NK Immunotherapy: Success and Challenges

NK Immunotherapy: Success and Challenges

Editor: Maverick Davies

MURPHY & MOORE
www.murphy-moorepublishing.com

www.murphy-moorepublishing.com

MURPHY & MOORE

Cataloging-in-Publication Data

NK immunotherapy : success and challenges / edited by Maverick Davies.
p. cm.
Includes bibliographical references and index.
ISBN 978-1-63987-398-2
1. Killer cells. 2. Killer cells--Therapeutic use. 3. Cancer--Immunotherapy. 4. Immunocompetent cells. I. Davies, Maverick.
QR185.8.K54 N53 2022
616.079--dc23

Murphy & Moore Publishing,
1 Rockefeller Plaza,
New York City, NY 10020, USA

ISBN 978-1-63987-398-2 (Hardback)

Contents

Preface

Over the recent decade, advancements and applications have progressed exponentially. This has led to the increased interest in this field and projects are being conducted to enhance knowledge. The main objective of this book is to present some of the critical challenges and provide insights into possible solutions. This book will answer the varied questions that arise in the field and also provide an increased scope for furthering studies.

Natural killer cells (NK cells) are the lymphocytes which belong to the same family as B and T cells, because they differentiate from a common progenitor. These cells detect and control early signs of cancer, and also recognize and kill virally infected cells. Specialized natural killer cells are found in placenta and are also known to play an important role in pregnancy. To enhance the immune system, natural killer cells secrete cytokines such as IFNγ and TNFα which act on other immune cells. Activating receptors recognize the molecules which are present on the surface of infected and cancer cells and switch on the natural killer cells. This book explores all the important aspects of natural killer cells. It brings forth the role of natural killer cells in disease response. In this book, using case studies and examples, constant effort has been made to make the understanding of the difficult concepts of this field as easy and informative as possible, for the readers.

I hope that this book, with its visionary approach, will be a valuable addition and will promote interest among readers. Each of the authors has provided their extraordinary competence in their specific fields by providing different perspectives as they come from diverse nations and regions. I thank them for their contributions.

Editor

1

Modulation of Human Leukocyte Antigen-C by Human Cytomegalovirus Stimulates KIR2DS1 Recognition by Natural Killer Cells

Kattria van der Ploeg[1], Chiwen Chang[1], Martin A. Ivarsson[1], Ashley Moffett[1], Mark R. Wills[2] and John Trowsdale[1]**

[1] Department of Pathology, University of Cambridge, Cambridge, UK, [2] Department of Medicine, University of Cambridge, Cambridge, UK

Edited by:
Ulrike Koehl,
Hannover Medical School, Germany

Reviewed by:
Ralf Dressel,
Universitätsmedizin Göttingen, Germany
Dagmar Stoiber,
Medical University of Vienna, Austria

***Correspondence:**
Mark R. Wills
mrw1004@cam.ac.uk;
John Trowsdale
jt233@cam.ac.uk

The interaction of inhibitory killer cell Ig-like receptors (KIRs) with human leukocyte antigen (HLA) class I molecules has been characterized in detail. By contrast, activating members of the KIR family, although closely related to inhibitory KIRs, appear to interact weakly, if at all, with HLA class I. KIR2DS1 is the best studied activating KIR and it interacts with C2 group HLA-C (C2-HLA-C) in some assays, but not as strongly as KIR2DL1. We used a mouse 2B4 cell reporter system, which carries NFAT-green fluorescent protein with KIR2DS1 and a modified DAP12 adaptor protein. KIR2DS1 reporter cells were not activated upon coculture with 721.221 cells transfected with different HLA-C molecules, or with interferon-γ stimulated primary dermal fibroblasts. However, KIR2DS1 reporter cells and KIR2DS1$^+$ primary natural killer (NK) cells were activated by C2-HLA-C homozygous human fetal foreskin fibroblasts (HFFFs) but only after infection with specific clones of a clinical strain of human cytomegalovirus (HCMV). Active viral gene expression was required for activation of both cell types. Primary NKG2A$^-$KIR2DS1$^+$ NK cell subsets degranulated after coculture with HCMV-infected HFFFs. The W6/32 antibody to HLA class I blocked the KIR2DS1 reporter cell interaction with its ligand on HCMV-infected HFFFs but did not block interaction with KIR2DL1. This implies a differential recognition of HLA-C by KIR2DL1 and KIR2DS1. The data suggest that modulation of HLA-C by HCMV is required for a potent KIR2DS1-mediated NK cell activation.

Keywords: natural killer cells, human cytomegalovirus, killer Ig-like receptor, KIR2DS1, HLA-C

INTRODUCTION

Since the discovery of natural killer (NK) cells more than 40 years ago (1–3), the interaction of inhibitory NK cell receptors with human leukocyte antigen (HLA) class I molecules has been characterized in detail. This led to new insights into NK cell differentiation, education, and function. However, ligands of most activating receptors, including activating killer cell Ig-like receptors (KIRs), are yet to be discovered.

KIR genes are members of the immunoglobulin (Ig) superfamily, encoded in the leukocyte receptor complex (LRC) on chromosome 19q14.3 (4). KIR molecules express either two or three extracellular Ig-like domains (2D or 3D) and consist of either a long (designated "L") or short (designated "S") cytoplasmic domain. KIRs with long cytoplasmic domains are inhibitory (iKIRs) and contain ITIMs. Activating KIRs (aKIRs) have a short cytoplasmic tail and transmit activating signals through the interaction with DAP12, which contains an ITAM (4).

Most iKIRs recognize certain allotypes of HLA class I. In general, allelic products of *KIR2DL1* bind to the C2 group of HLA-C molecules (C2-HLA-C) characterized by Asn77 and Lys80 (5), while KIR2DL2 and -2DL3, which are alleles at the same locus, recognize the C1 group (C1-HLA-C, Ser77, and Asn80) (6–8). These structural motifs were originally thought to be essential for the engagement of KIRs only on HLA-C. However, KIR2DL2 can also bind HLA-B46:01 and -B73:01 alleles, which have C1-related motifs at residues 77–83 (9). Furthermore, KIR2DL2 and -L3 receptors can bind many HLA-C alleles irrespective of -C1 or -C2 group (10, 11).

The extracellular parts of iKIRs and aKIRs are highly homologous and share conserved amino acid sequences, as "paired" receptors (11, 12). The balance between inhibitory and activating signaling through these paired receptors is tightly regulated by NK cells. Dysregulation of this balance might lead to autoimmunity or infectious diseases (13, 14). How the signaling is controlled by NK cells, however, is not completely understood, mainly due to uncertainty over the ligands and functions of aKIRs. The aKIR members seem to have evolved more rapidly than iKIRs, possibly through selection pressure imposed by pathogens (15, 16). If this hypothesis is true, it suggests that aKIR binding may be influenced by pathogen-derived proteins. Notably, KIR2DS1 and -2DS2 counterparts in chimpanzees, respectively, bind C2- and C1-HLA-C with high avidity compared to their inhibitory paired receptors (17). This indicates that the loss of binding by KIR2DS2, or highly reduced binding of KIR2DS1, to HLA-C is a product of human-specific evolution.

Most interactions of aKIRs and HLA class I molecules are very weak or undetectable (17–23). The best studied aKIR is KIR2DS1 and many studies have found that it binds C2-HLA-C (10, 11, 17, 24–35). However, this binding is much weaker compared to KIR2DL1 (10, 25, 27). Using surface plasmon resonance (SPR) analysis, Stewart and colleagues demonstrated that KIR2DS1 tetramer-binding avidity to the soluble HLA-Cw4/beta-2 microglobulin (β_2M)/peptide complex is approximately four times lower than KIR2DL1: dissociation constants (K_d) of 7.2 and 30 μM, respectively. In addition, amino acid substitution of the peptide at amino acid 7 or 8 drastically reduced KIR2DS1 tetramer binding to the HLA-Cw4 complex, indicating that the binding is peptide dependent (27). Both Moesta et al. and Hilton et al. demonstrated that KIR2DS1-Fc binds a range of C2-HLA-C allotypes linked to microbeads with different avidity (11, 17). Biassoni and colleagues demonstrated that the amino acid at position 70, a threonine in KIR2DL1 and a lysine in KIR2DS1, is the key residue that differentiates binding between KIR2DL1 and -2DS1. Substituting the threonine to a lysine in KIR2DL1-Fc prevented binding to 721.221 transfected with HLA-Cw4 (221-Cw4). When lysine was substituted for a threonine in KIR2DS1-Fc, the binding to 221-Cw4 was restored (25).

The rationale for KIR2DL1 and KIR2DS1 both binding C2-HLA-C would be understandable if the activating receptor was sensitive to structural changes in the HLA molecule, or bound an alternative molecule, induced by viral infection (18, 27, 36–42). For example, in mice, the cytomegalovirus (CMV)-encoded MHC class I homolog m157 is directly recognized by the activating Ly49H receptor (43). The mouse Ly49 receptor family serves a similar role to KIRs in humans, although KIRs and Ly49 receptors are from different molecular families. KIRs play an important role in human cytomegalovirus (HCMV) infections (44–47). For instance, a recent study demonstrated that KIR2DS1$^+$ decidual NK (dNK) cells degranulated after engaging with HCMV-infected decidual stromal cells (DSC), suggesting an increased ability of KIR2DS1-expressing dNK cells to respond to placental HCMV infection (47). Della Chiesa et al. have reported that HCMV can drive NK cell maturation in the absence of NKG2C in patients with hematological malignancies. These patients received umbilical cord blood transplantation from NKG2C$^{-/-}$ donors and when HCMV reactivation occurred, an expansion of NKG2A$^-$ NK cells expressing aKIRs was measured, particularly KIR2DS1 and KIR3DS1 (45). This finding is consistent with KIR2DS1 recognizing a ligand on HCMV-infected cells.

To probe the potential influence of HCMV on KIR2DS1 recognition, we designed a mouse 2B4 T cell hybridoma carrying an NFAT-green fluorescent protein (GFP) reporter. Our results suggest that modulation of HLA-C by HCMV is required for a potent KIR2DS1-mediated NK cell activation.

MATERIALS AND METHODS

Cell Lines and Cell Culture

Reporter cells, 721.221 cells, K562 cells, and primary NK cells were cultured in RPMI-1640 (Sigma-Aldrich, Steinheim, Germany) supplemented with 100 U/ml penicillin, 100 μg/ml streptomycin (Gibco, Paisley, UK), and 10% heat-inactivated Fetal Bovine Serum (Gibco). Human fetal foreskin fibroblasts (HFFFs, Culture Collections—ECACC, UK) and primary dermal fibroblasts (DFs) were cultured in Dulbecco's Modified Eagle's Medium (DMEM, Sigma-Aldrich) supplemented with 100 U/ml penicillin, 100 μg/ml streptomycin (Gibco), and 10% heat-inactivated fetal bovine serum. Adherent cells were harvested by washing the cells once with phosphate buffered saline (PBS, Sigma-Aldrich). Then the cells were either detached from the plastic using 0.05% Trypsin/EDTA (Gibco) or Accutase™ (Biolegend, San Diego, CA, USA) for 5 min at 37°C. All cellular experiments were performed at 37°C in 5% CO_2.

721.221 cells transfected with -A23:01, -B58:01 (Bw4), -B35:01 (Bw6), -C01:02 (C1), -C02:02 (C2), -C03:02 (C1), -C04:01 (C2), -C06:02 (C2), -C07:01 (C1), and -HLA-G were generated in house. 721.221-HLA-A11:02 was provided by Parham (22) (Stanford University, Palo Alto, CA, USA).

Establishment of the Reporter Cells

The 2B4 T cell hybridoma containing an NFAT-GFP reporter gene (2B4 reporter cells) was kindly provided by Lewis Lanier (43) (University of California San Francisco, USA). KIR2DL1*003, -2DL2*001, -2DS1*002, and -2DS1$^{(K70T)}$ reporter cells were generated as follows. First, pMX-neo constructs containing cDNA from a chimeric adaptor protein recombinant was used to transduce the 2B4 reporter cells. The chimeric adaptor consists of DAP12 and a cytoplasmic tail of DAP10 with spacer sequences in between. Then cDNA of the indicated KIRs was subcloned into a pMX-puro construct. For constructing 2DL1–2DS1TM (KIR2DL1 reporter) and 2DL2–2DS1TM (KIR2DL2 reporter) chimeric molecules, 5′-CCTGCACGTTCTGATTGGGACCTCAGT-3′ and 5′-CCCAATCAGAACGTGCAGGTGTCGGGGGTT-3′ primers were used. 5′-AGTCGCATGACGCAAGACCTGGCAGGG-3′ and 5′-GGTCTTGCGTCATGCGACTGATGGAG-3′ primers were used for constructing the KIR2DS1$^{(K70T)}$ reporter cell. Retroviruses were packaged in Phoenix-eco cells (generously provided by Lewis Lanier) using the non-modified polyethyleneimine (PEI, Sigma-Aldrich) reagent as described by Ehrhardt et al. (48). After 48 h, supernatant-containing retroviral particles was used to transduce the 2B4 reporter cells by adding Polybrene (8 ng/ml, Sigma-Aldrich) and by spin-infecting the cells at 2,500 rpm (AccuSpin 3R centrifuge, Fisher Scientific, Waltham, MA, USA) for 2 h. Cells expressing the KIRs were purified by surface staining using the PAN2D antibody (clone NKVFS1, Bio-Rad, Hercules, CA, USA), followed by single-cell sorting using the FACS sorter (BD Bioscience, Oxford, UK). The transduction success and the function of the reporter cells were analyzed by immunofluorescent staining and antibody crosslinking, as described below. The LILRB1 reporter cell was provided by Des Jones (Department of Pathology, University of Cambridge) and was constructed as described in Ref. (49).

Primary Cells

Primary NK cells and DFs were obtained from healthy individuals. Ethical approval for the use of these tissues was given by Addenbrookes National Health Service Hospital Trust institutional review board (Cambridge Research Ethics Committee) and informed written consent was obtained from all volunteers in accordance with the Declaration of Helsinki (LREC 97/092). The primary fibroblasts were fully HLA typed. HFFFs express HLA-A11:01, -A24:02, -B35:02 (Bw6), -B40:02 (Bw6), -C02:02 (C2), and -C04:01 (C2). Donor CMV307 expresses HLA-A01:01, -A26:01, -B08:01 (Bw6), -B27:05 (Bw4), -C07:01 (C1), and -C01:02 (C1). Donor CMV0005 expresses HLA-A02, -A03, -B07 (Bw6), -B13 (Bw4), -C07 (C1), and -C06 (C2). Primary NK cell donors both express KIR2DL1, -2DS1, 2DL3, NKG2A, and were C1/C1-HLA-C. In addition, donor 016 expresses NKG2C and donor 111 expresses KIR3DL1.

For functional NK cell studies, peripheral blood mononuclear cells (PBMCs) were extracted from 30 to 40 ml blood of donor 016 and 111 on a Ficoll-Hypaque density gradient (Lympoprep, Axis-Shield, Dundee, Scotland). The PBMCs were removed from the interface of the plasma and Lymphoprep layers and washed three times with PBS before further use. NK cells were separated from the PBMCs by negative selection using the EasySep™ Human NK cell Enrichment Kit from Stemcell Technologies (Vancouver, BC, Canada).

Dermal biopsies were taken from healthy individuals by Andrew Carmichael (Department of Medicine, University of Cambridge). They were sectioned with a scalpel and were grown beneath cover slips in a six-well culture plate containing Eagle's Minimum Essential Media (EMEM, GE Healthcare, Little Chalfont, UK) supplemented with 100 U/ml penicillin, 100 µg/ml streptomycin (Gibco), and 10% heat-inactivated fetal calf serum (Life Technologies, Carlsbad, CA, USA). The cells were grown until sufficient cell number was reached and were stored in liquid nitrogen at low passage numbers.

Coculture Experiments

Human fetal foreskin fibroblasts (10×10^3 cells/well) and DFs (10×10^3 cells/well) were seeded in 96-well flat-bottom culture plates with or without interferon (IFN)-γ (500 U/ml, PeproTech, Rocky Hill, CT, USA) for 72 h or infected with HCMV as described below. After appropriate stimulation/infection time was reached, reporter cells were added to the adherent cells at a concentration of 2×10^4 cells per well, and reporter cells were added directly. Cocultures using K562 cells and 721.221 cells were performed at an E:T ratio of 1:1 and 1:3, respectively. After an overnight coculture, the reporter cells were harvested, and GFP expression was analyzed by flow cytometry.

Antibody Cross-linking

For antibody crosslinking experiments, anti-mouse IgG-coated microplates (R&D systems, Minneapolis, MN, USA) were used. The plates were incubated for 30 min with 0.2–1 µg per well of PAN2D (clone NKVFS1) or anti-HA (clone HA-7, Sigma-Aldrich) antibodies in PBS at room temperature. After two PBS washes, the reporter cells were added.

Antibody-Blocking Assays

Unconjugated W6/32 (LEAF™ purified BioLegend), 6A4 (gift from Daniela Pende, Azienda Ospedaliera Universitaria San Martino di Genova), B1.23.2 (eBiosciences, San Diego, CA, USA), IgG2a (LEAF™ purified BioLegend), IgG2b (LEAF™ purified BioLegend), and IgG1 (LEAF™ purified BioLegend) isotype antibodies were added to pre-treated HFFFs. After an incubation of 10–15 min at room temperature and saturating concentration of 1–4 µg per well, the reporter cells were added.

Flow Cytometry

Immunofluorescence Cell-Surface Staining

The cells were harvested and rested at 37°C in 5% CO_2 for at least 30 min for the recovery of HLA class I molecules on the cell surface. Cells expressing Fc receptors or HCMV-encoded Fc receptors were first blocked with 40% human serum (Sigma-Aldrich) in PBS for at least 5 min before staining. Fractions of 721.221 cells were collected and directly incubated with the appropriate antibody. Cell-surface expression of different receptors was analyzed by immunofluorescent staining using unconjugated monoclonal primary antibodies listed in **Table 1**, unconjugated anti-HLA-E (3D12, IgG1, BioLegend), biotinylated anti-HLA-A11 (Abcam,

TABLE 1 | List of antibodies and their target epitopes where known.

Antibody	Isotype/*supplier*	Recognition	Specific HLA molecules	Reference
W6/32	IgG2a/*BioLegend, Hybridoma*	β_2M bound, fully assembled human leukocyte antigen (HLA) class I molecules. The epitope is not known. It includes residues on β_2M, α2, and α3 domains	HLA-A, -B, -C and HLA-E, -G	(50, 51)
B1.23.2	IgG2b/*eBiosciences*	β_2M bound, fully assembled HLA-B and -C molecules. Precise epitope unknown	HLA-B and HLA-C	(52)
6A4	IgG1/*Hybridoma*	β_2M bound, fully assembled HLA-B and -C molecules. Precise epitope unknown	HLA-B and HLA-C	(53)
HC10	IgG2a/*Hybridoma*	Free heavy chain of HLA class I. Residues 57–62, specifically residue 60 in the α1 domain in HLA-C. The epitope is blocked by peptide binding	HLA-B and HLA-C and some HLA-A (A10, A28, A29, A30, A31, A32, A33)	(54, 55)
DT9	IgG2b/*Hybridoma*	Fully assembled HLA-C and HLA-E bound to β_2M. Precise epitope unknown	HLA-C and HLA-E	(56)
L31	IgG1/*MediaParma*	Free heavy chain of HLA-C. Residues 66–68, with F or Y at position 67	Most HLA-C and a few HLA-B (B08, B07, B35, B51, B54, B56)	(57)

IgM, Cambridge, UK), APC-conjugated anti-HLA-Bw6 (IgG1, Miltenyi Biotec, Bergisch Gladbach, Germany), and the appropriate isotype control antibodies. The primary antibodies were added at a concentration of 0.5–1 µg per well in 5% FCS in PBS or 40% human serum in PBS and incubated on ice for 45 min. Following washing, the cells were incubated with secondary antibodies (0.4–1 µg per well), AlexaFluor® 647 (AF647)-conjugated Streptavidin (Life Technologies) and polyclonal anti-mouse IgG conjugated to FITC (BD Pharmingen, San Diego, CA, USA), AF647 (Life Technologies), or R-PE (Thermo Scientific), in 5% FCS with PBS or in 40% human serum with PBS for 30 min on ice. For double staining experiments, the W6/32 antibody or IgG2a isotype control conjugated to AF647 or FITC (BioLegend) was added to the cells for 30 min on ice. The cells were fixed with 1% formaldehyde in PBS before flow cytometry analysis.

Functional Assay with Primary NK Cells Including Subsequent Multicolor Immunofluorescence Staining

Natural killer cells were stimulated with medium alone or with IL-12 (10 ng/ml, R&D systems) and IL-15 (50 ng/ml, R&D systems) for 12 h. The following day, target cells were added at an E:T ratio of 1:1 or 10:1 (58). Cytokine-stimulated NK cells in medium alone, or cocultured with K562 cells (a gold standard NK cell target cell line, devoid of MHC class I) served as a positive control for function when HCMV-infected HFFFs were investigated for their capacity to trigger primary NK cells. After 1 h of coculture, GolgiPlug (Brefaldin A, BD Bioscience) and GolgiStop (Monensin, BD Bioscience) were added to the wells according to manufacturer's instructions, and incubated for an additional 4 h. Non-adherent cells were subsequently transferred to V-bottom well plates where multicolor immunofluorescence staining was performed (32). The following conjugated monoclonal antibodies were used: anti-KIR2DL3 FITC (REA147, Miltenyi), anti-NKG2A APC (Z199, Beckman Coulter, Brea, CA, USA), anti-CD107a APC-H7 (H4A3, BD Bioscience), anti-KIR3DL1 Brilliant Violet 421 (DX9, BioLegend), biotinylated anti-KIR2DL1 (143211, R&D Systems), anti-CD56 ECD (N901, Beckman Coulter), anti-CD3 PE-Cy5 (UCHT-1, BioLegend), and anti-KIR2DL1/S1 PE-Cy7 (EB6, Beckman Coulter). Briefly, the cells were stained with the primary antibody cocktail for 20 min. The anti-KIR2DL1/S1 antibody was added directly to the wells (dilution 1:20) for 10 min incubation. Following washing, the secondary antibody cocktail containing Live/Dead Aqua (Invitrogen, Carlsbad, CA, USA) and Qdot-605-conjugated Streptavidin (Invitrogen) was incubated for 20 min. Cells were next washed and fixed.

Data acquisition for the cellular and immunofluorescence cell-surface staining experiments was performed on FACScan (Department of Pathology), FACSCalibur or Accuri C6 flow cytometer (NIHR Cambridge BRC Cell Phenotyping Hub, BD Bioscience) depending on the experiment. The data acquisition of the multicolor primary NK cell staining experiments was performed on a BD LSR Fortessa flow cytometer, equipped with 4 lasers and 13 PMTs (Cambridge Stem Cell Institute). Flow cytometry data were analyzed using FlowJo software (TreeStar, OR, USA).

Viruses

Human cytomegalovirus strains TB40/E [isolated from a throat wash of a bone marrow transplant recipient (59)], AD169 (ATCC VR-538) and Merlin (a gift from Richard Stanton, University of Cardiff, UK) were grown, concentrated, and titrated as described previously (60). Confluent plates of HFFFs or DFs were infected with concentrated virus (TB40/E, AD169, Merlin) at a multiplicity of infection (MOI) as indicated in the experiment or with diluted (1:10) supernatant-containing virus (TB40/E-derived clones). The virus preparations were diluted in complete DMEM to obtain the required MOI. After 1 h of infection at room temperature, the cells were washed with PBS, and fresh complete DMEM was added. A coculture experiment was performed, or the cells were harvested for cell-surface staining, or western blotting, 24, 48, and/or 72 h p.i. as indicated in the experiment. Time points were chosen to reflect the temporal cascade of expression of HCMV viral proteins (immediate early, early, and late genes). Late genes, which are structural viral components, are last expressed at ~72 h post infection (p.i.) (61). UV inactivation of the virus was obtained by exposing the virus 30 min to UV light of a 30 W germicidal lamp.

Plaque Purification Assay

Human fetal foreskin fibroblasts were infected with serially diluted HCMV strain TB40/E. After 1 h of infection at room temperature, the virus-containing medium was replaced with 5 ml per well of 2% agarose mixed 1:1 with two times concentrated DMEM medium (Millipore, Bedford, MA, USA) containing 20% FCS, 200 U/ml penicillin, 200 μg/ml streptomycin, 8 g/L sodium bicarbonate (GE Healthcare), and 1 mM of sodium pyruvate (Sigma-Aldrich). After 3 weeks, plaques (areas of dead cells) were visible by eye. Well-separated plaques, representing different viral clones of the TB40/E wild-type strain, were picked using a glass Pasteur pipette by removing the agar and plaque as a plug. The agarose plug containing viral clones were disrupted and added to freshly plated HFFFs in 96-well flat-bottom culture plates to expand the viral clones. Selected clones were expanded up to a 75 cm^2 culture flask. After 100% infection was reached, as determined visually by microscopy, the supernatant-containing virus was harvested and stored at −80°C. Fresh complete DMEM was added to the flasks and further harvests were carried out every 72 h until approximately 90% cell lysis was visually determined. All the harvested supernatant was pooled, spun at 2,000 rpm to remove cell debris, aliquoted, and stored at −80°C.

Western Blot

Cells were washed with PBS and lysed in lysis buffer [1% Triton X-100, 20 mM Tris–HCl (pH 7.4), 150 mM NaCl, 1 mM EDTA, 5 mM $MgCl_2$, protease inhibitors (Roche, Mannheim, Germany), and 1 mM Phenylmethanesulfonyl fluoride (PMSF, Sigma-Aldrich) and 1.25 mg/ml N-ethylmaleimidine (Sigma-Aldrich)] for 20 min at 4°C. Debris and nuclei were removed by centrifugation at 13,000 rpm (centrifuge HAWK 15/05, MSE), and a bicinchoninic acid (BCA) assay (Thermo Scientific) was used to determine protein concentrations.

Supernatants of TB40/E-derived clones were pelleted by centrifuging at 15,000 rpm for 2 h at 4°C using an Avanti J-25 Ultracentrifuge (Beckman Coulter). The virus pellets were gently washed with PBS, and 200 μl of lysis buffer was added to lyse the particles. Lysates from positive and negative virus particles were paired based on approximately the same infectivity titer. Virus titers were calculated by $TCID_{50}$.

Total cell or virus particle lysates were loaded onto a 10–13% SDS-PAGE gel, and proteins were transferred onto Immobilon-P PVDF membranes (Millipore). The membranes were blocked 1 h in PBS, 5% dried milk, and 0.05% Tween 20 at room temperature. The membranes were incubated overnight at 4°C or 3 h at room temperature with primary antibodies [anti-pp28 (Abcam), L31, HC10, anti-I.E.1 (Merck Millipore, Billerica, MA, USA), anti-Calnexin (Enzo Life Sciences, Farmingdale, NY, USA), and anti-Flag (M2, Sigma-Aldrich)] at concentration of 0.1–0.2 μg. The membranes were washed thoroughly, and polyclonal HRP-conjugated goat anti-mouse IgG or goat anti-rabbit IgG (Dako, 1:4,000 dilution) secondary antibody was added for 30 min at room temperature. Chemiluminescence was performed according to the manufacturer's instructions using ECL Prime (GE Healthcare) or home-made ECL (62).

Establishment CRISPR-CAS9 Knockout (KO) for β$_2$M in HFFFs

Streptococcus pyogenes CAS9 and short-guide RNA (sgRNA) were expressed in separate lentivirus constructs: pHRSIN containing the SFFV promoter, FLAG tag, nuclear localization signals (NLS), CAS9 and pGK Hygro (kind gift from Lehner's group, CIMR, University of Cambridge), and pKLV-containing U6 promoter, sgRNA (modified *Bbs*I), pGK Puro, 2A, and BFP tag [Addgene plasmid #50946; created by Kosuke Yusa, Wellcome Trust Sanger Institute, Cambridge, UK (63)]. The pKLV construct containing sgRNA-targeting β_2M was a gift from Dick van den Boomen (CIMR, University of Cambridge), using the sgRNA sequence 5′-GGCCGAGATGTCTCGCTCCG-3′.

The pHRSIN and pKLV lentivirus constructs (6 μg) were packaged together with pMDG and pCMV9.81 packaging vectors (4 μg) in HEK 293T cells in 75 cm^2 culture flasks using Opti-MEM™ and GlutaMax™ media (Gibco), and *Trans*IT® Transfection reagent (Mirus, Madison, WI, USA). After 48 h, supernatant-containing lentiviral particles were used to transduce HFFFs in 75 cm^2 culture flasks by adding Polybrene (8 ng/ml). The transduced cells were selected with 200 μg/ml hygromycin and/or 2 μg/ml puromycin and grown until a confluent 75 cm^2 culture flask was reached. Cells negative for total HLA class I were purified by surface staining using W6/32 antibody, respectively, followed by single-cell sorting using the FACS sorter. The CAS9 is FLAG-tagged and a HRP-conjugated anti-M2 FLAG antibody was used to detect it by western blot.

Statistical Analysis

Non-parametric one-way analysis of variance (ANOVA) using the Kruskal–Wallis test and Dunn's multiple comparisons test was used to determine the statistical significance. In these tests, a p value of less than 0.05 was considered significant ($^*p < 0.05$, $^{**}p < 0.01$). The tests were done with GraphPad Prism version 6.00 (GraphPad software).

RESULTS

The Function and Specific Recognition of KIR2DS1 by Reporter Cells

Since KIR2DS1 binds C2-HLA-C only weakly, we aimed to investigate what might influence stronger binding. To accomplish this, we used a specifically designed mouse 2B4 T cell hybridoma carrying an NFAT-GFP reporter system, similar to that described by Arase et al. (43). KIR2DS1 was transduced into these 2B4 reporter cells, together with a modified adaptor protein. Once KIR2DS1 binds its cognate ligand, a signaling cascade is triggered through the adaptor protein, which then transcribes NFAT, resulting in GFP expression. KIR2DL1, -2DL2, and LILRB1 reporter cells were also generated. We confirmed that the reporter cells were constructed successfully by staining the surface expression of the different receptors with the relevant antibodies and by engaging the reporter cells with relevant plate-bound antibodies (**Figure 1A**).

There are seven amino acid differences between KIR2DL1 and -2DS1 alleles, including the threonine to lysine in KIR2DL1 and -2DS1 at position 70, respectively (**Figure 1B**). We wanted to investigate whether substituting the lysine to a threonine at position 70 in KIR2DS1 would result in the activation of the reporter cells as a result of the interaction with C2-HLA-C, as demonstrated previously by Biassoni and colleagues using Fc proteins (25). We made the KIR2DS1$^{(K70T)}$ reporter cell and cocultured these cells with 721.221 transfected with HLA-C03:02 (C1) or -C06:02 (C2) (221-Cw3 or 221-Cw6). Indeed, both KIR2DL1 and KIR2DS1$^{(K70T)}$ reporter cells were activated after coculture with 221-Cw6 (15 and 11% GFP-positive cells, respectively) and not with 221-Cw3 (**Figure 1C**). This confirms the observations made by Biassoni et al. in our cellular reporter system and shows that the reporter cells are functional, specific, and sensitive.

A

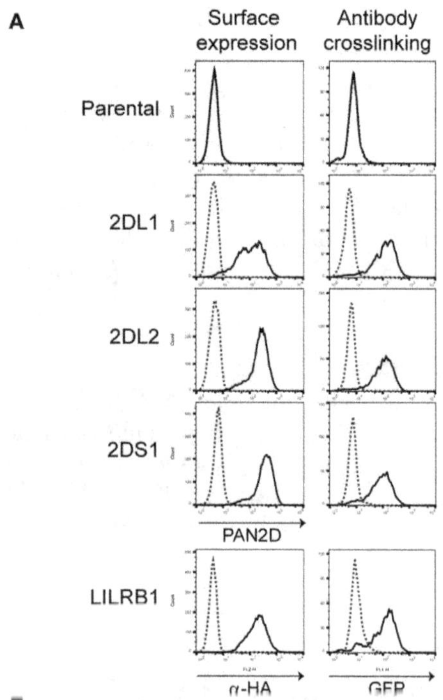

B

2DL1	T	V,L	P, L	T, P	N, D	R, H	E, K
#a.a.	70	90	114	154	163	182	216
2DS1	K,R	L,V	P	T	N	R	E,K

C

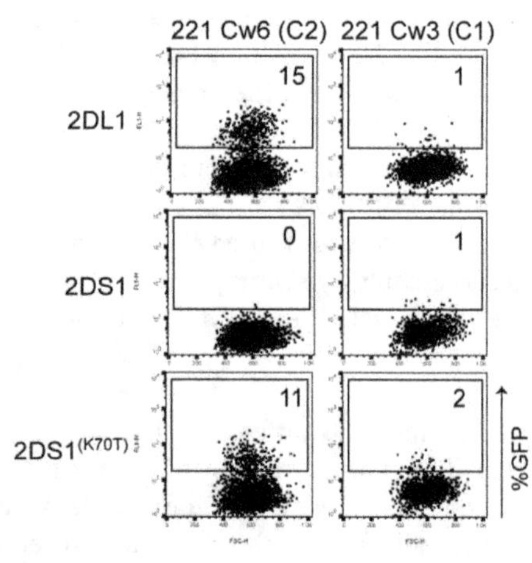

FIGURE 1 | Continued

FIGURE 1 | Continued
2B4 reporter system for activating killer cell Ig-like receptor (KIR). **(A)** Cell-surface expression on the indicated reporter cells was measured by flow cytometry after staining the cells with PAN2D (clone NKVFS1, recognizes all KIRs) or anti-HA antibody (black line, left panel). Parental 2B4 cells containing only the adaptor protein were used as a negative control. To test the function of these reporter cells, plate-bound PAN2D (clone NKVFS1) or anti-HA (LILRB1 reporter only) antibody crosslinking (black line) was used to engage KIR or LILRB1 molecules on the reporter cells during an overnight incubation (right panel). IgG1 antibody was used as isotype control (dotted line). **(B)** Table depicting amino acid positions of polymorphic sites in the extracellular domain of KIR2DL1 and -2DS1 alleles. Large letters indicate amino acids present in the majority of alleles, and small letters indicate amino acids that are present in few alleles. At position 70, a unique amino acid in KIR2DL1 and -2DS1, threonine (T) and lysine (K), respectively, was found. Allele sequences were aligned using the alignment tool in the *IPD-KIR database*. **(C)** KIR2DS1$^{(K70T)}$ was transfected into the 2B4 reporter system and an overnight coculture was performed using KIR2DL1, -2DS1, or -2DS1$^{(K70T)}$ reporter cells together with 721.221 cells expressing human leukocyte antigen (HLA)-Cw3 (C1) or HLA-Cw6 (C2). The next day green fluorescent protein (GFP) expression was measured by flow cytometry. The E:T ratio of the coculture was 1:3.

KIR2DS1 Reporter Cells Do Not Recognize Conventional HLA Class I

To verify whether KIR2DS1 reporter cells are activated by conventional HLA class I molecules, cocultures were performed using the KIR2DS1 reporter cells along with appropriate controls. The cocultures were done using a range of 221 cells transduced with different HLA molecules: HLA-A11:02, -A23:01, -Bw6 (B35:01), -Bw4 (B58:01), -G, -C01:02 (C1), -C02:02 (C2), -C03:02 (C1), -C04:01 (C2), -C06:02 (C2), and -C07:01 (C1). HLA expression levels on these 221 cells were confirmed by staining with W6/32 antibody (**Figure 2A**). KIR2DS1 reporter cells were not activated in cocultures with different 221-HLA-C or other 221-HLA class I cells. However, the positive controls KIR2DL1, -L2, and LILRB1 were activated after engaging their documented ligands (**Figures 2B,C**). As expected, KIR2DL1, -L2, and LILRB1 reporter cells were differentially activated, depending on the HLA-C allele they engaged. For instance, the KIR2DL2 reporter cell was weakly activated after binding HLA-C01:02 (<5% GFP-positive cells) compared to HLA-C07:01 (>20% GFP cells), as depicted in **Figure 2C**, lower panel. This is in line with binding studies using Fc proteins and HLA-coated beads (11, 64).

In our hands, the KIR2DS1 reporter cells did not recognize endogenously expressed HLA class I molecules on 221 cells. To investigate this further, they were cocultured with primary cells: HFFFs, CMV307, and CMV0005 DFs, which were stimulated with or without IFN-γ for 72 h. Before the coculture experiments, HLA class I surface expression levels of untreated and IFN-γ stimulated cells were measured. The primary fibroblasts all expressed high levels of folded HLA class I and HLA-C/-E molecules, revealed by W6/32 and DT9 antibody staining, respectively. These HLA class I levels were further increased after 72 h of IFN-γ stimulation (**Figure 3A**). KIR2DS1 reporter cells remained GFP-negative when cocultured with the different untreated or IFN-γ stimulated primary fibroblasts (**Figure 3B**). By contrast, LILRB1 reporter cells were activated in all coculture experiments with untreated cells and were further activated in

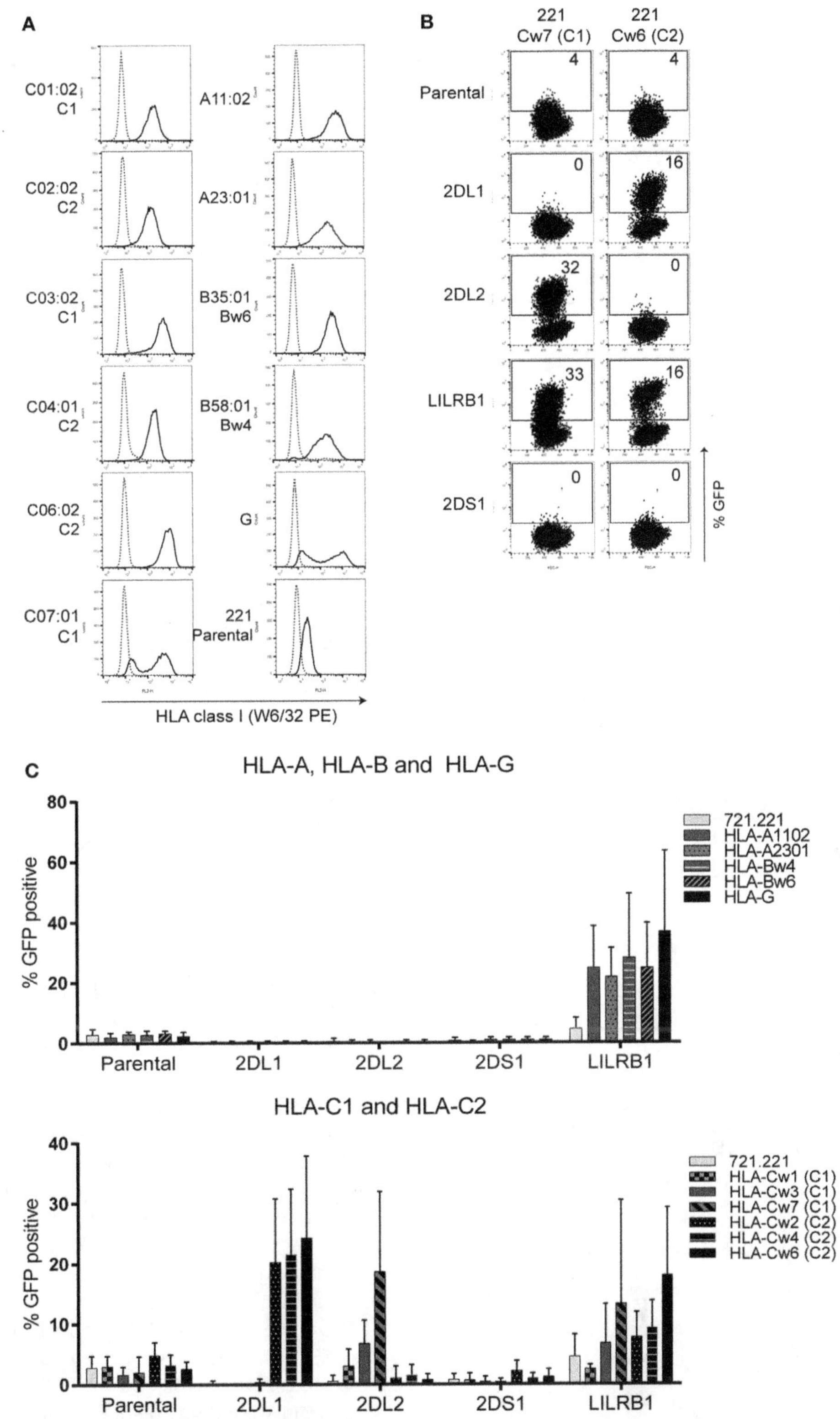

FIGURE 2 | KIR2DS1 reporter cells are not activated by conventional human leukocyte antigen (HLA) class I molecules. (A) The 721.221-HLA transfectants were stained with W6/32 (black line), and measured by flow cytometry. IgG2a antibody was used as isotype control (dotted line). **(B)** Dot plots of a selection of representative data from the same experiment. A coculture of the indicated reporter cells together with 721.221 cells containing HLA-C07:01 (221 Cw7, C1) or HLA-C06:02 (221-Cw6, C2) is shown. GFP expression was determined by flow cytometry. The E:T ratio was 1:3. **(C)** The data are depicted as the mean ± SD of individual samples collected from five independent experiments.

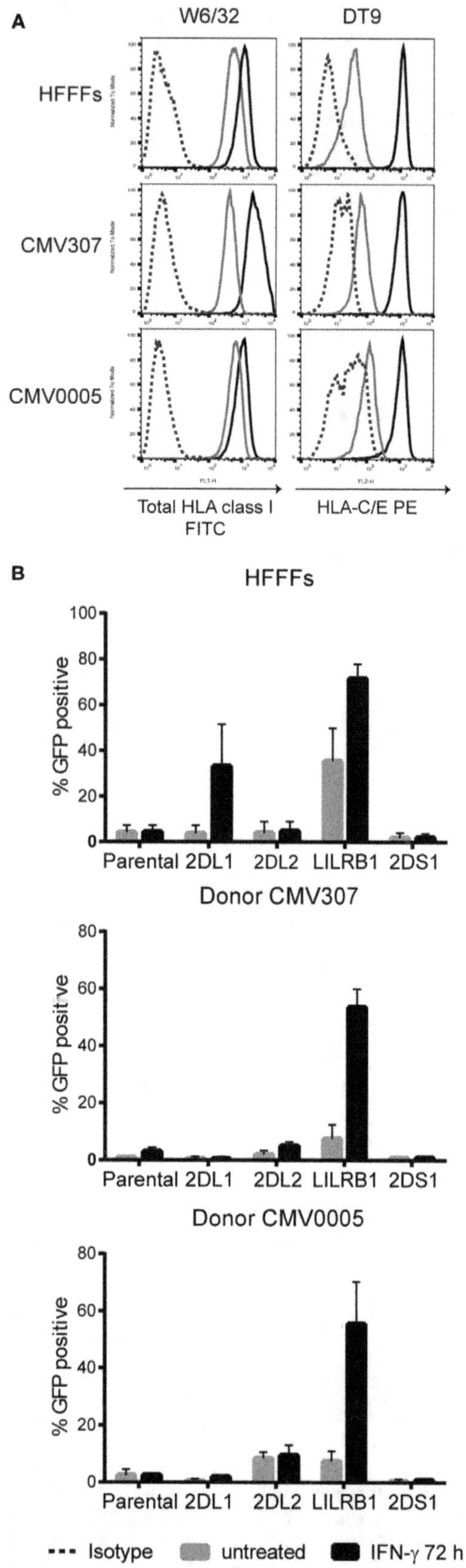

FIGURE 3 | Continued

FIGURE 3 | Continued
KIR2DS1 reporter cells are not activated by interferon (IFN)-γ stimulated human fetal foreskin and primary dermal fibroblasts of healthy individuals. (A) Total human leukocyte antigen (HLA) class I (W6/32) and HLA-C/-E (DT9) expression levels of untreated and IFN-γ stimulated HFFFs (C2/C2) and DFs of donor CMV307 (C1/C1) and CMV0005 (C1/C2) were measured by flow cytometry. Gray = untreated, black = IFN-γ stimulated (500 U/ml, 72 h) and dotted line = isotype control. **(B)** The untreated and IFN-γ-stimulated fibroblasts were cocultured overnight with the indicated reporter cells, and green fluorescent protein (GFP) expression was measured by flow cytometry. Three independent experiments are depicted in the bar graphs as the mean ± SD of individual samples.

settings with IFN-γ stimulated cells. KIR2DL2 and KIR2DL1 reporter cells were activated, in some conditions only minimally, depending on the particular HLA-C alleles expressed by the cells (**Figure 3B**). KIR2DS1 reporter cells were similarly unresponsive to tumor cell lines including HeLa, Meljuso, Caski, and JEG-3 cells (data not shown). In conclusion, KIR2DS1 reporter cells were not activated after engaging with conventional C2-HLA-C molecules.

KIR2DS1 Reporter Cells Bind a Ligand on HFFFs Infected with Specific HCMV Clones

Since KIR2DS1 reporter cells were not activated by conventional HLA molecules, we considered that KIR2DS1 might recognize a pathogen-induced ligand. Several studies have suggested a role for aKIR in HMCV infection (44–47). We therefore investigated whether the KIR2DS1 ligand might be upregulated after HCMV infection. HFFFs were infected with the HCMV TB40/E strain for 24, 48, and 72 h and cocultured with the panel of reporter cells. HFFFs were used because they support full lytic HCMV infection (65) and express HLA-C02:02 and -C04:01, both C2-HLA-C.

HFFFs Infected with the TB40/E Wild-Type Strain Express a Ligand for KIR2DS1

In coculture, KIR2DS1 reporter cells increased in GFP positivity from 1% (uninfected and 24 h) to 5% (48 h) and to 21% positive cells (72 h) (**Figure 4A**). This indicates that KIR2DS1 recognized a ligand, which was detected by the reporter cells 48 h p.i. with HCMV. In addition, there was increased triggering of both KIR2DL1 and LILRB1 reporter cells over time (**Figure 4A**). This was of interest as HCMV downregulates HLA class I molecules (66–71), and a decrease in KIR2DL1 reporter activation over infection time was expected. The increase in the LILRB1 reporter cell activation might be explained by the recognition of UL18 protein, which HCMV produces as an immune evasion strategy (72, 73). Subsequently, in a total of 10 KIR2DS1 reporter cell coculture experiments with HFFFs infected with HCMV TB40/E strain were done, demonstrating the reproducibility of the KIR2DS1 reporter cell activation after infection (Figure S1 in Supplementary Material). The KIR2DS1 ligand was only expressed on HFFFs infected with the TB40/E clinical strain and not with other strains tested, such as Merlin (another clinical strain) and AD169 (laboratory strain with deletion in the ULb' region) (data not shown).

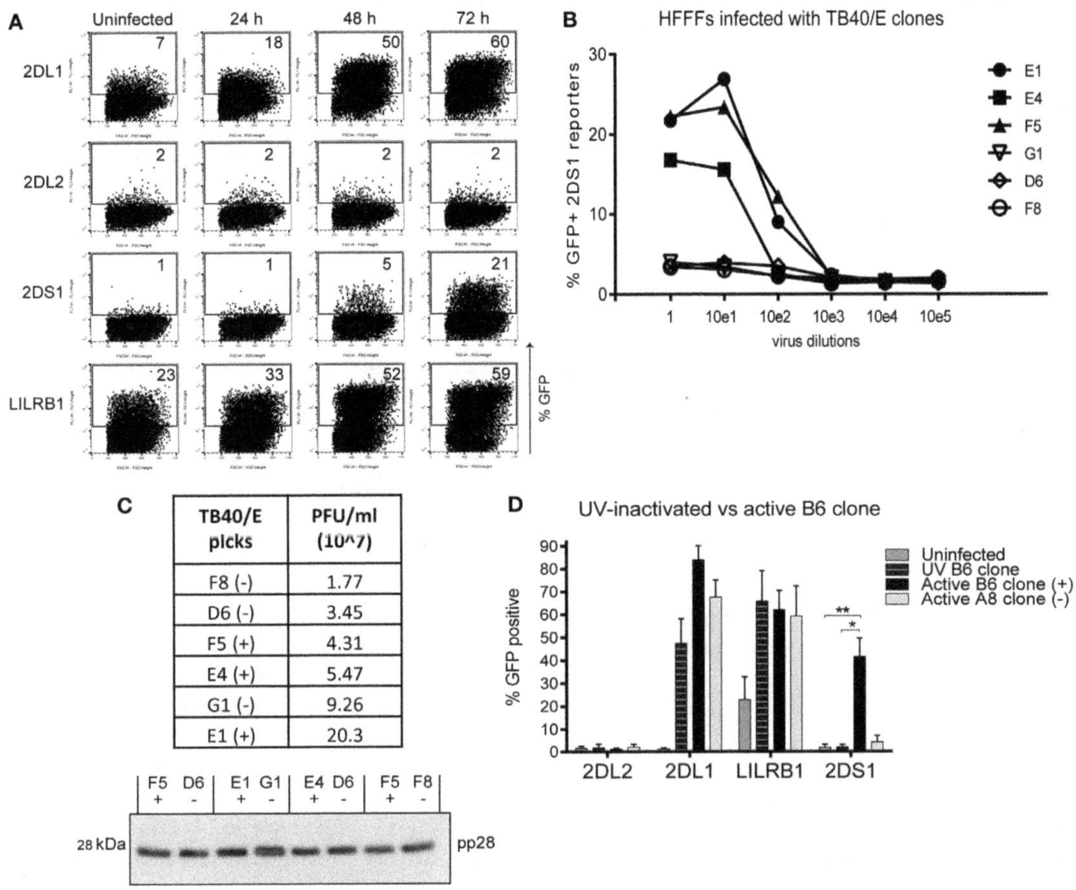

FIGURE 4 | The KIR2DS1 reporter cell recognizes a ligand on human fetal foreskin fibroblasts (HFFFs) infected with specific human cytomegalovirus (HCMV) strains. (A) Uninfected HFFFs and 24, 48, and 72 h infected HFFFs with HCMV TB40/E strain with an MOI of 10 were cocultured overnight with the reporter cells as indicated. A representative experiment of two independent time course cocultures is shown. **(B)** HFFFs were infected with six different TB40/E-isolated clones (E1, E4, F5, G1, D6, and F8) in 10-fold serial dilutions for 72 h and cocultured with KIR2DS1 reporter cells. A representative experiment from over three independent experiments is depicted. **(C)** The infectivities of three positive and three negative TB40/E clones were calculated by $TCID_{50}$ assay and are depicted in the table (virus titers in PFU/ml). Viral particles were isolated from the supernatant, lysed, and loaded onto a 12% SDS-PAGE gel for western blotting with anti-pp28 antibody. Positive and negative clones were paired based on approximately the same infectivity. Equal amounts of lysate from these pairs were loaded onto the gel. (−) indicates a negative clone, (+) positive clone. **(D)** HFFFs were stimulated with UV-inactivated TB40/E, infected with the positive B6 or negative A8 clone for 72 h. Forty-eight hours p.i. KIR2DL2 (negative control), KIR2DL1, LILRB1 (positive control), and KIR2DS1 reporter cells were added. After an overnight coculture, the GFP expression was measured using flow cytometry. The data are depicted as the mean ± SD of individual samples collected from five independent experiments. $^*p < 0.05$ and $^{**}p < 0.01$ are calculated by non-parametric one-way ANOVA using the Kruskal–Wallis test and Dunn's multiple comparisons test.

Specific TB40/E Clones Activate the KIR2DS1 Reporter Cells

Tomasec and colleagues isolated two viral clones, called Lisa and Bart, from the TB40/E strain, which had different functional properties (74). We wanted to investigate whether different viral clones were also present in our TB40/E strain. Multiple viral clones were isolated using a plaque purification assay. Twelve clones were selected and used to infect HFFFs followed by coculture with KIR2DS1 reporter cells. Six clones (B6, D7, E1, E4, E5, and F5) activated KIR2DS1 reporter cells (referred to as positive clones in what follows), while the other "negative" clones (A6, A8, D6, F2, F8, and G1) did not. **Figure 4B** depicts a representative coculture experiment where HFFFs were infected with six randomly selected clones in a 10-fold serial dilution for 72 h and then cocultured with KIR2DS1 reporter cells. The reporter cells were highly activated by HFFFs infected with the positive clones; E1, E4, and F5. They were not activated after infecting with negative clones; G1, D6, and F8 (**Figure 4B**). Infecting HFFFs with the positive clones resulted in a higher percentage of GFP-positive KIR2DS1 reporter cells (ranging from 16% by E4 clone to 50% by B6 clone) compared to the wild-type TB40/E strain (ranging from 4 to 21%, Figure S1 in Supplementary Material), indicating that the positive clones induced the KIR2DS1 ligand more efficiently. This is in line with the idea that the parental TB40/E strain contains a mixture of positive and negative viruses with respect to KIR2DS1 ligand expression.

Human cytomegalovirus infection indeed led to reduction in total HLA class I expression on most HFFFs as monitored by binding of W6/32 antibody and this was true of both positive and negative clones (**Figure 6A**). This indirectly indicates that the positive and negative clones were equally infectious and consistently infected over 90% of the HFFFs. In addition, the differential response of the KIR2DS1 reporter cells to the positive and negative clones was not due to differences in overall viral particle numbers of the different clones. This was demonstrated by comparing the infectivity (functional virus particles) with the total number of viral particles (functional and empty/non-functional particles) by detecting a structural tegument protein pp28 by western blot after pairing positive and negative virus clones based on approximately the same infectivity titer (**Figure 4C**).

KIR2DS1 Ligand Is Only Expressed on HFFFs after Infecting with Infectious Virus

Within the investigated infection timeframe, non-infected cells could be refractory for HCMV infection, yet exposed to pro-inflammatory cytokines, such as type I interferons (IFNs), which subsequently induce HLA class I surface expression. It was therefore critically important to investigate whether KIR2DS1 reporter cells were activated by the infected cells or by the surrounding non-infected cells. To examine this, TB40/E viruses were exposed to UV light for 30 min to inactivate the virus. After exposure, viral particles will be present and able to enter the cell, but the viral genes will be inactivated and will not be transcribed. Virus inactivation was confirmed by immunohistochemistry staining of Immediate Early 1 (I.E.1) viral proteins on the treated HFFFs, as shown in Figure S2 in Supplementary Material. These UV-inactivated viruses are called "UV virus."

After stimulating HFFFs with UV B6 clone and infecting with the positive B6 and negative A8 clones for 72 h, a coculture with parental, KIR2DL1, LILRB1, and KIR2DS1 reporter cells was performed. KIR2DS1 reporter cells were significantly activated in coculture with the active B6 clone-infected HFFFs, but not with UV B6-stimulated HFFFs nor in any other conditions, as shown in **Figure 4D**. By contrast, the KIR2DL1 reporter cells were highly activated after encountering UV B6-stimulated HFFFs, as well as with B6 clone-infected HFFFs (**Figure 4D**). Total HLA class I and HLA-C/-E cell-surface levels using W6/32 and DT9 antibodies were assessed and both HLA-A, -B, -C and HLA-C/-E surface expression levels were highly increased on UV virus-stimulated HFFFs (**Figure 6**). Since fibroblasts are known to produce IFN-α and -β (75, 76), it was expected to see such an increase in HLA class I cell-surface levels, after exposure to viral particles. This is the host response to HCMV particles without the interference of HCMV genes downregulating HLA class I molecules.

In conclusion, from the different HCMV strains tested, only the TB40/E strain activated KIR2DS1 reporter cells after infecting HFFFs. Furthermore, the TB40/E strain consists of different virus clones and these virus clones differentially activated KIR2DS1 reporter cells. Since the clones have comparable amounts of functional viral particles, the levels of activation were not governed by the number of viral particles. We may also conclude that active viral gene expression is necessary to induce the KIR2DS1 ligand.

Primary, Single-Positive KIR2DS1 NK Cells Are Only Activated in Coculture when HFFFs Are Infected with Specific HCMV Clones

After establishing that KIR2DS1 reporter cells recognize a ligand on HFFFs infected with specific clones, we asked whether primary NK cells would also interact with these infected cells. Coculture experiments were performed using freshly isolated peripheral blood NK cells from healthy individuals. NK cells expressing KIR2DS1 are hyporesponsive, if the donor is homozygous for C2-HLA-C (32). We therefore chose donors that were C1-HLA-C homozygous, bearing fully functional KIR2DS1-positive NK cells. General NK cell functionality was verified using IL-12/15 primed NK cells cocultured with the prototypic HLA class I negative NK cell target cell line K562 (positive control), or with no target cells (negative control). The KIR$^-$ NK cell subset represents the KIR-independent activation of NK cells, which is the background NK cell activation in this experiment.

Forty-nine percent of the cytokine-primed KIR$^-$ NK cells degranulated after NK cells of donor 016 encountered K562 cells, and degranulation of 4% was observed in culture without target cells, indicating that the NK cells are functional, and little background activation was observed (**Figure 5A**). In the same experiment, rested NK cells (without cytokine stimulation) were cocultured with HFFFs. NKG2A$^-$ cells were separated into KIR-negative (KIR$^-$), KIR2DL1 single-positive (2DL1sp), and KIR2DS1 single-positive (2DS1sp) NK cells. The gating strategy is described in Figure S3 in Supplementary Material. No response (1% CD107a expression) was seen from 2DS1sp NK cells cocultured with uninfected or UV B6 clone-stimulated HFFFs (**Figure 5B**). Similarly, minimal functional response from 2DS1sp NK cells was observed after coculture with HFFFs infected with the negative A8 clone (5% CD107a expression). Notably, 2DS1sp NK cells engaging HFFFs infected with the positive B6 clone showed 20% CD107a expression (**Figure 5B**). Only a slight background activation of 3% was observed in the KIR$^-$ population in coculture with positive B6 clone-infected HFFFs. In all the other conditions, no degranulation in the KIR$^-$ population was observed. Furthermore, in every condition, including the positive B6 clone condition, no degranulation of 2DL1sp NK cells was observed (**Figure 5B**). The experiment was repeated with isolated NK cells of donor 111, and these NK cells responded similarly, though slightly weaker, to the different types of treated HFFFs (**Figure 5C**). These results were reproducible in two independent coculture experiments using either isolated NK cells (Figure S4A in Supplementary Material) or PBMCs (Figure S4B in Supplementary Material).

In conclusion, similar to the reporter cells, primary NK cells expressing KIR2DS1 recognize a ligand on HFFFs infected with specific clones of the TB40/E strain.

Specific HCMV Clones Are Less Effective in Targeting HLA-C

After confirming that the KIR2DS1 ligand is expressed on HFFFs infected with positive clones of HCMV, different HLA class I surface levels of uninfected HFFFs, UV B6 clone-stimulated

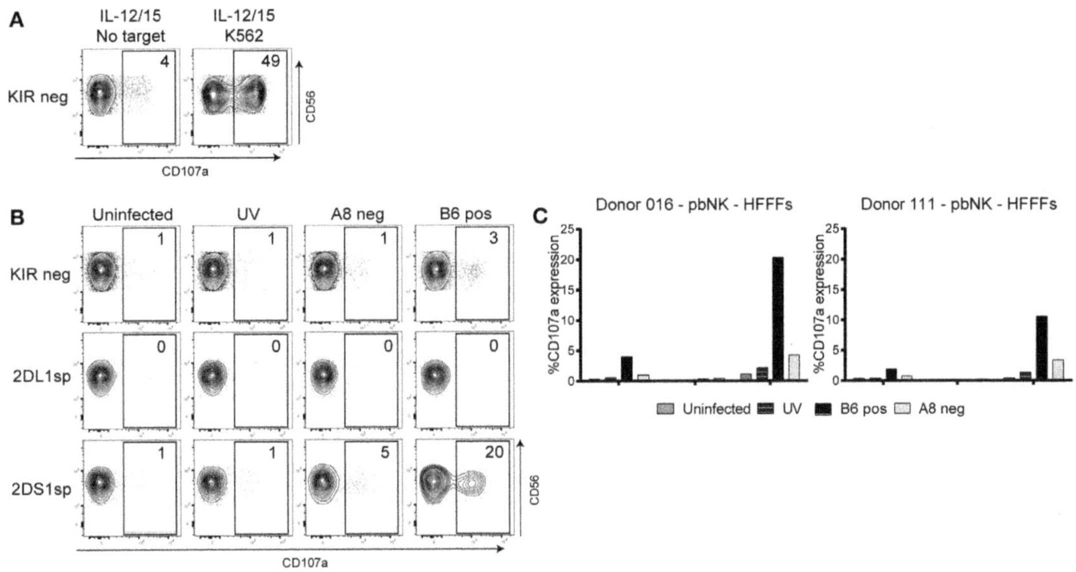

FIGURE 5 | Primary natural killer (NK) cells degranulated after coculture with human fetal foreskin fibroblasts (HFFFs) infected with the positive B6 TB40/E clone. **(A)** After NK cell isolation, the NK cells of donor 016 were rested or stimulated with IL-12 and IL-15 overnight. IL12/IL15-stimulated NK cells were cocultured with and without K562 cells. **(B)** Unstimulated NK cells were cocultured with uninfected HFFFs, UV B6 clone-stimulated HFFFs (UV), negative A8 clone-(A8 neg), and positive B6 clone-infected HFFFs (B6 pos). After harvesting, the NK cells were stained with antibodies targeting different surface markers and the degranulation marker CD107a. NK cells from the KIR-negative (KIR), KIR2DL1 single-positive (2DL1sp), and KIR2DS1 single-positive (2DS1sp) subsets are shown. The percentage of CD107a expression is depicted. The NK subsets are grouped on NKG2A⁻ NK cells (A⁻). **(C)** The same coculture experiment with NK cells of donor 016 is represented in a bar graph (left) including coculture data of NK cells of donor 111 (right). The percentage of CD107a expression is shown. All cocultures were performed for 5 h at an E:T ratio of 1:1. A representative data from three independent coculture experiments are shown. The other two independent experiments are illustrated in Figure S4 in Supplementary Material.

HFFFs, and both positive and negative clone-infected HFFFs were compared. Surface expression of HLA-E (3D12), HLA-C/-E (DT9), total HLA class I (W6/32), HLA-A11, HLA-Bw6, total free heavy chain (FHC) of HLA class I (HC10), and FHC of HLA-C (L31) was analyzed. The antibodies used are listed in **Table 1** with a description of their specific recognition patterns and are discussed below.

Free Heavy Chain of HLA-C and Assembled HLA-C Remain on the Cell Surface of HFFFs Infected with Positive Clones

Positive and negative clones both reduced total HLA class I, HLA-E, HLA-Bw6, and HLA-A11 cell-surface levels on infected HFFFs compared to the UV B6 clone-stimulated HFFFs ($p < 0.05$ and $p < 0.01$, **Figures 6A,B**). However, using antibodies to FHC forms of HLA, such as HC10 and, in particular, L31 staining was higher in HFFFs infected with the positive clones compared to the negative clones. FHC HLA-C levels, as detected by L31, were close to the levels of UV B6-stimulated HFFFs, indicating that the positive clones did not effectively downregulate the FHC HLA-C surface levels (**Figures 6A,B**). This difference observed in L31 staining was reproducible in every experiment performed ($n = 11$) and in an experiment where FHC HLA-C surface levels were compared in HFFFs infected with 12 additionally isolated clones comprising six positive and six negative clones (Figure S5A in Supplementary Material). Similar differences between the clones were observed with HC10 staining (FHC of HLA class I molecules, **Figures 6A,B**). This difference could be due to the antibody detecting the elevated FHC of HLA-C specifically. However, this remains uncertain, because specific antibodies against FHC of HLA-A and -B are not available.

The DT9 staining (detects conformational HLA-C/-E containing β_2M) was slightly weaker and less consistent than L31, but similar differences between the clones were measurable (**Figures 6A,B**). Since DT9 antibody cross-reacts with HLA-E (56), an anti-HLA-E monoclonal antibody (3D12) was included. HLA-E surface expression in general was very low on HFFFs (**Figure 6A**) and no difference was observed between HFFFs infected with positive and negative HCMV clones (**Figure 6B**). Low HLA-E expression in the infected HFFFs is most likely due to a mutation at position 2 (Met to Val) in the canonical sequence (VMAPRTLIL) of UL40 expressed in all the TB40/E clones, as described previously (77). This result implies that the difference observed in the DT9 staining is, most likely, due to higher HLA-C expression and not HLA-E expression. In addition, assembled HLA-C and FHC of HLA-C surface levels on HFFFs infected with other viral strains, such as Merlin and AD169, were downregulated comparable to the surface levels found with the negative clones (Figure S6 in Supplementary Material). Together, these results indicate that the positive clones are less effective in downregulating assembled HLA-C and, in particular, FHC of HLA-C in HFFFs, compared to the negative clones and other HCMV strains. However, both sets of clones downregulated other HLA molecules to similar levels.

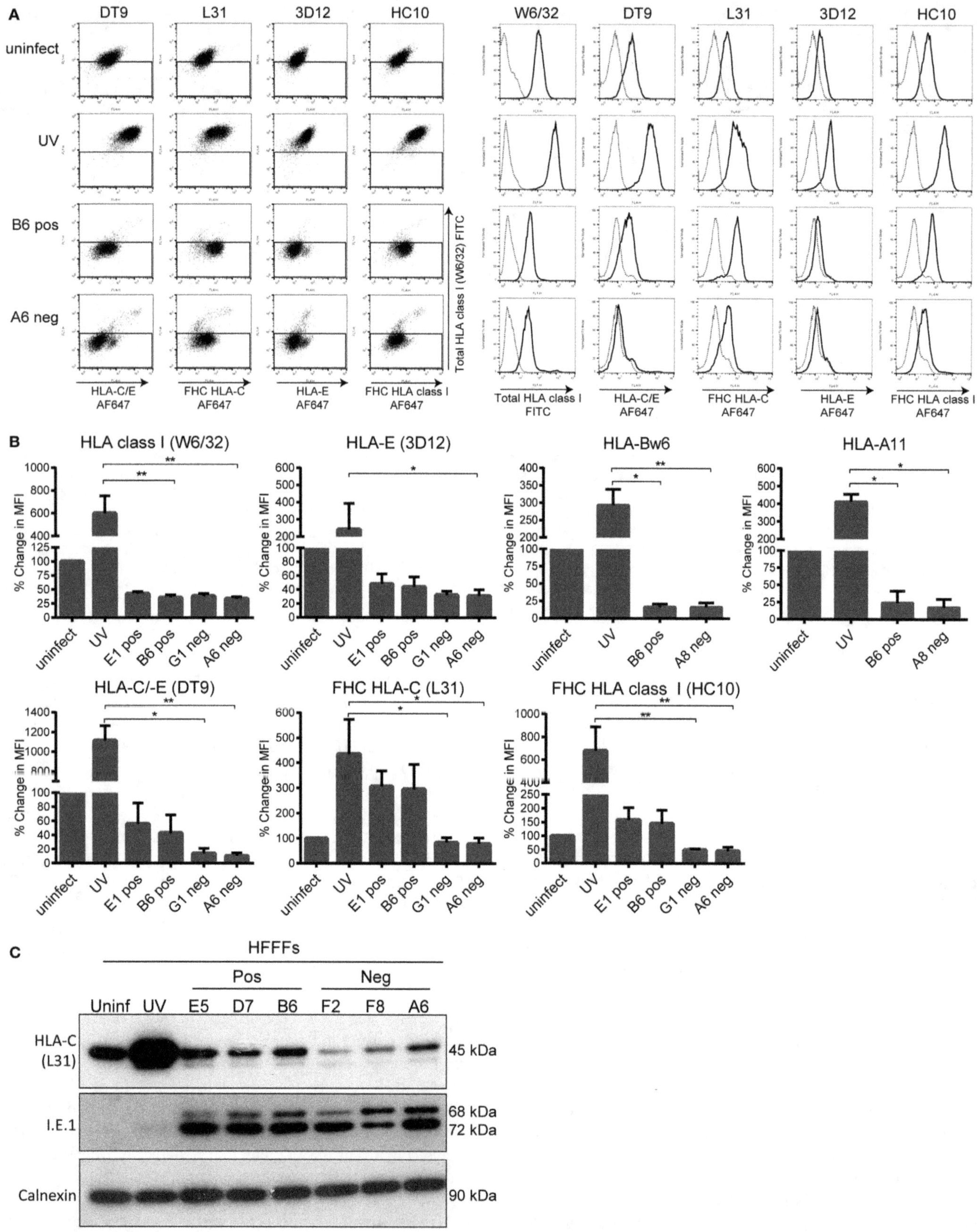

FIGURE 6 | Continued

FIGURE 6 | Continued
Human leukocyte antigen (HLA) class I surface and total HLA-C protein expression of infected human fetal foreskin fibroblasts (HFFFs). (A) Dot plots (left) and histograms (right) from the same representative experiment are shown. The HFFFs infected with the indicated positive or negative clones and the UV B6 clone-stimulated HFFFs (UV) were treated for 72 h. The indicated treated HFFFs were stained subsequently with W6/32 (total HLA class I), DT9 (HLA-C/-E), 3D12 (HLA-E), L31 (FHC HLA-C), and HC10 (FHC HLA class I) antibodies (black line). Cells stained with the appropriate isotype control (dotted line) were included. **(B)** Collection of up to four independent cell-surface staining experiments is depicted in bar graphs as the mean ± SD of individual samples. Anti-Bw6 and anti-A11 antibodies were also included. The percentage change in MFI is depicted with the uninfected condition at 100%. $^{*}p < 0.05$ and $^{**}p < 0.01$ are calculated by non-parametric one-way ANOVA using the Kruskal–Wallis test and Dunn's multiple comparisons test. **(C)** Total lysate of uninfected, UV B6 clone-stimulated (UV), positive clones- (E5, D7, B6), and negative clones (F2, F8, A6)-infected HFFFs were loaded onto a reduced 10% SDS-PAGE gel. The membrane was blotted with the L31, anti-I.E.1 viral protein (measure of infection) and anti-calnexin (loading control) antibodies.

More Total HLA-C Protein Is Expressed on HFFFs Infected with Selected Clones of HCMV

After having found that HLA-C and, in particular, FHC of HLA-C cell-surface levels were higher on positive clone-infected HFFFs, we investigated whether differential expression of total HLA-C protein levels could be tracked by western blot. HFFFs infected with three positive clones had higher amounts of total HLA-C compared to the negative clones, although the levels in general were lower than for the uninfected and UV B6 clone samples. As expected, HFFFs stimulated with UV B6 clone contained high amounts of HLA-C protein (**Figure 6C**). Additionally, we tested HFFFs infected with a further 12 clones (6 positive and 6 negative, Figure S5B in Supplementary Material). From the 18 clones tested in total, 7 positive clones had high amounts of total HLA-C, while 7 negative clones had lower amounts.

In conclusion, positive TB40/E clones are defective in down-regulating HLA-C and, in particular, FHC of HLA-C in HFFF cells. Higher amounts of total HLA-C protein were found in HFFFs infected with the positive TB40/E clone compared to infection with the negative TB40/E clone. This may reflect a deficiency in degradation/turnover of HLA-C by positive TB40/E clones.

Pan HLA Class I Antibodies Block the KIR2DS1–Ligand Interaction

Antibody-blocking experiments were performed using various anti-HLA class I antibodies to obtain a better understanding of the KIR2DS1 interaction. Previously, Stewart and colleagues tested a large panel of pan HLA class I antibodies in a blocking experiment to analyze the interaction of KIR2DL1 and KIR2DS1 tetramers with 221-HLA-C transfectants (27). They concluded that W6/32 and HC10 antibodies were not able to block both KIR2DL1 and KIR2DS1 tetramer interactions with C2-HLA-C. However, other pan HLA class I antibodies, such as 6A4 and B1.23.2, did block (27). To confirm the previous findings, the interaction of KIR2DL1 reporter cells with UV B6-stimulated HFFFs, which highly expressed all HLA class I molecules, was tested. As expected, W6/32 antibody did not block the KIR2DL1 interaction with the HFFFs, but the 6A4 and B1.23.2 antibodies did (**Figures 7A,B**, left panels). Next, the HFFFs were infected with the positive B6 clone and the antibody blocking experiment was repeated. W6/32, 6A4, and B1.23.2 antibodies blocked the KIR2DS1 reporter cell interaction with B6 clone-infected HFFFs. Notably, the W6/32 antibody did not block and the 6A4 and B1.23.2 antibodies only partially blocked the KIR2DL1 reporter cell interaction (**Figures 7A,B**, right panels). Furthermore, DT9, L31, and HC10 antibodies were also not able to block the KIR2DS1–ligand interaction and the anti-β_2M only partially (data not shown). Together, these data are consistent with a differential mode of recognition of C2-HLA-C by KIR2DL1 and KIR2DS1.

HFFFs with a β_2M KO Infected with Positive HCMV Clones Do Not Induce the KIR2DS1 Ligand

To confirm whether the KIR2DS1 ligand is a HLA class I molecule and specifically HLA-C, the β_2M gene was knocked out of HFFFs. The HLA class I complex cannot be formed without β_2M and therefore, HLA class I molecules, but also FHC of HLA class I, will not be transported efficiently to the cell surface (78). The *β_2M* gene was knocked out by using the CRISPR/CAS9 genome editing tool (79). After selection and single-cell sorting, β_2M KO HFFFs were checked for β_2M, total HLA class I (W6/32), and FHC of HLA-C (L31) surface expression by flow cytometry and total protein expression by western blot. β_2M KO HFFFs did not express surface β_2M, total HLA class I, and FHC of HLA-C. B6 and A8 clone infection did not alter these expression levels (**Figure 8A**). Total β_2M, total HLA class I (detected with HC10), and most HLA-C protein were also absent in the β_2M KO HFFFs, compared to the untreated HFFFs (WT) and HFFFs containing CAS9 without the sgRNA (CAS9), indicating that the β_2M KO was successful (**Figure 8B**).

Subsequently, β_2M KO HFFFs and WT HFFFs, either stimulated with UV B6 clone or infected with a positive and negative clone, were cocultured with the KIR2DS1, -L1, -L2, and LILRB1 reporter cells. The KIR2DS1 reporter cell was not activated after encountering β_2M KO HFFFs infected with the positive B6 clone, while, in the same experiment, 44% of KIR2DS1 reporter cells were GFP positive after coculture with B6 clone-infected WT HFFFs (**Figure 8C**). KIR2DL1 and LILRB1 reporter cells were not significantly triggered by β_2M KO HFFFs in all the conditions (**Figure 8C**).

This result is consistent with KIR2DS1 recognizing an HLA class I molecule, including HLA-C. Together with the cell-surface staining and the antibody-blocking experiment, the data suggest that modulation of C2-HLA-C by HCMV induces a potent KIR2DS1-mediated NK cell activation.

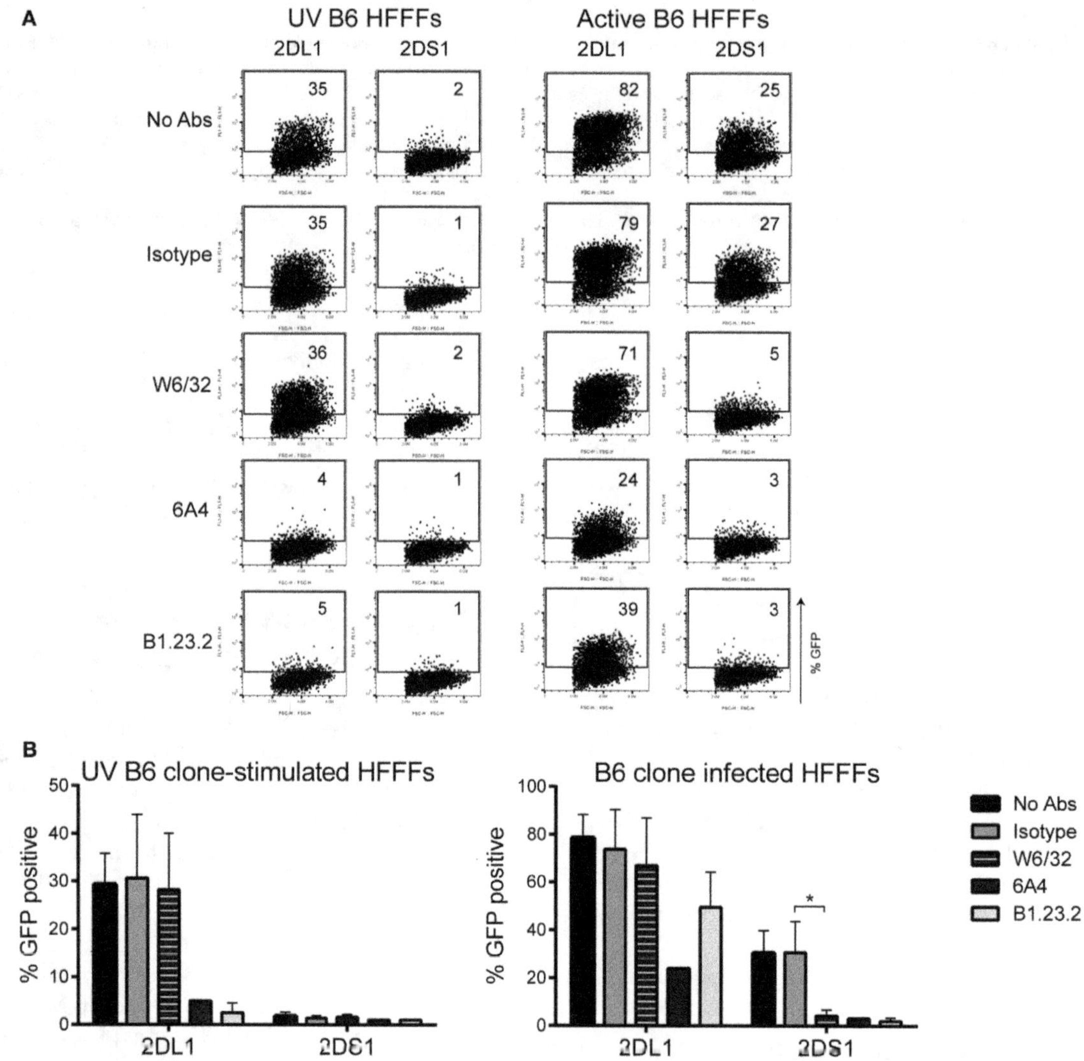

FIGURE 7 | Antibody-blocking experiment of reporter cells and HFFFs stimulated with UV B6 clone or infected with the B6 clone. W6/32, 6A4, B1.23.2, and isotype antibodies were added to 72 h UV-stimulated or infected HFFFs, before an overnight coculture with KIR2DS1 and KIR2DL1 reporter cells was performed. GFP expression was measured by flow cytometry. **(A)** The dot plots are data from a representative experiment. **(B)** Collection of up to five independent blocking experiments is depicted in bar graphs as the mean ± SD of individual samples with $^*p < 0.05$ calculated by non-parametric one-way ANOVA using the Kruskal–Wallis test and Dunn's multiple comparisons test.

DISCUSSION

We found that KIR2DS1 recognizes a ligand on HFFFs infected with the TB40/E strain of HCMV. This wild type strain consists of at least two sets of virus clones: one set that, after HFFF infection, activates KIR2DS1-expressing cells (positive clones) and one that does not (negative clones). This specific KIR2DS1 recognition was only present when the HFFFs were infected with fully functioning viruses, indicating that the virus plays a direct role in expressing the KIR2DS1 ligand. In addition, KIR2DS1 single-positive (2DS1sp) primary NK cells degranulated after engaging with positive clone-infected HFFFs. Together, this indicates that KIR2DS1 reporter cell activation correlates with physiological KIR2DS1 binding to its ligand.

The Reporter System Is a Valid Way of Examining KIR Specificity

Using the reporter system, we confirmed that a single amino acid substitution K70T in KIR2DS1 altered the recognition from no binding to binding the same cognate ligands as KIR2DL1 (**Figure 1C**). This confirms the important role of residue 70 in the binding avidity to HLA-C by KIR2DL1/2DS1. These findings together with the differential recognition of HLA-C alleles by LILRB1, KIR2DL1, and -2DL2 reporter cells (**Figure 2C**) indicate that the reporter system is sensitive to subtle differences. One possible explanation of KIR2DL1 and -L2 reporter cells responding minimally to DFs (**Figure 3B**) is that there is binding of KIR to its ligand, but this binding is not strong enough to trigger a signaling cascade to activate the reporter cell. The signaling cascade

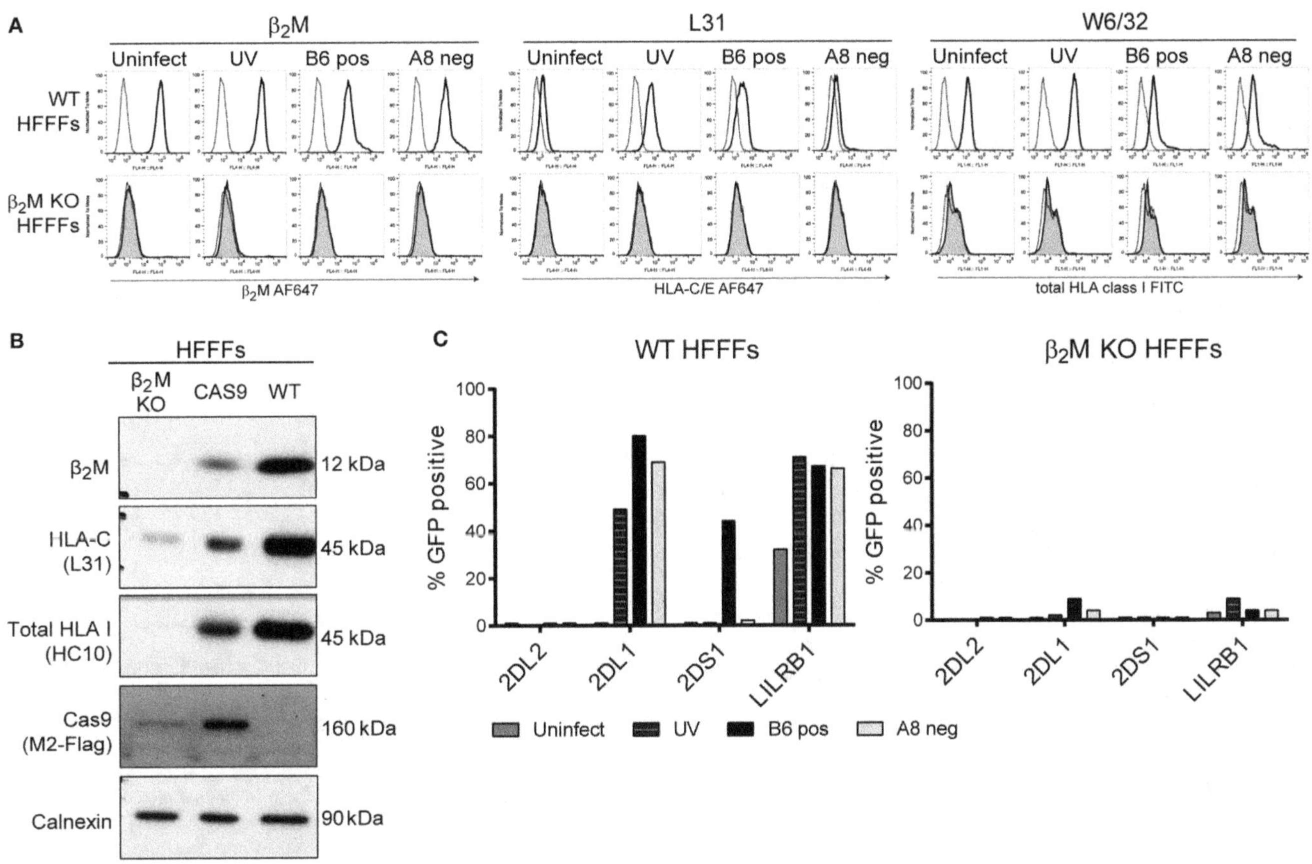

FIGURE 8 | β_2M KO HFFFs do not activate KIR2DS1 reporter cells. (A) To control the success of the β_2M knockout, untreated (WT) HFFFs (black line) and β_2M KO HFFFs (shaded gray line) were stained with anti-β_2M, W6/32, and L31 antibodies. Before staining, the cells were stimulated with UV clone (UV) or infected for 72 h with the positive B6 clone (B6 pos) or the negative A8 clone (A8 neg). Unstained cells were included (dotted line). **(B)** Total lysate of β_2M KO HFFFs, HFFFs containing CAS9 without sgRNA (CAS9) and WT HFFFs were loaded onto a reduced 10% SDS-PAGE gel. The membrane was blotted with the anti-β_2M, L31, HC10, anti-M2 Flag, and anti-calnexin (loading control) antibodies. **(C)** β_2M KO HFFFs and WT HFFFs were stimulated with UV B6 clone (UV) and infected with the positive B6 clone (B6 pos) or negative A8 clone (A8 neg) for 72 h. An overnight coculture was performed with KIR2DL2 (negative control), KIR2DL1, KIR2DS1, and LILRB1 reporter cells. GFP expression was measured by flow cytometry. This experiment was performed twice.

will only be triggered if there is a true interaction. The degree of receptor/ligand clustering might also influence the downstream signaling, as shown by Oszmiana et al. (80). Therefore, the reporter cells may be a more physiological system compared to Fc proteins or tetramers.

KIR2DS1 Interacts with HLA-C but Not in the Same Way as KIR2DL1

There are several examples of "paired" immunoreceptors consisting of almost identical external moieties with positive and negative signaling tails, respectively (13). It is believed, but by no means proven, that this situation is driven by host–pathogen interaction. Our data are broadly consistent with this proposal. Our data fit with KIR2DL1, the inhibitory receptor, interacting with C2-HLA-C for recognition of a self-ligand in order to promote education/licensing of NK cells and subsequent loss of inhibition when the ligand is missing. By contrast, the role of aKIR has been unclear. Some groups reported weak binding, particularly for KIR2DS1, but these effects are inconsistent. In general, aKIR appear to interact with HLA molecules weakly except in certain circumstances (27, 41, 42). We found that KIR2DS1 reporter cells were not activated after engaging conventional HLA class I molecules (**Figures 2** and **3**), as shown previously by others (10, 11, 17, 24–35). There is a possibility that the target cells used in previous KIR2DS1 studies did not only express conventional HLA-C molecules. These cells were EBV positive (721.221 cells, BLCLs, C1R cells), and/or tumor-derived cells, such as leukemia blasts and lymphomas. In addition, other studies included primary cells such as DCs (33), T cells (30, 33), B cells (27, 29), and MRC-5 fibroblasts (supplementary data of Stewart et al.) KIR2DS1 did not interact with or bind these primary cells, unless they were stimulated: in the case of DCs, stimulated with LPS and T cells, stimulated with PHA to form T cell blasts. Furthermore, Crespo et al. found that HCMV-infected JEG-3 cells and fetal extravillous trophoblasts (EVT) did not induce degranulation and cytokine production of dNK cells. They only found a cytotoxic response when dNK cells were exposed to HCMV-infected DSC specifically, indicating differential recognition of dNK cells to HCMV-infected cells (47). One explanation could be that KIR2DS1-mediated NK cell activation could only occur through an unknown synergistic engagement of other activating receptors, as proposed by Bryceson et al. (81). Additionally, perhaps a high level of C2-HLA-C is needed for a potent KIR2DS1-mediated NK cell activation, which is the case for these target cells. Primary cells might express too low levels of C2-HLA-C and in combination with the weak binding of KIR2DS1 to C2-HLA-C might result in the absence of NK cell activation. Crespo et al. also found reduced levels of HLA-C on HCMV-infected DSC leading the authors to speculate that an unknown activating ligand for KIR2DS1 is upregulated by HCMV infection which is recognized by dNK cells (47). Alternatively, our findings indicate that KIR2DS1 might recognize a modified form of C2-HLA-C, which is induced by selected TB40/E clones in HFFFs. Compared to HCMV from other sources, positive TB40/E clones were less effective in controlling FHC of HLA-C and, to a certain extent, assembled HLA-C (**Figure 6**). W6/32, 6A4, and B1.23.2 antibodies were able to block the KIR2DS1–ligand interaction on positive clone-infected HFFFs, while these antibodies could not block, or only partially block, the KIR2DL1 interaction (**Figure 7**). β_2M KO experiments confirmed that KIR2DS1 binds a HLA class I molecule, most likely HLA-C, on these infected cells (**Figure 8**). It is unlikely that KIR2DS1 was binding directly to an HCMV-encoded protein, since KIR2DS1 reporter cells were not activated in coculture with β_2M KO HFFFs infected with the positive clone. The most parsimonious interpretation is that the virus influences the balance of recognition directly of HLA-C by the KIR2DL1/S1 pair. At this stage, however, other possibilities cannot be ruled out, such as a combination of HLA class I with another protein.

What Is the Difference Between HLA-C Recognized by KIR2DS1 and by KIR2DL1?

Since all the data indicate that HLA-C forms at least part of the KIR2DS1 ligand the question that arises is how does HLA-C differ upon viral infection such that KIR2DS1 is brought into play? **Figure 9** suggests various models which may be tested, namely:

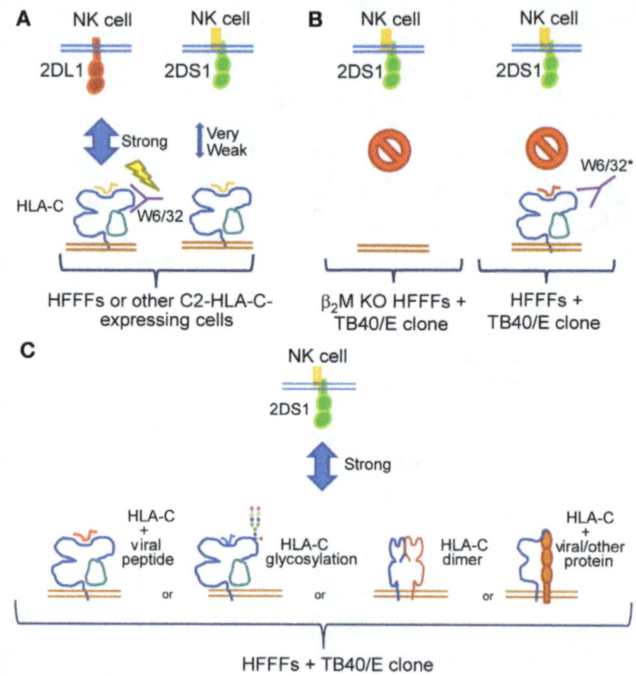

FIGURE 9 | Summary of the findings and working models. (A) Both KIR2DL1 (red) and -2DS1 (green) bind conventional group C2 human leukocyte antigen (HLA)-C; however, KIR2DS1 binds weakly. Our data were consistent with previous studies demonstrating that the W6/32 antibody does not block the KIR2DL1–HLA-C interaction. The lightning bolt indicates KIR2DL1 triggering even after adding W6/32 to the coculture. **(B)** Strong KIR2DS1 activation was observed after coculture with HFFFs-infected selected TB40/E clones. After infecting β_2M KO HFFFs with selected TB40/E clones, KIR2DS1 interaction was diminished (left). In addition, W6/32 blocked the KIR2DS1–ligand interaction after coculture with HFFFs-infected selected TB40/E clones (right). *Other pan class I antibodies, such as 6A4 and B1.23.2, blocked both KIR2DL1 and KIR2DS1 interaction. The red symbol indicates no binding of KIR2DS1 after coculture in the indicated setting. **(C)** Remaining FHC of HLA-C and, to certain extent, assembled HLA-C surface expression was found on infected HFFFs. Together with the other findings, we hypothesize that a modification of C2-HLA-C is induced by HCMV, which influences KIR2DS1 recognition. This modification could relate to the following: presentation of HCMV-derived peptides; alteration of glycosylation patterns on the HLA-C molecule; formation of HLA-C homodimers; or heterodimers association with another protein.

1. *Bound peptide*: there have been reports of KIRs recognizing certain peptide motifs presented by HLA class I molecules. Stewart and colleagues have demonstrated that amino acids at position 7 and 8 of the peptide play a role in KIR2DS1 binding. KIR2DS1 has similar peptide selectivity to KIR2DL1 (27). Work from Khakoo's group has shown that peptides with certain motifs have either strong inhibitory, low inhibitory, or antagonistic effects on KIR2DL2 and -2DL3$^+$ NK cells (82). These data suggest that NK cells are able to sense alterations of target cells, through selective peptide recognition.
2. *Modified glycosylation*: KIR3DL1 binding to HLA-B57:01 is dependent on the N-glycan on HLA-B57:01. Removing the N-glycan resulted in reduced inhibition, thus increasing degranulation of KIR3DL1$^+$ NK cell clones (38). The N-glycosylation site on HLA class I is highly conserved and the glycan structures on HLA-C are relatively uniform between HLA-C allotypes (83). Perhaps viral infections could alter these glycosylation patterns and break the uniformity, resulting in the recognition of KIR2DS1. HIV and HCV infection has been shown to alter glycosylation in host cells, due to ER stress (84, 85). This might change the glycosylation pattern of HLA class I.
3. *Formation of HLA-C homodimers or heterodimers with other HLA class I molecules*: KIR3DL2 binds FHC forms of HLA-B27, including HLA-B27 dimers (86). LILRB1 interacts with different HLA class I molecules, which are able to dimerize via a cysteine residue in the cytoplasmic tail (49). Recently, it is described that the ligand for KIR3DS1 is FHC of HLA-F (41, 42). These data are evidence of the capability of Ig-like receptors to bind FHC HLA molecules or HLA homodimers.
4. *Modification by formation of HLA-C heterodimers with a virus-derived protein*: in mice, activating receptor Ly49P binding to an MHC-CMV protein heterodimer has been reported.

Ly49P binds CMV-infected cells expressing a complex of the m04 CMV protein together with the MHC class I molecule H-2D^k (87). Very recently, Serena et al. proposed that HIV-1 Env protein associates with FHC of HLA-C and that HIV-1 specifically upregulates FHC of HLA-C at the cell surface of infected cells (88). Their work has some parallels with our findings. Our preliminary experiments, however, did not show any evidence of HLA-C forming homo- or heterodimers (data not shown).

Based on these observations together with our findings, we argue that KIR2DS1 most likely binds C2-HLA-C either through recognizing HCMV-derived peptide or changes in glycosylation patterns. This will be the focus of future experiments.

Have Activating KIRs Evolved to Recognize Infected Cells?

Studies investigating the evolution of KIRs and other paired receptors have proposed that the activating members may be evolving more rapidly than the inhibitory members through selection imposed by pathogens (15, 16). The positively charged lysine at position 70 in KIR2DS1 is critical for the diminished binding to C2-HLA-C compared to KIR2DL1. This amino acid is conserved in all KIR2DS1 allotypes (with the exception of KIR2DS1*001, which has a charged arginine) (11). Conversely, the lysine at position 70 in KIR2DS1 could be crucial for binding modified HLA-C induced by pathogens. KIR2DL1 might have evolved to recognize HLA-C on healthy cells (induced-self), while KIR2DS1 might recognize slight structural changes on HLA-C induced by pathogens (altered-self). KIR2DS1 may still bind conventional HLA-C weakly to secure tolerance, yet recognition by KIR2DS1 of a modified form of HLA-C could overcome this tolerance. Our findings favor this hypothesis.

Why Were KIR2DL1 Reporter Cells More Activated after Encountering TB40/E-Infected HFFFs?

The GFP expression of KIR2DL1 reporter cells, after encountering TB40/E-infected HFFFs, was even higher than in coculture with UV TB40/E-stimulated HFFFs, which contain high expression levels of HLA-C (**Figure 4A**). HCMV downregulates HLA class I molecules, thus these findings were unexpected. A possible explanation is that KIR2DL1 binds the remaining HLA-C on the infected cell surface (**Figure 6B**). Ameres et al. reported that HCMV downregulates certain HLA-A and -B alleles more efficiently than HLA-C alleles (89, 90). Another explanation could be that KIR2DL1 recognizes an alternative ligand, comparable to how both inhibitory Ly49I and activating Ly49H-binding m157 (43, 91).

Differences in HCMV Isolates

Natural killer cell responses appear to differ when encountering cells infected with various HCMV strains and even clones within strains. Chen et al. also demonstrated that the ability of NK cells to control virus spread through LILRB1 was variable between HCMV viral strains, depending on the amino acid sequence within UL18 (92). Thus, the variable effect between HCMV strains on NK cell activity and vice versa should be taken into account when setting up experiments and interpreting published data. This will also count for cytotoxic T cells and other immune cell responses.

TB40/E-derived positive clones were the only viruses that, upon infecting HFFFs, expressed the KIR2DS1 ligand. Together, these findings imply that after infection the positive clones are less successful in downregulating the KIR2DS1 ligand than TB40/E-derived negative clones or other HCMV strains. As a result, KIR2DS1 reporter cells and 2DS1sp NK cells are specifically detecting the ligand on positive clone-infected HFFFs. This suggests that the other HCMV strains are capable of downregulating the KIR2DS1 ligand as an immune evasion strategy to NK cells. Our findings could explain why this interaction has not been detected in previous studies. Identification of the differences between the positive and negative clones by whole virus genome sequencing should help to resolve this issue.

CONCLUSION

Our findings indicate that activating KIRs do not bind the same conventional HLA molecules as their inhibitory counterparts. They suggest that pathogenic infections are required for strong activating KIR binding, discriminating between healthy and unhealthy cells. To our knowledge, this is the first time that the role of HCMV on activating KIR recognition has been conclusively shown. Future work will provide new insights into the role of NK cells in HCMV infection and transplantation. This could lead to more targeted and effective therapeutic avenues in the treatments for HCMV infection in new-born babies, immunosuppressed individuals, and patients undergoing solid organ or HSCT transplantation.

ETHICS STATEMENT

This study was carried out in accordance with the recommendations of Addenbrookes National Health Service Hospital Trust institutional review board, Cambridgeshire 2 Research Ethics Committee (REC 97/092) with written informed consent from all subjects. All subjects gave written informed consent in accordance with the Declaration of Helsinki. The protocol was approved by the Cambridgeshire 2 Research Ethics Committee.

AUTHOR CONTRIBUTIONS

KP, MW, and JT conceived and designed the experiments. KP and MI performed the experiments and analyzed the data. CC designed and provided the reporter cells. KP, MW, JT, CC, MI, and AM discussed the data and commented on the manuscript. KP and JT wrote the manuscript.

ACKNOWLEDGMENTS

The authors would like to thank Nigel Miller at the Department of Pathology, colleagues at the NIHR Cambridge BRC Cell Phenotyping Hub for the assistance with the flow cytometry and the cell sorting, and NIHR UK Biomedical Research Centre for their support. The authors are grateful to all the blood and DF donors for their contribution to this project. Also, many thanks go to Georgina Okecha for all the experimental support. Moreover, their gratitude goes to Andrew Davison at the University of Glasgow for the collaboration and helpful discussions. Finally, the authors are thankful for all the researchers that provided them with the necessary cells and reagents.

Training Network, FP7-PEOPLE-2012-ITN-317013, grant agreement No: 317013), the European Research Council (ERC) under the European Union's Horizon 2020 research and innovation programme (Grant agreement No: 695551), the Wellcome Trust (Grant agreement No: WT094107/Z/10/Z) and the UK Medical Research Council (Grant agreement No: 0701279 and MR/K021087/1).

REFERENCES

1. Kiessling R, Klein E, Wigzell H. 'Natural' killer cells in the mouse. I. Cytotoxic cells with specificity for mouse Moloney leukemia cells. Specificity and distribution according to genotype. *Eur J Immunol* (1975) 5(2):112–7. doi:10.1002/eji.1830050208
2. Kiessling R, Klein E, Pross H, Wigzell H. 'Natural' killer cells in the mouse. II. Cytotoxic cells with specificity for mouse Moloney leukemia cells. Characteristics of the killer cell. *Eur J Immunol* (1975) 5(2):117–21. doi:10.1002/eji.1830050209
3. Herberman RB, Nunn ME, Lavrin DH. Natural cytotoxic reactivity of mouse lymphoid cells against syngeneic and allogeneic tumors. I. Distribution of reactivity and specificity. *Int J Cancer* (1975) 16(2):216–29. doi:10.1002/ijc.2910160204
4. Carrington M, Norman P. *The KIR Gene Cluster. Bookshelf NCBI*. Bethesda (MD): National Center for Biotechnology Information (2003). Bookshelf ID: NBK45311.
5. Fan QR, Long EO, Wiley DC. Crystal structure of the human natural killer cell inhibitory receptor KIR2DL1-HLA-Cw4 complex. *Nat Immunol* (2001) 2(5):452–60. doi:10.1038/87766
6. Snyder GA, Brooks AG, Sun PD. Crystal structure of the HLA-Cw3 allotype-specific killer cell inhibitory receptor KIR2DL2. *Proc Natl Acad Sci U S A* (1999) 96(7):3864–9. doi:10.1073/pnas.96.7.3864
7. Winter CC, Gumperz JE, Parham P, Long EO, Wagtmann N. Direct binding and functional transfer of NK cell inhibitory receptors reveal novel patterns of HLA-C allotype recognition. *J Immunol* (1998) 161(2):571–7.
8. Moesta AK, Parham P. Diverse functionality among human NK cell receptors for the C1 epitope of HLA-C: KIR2DS2, KIR2DL2, and KIR2DL3. *Front Immunol* (2012) 3:1–13. doi:10.3389/fimmu.2012.00336
9. Moesta AK, Norman PJ, Yawata M, Yawata N, Gleimer M, Parham P. Synergistic polymorphism at two positions distal to the ligand-binding site makes KIR2DL2 a stronger receptor for HLA-C than KIR2DL3. *J Immunol* (2008) 180(6):3969–79. doi:10.4049/jimmunol.180.6.3969
10. David G, Djaoud Z, Willem C, Legrand N, Rettman P, Gagne K, et al. Large spectrum of HLA-C recognition by killer Ig-like receptor (KIR)2DL2 and KIR2DL3 and restricted C1 specificity of KIR2DS2: dominant impact of KIR2DL2/KIR2DS2 on KIR2D NK cell repertoire formation. *J Immunol* (2013) 191(9):4778–88. doi:10.4049/jimmunol.1301580
11. Hilton HG, Guethlein LA, Goyos A, Nemat-Gorgani N, Bushnell DA, Norman PJ, et al. Polymorphic HLA-C receptors balance the functional characteristics of KIR haplotypes. *J Immunol* (2015) 195:3160–70. doi:10.4049/jimmunol.1501358
12. Biassoni R, Cantoni C, Falco M, Verdiani S, Bottino C, Vitale M, et al. The human leukocyte antigen (HLA)-C-specific 'activatory' or 'inhibitory' natural killer cell receptors display highly homologous extracellular domains but differ in their transmembrane and intracytoplasmic portions. *J Exp Med* (1996) 183(2):645–50. doi:10.1084/jem.183.2.645
13. Akkaya M, Barclay AN. How do pathogens drive the evolution of paired receptors? *Eur J Immunol* (2013) 43(2):303–13. doi:10.1002/eji.2012 42896
14. Ivarsson MA, Michaëlsson J, Fauriat C. Activating killer cell Ig-like receptors in health and disease. *Front Immunol* (2014) 5:1–9. doi:10.3389/fimmu.2014.00184
15. Abi-Rached L, Parham P. Natural selection drives recurrent formation of activating killer cell immunoglobulin-like receptor and Ly49 from inhibitory homologues. *J Exp Med* (2005) 201(8):1319–32. doi:10.1084/jem.20042558
16. Barclay AN, Hatherley D. The counterbalance theory for evolution and function of paired receptors. *Immunity* (2008) 29(5):675–8. doi:10.1016/j.immuni.2008.10.004
17. Moesta AK, Graef T, Abi-Rached L, Older Aguilar AM, Guethlein LA, Parham P. Humans differ from other hominids in lacking an activating NK cell receptor that recognizes the C1 epitope of MHC class I. *J Immunol* (2010) 185(7):4233–7. doi:10.4049/jimmunol.1001951
18. Liu J, Xiao Z, Ko HL, Shen M, Ren EC. Activating killer cell immunoglobulin-like receptor 2DS2 binds to HLA-A*11. *Proc Natl Acad Sci U S A* (2014) 111(7):2662–7. doi:10.1073/pnas.1322052111
19. Valés-Gómez M, Reyburn HT, Erskine RA, Strominger J. Differential binding to HLA-C of p50-activating and p58-inhibitory natural killer cell receptors. *Proc Natl Acad Sci U S A* (1998) 95(24):14326–31. doi:10.1073/pnas.95.24.14326
20. Campbell KS, Cella M, Carretero M, López-Botet M, Colonna M. Signaling through human killer cell activating receptors triggers tyrosine phosphorylation of an associated protein complex. *Eur J Immunol* (1998) 28(2):599–609. doi:10.1002/(SICI)1521-4141(199802)28:02<599:AID-IMMU599>3.0.CO;2-F
21. Katz G, Gazit R, Arnon TI, Gonen-Gross T, Tarcic G, Markel G, et al. MHC class I-independent recognition of NK-activating receptor KIR2DS4. *J Immunol* (2004) 173(3):1819–25. doi:10.4049/jimmunol.173.3.1819
22. Graef T, Moesta AK, Norman PJ, Abi-Rached L, Vago L, Older Aguilar AM, et al. KIR2DS4 is a product of gene conversion with KIR3DL2 that introduced specificity for HLA-A*11 while diminishing avidity for HLA-C. *J Exp Med* (2009) 206(11):2557–72. doi:10.1084/jem.20091010
23. VandenBussche CJ, Mulrooney TJ, Frazier WR, Dakshanamurthy S, Hurley CK. Dramatically reduced surface expression of NK cell receptor KIR2DS3 is attributed to multiple residues throughout the molecule. *Genes Immun* (2009) 10(2):162–73. doi:10.1038/gene.2008.91
24. Moretta A, Sivori S, Vitale M, Pende D, Morelli L, Augugliaro R, et al. Existence of both inhibitory (p58) and activatory (p50) receptors for HLA-C molecules in human natural killer cells. *J Exp Med* (1995) 182(3):875–84. doi:10.1084/jem.182.3.875
25. Biassoni R, Pessino A, Malaspina A, Cantoni C, Bottino C, Sivori S, et al. Role of amino acid position 70 in the binding affinity of p50.1 and p58.1 receptors for HLA-Cw4 molecules. *Eur J Immunol* (1997) 27(12):3095–9. doi:10.1002/eji.1830271203
26. Mandelboim O, Kent S, Davis DM, Wilson SB, Okazaki T, Jackson R, et al. Natural killer activating receptors trigger interferon secretion from T cells and natural killer cells. *Proc Natl Acad Sci U S A* (1998) 95(7):3798–803. doi:10.1073/pnas.95.7.3798
27. Stewart CA, Laugier-Anfossi F, Vély F, Saulquin X, Riedmuller J, Tisserant A, et al. Recognition of peptide-MHC class I complexes by activating killer immunoglobulin-like receptors. *Proc Natl Acad Sci U S A* (2005) 102(37):13224–9. doi:10.1073/pnas.0503594102
28. Chewning JH, Gudme CN, Hsu KC, Selvakumar A, Dupont B. KIR2DS1-positive NK cells mediate alloresponse against the C2 HLA-KIR ligand group in vitro. *J Immunol* (2007) 179(2):854–68. doi:10.4049/jimmunol.179.2.854

29. Morvan M, David G, Sébille V, Perrin A, Gagne K, Willem C, et al. Autologous and allogeneic HLA KIR ligand environments and activating KIR control KIR NK-cell functions. *Eur J Immunol* (2008) 38(12):3474–86. doi:10.1002/eji.200838407
30. Foley B, De Santis D, Lathbury L, Christiansen F, Witt C. KIR2DS1-mediated activation overrides NKG2A-mediated inhibition in HLA-C C2-negative individuals. *Int Immunol* (2008) 20(4):555–63. doi:10.1093/intimm/dxn013
31. Pende D, Marcenaro S, Falco M, Martini S, Bernardo ME, Montagna D, et al. Anti-leukemia activity of alloreactive NK cells in KIR ligand-mismatched haploidentical HSCT for pediatric patients: evaluation of the functional role of activating KIR and redefinition of inhibitory KIR specificity. *Blood* (2009) 113(13):3119–29. doi:10.1182/blood-2008-06-164103
32. Fauriat C, Ivarsson MA, Ljunggren HG, Malmberg KJ, Michaëlsson J. Education of human natural killer cells by activating killer cell immunoglobulin-like receptors. *Blood* (2010) 115(6):1166–74. doi:10.1182/blood-2009-09-245746
33. Sivori S, Carlomagno S, Falco M, Romeo E, Moretta L, Moretta A. Natural killer cells expressing the KIR2DS1-activating receptor efficiently kill T-cell blasts and dendritic cells: implications in haploidentical HSCT. *Blood* (2011) 117(16):4284–92. doi:10.1182/blood-2010-10-316125
34. Xiong S, Sharkey AM, Kennedy PR, Gardner L, Farrell LE, Chazara O, et al. Maternal uterine NK cell–activating receptor KIR2DS1 enhances placentation. *J Clin Invest* (2013) 123(10):4264–72. doi:10.1172/JCI68991
35. Pittari G, Liu XR, Selvakumar A, Zhao Z, Merino E, Huse M, et al. NK cell tolerance of self-specific activating receptor KIR2DS1 in individuals with cognate HLA-C2 ligand. *J Immunol* (2013) 190(9):4650–60. doi:10.4049/jimmunol.1202120
36. Cassidy SA, Cheent KS, Khakoo SI. Effects of peptide on NK cell-mediated MHC I recognition. *Front Immunol* (2014) 5:133. doi:10.3389/fimmu.2014.00133
37. Das J, Khakoo SI. NK cells: tuned by peptide? *Immunol Rev* (2015) 267(1):214–27. doi:10.1111/imr.12315
38. Salzberger W, Garcia-Beltran WF, Dugan H, Gubbala S, Simoneau C, Gressens SB, et al. Influence of glycosylation inhibition on the binding of KIR3DL1 to HLA-B*57:01. *PLoS One* (2015) 10(12):1–14. doi:10.1371/journal.pone.0145324
39. Steiner NK, Dakshanamurthy S, Nguyen N, Hurley CK. Allelic variation of killer cell immunoglobulin-like receptor 2DS5 impacts glycosylation altering cell surface expression levels. *Hum Immunol* (2014) 75(2):124–8. doi:10.1016/j.humimm.2013.11.012
40. Kollnberger S, Chan A, Sun MY, Chen LY, Wright C, di Gleria K, et al. Interaction of HLA-B27 homodimers with KIR3DL1 and KIR3DL2, unlike HLA-B27 heterotrimers, is independent of the sequence of bound peptide. *Eur J Immunol* (2007) 37(5):1313–22. doi:10.1002/eji.200635997
41. Garcia-Beltran WF, Hölzemer A, Martrus G, Chung AW, Pacheco Y, Simoneau CR, et al. Open conformers of HLA-F are high-affinity ligands of the activating NK-cell receptor KIR3DS1. *Nat Immunol* (2016) 17(9):1067–74. doi:10.1038/ni.3513
42. Burian A, Wang KL, Finton KAK, Lee N, Ishitani A, Strong K, et al. HLA-F and MHC-I open conformers bind natural killer cell Ig-like receptor KIR3DS1. *PLoS One* (2016) 11(9):e0163297. doi:10.1371/journal.pone.0163297
43. Arase H, Mocarski ES, Campbell AE, Hill AB, Lanier LL. Direct recognition of cytomegalovirus by activating and inhibitory NK cell receptors. *Science* (2002) 296(5571):1323–6. doi:10.1126/science.1070884
44. Béziat V, Liu LL, Malmberg J-A, Ivarsson MA, Sohlberg E, Björklund AT, et al. NK cell responses to cytomegalovirus infection lead to stable imprints in the human KIR repertoire and involve activating KIRs. *Blood* (2013) 121(14):2678–88. doi:10.1182/blood-2012-10-459545
45. Della Chiesa M, Falco M, Bertaina A, Muccio L, Alicata C, Frassoni F, et al. Human cytomegalovirus infection promotes rapid maturation of NK cells expressing activating killer Ig-like receptor in patients transplanted with NKG2C-/- umbilical cord blood. *J Immunol* (2014) 192(4):1471–9. doi:10.4049/jimmunol.1302053
46. Charoudeh HN, Terszowski G, Czaja K, Gonzalez A, Schmitter K, Stern M. Modulation of the natural killer cell KIR repertoire by cytomegalovirus infection. *Eur J Immunol* (2013) 43(2):480–7. doi:10.1002/eji.201242389
47. Crespo ÂC, Strominger JL, Tilburgs T. Expression of KIR2DS1 by decidual natural killer cells increases their ability to control placental HCMV infection. *Proc Natl Acad Sci U S A* (2016) 113(52):15072–7. doi:10.1073/pnas.1617927114
48. Ehrhardt C, Schmolke M, Matzke A, Knoblauch A, Will C, Wixler V, et al. Polyethylenimine, a cost-effective transfection reagent. *Signal Transduct* (2006) 6(3):179–84. doi:10.1002/sita.200500073
49. Baía D, Pou J, Jones DC, Mandelboim O, Trowsdale J, Muntasell A, et al. Interaction of the LILRB1 inhibitory receptor with HLA class Ia dimers. *Eur J Immunol* (2016) 46(7):1681–90. doi:10.1002/eji.201546149
50. Tran T, Ivanyi P, Hilgert I, Brdička T, Pla M, Breur B, et al. The epitope recognized by pan-HLA class I-reactive monoclonal antibody W6/32 and its relationship to unusual stability of the HLA-B27/β_2-microglobulin complex. *Immunogenetics* (2001) 53(6):440–6. doi:10.1007/s002510100353
51. Barnstable CJ, Bodmer WF, Brown G, Galfre G, Milstein C, Williams AF, et al. Production of monoclonal antibodies to group A erythrocytes, HLA and other human cell surface antigens-new tools for genetic analysis. *Cell* (1978) 14(1):9–20. doi:10.1016/0092-8674(78)90296-9
52. Rebaï N, Malissen B. Structural and genetic analyses of HLA class I molecules using monoclonal xenoantibodies. *Tissue Antigens* (2008) 22(2):107–17. doi:10.1111/j.1399-0039.1983.tb01176.x
53. Ciccone E, Pende D, Nanni L, Di Donato C, Viale O, Beretta A, et al. General role of HLA class I molecules in the protection of target cells from lysis by natural killer cells: evidence that the free heavy chains of class I molecules are not sufficient to mediate the protective effect. *Int Immunol* (1995) 7(3):393–400. doi:10.1093/intimm/7.3.393
54. Stam NJ, Spits H, Ploegh HL. Monoclonal antibodies raised against denatured HLA-B locus heavy chains permit biochemical characterization of certain HLA-C locus products. *J Immunol* (1986) 137(7):2299–306.
55. Sernee MF, Ploegh HL, Schust DJ. Why certain antibodies cross-react with HLA-A and HLA-G: epitope mapping of two common MHC class I reagents. *Mol Immunol* (1998) 35(3):177–88. doi:10.1016/S0161-5890(98)00026-1
56. Braud VM, Allan DS, Wilson D, McMichael AJ. TAP- and tapasin-dependent HLA-E surface expression correlates with the binding of an MHC class I leader peptide. *Curr Biol* (1998) 8(1):1–10. doi:10.1016/S0960-9822(98)70014-4
57. Setini A, Beretta A, De Santis C, Meneveri R, Martayan A, Mazzilli MC, et al. Distinctive features of the α_1-domain alpha helix of HLA-C heavy chains free of β_2-microglobulin. *Hum Immunol* (1996) 46(2):69–81. doi:10.1016/0198-8859(96)00011-0
58. Bryceson YT, Fauriat C, Nunes JM, Wood SM, Björkström NK, Long EO, et al. Functional analysis of human NK cells by flow cytometry. *Methods Mol Biol* (2010) 612:335–52. doi:10.1007/978-1-60761-362-6_23
59. Sinzger C, Schmidt K, Knapp J, Kahl M, Beck R, Waldman J, et al. Modification of human cytomegalovirus tropism through propagation in vitro is associated with changes in the viral genome. *J Gen Virol* (1999) 80(11):2867–77. doi:10.1099/0022-1317-80-11-2867
60. Wills MR, Ashiru O, Reeves MB, Okecha G, Trowsdale J, Tomasec P, et al. Human cytomegalovirus encodes an MHC class I-like molecule (UL142) that functions to inhibit NK cell lysis. *J Immunol* (2005) 175(11):7457–65. doi:10.4049/jimmunol.175.11.7457
61. Stinski MF. Sequence of protein synthesis in cells infected by human cytomegalovirus: early and late virus-induced polypeptides. *J Virol* (1978) 26(3):686–701.
62. Haan C, Behrmann I. A cost effective non-commercial ECL-solution for western blot detections yielding strong signals and low background. *J Immunol Methods* (2007) 318(1–2):11–9. doi:10.1016/j.jim.2006.07.027
63. Koike-Yusa H, Li Y, Tan EP, Velasco-Herrera Mdel C, Yusa K. Genome-wide recessive genetic screening in mammalian cells with a lentiviral CRISPR-guide RNA library. *Nat Biotechnol* (2013) 32(3):267–73. doi:10.1038/nbt.2800
64. Jones DC, Kosmoliaptsis V, Apps R, Lapaque N, Smith I, Kono A, et al. HLA class I allelic sequence and conformation regulate leukocyte Ig-like receptor binding. *J Immunol* (2011) 186(5):2990–7. doi:10.4049/jimmunol.1003078
65. Sinzger C, Grefte A, Plachter B, Gouw AS, The TH, Jahn G. Fibroblasts, epithelial cells, endothelial cells and smooth muscle cells are major targets of human cytomegalovirus infection in lung and gastrointestinal tissues. *J Gen Virol* (1995) 76(4):741–50. doi:10.1099/0022-1317-76-4-741
66. Ahn K, Angulo A, Ghazal P, Peterson PA, Yang Y, Früh K. Human cytomegalovirus inhibits antigen presentation by a sequential multistep process.

Proc Natl Acad Sci U S A (1996) 93(20):10990–5. doi:10.1073/pnas.93.20.10990

67. Noriega VM, Tortorella D. Human cytomegalovirus-encoded immune modulators partner to downregulate major histocompatibility complex class I molecules. *J Virol* (2009) 83(3):1359–67. doi:10.1128/JVI.01324-08
68. Wiertz EJ, Jones TR, Sun L, Bogyo M, Geuze HJ, Ploegh HL. The human cytomegalovirus US11 gene product dislocates MHC class I heavy chains from the endoplasmic reticulum to the cytosol. *Cell* (1996) 84(5):769–79. doi:10.1016/S0092-8674(00)81054-5
69. Jones TR, Sun L. Human cytomegalovirus US2 destabilizes major histocompatibility complex class I heavy chains. *J Virol* (1997) 71(4):2970–9.
70. Ahn K, Gruhler A, Galocha B, Jones TR, Wiertz EJ, Ploegh HL, et al. The ER-luminal domain of the HCMV glycoprotein US6 inhibits peptide translocation by TAP. *Immunity* (1997) 6(5):613–21. doi:10.1016/S1074-7613(00)80349-0
71. Lehner PJ, Karttunen JT, Wilkinson GWG, Cresswell P. The human cytomegalovirus US6 glycoprotein inhibits transporter associated with antigen processing-dependent peptide translocation. *Proc Natl Acad Sci U S A* (1997) 94(13):6904–9. doi:10.1073/pnas.94.13.6904
72. Cerboni C, Achour A, Wärnmark A, Mousavi-Jazi M, Sandalova T, Hsu ML, et al. Spontaneous mutations in the human CMV HLA class I homologue UL18 affect its binding to the inhibitory receptor LIR-1/ILT2/CD85j. *Eur J Immunol* (2006) 36(3):732–41. doi:10.1002/eji.200425220
73. Prod'homme V, Griffin C, Aicheler RJ, Wang EC, McSharry BP, Rickards CR, et al. The human cytomegalovirus MHC class I homolog UL18 inhibits LIR-1+ but activates LIR-1- NK cells. *J Immunol* (2007) 178(7):4473–81. doi:10.4049/jimmunol.178.7.4473
74. Tomasec P, Wang EC, Davison AJ, Vojtesek B, Armstrong M, Griffin C, et al. Downregulation of natural killer cell-activating ligand CD155 by human cytomegalovirus UL141. *Nat Immunol* (2005) 6(2):181–8. doi:10.1038/ni1156
75. Fitzgerald-Bocarsly P. Human natural interferon-α producing cells. *Pharmacol Ther* (1993) 60(1):39–62. doi:10.1016/0163-7258(93)90021-5
76. Van Damme J, Schaafsma MR, Fibbe WE, Falkenburg JH, Opdenakker G, Billiau A. Simultaneous production of interleukin 6, interferon-β and colony-stimulating activity by fibroblasts after viral and bacterial infection. *Eur J Immunol* (1989) 19(1):163–8. doi:10.1002/eji.1830190126
77. Magri G, Muntasell A, Romo N, Sáez-Borderías A, Pende D, Geraghty DE, et al. NKp46 and DNAM-1 NK cell receptors drive the response to human cytomegalovirus infected myeloid dendritic cells overcoming viral immune evasion strategies. *Blood* (2011) 117(3):848–56. doi:10.1182/blood-2010-08-301374
78. Blum JS, Wearsch PA, Cresswell P. Pathways of antigen processing. *Annu Rev Immunol* (2013) 31:443–73. doi:10.1146/annurev-immunol-032712-095910
79. Ran FA, Hsu PD, Wright J, Agarwala V, Scott DA, Zhang F. Genome engineering using the CRISPR-Cas9 system. *Nat Protoc* (2013) 8(11):2281–308. doi:10.1038/nprot.2013.143
80. Oszmiana A, Williamson DJ, Cordoba SP, Morgan DJ, Kennedy PR, Stacey K, et al. The size of activating and inhibitory killer Ig-like receptor nanoclusters is controlled by the transmembrane sequence and affects signaling. *Cell Rep* (2016) 15(9):1957–72. doi:10.1016/j.celrep.2016.04.075
81. Bryceson YT, March ME, Ljunggren H-G, Long EO. Synergy among receptors on resting NK cells for the activation of natural cytotoxicity and cytokine secretion. *Blood* (2006) 107(1):159–66. doi:10.1182/blood-2005-04-1351
82. Fadda L, Borhis G, Ahmed P, Cheent K, Pageon SV, Cazaly A, et al. Peptide antagonism as a mechanism for NK cell activation. *Proc Natl Acad Sci U S A* (2010) 107(22):10160–5. doi:10.1073/pnas.0913745107
83. Barber LD, Patel TP, Percival L, Gumperz JE, Lanier LL, Phillips JH, et al. Unusual uniformity of the N-linked oligosaccharides of HLA-A, -B, and -C glycoproteins. *J Immunol* (1996) 156(9):3275–84.
84. Tardif KD, Siddiqui A. Cell surface expression of major histocompatibility complex class I molecules is reduced in hepatitis C virus subgenomic replicon-expressing cells. *J Virol* (2003) 77(21):11644–50. doi:10.1128/JVI.77.21.11644-11650.2003
85. Lantéri M, Giordanengo V, Hiraoka N, Fuzibet JG, Auberger P, Fukuda M, et al. Altered T cell surface glycosylation in HIV-1 infection results in increased susceptibility to galectin-1-induced cell death. *Glycobiology* (2003) 13(12):909–18. doi:10.1093/glycob/cwg110
86. Kollnberger S, Bird L, Sun MY, Retiere C, Braud VM, McMichael A, et al. Cell-surface expression and immune receptor recognition of HLA-B27 homodimers. *Arthritis Rheum* (2002) 46(11):2972–82. doi:10.1002/art.10605
87. Kielczewska A, Pyzik M, Sun T, Krmpotic A, Lodoen MB, Munks MW, et al. Ly49P recognition of cytomegalovirus-infected cells expressing H2-Dk and CMV-encoded m04 correlates with the NK cell antiviral response. *J Exp Med* (2009) 206(3):515–23. doi:10.1084/jem.20080954
88. Serena M, Parolini F, Biswas P, Sironi F, Blanco Miranda A, Zoratti E, et al. HIV-1 env associates with HLA-C free-chains at the cell membrane modulating viral infectivity. *Sci Rep* (2017) 7:40037. doi:10.1038/srep40037
89. Ameres S, Besold K, Plachter B, Moosmann A. CD8 T cell-evasive functions of human cytomegalovirus display pervasive MHC allele specificity, complementarity, and cooperativity. *J Immunol* (2014) 192(12):5894–905. doi:10.4049/jimmunol.1302281
90. Ameres S, Mautner J, Schlott F, Neuenhahn M, Busch DH, Plachter B, et al. Presentation of an immunodominant immediate-early CD8+ T cell epitope resists human cytomegalovirus immunoevasion. *PLoS Pathog* (2013) 9(5):e1003383. doi:10.1371/journal.ppat.1003383
91. Arase H, Lanier LL. Virus-driven evolution of natural killer cell receptors. *Microbes Infect* (2002) 4(15):1505–12. doi:10.1016/S1286-4579(02)00033-3
92. Chen KC, Stanton RJ, Banat JJ, Wills MR. Leukocyte immunoglobulin-like receptor 1-expressing human natural killer cell subsets differentially recognize isolates of human cytomegalovirus through the viral major histocompatibility complex class I homolog UL18. *J Virol* (2016) 90(6):3123–37. doi:10.1128/JVI.02614-15

2

Regulation of Murine Natural Killer Cell Development

*Wilford Goh[1,2] and Nicholas D. Huntington[1,2]**

[1] The Walter and Eliza Hall Institute of Medical Research, Parkville, VIC, Australia, [2] Department of Medical Biology, University of Melbourne, Melbourne, VIC, Australia

Edited by:
Antoine Toubert,
Paris Diderot University, France

Reviewed by:
Evelyn Ullrich,
Goethe University Frankfurt, Germany
Reem Al-Daccak,
Institut national de la santé et de la recherche médicale, France
Jean Villard,
University of Geneva, Switzerland

***Correspondence:**
Nicholas D. Huntington
huntington@wehi.edu.au

Natural killer (NK) cells are effector lymphocytes of the innate immune system that are known for their ability to kill transformed and virus-infected cells. NK cells originate from hematopoietic stem cells in the bone marrow, and studies on mouse models have revealed that NK cell development is a complex, yet tightly regulated process, which is dependent on both intrinsic and extrinsic factors. The development of NK cells can be broadly categorized into two phases: lineage commitment and maturation. Efforts to better define the developmental framework of NK cells have led to the identification of several murine NK progenitor populations and mature NK cell subsets, each defined by a varied set of cell surface markers. Nevertheless, the relationship between some of these NK cell subsets remains to be determined. The classical approach to studying both NK cell development and function is to identify the transcription factors involved and elucidate the mechanistic action of each transcription factor. In this regard, recent studies have provided further insight into the mechanisms by which transcription factors, such as ID2, FOXO1, Kruppel-like factor 2, and GATA-binding protein 3 regulate various aspects of NK cell biology. It is also becoming evident that the biology of NK cells is not only transcriptionally regulated but also determined by epigenetic alterations and posttranscriptional regulation of gene expression by microRNAs. This review summarizes recent progress made in NK development, focusing primarily on transcriptional regulators and their mechanistic actions.

Keywords: NK cell, transcription factors, ontogeny, maturation, homeostasis, IL-15

INTRODUCTION TO NATURAL KILLER (NK) CELLS

Natural killer cells in mice were first described in 1975 (1–3), following further investigation into splenocytes that were able to kill tumor and virus-infected cells without prior sensitization (4–6). NK cells exert their cytotoxic effect on target cells by inducing apoptosis. Upon formation of an immunological synapse with the target cell, NK cells become activated and release cytolytic granules containing perforin and granzymes (7–9). Perforin forms pores in the membrane of target cells, thereby allowing granzymes to enter the cell, activate caspases, and initiate apoptosis (8). In a similar process known as antibody-dependent cell cytotoxicity, NK cells are able to release cytolytic granules and initiate apoptosis in opsonized cells, following recognition of the opsonized cells *via* low-affinity Fc receptors (CD16) expressed on the surface of NK cells (10). NK cells can also initiate apoptosis in target cells through the respective engagement of Fas ligands and tumor necrosis factor-related apoptosis-inducing ligand (TRAIL) on their cell surface with Fas and TRAIL receptors on the target cells (11, 12). In addition to inducing apoptosis, NK cells can indirectly mediate the clearance of target cells by producing pro-inflammatory cytokines [e.g., interferon-gamma (IFN-γ)], which boost the innate response and recruit adaptive immune responses (13–15).

The surface markers that are commonly used to identify murine NK cells by flow cytometry vary depending on the mouse strain. C57B/6 and SJL mice express the surface markers NK1.1, NKp46, and CD49b, but not CD3, which is a surface marker of T cells. CD3 is used to exclude contaminating T cell subsets, such as natural killer T cells and NK-like T cells, that, respectively, express NK1.1 and NKp46 (16). As for other mouse strains, such as BALB/c, NK cells are identified with only CD49b and NKp46 as these strains possess allelic variants of NK1.1 that cannot be detected with the widely used PK136 antibody (16, 17).

MURINE NK CELL DEVELOPMENT

Murine NK cells can be found in all lymphoid organs and many non-lymphoid tissues, such as salivary glands, liver, and kidney. The more recent discovery of related innate lymphoid cells (ILCs) places NK cells within this family, specifically in the IL-15 dependent, IFN-γ producing group 1 ILCs. ILCs are lymphoid cells that lack rearranged antigen receptors and are dependent on the transcription factors inhibitor of DNA-binding 2 (ID2) and nuclear factor, interleukin 3 regulated (NFIL3) for their development. While NK cells are phenotypically heterogeneous and previously categorized based on their tissue of origin or location (bone marrow, thymus, fetal liver, adult liver), we appreciate that some of this heterogeneity stems from NK cells (Eomes$^+$) and other ILC1s (Eomes$^-$) being viewed as the same cell type. As much of our current understanding of murine NK cell development is built upon studies on bone marrow-derived NK cells [referred to here as conventional NK (cNK) cells], which represent the majority of NK cells within the body, this review will focus primarily on progress made in our understanding of cNK development.

cNK Development in the Bone Marrow—Lineage Commitment

Conventional NK cells develop from HSCs in the bone marrow, through a sequential order of intermediate progenitors. The first progenitor to arise from HSCs is the lymphoid-primed multipotent progenitor, which then gives rise to the common lymphoid progenitor (CLP) (18). The earliest NK lineage committed progenitor to arise from CLPs is known as pre-pro NK (19), which was subdivided into pre-pro A and pre-pro B (19, 20). Differing only in c-kit (CD117) expression, the relationship between pre-pro A and B remains unclear and requires further investigation. Pre-pro NK cells then differentiate into the NK progenitor (NKP) (19, 21). NKPs give rise to immature NK (iNK) cells that either undergo further development within the bone marrow (22) or enter the periphery and develop into mature NK cells (23, 24).

As the early stages of murine NK development are still poorly defined, the developmental pathway outlined above is by no means the definitive model. Heterogeneity within existing progenitor populations, along with the discovery of new distinguishing cell markers, have led to the identification of new sub-populations and, therefore, refinements to the developmental pathway of NK cells. For instance, the common innate lymphoid progenitor (CILP) was found to possess the capacity to give rise to all lineages within the ILC family, of which NK cells are the founding member, but not B and T cells, thus making it an earlier progenitor than the pre-pro NK (25). As the CILP is expresses $\alpha_4\beta_7$, it is also alternatively referred to as the α-lymphoid precursor (αLP) (25). However, it has been postulated that there could be an even earlier progenitor population, as there were only about 50 CILPs per mice, and only 2.5% of CILPs efficiently developed into all ILC lineages (25).

The early innate lymphoid progenitor (EILP) is proposed to be the earliest known progenitor for ILCs and was identified using a TCF1 (T-cell factor 1, encoded by *Tcf7*) transgenic mouse strain that expresses a green fluorescent protein reporter (26). Like CILPs, EILPs give rise to all ILC lineages both *in vivo* and *in vitro*, albeit more efficiently (26). Nonetheless, EILPs have not been shown to differentiate into CILPs and hence, the relationship between these two progenitors remains unresolved. As most of the surface markers used to identify the EILP and CILP were different, a detailed comparison of the surface marker phenotype between the two progenitors may also provide further insight.

The discovery of pre–pro NKPs was the outcome of efforts to better understand why only 8–40% of NKPs had solely NK cell potential (19). Similar studies resulted in the identification of a pre-NKP population that preceded a "streamlined" population of NKPs known as refined-NKP (rNKP) (27). Even though many parallels have been drawn between the pre-NKP and pre-pro NK, likewise with NKP and rNKP, it remains to be determined if these populations are exactly identical (20). A summary of the surface markers that are expressed on the various progenitors are provided in **Table 1**.

cNK Development in the Periphery—NK Cell Maturation

Natural killer cell maturation is a process by which lineage committed NK cells acquire their full effector functions. This process is also accompanied by the expression of different cell surface markers, which have helped in the identification of different NK cell maturation subsets. At present, most studies use CD11b and CD27 to divide cNK cells into three maturation subsets: immature (Imm), mature 1 (M1), and mature 2 (M2). Low (lo) and high (hi) CD11b expression divides cNK cells into immature and mature subsets, respectively (22). Heterogeneity in CD27 expression further delineates the mature NK compartment into CD27^{hi} and CD27^{lo} subsets, which has also been referred to as M1 and M2 NK subsets (24, 28). CD27 and CD11b expressing M1 NK cells have also been termed double-positive NK cells (24). The three subsets differ in proliferative and cytotoxic capacity. In general, NK cells lose proliferative potential and produce less cytokine, but become more cytotoxic against target cells as they mature (22, 24, 28).

Apart from CD27 and CD11b, markers such as KLRG1 (23, 29), CD62L (30), MCAM (31), CD49b (21), CD43 (32, 33), Ly6C (34), DNAM1 (35), and CD160 (36) have further dissected maturing NK cells into various phenotypic subsets (37, 38). NK cells that express CD160 exhibit enhanced IFN-γ

TABLE 1 | Surface markers expressed by different natural killer (NK) cell populations reported in the literature.

Surface markers	Common lymphoid progenitor	Early innate lymphoid progenitor	Common innate lymphoid progenitor (CILP)/αLP	Pre-pro A NK	Pre-pro B NK	NK progenitor (NKP)	pre-NKP	refined-NKP (rNKP)	Imm NK	M1 NK	M2 NK
NK1.1	–	–	–	–	–	–	–	–	+	+	+
CD11b (MAC-1)	–	–	–	–	–	–	–	–	–	+	+
CD127 (IL-7Rα)	hi	lo	hi	hi	hi	hi	hi	int	int	lo	lo
CD117 (c-kit)	int	lo	int	int	lo	lo	int	lo	int	lo	lo
Sca-1	int	–	int	+	+	+	+	+	–	–	–
CD49b	–	–	–	–	–	–	–	–	–	+	+
CD27	+	ND	+	+	+	+	+	+	+	+	–
CD244 (2B4)	+	ND	+	+	+	+	+	+	+	+	+
CD25 (IL-2Rα)	–	–	–	ND	ND	–	–	+	–	–	–
CD122 (IL-2Rβ)	–	–	–	–	–	+	–	+	+	+	+
CD132 (IL-2Rγ_c)	+	+	+	+	+	+	+	+	+	+	+
CD314 (NKG2D)	–	–	–	+	+	+	+	+	+	+	+
CD226 (DNAM1)	–	ND	int	ND	ND	+	ND	ND	hi	int	lo
CD279 (PD1)	–	ND	–/+	ND	ND	hi	ND	ND	–	–	–
CD43 (Leukosialin)	–	–	–	–	–	–	–	–	lo	int	hi
CD335 (NKp46)	–	–	–	–	–	–	–	–	+	+	+
CD253 (tumor necrosis factor-related apoptosis-inducing ligand)	–	ND	ND	ND	ND	ND	ND	ND	int	lo	lo
KLRG1	–	ND	ND	–	–	–	ND	ND	–	–	+
$\alpha_4\beta_7$ (LPAM)	+	+	+	ND	ND	–	ND	ND	–	–	–
Ly49s	–	–	–	–	–	–	–	–	–/+	–/+	–/+
CD94-NKG2	–	ND	ND	–	–	–	–	–	+	+	+
CD62L (L-selectin)	–	ND	ND	–	–	–	ND	ND	–/+	+	+
CD146 (MCAM)	–	ND	ND	–	–	–	ND	ND	lo	int	hi
CXCR3	–	ND	ND	–	–	+	ND	ND	hi	int	lo
CXCR6	–	–	+	+	+	+	ND	ND	–	–	–
Ly6C	–	ND	ND	–	–	–	ND	ND	lo	int	hi

production (36), while the opposite is true for mature NK cells that express higher levels of Ly6C (34), and KLRG1 (39). Like Ly6C, MCAM is also more highly expressed on mature NK cells, although MCAM$^+$ and MCAM$^-$ NK cells differ in their ability to kill target cells rather than cytokine production (31). Interestingly, the expression of DNAM1 appears to be independent of NK cell maturation that is defined by CD27 and CD11b, as DNAM-1$^+$ and DNAM-1$^-$ NK cells were observed in both the immature and mature NK compartments (35). As the correlation between these markers have not been studied in detail, the relationships between these phenotypic subsets remain unclear and warrants further investigation. A summary of the surface markers that are expressed on the various mature NK cell subsets are provided in **Table 1**.

TRANSCRIPTIONAL REGULATION OF MURINE cNK CELL DEVELOPMENT

Transcription factors control gene expression by either activating or repressing gene transcription. This is achieved by first binding to specific DNA sequences in the enhancer or promoter regions that lay upstream of target genes, then promoting or blocking the recruitment of RNA polymerases that transcribe those genes (40). In terms of murine NK cell development, several TFs have been shown to play crucial roles in regulating NK cell lineage specification, NK cell maturation, or even both. Conventional NK cell development occurs mostly in the bone marrow, under the coordinated control of the TFs and cytokines. TFs like ID2, NFIL3, T-box brain protein 2 (EOMES), and T-box protein 21 (TBET) to name a few, fall into the category of intrinsic factors that regulate NK cell development. A summary of the transcription factors that are implicated in NK cell development is provided in **Table 2**.

Transcription Factors Regulating NK Cell Lineage Specification

The TFs that are involved in regulating NK cell lineage specification include ETS proto-oncogene 1 (ETS1), NFIL3, and TCF1. ETS1 is a key regulator of early NK cell development as ETS1-deficient mice have normal CLP numbers but lack NK cells (45). Further investigation into the impact of ETS1 deficiency on NK lineage specification revealed a reduction in pre-pro NK, pre-NKP and

TABLE 2 | Transcription factors implicated in natural killer (NK) cell development and function.

	Phenotype of germline (KO) or conditional (cKO) deficiency				
Transcription factor (gene symbol)	**Bone marrow NK precursor #**	**NK cells #**	**IL-15 responsiveness**	**Interferon-gamma (IFN-γ) production/ degranulation**	**Cytotoxicity (tumor/target cells)**
Ikaros family zinc finger 3, Aiolos (*Ikzf3*)	ND	KO: normal; accumulation of immature NKs (iNKs)	KO: hyperresponsive to IL-2/anti-IL-2 mAB complex *in vitro*	KO: slightly impaired IFN-γ production	KO: normal *in vitro* killing, augmented *in vivo* killing
B lymphocyte-induced maturation protein 1 (*Prdm1*)	ND	KO: reduced in spleen, liver, and lung; accumulation in bone marrow and lymph nodes; loss of mNKs	KO: hyperresponsive	KO: normal IFN-γ production	KO: normal *in vitro* killing, augmented *in vivo* killing
T-box brain protein 2, EOMES (*eomesodermin*)	ND	cKO: reduced; loss of mNKs	ND	cKO: slightly impaired IFN-γ production	ND
ETS proto-oncogene 1, ETS1 (*Ets1*)	KO: lack pre-NKPs and rNKPs	KO: reduced	KO: hyperresponsive	ND	KO: impaired *in vitro* killing and degranulation
Forkhead box protein O1, FOXO1 (*Foxo1*)	cKO: normal NKP (41)	cKO: reduced; loss of mNKs (41) cKO: normal; accumulation of mNKs (42)	ND	cKO: augmented IFN-γ production (42)	cKO: augmented *in vitro* and *in vivo* killing (42)
GATA-binding protein 3 (*Gata3*)	ND	cKO: reduced in bone marrow; accumulation in spleen and liver, systemic accumulation of iNKs	ND	cKO: impaired IFN-γ production; normal degranulation	cKO: normal *in vitro* killing
Inhibitor of DNA-binding 2, ID2 (*Id2*)	KO: normal NKP	KO: reduced; loss of mNKs; cKO: systemic reduction	cKO: hyporesponsive	KO: impaired IFN-γ production cKO: normal IFN-γ production	KO: impaired *in vitro* killing cKO: impaired *in vivo* killing
Interferon Regulatory Factor 2 (*Irf2*)	ND	KO: reduced; loss of mNKs	KO: hyporesponsive	KO: impaired IFN-γ production	KO: normal *in vitro* killing (43); impaired *in vitro killing* (44)
Kruppel-like factor 2 (*Klf2*)	ND	cKO: reduced in spleen, blood, and lung; accumulation in bone marow and liver; loss of mNKs	ND	ND	ND
Myeloid elf-1-like factor (*Mef*)	ND	KO: reduced	ND	KO: impaired IFN-γ production	KO: impaired *in vitro* killing
Nuclear factor, interleukin 3 regulated, NFIL3 (*E4bp4*)	KO: lack CILP and NKPs, pre-NKPs and rNKPs	KO: reduced	ND	ND	ND
T-box protein 21, TBET (*Tbx21*)	ND	cKO and KO: reduced in spleen and liver; accumulation in bone marrow; loss of mNKs	ND	KO: impaired IFN-γ production	KO: impaired *in vitro* killing
T-cell-specific transcription factor 1, TCF1 (*Tcf7*)	KO: lack NKPs	KO: reduced in bone marrow; normal in periphery	ND	ND	ND
Thymocyte selection-associated high mobility group box protein (*Tox*)	KO: normal NKP	KO: reduced; loss of mNKs	ND	ND	KO: impaired *in vivo* killing
Zinc finger E-box binding homeobox 2 (*Zeb2*)	ND	KO: reduced in periphery; normal in bone marrow; loss of mNKs	cKO: hyporesponsive	cKO: normal to slightly augmented IFN-γ production	cKO: impaired killing *in vivo*

rNKP, and mature NK cell numbers, thereby supporting the role of ETS1 in NK cell lineage specification (46). ETS1 is believed to impact early NK cell development by regulating the expression of ID2 and TBET, which are also important TFs for NK cell development (46).

TCF1 and NFIL3 are also key regulators of NK cell lineage specification, as marked reductions in pre-pro NK, pre-NKP, and rNKP numbers were observed within the bone marrow of NFIL3- and TCF1-deficient mice, although only the former mouse strain exhibited an additional reduction in CILP numbers (25, 26, 47, 48). NFIL3 appears to be dispensable for mature NK cell as their numbers remained unchanged after its deletion in mature NKp46$^+$ NK cells (49). Despite reduced progenitor and mature NK cell numbers in the bone marrow of TCF1-deficient mice, mature NK cells have been found in the periphery at frequencies comparable to littermate controls (26, 50). As most NK cells are derived from bone marrow precursors at steady state, an investigation into the source of NK cells in TCF1-deficient mice might perhaps shed some light on alternative pathways of NK development. The mechanism by which TCF1 mediates NK lineage specification remains poorly understood. However, TCF1 has been shown to regulate T-lineage specification by promoting the expression of genes, such as *Gata3*, *Bcl11b*, *Il2ra*, and *Cd3e*, that are critical for T cell development (51). Similarly, the mechanism by which NFIL3 specifically mediates lineage specification in NK cells remains unclear, although NFIL3 was found to promote the expression of ID2 and EOMES (47), transcription factors known to be involved in the later stages of NK cell development (52, 53). The role of NFIL3 in ID2 expression remains to be clarified given a contradicting report that ID2 expression is normal in NK cells lacking NFIL3 (48).

The role of ID2 in the lineage specification of NK cells is unclear, due to contradicting reports following deletion of the encoding gene *Id2*. While Yokota et al. (54). reported poor reconstitution of NK cells following bone marrow transplantation, implying a defect early on during NK cell development, Boos et al. (55) did not observe any reduction in NKP and iNK cell numbers. A recent demonstration of low ID2 levels in CLPs but consistently high levels in pre-pro NK, NKP, and immature and mature NK cells lends further support for the hypothesis that ID2 could indeed be important for NK lineage specification/maintenance (53). Interestingly, ID2 has been found to suppress T and B cell development through heterodimer formation with the E-box protein E2A (55–57), although how the interaction specifically promotes commitment to the NK cell lineage remains unknown.

Transcription Factors Regulating NK Cell Maturation

A greater number of TFs have been shown to play a role in NK cell maturation. These factors include ID2, TBET, EOMES, Zinc finger E-box-binding homeobox 2 (ZEB2), Thymocyte selection-associated high mobility group box (TOX), IKAROS family zinc finger 3 (AIOLOS), Interferon regulatory factor 2 (IRF2), B lymphocyte-induced maturation protein 1 (BLIMP1), Forkhead box O1 (FOXO1), Kruppel-like factor 2 (KLF2), and GATA-binding protein 3 (GATA3). An overview of the expression these TFs during NK cell maturation is presented in **Figure 1**.

The role of ID2 in NK cell maturation is better understood than its role in early NK cell development. A recent study provided additional insight into the underlying mechanism of ID2 by demonstrating that it modulates the expression of E2A target genes (i.e., *Socs3*, *Tcf7*, and *Cxcr5*) by titrating E-protein activity, thereby controlling the responsiveness of NK cells to IL-15 that is crucial for survival (53, 58, 59).

TBET and EOMES are members of the T-box family of transcription factors that appear to regulate distinct checkpoints in NK cell maturation. TBET- and EOMES-deficient mice exhibited a similar phenotype where NK cell numbers were reduced in the all lymphoid tissues (52, 60, 61), except in the bone marrow of the former where there was an increase in NK cell numbers (60, 61). Detailed analyses of the bone marrow from TBET-deficient mice revealed that the increase in NK cell numbers was due to an accumulation of iNK cells, which the authors attributed to a developmental block (60). However, the possibility of a defect in NK cell trafficking remains unaddressed, given that TBET plays a role in the expression of sphigosine-1-phosphate receptor 5 ($S1P_5$) that is required for NK cell egression from the bone marrow (62). Nonetheless, both TFs are crucial for maturation, as mice that are deficient for both have a systemic lack of NK cells despite normal progenitor numbers (52, 61). As TBET and EOMES are, respectively, required by immature and mature NK cells, the two TFs are believed to function in a sequential manner for NK cell maturation (52, 60). Quite fittingly, both TBET and EOMES were shown to be essential for IL-15 responsiveness by enforcing high CD122 expression, with *Il2rb* (gene encoding CD122) being a direct target of EOMES (63).

Similar to that of TBET-deficient mice, NK cell numbers in the bone marrow of mice deficient for ZEB2 or BLIMP1 were higher than littermate controls (64, 65). The unusual accumulation of NK cells within the bone marrow of ZEB2-deficient mice was due to reduced $S1P_5$ expression (64), while enhanced proliferation was found to be the underlying cause in BLIMP1-deficient mice (65). Profound losses in terminally mature NK cells were observed on closer examination of the bone marrow NK cells from both strains (64, 65). The lack of mature NK cells in ZEB2-deficient was further attributed to poor responsiveness to IL-15, which resulted in poor survival (64).

The lack of mature NK cells was reported in mice that were deficient for TOX (66), AIOLOS (67), IRF2 (43, 44), KLF2 (68), or GATA3 (69, 70). A similar lack of mature NK cells in FOXO1-deficient mice was reported most recently (41), although this remains to be clarified against an earlier study, which instead found an accumulation of mature NK cells (42). Interestingly, the iNK cells in AIOLOS-deficient mice retained their expression of KLRG1, which is typically expressed on terminally mature NK cells (67). NK cells from IRF2-deficient mice have been shown to undergo accelerated apoptosis, indicating a role for IRF2 in regulating NK cell survival as well as maturation (43). Unlike the foregoing TFs, the temporal requirement for myeloid elf-1-like factor (MEF) by maturing NK cells has not been determined, as only an overall reduction in NK cell numbers was reported (71).

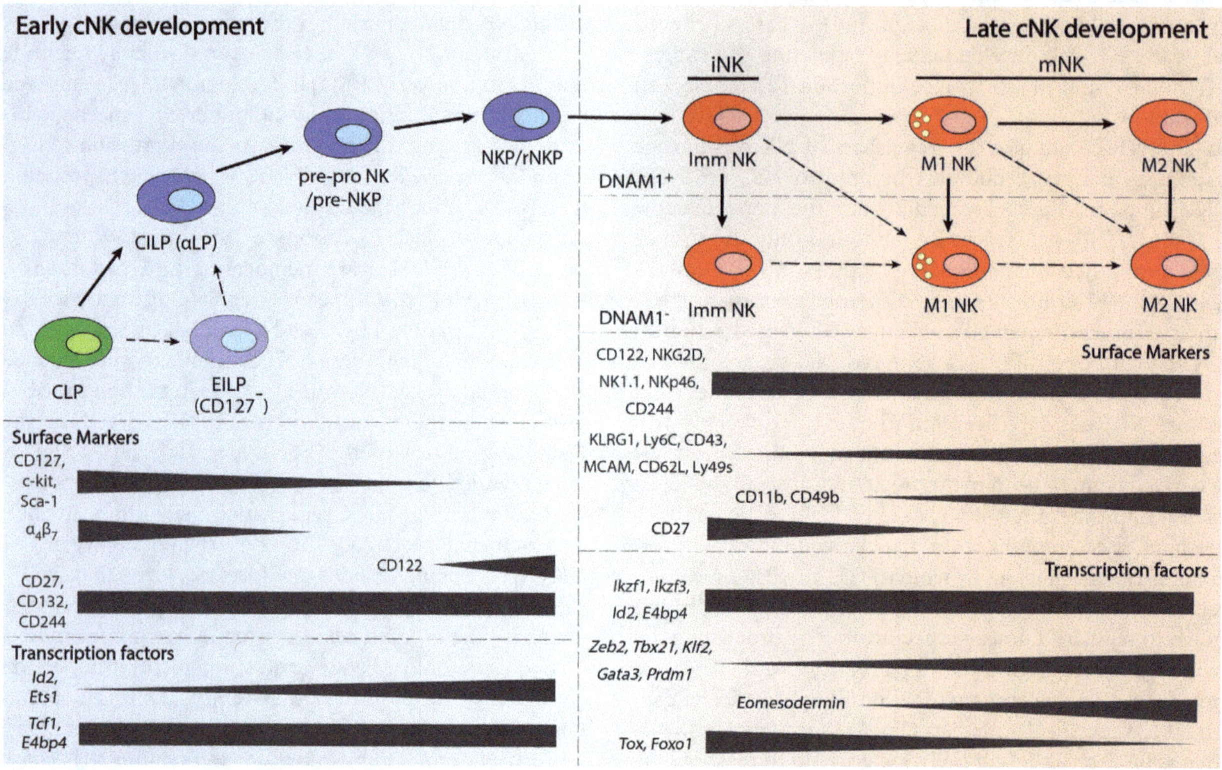

FIGURE 1 | Overview of the early and late stages of natural killer cell development.

FOXO1 was recently found to be directly involved in the initiation of autophagy in iNK cells, most likely *via* cytosolic interactions between FOXO1 and the autophagy protein ATG7 (41). KLF2 appears to regulate NK cell maturation *via* a different mechanism, influencing the expression of homing receptors such as CD62L (i.e., L-selectin) on maturing NK cells, thereby dictating their access to IL-15 signaling that is essential for survival (68). Current knowledge of how GATA3 regulates NK cell maturation is limited to disturbances in expression of the TFs ID2, TBET, and NFIL3 (70). Nevertheless, GATA3 has also been shown to regulate NK cell egression from the bone marrow in a CXCR4-dependent manner, and also NK cell proliferation in response to IL-15 *via* CD25 expression (70).

Transcription Factors Regulating NK Cell Effector Function

Natural killer cell effector function is also regulated by many of the TFs outlined above. FOXO1 has also been proposed to negatively regulate NK cell effector function, as its absence was correlated with augmented IFN-γ production in response to murine cytomegalovirus (MCMV) infections and anti-metastatic activity against the B16F10 mouse melanoma cell line (42). Augmented anti-metastatic activity against the same melanoma cell line also was reported in mice deficient for either BLIMP1 or AIOLOS, despite the lack of any significant impact on cytokine production (65, 67). On the other hand, MEF is required for normal cytokine production and cytotoxicity, as it positively regulates IFN-γ and perforin expression, which corresponds to poorer cytotoxicity against tumor cell targets in MEF-deficient mice (71). Normal cytotoxicity but reduced IFN-γ production has been observed in IRF2-deficient mice and mice specifically lacking GATA3 in NK cells (43, 70). NK cell function is also regulated by TBET and EOMES, as TBET has been shown to bind to the regulatory regions of genes encoding granzyme B and perforin, while the expression of EOMES as NK cells mature is associated with increased transcription of mRNA (52, 60).

POSTTRANSCRIPTIONAL REGULATION OF NK CELL DEVELOPMENT BY microRNAs (miRs)

microRNAs are short non-coding RNAs (19–26 nt) that modulate gene expression at a posttranscriptional level. Recent studies have shown that miRs are also important for NK cell development and function. Using a Dicer1-deficient mouse model that abrogates miR biogenesis in NK cells, Degouve et al. (72). showed that a 10-fold global reduction in miR expression resulted in reduced NK cell numbers, aberrant NK cell maturation, along with attenuated IFN-γ production and cytotoxicity against target cells. Given that IL-15 signaling *via* the STAT5 and mTOR pathways was significantly affected, it was proposed that miRs regulate NK cell survival by modulating IL-15 sensitivity (72). Although it remains unclear as to whether NK cell survival is dependent on specific miRs, miR-155 and miR-15/16 are unlikely candidates since mice that are deficient for either miR have normal NK cell numbers (73, 74).

Rather than NK cell survival, mIR-155 and miR15/16 appear to be essential for normal NK cell maturation, as NK cells lacking miR-15/16 are unable to fully mature into M2 NK cells (74), much like Dicer1-deficient mice, while miR-155-deficient NK cells undergo accelerated maturation (73). In contrast to the accumulation of M2 NK cells in miR-155 deficient mice, an accumulation of Imm NK cells was observed in mice that over-expressed miR-155, thereby providing additional evidence for the role of miR-155 in NK cell maturation (75). More importantly, the opposing effect that miR-15/16 and miR-155 have on NK cell maturation highlights the pleiotropic effects of miRs and suggests that there is still much to learn about the role of miRs in NK cell biology, particularly about redundancies between miRs. Nevertheless, it has been shown that miR-15/16 controls NK maturation by directly regulating levels of the transcription factor MYB, since the overexpression of miR-15/16 or MYB deficiency in miR-15/16-deficient NK cells rescues the maturation defect (74).

The mechanistic action of miR-155 can be appreciated in the context of NK cell proliferation and homeostasis, as NK cells deficient for miR-155 were unable to proliferate in response to MCMV infections and were also outcompeted by wild-type NK cells when cotransferred into homeostatic or lymphopenic environments (73). This dependency on miR-155 for proliferation under both homeostatic and infectious conditions appears to be mediated through the direct suppression of its target genes suppressor of cytokine signaling 1 (*Socs1*) and pro-apoptotic molecule phorbol-12-myristate-13-acetate-induced protein 1 (*Pmaip1*; encoding NOXA) (73). Interestingly, an accumulation of NK cells was observed in transgenic mice that over-expressed miR-155, lending further support for the regulatory role of miR-155 in NK cell proliferation (75).

EPIGENETIC REGULATION OF NK CELL DEVELOPMENT

Histone modifications have previously been shown to be essential events in B and T cell development (76, 77). Recent studies have demonstrated that defects in histone modification also impacts NK cell development with respect to lineage commitment (78) and maturation (79). Mice deficient for enhancer of zeste homolog 2 (EZH2), a H3K27 methyltransferase, were observed to have higher numbers NKPs and NK cells. Microarray analysis revealed that the difference was associated with the upregulation of genes essential for NK cell development and function, thereby resulting in earlier lineage commitment and enhanced survival of NKPs (78). This increase in NK cell production was also observed when hematopoietic progenitors from human and wild-type mice were treated *in vitro* with EZH2 inhibitors (78).

A different type of histone modification, deubiquitination, also appears to be involved in the epigenetic regulation of NK cell maturation. The histone deubiquitinase, MYSM1 (Myb-like, SWIRM, and MPN domains-containing protein 1), was found to regulate NK cell maturation as MYSM1-deficient mice possessed fewer NK cells that were mostly immature (79). Given that no defects were observed in the NKP compartment, MYSM1 was proposed to be uniquely required during NK cell maturation. Mechanistically, MYSM1 regulates NK cell maturation by binding directly to the *Id2* gene locus, as revealed by chromatin immunoprecipitation, thereby maintaining expression of the TF (79). However, the mechanism by which MYSM1 is selectively directed to the *Id2* gene locus remains unclear and thus requires further investigation.

CONCLUSION

The discovery of new members within the ILC family has rekindled efforts to better understand the development of NK cells, the founding member of the ILC family. Many of the recent breakthroughs made in the transcriptional regulation of NK cell development have been aided by key tools and techniques such as single cell RNA-seq, *in vitro* differentiation conditions, transcription factor reporter mice and conditional alleles and lineage specific Cre-expressing mouse strains. As these techniques and tools become commonplace in the field of NK cell biology, our understanding of the temporal–spatial transcriptional regulation of NK cell development and the key target genes that govern NK cell fate, homeostasis, and function becomes increasingly more complete. Recent studies have advanced our understanding of how individual TFs may be regulating NK cell commitment and NK cell lineage maintenance. However, how these various TFs form a transcriptional network and act in concert to ensure NK cell homeostasis remains unclear.

On a translational front, defining the extrinsic cues and TFs that regulate NK cell maturation, proliferation, cytokine responsiveness, and priming of effector functions represents an area of therapeutic interest. Proteins that negatively regulate NK cell maturation and fitness are tangible drug targets in cancer immunotherapy as recently evidenced by our group. Building on these potential targets will increase the likelihood of developing specific inhibitors for clinical translation.

AUTHOR CONTRIBUTIONS

WG wrote the review. NH conceived and edited the review.

REFERENCES

1. Kiessling R, Klein E, Pross H, Wigzell H. "Natural" killer cells in the mouse. II. Cytotoxic cells with specificity for mouse Moloney leukemia cells. Characteristics of the killer cell. *Eur J Immunol* (1975) 5:117–21. doi:10.1002/eji.1830050209
2. Herberman RB, Nunn ME, Lavrin DH. Natural cytotoxic reactivity of mouse lymphoid cells against syngeneic and allogeneic tumors. I. Distribution of reactivity and specificity. *Int J Cancer* (1975) 16:216–29. doi:10.1002/ijc.2910160204
3. Kiessling R, Klein E, Wigzell H. Natural. *Eur J Immunol* (1975) 5:112–7. doi:10.1002/eji.1830050208
4. Hellström I, Hellström KE, Pierce GE, Yang JP. Cellular and humoral immunity to different types of human neoplasms. *Nature* (1968) 220:1352–4. doi:10.1038/2201352a0
5. Herberman RB, Nunn ME, Lavrin DH, Asofsky R. Effect of antibody to theta antigen on cell-mediated immunity induced in syngeneic mice by murine sarcoma virus. *J Natl Cancer Inst* (1973) 51:1509–12. doi:10.1093/jnci/51.5.1509
6. Kiessling R, Bataillon G, Lamon EW, Klein E. The lymphocyte response to primary Moloney sarcoma virus tumors: definition of a non-specific component of the in vitro cellular hyporeactivity of tumor-bearing hosts. *Int J Cancer* (1974) 14:642–8. doi:10.1002/ijc.2910140511
7. Fehniger TA, Cai SF, Cao X, Bredemeyer AJ, Presti RM, French AR, et al. Acquisition of murine NK cell cytotoxicity requires the translation of a pre-existing pool of granzyme B and perforin mRNAs. *Immunity* (2007) 26:798–811. doi:10.1016/j.immuni.2007.04.010
8. Thiery J, Keefe D, Boulant S, Boucrot E, Walch M, Martinvalet D, et al. Perforin pores in the endosomal membrane trigger the release of endocytosed granzyme B into the cytosol of target cells. *Nat Immunol* (2011) 12:770–7. doi:10.1038/ni.2050
9. Henkart MP, Henkart PA. Lymphocyte mediated cytolysis as a secretory phenomenon. In: Clark WR, Golstein P, editors. *Mechanisms of Cell-Mediated Cytotoxicity Advances in Experimental Medicine and Biology*. Boston, MA: Springer (1982). p. 227–47.
10. Ojo E, Wigzell H. Natural killer cells may be the only cells in normal mouse lymphoid cell populations endowed with cytolytic ability for antibody-coated tumour target cells. *Scand J Immunol* (1978) 7:297–306. doi:10.1111/j.1365-3083.1978.tb00457.x
11. Takeda K, Cretney E, Hayakawa Y, Ota T, Akiba H, Ogasawara K, et al. TRAIL identifies immature natural killer cells in newborn mice and adult mouse liver. *Blood* (2005) 105:2082–9. doi:10.1182/blood-2004-08-3262
12. Arase H, Arase N, Saito T. Fas-mediated cytotoxicity by freshly isolated natural killer cells. *J Exp Med* (1995) 181:1235–8. doi:10.1084/jem.181.3.1235
13. Stetson DB, Mohrs M, Reinhardt RL, Baron JL, Wang Z-E, Gapin L, et al. Constitutive cytokine mRNAs mark natural killer (NK) and NK T cells poised for rapid effector function. *J Exp Med* (2003) 198:1069–76. doi:10.1084/jem.20030630
14. Martín-Fontecha A, Thomsen LL, Brett S, Gerard C, Lipp M, Lanzavecchia A, et al. Induced recruitment of NK cells to lymph nodes provides IFN-|[gamma]| for TH1 priming. *Nat Immunol* (2004) 5:1260–5. doi:10.1038/ni1138
15. Moretta L, Ferlazzo G, Bottino C, Vitale M, Pende D, Mingari MC, et al. Effector and regulatory events during natural killer-dendritic cell interactions. *Immunol Rev* (2006) 214:219–28. doi:10.1111/j.1600-065X.2006.00450.x
16. Walzer T, Bléry M, Chaix J, Fuseri N, Chasson L, Robbins SH, et al. Identification, activation, and selective in vivo ablation of mouse NK cells via NKp46. *Proc Natl Acad Sci U S A* (2007) 104:3384–9. doi:10.1073/pnas.0609692104
17. Walzer T, Jaeger S, Chaix J, Vivier E. Natural killer cells: from CD3− NKp46+ to post-genomics meta-analyses. *Curr Opin Immunol* (2007) 19:365–72. doi:10.1016/j.coi.2007.04.004
18. Kondo M, Weissman IL, Akashi K. Identification of clonogenic common lymphoid progenitors in mouse bone marrow. *Cell* (1997) 91:661–72. doi:10.1016/S0092-8674(00)80291-3
19. Carotta S, Pang SHM, Nutt SL, Belz GT. Identification of the earliest NK-cell precursor in the mouse BM. *Blood* (2011) 117:5449–52. doi:10.1182/blood-2010-11-318956
20. Huntington ND, Nutt SL, Carotta S. Regulation of murine natural killer cell commitment. *Front Immunol* (2013) 4:14. doi:10.3389/fimmu.2013.00014
21. Rosmaraki EE, Douagi I, Roth C, Colucci F, Cumano A, Di Santo JP. Identification of committed NK cell progenitors in adult murine bone marrow. *Eur J Immunol* (2001) 31:1900–9. doi:10.1002/1521-4141(200106)31:6<1900::AID-IMMU1900>3.0.CO;2-M
22. Kim S, Iizuka K, Kang H-SP, Dokun A, French AR, Greco S, et al. In vivo developmental stages in murine natural killer cell maturation. *Nat Immunol* (2002) 3:523–8. doi:10.1038/ni796
23. Huntington ND, Tabarias H, Fairfax K, Brady J, Hayakawa Y, Degli-Esposti MA, et al. NK cell maturation and peripheral homeostasis is associated with KLRG1 up-regulation. *J Immunol* (2007) 178:4764–70. doi:10.4049/jimmunol.178.8.4764
24. Chiossone L, Chaix J, Fuseri N, Roth C, Vivier E, Walzer T. Maturation of mouse NK cells is a 4-stage developmental program. *Blood* (2009) 113:5488–96. doi:10.1182/blood-2008-10-187179
25. Yu X, Wang Y, Deng M, Li Y, Ruhn KA, Zhang CC, et al. The basic leucine zipper transcription factor NFIL3 directs the development of a common innate lymphoid cell precursor. *Elife* (2014) 3:945. doi:10.7554/eLife.04406
26. Yang Q, Li F, Harly C, Xing S, Ye L, Xia X, et al. TCF-1 upregulation identifies early innate lymphoid progenitors in the bone marrow. *Nat Immunol* (2015) 16:1044–50. doi:10.1038/ni.3248
27. Fathman JW, Bhattacharya D, Inlay MA, Seita J, Karsunky H, Weissman IL. Identification of the earliest natural killer cell-committed progenitor in murine bone marrow. *Blood* (2011) 118:5439–47. doi:10.1182/blood-2011-04-348912
28. Hayakawa Y, Smyth MJ. CD27 dissects mature NK cells into two subsets with distinct responsiveness and migratory capacity. *J Immunol* (2006) 176:1517–24. doi:10.4049/jimmunol.176.3.1517
29. Robbins SH, Tessmer MS, Mikayama T, Brossay L. Expansion and contraction of the NK cell compartment in response to murine cytomegalovirus infection. *J Immunol* (2004) 173:259–66. doi:10.4049/jimmunol.173.1.259
30. Peng H, Sun R, Tang L, Wei H, Tian Z. CD62L is critical for maturation and accumulation of murine hepatic NK cells in response to viral infection. *J Immunol* (2013) 190:4255–62. doi:10.4049/jimmunol.1202395
31. Despoix N, Walzer T, Jouve N, Blot-Chabaud M, Bardin N, Paul P, et al. Mouse CD146/MCAM is a marker of natural killer cell maturation. *Eur J Immunol* (2008) 38:2855–64. doi:10.1002/eji.200838469
32. Zafirova B, Mandarić S, Antulov R, Krmpotić A, Jonsson H, Yokoyama WM, et al. Altered NK cell development and enhanced NK cell-mediated resistance to mouse cytomegalovirus in NKG2D-deficient mice. *Immunity* (2009) 31:270–82. doi:10.1016/j.immuni.2009.06.017
33. Shehata HM, Hoebe K, Chougnet CA. The aged nonhematopoietic environment impairs natural killer cell maturation and function. *Aging Cell* (2015) 14:191–9. doi:10.1111/acel.12303
34. Omi A, Enomoto Y, Kiniwa T, Miyata N, Miyajima A. Mature resting Ly6Chigh natural killer cells can be reactivated by IL-15. *Eur J Immunol* (2014) 44:2638–47. doi:10.1002/eji.201444570
35. Martinet L, De Andrade LF, Guillerey C, Lee JS, Liu J, Souza-Fonseca-Guimaraes F, et al. DNAM-1 expression marks an alternative program of NK cell maturation. *Cell Rep* (2015) 11:85–97. doi:10.1016/j.celrep.2015.03.006
36. Tu TC, Brown NK, Kim T-J, Wroblewska J, Yang X, Guo X, et al. CD160 is essential for NK-mediated IFN-γ production. *J Exp Med* (2015) 212:415–29. doi:10.1084/jem.20131601
37. Seillet C, Belz GT, Huntington ND. Development, homeostasis, and heterogeneity of NK cells and ILC1. *Curr Top Microbiol Immunol* (2015) 395:37–61. doi:10.1007/82_2015_474
38. Huntington ND, Martinet L, Smyth MJ. DNAM-1: would the real natural killer cell please stand up! *Oncotarget* (2015) 6:28537–8. doi:10.18632/oncotarget.5952
39. Robbins SH, Nguyen KB, Takahashi N, Mikayama T, Biron CA, Brossay L. Cutting edge: inhibitory functions of the killer cell lectin-like receptor G1 molecule during the activation of mouse NK cells. *J Immunol* (2002) 168:2585–9. doi:10.4049/jimmunol.168.6.2585
40. Lemon B, Tjian R. Orchestrated response: a symphony of transcription factors for gene control. *Genes Dev* (2000) 14:2551–69. doi:10.1101/gad.831000
41. Wang S, Xia P, Huang G, Zhu P, Liu J, Ye B, et al. FoxO1-mediated autophagy is required for NK cell development and innate immunity. *Nat Commun* (2016) 7:11023. doi:10.1038/ncomms11023
42. Deng Y, Kerdiles Y, Chu J, Yuan S, Wang Y, Chen X, et al. Transcription factor Foxo1 is a negative regulator of natural killer cell maturation and function. *Immunity* (2015) 42:457–70. doi:10.1016/j.immuni.2015.02.006
43. Taki S, Nakajima S, Ichikawa E, Saito T, Hida S. IFN regulatory factor-2 deficiency revealed a novel checkpoint critical for the generation of

peripheral NK cells. *J Immunol* (2005) 174:6005–12. doi:10.4049/jimmunol.174.10.6005
44. Lohoff M, Duncan GS, Ferrick D, Mittrücker HW, Bischof S, Prechtl S, et al. Deficiency in the transcription factor interferon regulatory factor (IRF)-2 leads to severely compromised development of natural killer and T helper type 1 cells. *J Exp Med* (2000) 192:325–36. doi:10.1084/jem.192.3.325
45. Barton K, Muthusamy N, Fischer C, Ting CN, Walunas TL, Lanier LL, et al. The Ets-1 transcription factor is required for the development of natural killer cells in mice. *Immunity* (1998) 9:555–63. doi:10.1016/S1074-7613(00)80638-X
46. Ramirez K, Chandler KJ, Spaulding C, Zandi S, Sigvardsson M, Graves BJ, et al. Gene deregulation and chronic activation in natural killer cells deficient in the transcription factor ETS1. *Immunity* (2012) 36:921–32. doi:10.1016/j.immuni.2012.04.006
47. Male V, Nisoli I, Kostrzewski T, Allan DS, Carlyle JR, Lord GM, et al. The transcription factor E4bp4/Nfil3 controls commitment to the NK lineage and directly regulates Eomes and Id2 expression. *J Exp Med* (2014) 211:635–42. doi:10.1084/jem.20132398
48. Seillet C, Huntington ND, Gangatirkar P, Axelsson E, Minnich M, Brady HJ, et al. Differential requirement for Nfil3 during NK cell development. *J Immunol* (2014) 192:2667–76. doi:10.4049/jimmunol.1302605
49. Firth MA, Madera S, Beaulieu AM, Gasteiger G, Castillo EF, Schluns KS, et al. Nfil3-independent lineage maintenance and antiviral response of natural killer cells. *J Exp Med* (2013) 210:2981–90. doi:10.1084/jem.20130417
50. Held W, Kunz B, Lowin-Kropf B, van de Wetering M, Clevers H. Clonal acquisition of the Ly49A NK cell receptor is dependent on the trans-acting factor TCF-1. *Immunity* (1999) 11:433–42. doi:10.1016/S1074-7613(00)80118-1
51. Weber BN, Chi AW-S, Chavez A, Yashiro-Ohtani Y, Yang Q, Shestova O, et al. A critical role for TCF-1 in T-lineage specification and differentiation. *Nature* (2011) 476:63–8. doi:10.1038/nature10279
52. Gordon SM, Chaix J, Rupp LJ, Wu J, Madera S, Sun JC, et al. The transcription factors T-bet and Eomes control key checkpoints of natural killer cell maturation. *Immunity* (2012) 36:55–67. doi:10.1016/j.immuni.2011.11.016
53. Delconte RB, Shi W, Sathe P, Ushiki T, Seillet C, Minnich M, et al. The helix-loop-helix protein ID2 governs NK cell fate by tuning their sensitivity to interleukin-15. *Immunity* (2016) 44:103–15. doi:10.1016/j.immuni.2015.12.007
54. Sathe P, Delconte RB, Souza-Fonseca-Guimaraes F, Seillet C, Chopin M, Vandenberg CJ, et al. Innate immunodeficiency following genetic ablation of Mcl1 in natural killer cells. *Nat Commun* (2014) 5:4539. doi:10.1038/ncomms5539
55. Boos MD, Yokota Y, Eberl G, Kee BL. Mature natural killer cell and lymphoid tissue-inducing cell development requires Id2-mediated suppression of E protein activity. *J Exp Med* (2007) 204:1119–30. doi:10.1084/jem.20061959
56. Murre C, McCaw PS, Vaessin H, Caudy M, Jan LY, Jan YN, et al. Interactions between heterologous helix-loop-helix proteins generate complexes that bind specifically to a common DNA sequence. *Cell* (1989) 58:537–44. doi:10.1016/0092-8674(89)90434-0
57. Quong MW, Romanow WJ, Murre C. E protein function in lymphocyte development. *Annu Rev Immunol* (2003) 20:301–22. doi:10.1146/annurev.immunol.20.092501.162048
58. Huntington ND, Puthalakath H, Gunn P, Naik E, Michalak EM, Smyth MJ, et al. Interleukin 15-mediated survival of natural killer cells is determined by interactions among Bim, Noxa and Mcl-1. *Nat Immunol* (2007) 8:856–63. doi:10.1038/ni1487
59. Sathe P, Delconte RB, Souza-Fonseca-Guimaraes F, Seillet C, Chopin M, Vandenberg CJ, et al. Innate immunodeficiency following genetic ablation of Mcl1 in natural killer cells. *Nat Commun* (2014) 5:1–10. doi:10.1038/ncomms5539
60. Townsend MJ, Weinmann AS, Matsuda JL, Salomon R. T-bet regulates the terminal maturation and homeostasis of NK and Vα14i NKT cells. *Immunity* (2004) 20(4):477–94. doi:10.1016/S1074-7613(04)00076-7
61. Pikovskaya O, Chaix J, Rothman NJ, Collins A, Chen Y-H, Scipioni AM, et al. Cutting edge: eomesodermin is sufficient to direct type 1 innate lymphocyte development into the conventional NK lineage. *J Immunol* (2016) 196:1449–54. doi:10.4049/jimmunol.1502396
62. Jenne CN, Enders A, Rivera R, Watson SR, Bankovich AJ, Pereira JP, et al. T-bet-dependent S1P5 expression in NK cells promotes egress from lymph nodes and bone marrow. *J Exp Med* (2009) 206:2469–81. doi:10.1084/jem.20090525
63. Intlekofer AM, Takemoto N, Wherry EJ, Longworth SA, Northrup JT, Palanivel VR, et al. Effector and memory CD8+ T cell fate coupled by T-bet and eomesodermin. *Nat Immunol* (2005) 6:1236–44. doi:10.1038/ni1268
64. van Helden MJ, Goossens S, Daussy C, Mathieu A-L, Faure F, Marçais A, et al. Terminal NK cell maturation is controlled by concerted actions of T-bet and Zeb2 and is essential for melanoma rejection. *J Exp Med* (2015) 212:2015–25. doi:10.1084/jem.20150809
65. Kallies A, Carotta S, Huntington ND, Bernard NJ, Tarlinton DM, Smyth MJ, et al. A role for Blimp1 in the transcriptional network controlling natural killer cell maturation. *Blood* (2011) 117:1869–79. doi:10.1182/blood-2010-08-303123
66. Aliahmad P, la Torre de B, Kaye J. Shared dependence on the DNA-binding factor TOX for the development of lymphoid tissue-inducer cell and NK cell lineages. *Nat Immunol* (2010) 11:945–52. doi:10.1038/ni.1930
67. Holmes ML, Huntington ND, Thong RP, Brady J, Hayakawa Y, Andoniou CE, et al. Peripheral natural killer cell maturation depends on the transcription factor Aiolos. *EMBO J* (2014) 33:2721–34. doi:10.15252/embj.201487900
68. Rabacal W, Pabbisetty SK, Hoek KL, Cendron D, Guo Y, Maseda D, et al. Transcription factor KLF2 regulates homeostatic NK cell proliferation and survival. *Proc Natl Acad Sci U S A* (2016) 113:5370–5. doi:10.1073/pnas.1521491113
69. Samson SI, Richard O, Tavian M, Ranson T, Vosshenrich CAJ, Colucci F, et al. GATA-3 promotes maturation, IFN-γ production, and liver-specific homing of NK cells. *Immunity* (2003) 19:701–11. doi:10.1016/S1074-7613(03)00294-2
70. Ali AK, Oh JS, Vivier E, Busslinger M, Lee SH. NK cell-specific Gata3 ablation identifies the maturation program required for bone marrow exit and control of proliferation. *J Immunol* (2016) 196:1753–67. doi:10.4049/jimmunol.1501593
71. Lacorazza HD, Miyazaki Y, Di Cristofano A, Deblasio A, Hedvat C, Zhang J, et al. The ETS protein MEF plays a critical role in perforin gene expression and the development of natural killer and NK-T cells. *Immunity* (2002) 17:437–49. doi:10.1016/S1074-7613(02)00422-3
72. Degouve S, Tavares A, Viel S, Walzer T, Marçais A. NKp46-mediated Dicer1 inactivation results in defective NK-cell differentiation and effector functions in mice. *Eur J Immunol* (2016) 46:1902–11. doi:10.1002/eji.201546163
73. Zawislak CL, Beaulieu AM, Loeb GB, Karo J, Canner D, Bezman NA, et al. Stage-specific regulation of natural killer cell homeostasis and response against viral infection by microRNA-155. *Proc Natl Acad Sci U S A* (2013) 110:6967–72. doi:10.1073/pnas.1304410110
74. Sullivan RP, Leong JW, Schneider SE, Ireland AR, Berrien-Elliott MM, Singh A, et al. microRNA-15/16 antagonizes Myb to control NK cell maturation. *J Immunol* (2015) 195:2806–17. doi:10.4049/jimmunol.1500949
75. Trotta R, Chen L, Costinean S, Josyula S, Mundy-Bosse BL, Ciarlariello D, et al. Overexpression of miR-155 causes expansion, arrest in terminal differentiation and functional activation of mouse natural killer cells. *Blood* (2013) 121:3126–34. doi:10.1182/blood-2012-12-467597
76. Su I-H, Basavaraj A, Krutchinsky AN, Hobert O, Ullrich A, Chait BT, et al. Ezh2 controls B cell development through histone H3 methylation and Igh rearrangement. *Nat Immunol* (2003) 4:124–31. doi:10.1038/ni876
77. Tumes DJ, Onodera A, Suzuki A, Shinoda K, Endo Y, Iwamura C, et al. The polycomb protein Ezh2 regulates differentiation and plasticity of CD4+ T helper type 1 and type 2 cells. *Immunity* (2013) 39:819–32. doi:10.1016/j.immuni.2013.09.012
78. Yin J, Leavenworth JW, Li Y, Luo Q, Xie H, Liu X, et al. Ezh2 regulates differentiation and function of natural killer cells through histone methyltransferase activity. *Proc Natl Acad Sci U S A* (2015) 112:15988–93. doi:10.1073/pnas.1521740112
79. Nandakumar V, Chou Y, Zang L, Huang XF, Chen SY. Epigenetic control of natural killer cell maturation by histone H2A deubiquitinase, MYSM1. *Proc Natl Acad Sci U S A* (2013) 110:E3927–36. doi:10.1073/pnas.1308888110

3

Hepatitis C Virus and Human Cytomegalovirus —Natural Killer Cell Subsets in Persistent Viral Infections

*Julia Pollmann[1], Alexander Rölle[2], Maike Hofmann[3] and Adelheid Cerwenka[1,4]**

[1] Research Group Innate Immunity, German Cancer Research Center (DKFZ), Heidelberg, Germany, [2] Clinical Cooperation Unit Applied Tumor-Immunity, German Cancer Research Center (DKFZ), National Center for Tumor Diseases (NCT), Heidelberg, Germany, [3] Faculty of Medicine, Department of Medicine II, University Hospital Freiburg, University of Freiburg, Freiburg im Breisgau, Germany, [4] Medical Faculty Mannheim, Division of Immunbiochemistry, University Heidelberg, Heidelberg, Germany

Edited by:
Ulrike Koehl, Hannover Medical School, Germany

Reviewed by:
Roland Jacobs, Hannover Medical School, Germany
Miguel López-Botet, Institut Hospital del Mar d'Investigacions Mèdiques, Spain

***Correspondence:**
Adelheid Cerwenka
a.cerwenka@dkfz-heidelberg.de, adelheid.cerwenka@medma.uni-heidelberg.de

Hepatitis C virus (HCV) and human cytomegalovirus (HCMV) are prominent examples of RNA and DNA viruses, respectively, that establish a persistent infection in their host. HCV affects over 185 million patients worldwide, who are at high risk for developing liver fibrosis, liver cirrhosis, and ultimately hepatocellular carcinoma. Recent breakthroughs in HCV therapy, using direct-acting antivirals have provided the opportunity to monitor natural killer (NK) cells after clearance of a chronic infection. There is now increasing evidence that the individual NK cell repertoire before infection is predictive for the course of disease. HCMV affects the majority of the global population. While being asymptomatic in healthy individuals, HCMV represents a severe clinical challenge in immunocompromised patients. Both viral infections, HCV and HCMV, lead to long-lasting and profound alterations within the entire NK cell compartment. This review article, will discuss the diverse range of changes in the NK cell compartment as well as potential consequences for the course of disease.

Keywords: natural killer cells, hepatitis C virus, human cytomegalovirus, chronic infection, natural killer subsets

INTRODUCTION

A wide range of viral infections challenge the immune system throughout the lifetime of its host exerting a substantial and often long-lasting impact on multiple immune parameters. Natural killer (NK) cells, vital players in the antiviral immune defense, have been shown to undergo substantial changes in phenotype, function, and subset distribution during persistent viral infections. Specific NK subsets have been associated with both efficient clearance of viruses and immune dysfunction.

Persistent viral infections can be latent or chronic. Latent infection is characterized by long periods of viral inactivity with no replication or production of new virions although stress stimuli can trigger episodes of reactivations. Prominent examples of viruses establishing latent infection are the herpes viruses [human cytomegalovirus (HCMV), herpes simplex virus (HSV), Epstein–Barr virus (EBV), varicella-zoster-virus (VZV)]. Other viruses, such as the majority of hepatitis viruses [hepatitis C Virus (HCV), hepatitis B Virus (HBV), hepatitis D Virus (HDV)] and human immunodeficiency virus (HIV), establish chronic infections in which constant replication takes place. This drives chronic inflammation, often resulting in severe tissue damage of the infected organ (1). In this review, we will focus primarily on the effects of latent HCMV and chronic HCV infection on NK cells.

Hepatitis C Virus is a hepatotropic, enveloped, (+)-strand RNA virus that is transmitted person-to-person *via* blood and establishes chronic infection in 55–85% of patients. The probability for spontaneous viral clearance depends on several factors such as age, sex, host genetic factors, coinfection with other viruses, and viral genotype (2, 3).

Currently, seven genotypes and multiple subgenotypes are described with distinct global distribution patterns. In developed countries, genotype 1 is the most common, accounting for around 50% of all HCV infections, even though it has the most favorable prognosis (4). Worldwide an estimated 2.5% of the world's population is chronically infected with HCV. Throughout the decade-long infection, the liver suffers from immunopathology, resulting in fibrosis, cirrhosis, often progressing to hepatocellular carcinoma. Each year around 500,000 people die from HCV-related liver diseases (5). To establish chronicity, the virus interferes with several innate and adaptive immune pathways, such as recognition by retinoic acid inducible gene I (RIG I), the primary sensor for HCV-RNA in the host cell's cytoplasm (6). The emergence of viral escape variants facilitates evasion from recognition by CD8 T cells, which are the main effector cells against HCV (7).

Until 2011, standard therapy for HCV consisted of pegylated IFN-α/ribavirin. However, only around half of the patients achieved a sustained virological response (SVR) defined by no detectable HCV-RNA 24 weeks after treatment and side effects were drastic. In 2011, the first direct-acting antivirals (DAAs) were approved targeting essential viral proteins which revolutionized therapy by reaching SVR rates of >90% (4, 8, 9).

Human cytomegalovirus has a linear double-stranded DNA genome of 236 kbp. The virus spreads vertically and horizontally *via* bodily fluids by infecting epithelial and endothelial cells, macrophages, and DCs wherein it establishes life-long latency. This leads to high global prevalence rates of 60–85%, depending on socioeconomic factors, geographical location, and age. HCMV is an opportunistic pathogen, causing disease only in immunocompromised people, e.g., during transplantations or in HIV patients. Furthermore, transplacental transmission of HCMV can cause severe, primarily neurological, damage to the fetus (10).

Human cytomegalovirus has developed a plethora of strategies and dedicates a large portion of its genome to interfere with the host immune system. Many of these escape mechanisms have evolved to avoid recognition by NK cells (11–13).

Natural killer cells are important effector cells in the antiviral immune response in during HCMV and HCV infection (14–17).

The importance of NK cells in human Herpes virus infections was initially highlighted in a patient with a very rare NK cell deficiency and his enhanced susceptibility to recurrent infections (18), a clinical phenotype corroborated in later reports (19, 20). Moreover, it is also indirectly evident in the multiple immune evasion strategies that HCMV utilizes to prevent NK cell recognition (11, 12).

The two main strategies employed by various viruses to evade NK cells are preventing the upregulation of activating ligands or enhancing the expression of ligands for inhibitory NK cell receptors. Upon viral infection, a variety of stress-induced molecules are expressed on the surface of infected cells, which can be recognized by activating receptors on NK cells. Important activating receptors belong to the natural cytotoxicity receptor (NCR) family, including NKp30, NKp44, and NKp46, which can recognize cellular as well as viral ligands. However, many of the NCR ligands still remain elusive. MHC class I molecules, in particular HLA-C, provide the main inhibitory signals for NK cells *via* interacting with killer immunoglobulin-like receptors (KIRs) (21). The KIRs, like their MHC ligands, are genetically highly polymorphic and expressed in a stochastic manner, leaving every NK cell with 0–4 KIR receptors (22). Furthermore, different KIR haplotypes—group A and B—have been identified. While the group A haplotype comprises almost exclusively inhibitory, the group B haplotypes also encode activating KIRs (23). Indeed, most receptors are expressed only on subsets of NK cells. Therefore, NK cells are not a uniform cell population but composed of many different subsets that differ in their mode of activation and their functional properties. In this review, we will discuss some of the NK cell subsets that have been studied in HCV and HCMV infections (**Figure 1**).

NK CELL SUBSETS CARRYING SPECIFIC KIRs

One receptor that has repeatedly been implicated in the anti-HCV immune response is the inhibitory KIR2DL3. In 2004, Khakoo and colleagues (24) reported a genetic association between the expression of KIR2DL3 and HCV clearance. They studied more than thousand patients infected with HCV, of whom 685 developed a chronic infection while 352 cleared the virus. Those patients, whose NK cells expressed the inhibitory KIR2DL3 homozygously on an HLA-C1/C1 background, were more likely to clear the virus spontaneously. Of note, when stratified in terms of route of infection, this association was only observed in patients with presumed low-dose viral inoculum, e.g., after a needle-stick, suggesting that the antiviral NK cell response is most efficient in situations with low viral load. Other studies in cohorts of injection drug users confirmed that exposed but uninfected individuals had a higher frequency of KIR2DL3 expression than drug users with chronic and resolved infections or healthy donors (25). This association was also reflected in the response to treatment, in which KIR2DL3–HLA-C1 expressing patients had a higher chance to achieve SVR after IFN-α-based treatment (26). However, in other studies, a correlation between KIR2DL3 and outcome of infection was not observed, albeit with smaller and different patient cohorts (27, 28). The current hypothesis for this association is that NK cells expressing KIR2DL3, which has a rather weak binding affinity to HLA-C (29), tend to receive less inhibitory signals and are, therefore, more easily activated than NK cells expressing KIRs with higher binding affinities.

Additionally, there is now increasing evidence that peptides presented in MHC class I complexes also influence binding affinities of KIRs. This has already been shown for HIV peptides (30–32) and could in 2016 for the first time be shown for an HCV peptide presented by an HLA-C1 molecule to KIR2DL3 on NK cells (33). Lunemann et al. (33) identified peptides derived from HCV NS3 and Core protein that stabilized expression of

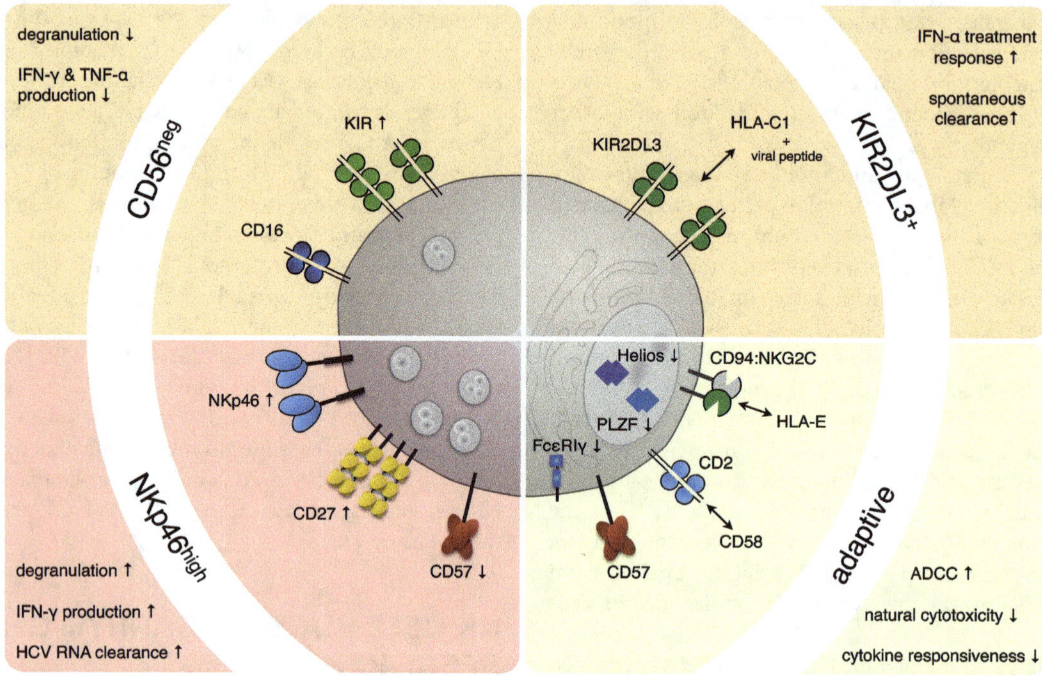

FIGURE 1 | NK cell subsets in chronic viral infections. Persistent infections exert a profound and long-lasting impact on the NK cell compartment. In hepatitis C virus (HCV) (red background) and human cytomegalovirus (HCMV) (yellow background) infection, a number of NK cell subsets have been shown to expand and/or correlate with disease progression and outcome. During chronic HCV infection, an NK cell subset displaying increased expression of the natural cytotoxicity receptor NKp46 was observed, which proved superior in clearing HCV-RNA from infected hepatocytes. On the other hand, expansions of functionally impaired CD56neg NK cells ($CD56^-CD16^+CD3^-CD14^-CD19^-$) were described during chronic HCV and human immunodeficiency virus and sporadically during reactivation episodes of HCMV. HCV patients with high numbers of NK cells expressing the inhibitory receptor KIR2DL3 on an HLA-C1/C1 background are more likely to spontaneously clear the virus or respond favorably to IFN-α treatment. Furthermore, certain killer immunoglobulin-like receptors (KIRs), including KIR2DL3 seem to be overrepresented on expanded adaptive NK cell subsets in HCMV infection. These adaptive NK cell subsets often express NKG2C and CD57 and are characterized by low/absent levels of key signaling molecules (e.g., FcεRIγ) and transcription factors (e.g., PLZF and/or Helios).

HLA-C*03:04 on transfected 721.221 cells and facilitated binding of KIR2DL3–Fc proteins. One of the peptides derived from HCV Core protein was furthermore able to inhibit degranulation of primary $KIR2DL3^+$ NK cells. When comparing core peptide sequences from different HCV genotypes, the genotype 1-derived peptide provided the strongest inhibitory signal, while peptides from other genotypes were much less effective. These results encourage speculations about possible viral escape mechanisms by modulation of KIR binding and the implication that NK cells exert evolutionary pressure on the HCV genome. Future studies should address if the identified peptides can also be presented on primary hepatocytes and if other KIRs/HLA interactions are involved as well.

Moreover, there is substantial evidence for an overrepresentation of certain KIRs in expanded adaptive NK cell subpopulations (see below) in HCMV. The currently available data yield a complex picture with some reports highlighting KIR2DL2/3 (34–36) others KIR3DL1 (37) or activating KIRs (38).

THE $CD56^{neg}$ NK CELL SUBSET

Traditionally, NK cells are classified as $CD3^-CD56^+$ lymphocytes, which are further divided into a $CD56^{dim}CD16^+$ and a $CD56^{bright}CD16^-$ subset. During chronic viral infections, especially in HIV and HCV, a subset of $CD3^-CD56^-CD16^+$ NK cells is detectable (39). These cells miss expression of lineage markers, such as CD14 and CD19 or markers of other cell types positive for CD16, while expressing a variety of NK cell receptors (NCR, KIR, and NKG2). They are, therefore, classified as $CD56^{neg}$ NK cells. This NK cell subset is found at low percentages (around 5% of all NK cells) in healthy adults and even in neonates, but can expand to 10–40% of all NK cells during HIV, acute, chronic, and resolved HCV infections or HIV/HCV coinfection. Concurrently, a drop in the percentage of $CD56^{dim}$ NK cells is observed (40–42).

Phenotypically, $CD56^{neg}$ NK cells from healthy donors and chronic HCV patients are similar and expression of many receptors is comparable between $CD56^{neg}$ and $CD56^{dim}$ NK cells from chronic HCV patients. Only CD57 and to a lesser extent NKp30 were found to be expressed at lower levels in the $CD56^{neg}$ NK subset (40). Functionally, however, $CD56^{neg}$ NK cells appear impaired. In response to different stimuli, the $CD56^{neg}$ NK subset of HCV patients failed to secrete significant amounts of IFN-γ and TNF-α and displayed low perforin expression and degranulation (40, 42). Yet, they showed higher TRAIL expression compared to $CD56^{neg}$ NK cells from healthy controls (41). This functional impairment seems to be a general feature of the $CD56^{neg}$ NK cell subset, as it was observed in various chronic inflammatory situations. However, overnight *in vitro* culture of CD56^{neg} NK cells with IL-2, IL-12, or IL-15 resulted in cytotoxicity levels comparable to stimulated CD56^{+} NK cells, indicating that upon stimulation CD56^{neg} NK cells can effectively function (43). It was shown in HCV, as well as in HIV/HCV coinfection, that high pre-treatment levels of CD56^{neg} NK cells correlated with treatment failure (40, 44). Furthermore, after successful IFN-α/ribavirin treatment of HIV/HCV coinfected patients, absolute levels and percentage of CD56^{neg} NK cells normalized after 4 weeks (45).

So far, the role of this particular NK cell subset was not evaluated during DAA treatment of patients, but several lines of evidence suggest that the decline in CD56^{neg} NK cells after treatment is not induced by IFN-α, but results from a decrease in viral load (39). Until now, it was shown that the CD56^{dim} and CD56^{bright} NK cell subsets quickly normalize after DAA treatment in regard to numbers and functionality (46–48). This observation could indicate that the CD56^{neg} subset also normalizes. However, further studies are needed to address this and other open questions: does the CD56^{neg} subset represent a terminally differentiated or exhausted cell type or rather a specific lineage? What is the degree of plasticity in terms of other NK cell subsets becoming CD56^{neg} or CD56^{neg} NK cells acquiring CD56^{pos} phenotypes? The fact that neonates and healthy individuals already have this CD56^{neg} subset, might argue against an exhausted phenotype, although many of their properties resemble exhausted T cells.

In contrast to HCV and HIV, increased levels of CD56^{neg} cells have only been reported sporadically in HCMV infection and were observed only in a subset of patients experiencing viral reactivation (49).

THE NKp46^{high} NK CELL SUBSET

Expression of NK cell receptors was analyzed in different HCV patient cohorts (e.g., in acute or chronic infection and during treatment), but the obtained results have been highly controversial (50, 51). With regards to NKp46, however, several independent studies observed the involvement of an NKp46^{high} NK cell subset expressing multiple markers of immature NK cells (52) in protection from infection, spontaneous clearance, liver inflammation, progression of fibrosis, and outcome of treatment.

In prospective studies of injection drug users, a higher percentage of CD56^{dim} NK cells with increased levels of NKp46 was correlated with a higher percentage of individuals remaining seronegative, suggesting that high expression of NKp46 might be predictive for protection from infection (25). However, Alter et al. (42) reported lower expression of activating receptors, including NKp46, in patients with acute infection who subsequently cleared the virus than in those, who progressed to chronic infection.

In chronically infected patients, the majority of studies report elevated NKp46 expression on peripheral blood NK cells compared to healthy donors (51). Of note, this is even more pronounced in the liver (52). NKp46^{high} peripheral NK cells from healthy donors and HCV patients perform better in reducing HCV-RNA from *in vitro* infected hepatocytes, produce higher levels of IFN-γ, and degranulate more *ex vivo* in response to different stimuli (52, 53). Likewise, intrahepatic NKp46^{high} NK cells from HCV patients were shown to degranulate more *ex vivo* than NKp46^{dim} cells (54). Furthermore, staining with an NKp46-Ig fusion protein revealed higher expression of a yet unknown NKp46-ligand on HCV-infected Huh7.5 hepatoma cells than on uninfected cells (53).

Accordingly, NKp46 levels in patients correlate positively with liver inflammation scores (54) and inversely with HCV serum levels (52), suggesting that NKp46^{high} NK cells can kill infected hepatocytes and contribute to viral control during chronic infection. The NKp46^{high} subset correlated with low fibrosis stages, possibly due to NKp46-dependent killing of hepatic stellate cells, the main drivers of fibrosis (54–56).

Even though the NKp46^{high} NK cell subset might be beneficial in reducing viral load and liver fibrosis during chronic infection, it also predicts failure to IFN-α therapy (54, 57). After successful DAA treatment, previously elevated NKp46 levels in liver and blood normalize, concomitantly with a normalization of many other NK cell receptors (47).

In contrast, in HCMV infection, there is little evidence for a direct modulation or involvement of NCRs. One report observed the dissociation of the CD3zeta chain from NKp30 after engaging the HCMV tegument protein pp65, leading to greatly reduced NKp30-mediated killing (58).

ADAPTIVE NK CELL SUBSETS

The most striking example for a long-lasting impact of a pathogenic challenge on distinct NK subsets (59) was initially identified in two key reports by Miguel Lopez-Botet's group (60, 61). In HCMV-seropositive individuals, a higher proportion of NK cells expressing the activating receptor CD94/NKG2C was detected. This expanded subset displays lower NCR levels and increased expression of CD85j/LIR-1 (60) and CD2 (38). Similar observations were made in transplant recipients who suffered from acute CMV infection/reactivation (34, 37, 49, 62, 63). *In vitro* studies recapitulated subset expansion suggesting that exposure of NKG2C^{+} NK cells to infected cells was critical for this process (61). Moreover, the interaction between HLA-E and CD94/NKG2C was defined as a critical event for subset expansion (64, 65). To date, several additional factors have been reported to contribute to the expansion and activation of NKG2C^{+} NK cells in response to HCMV infection, such as IL-15 (61), IL-12 (64), and CD14^{+} monocytes (64, 66), as well as the interaction between CD2 and upregulated CD58 on infected cells (67). As expansion of subpopulations and their subsequent longevity resemble hallmarks of adaptive immune responses, the term "adaptive NK cells" was coined for human NK cells displaying these characteristics. From here on, we will use the term in this broadly defined sense, comprising multiple subsets.

Intriguingly, while a large number of studies describe NK subset expansions in other infections, e.g., HIV (68–70) Hantavirus (65), Chikungunya virus (71), EBV (72), and HBV/HCV (35, 73), seropositivity for HCMV seems to be a necessary pre-requirement. Altered HLA-E levels and/or a certain inflammatory cytokine milieu could be common denominators permitting the (re)expansion of NK cells "primed" initially by HCMV. The initial events, however, underlying the formation of this NK cell subpopulation in primary HCMV infection, remain enigmatic and represent a field of intense interest.

In recent years, it became clear that the initial definition as NKG2C$^+$ (and CD57$^+$) was not sufficient to encompass all adaptive NK cell subsets.

A study by Hwang and coworkers in 2012 (74) identified NK cell subpopulations with low or absent expression of the adaptor protein FcεRIγ in about one-third of all individuals tested. Zhang et al. (75) then established that the presence of FcεRIγ-deficient NK cells was strictly associated with prior exposure to HCMV. Expansion of an FcεRIγ^- subset was also observed in HCMV$^+$ chronic HCV patients and correlated with low liver damage and fibrosis levels, possibly implying an involvement of this subset in protection from immunopathology (76). Further reports extended the concept of HCMV driving the expansion of adaptive NK cell populations with deficiencies in key signaling molecules to Syk, EAT-2, and DAB2 and the transcription factors PLZF and Helios (77, 78). These features are not necessarily combined at a single-cell level and instead found in different combinations creating a previously unappreciated heterogeneity among adaptive NK cells.

Intriguingly, this molecular signature partially resembles exhausted T cells and in fact a recent study described high PD-1 expression on a subset of CD57$^+$ NK cells also displaying increased LIR-1 levels as well as higher NKG2C expression in some donors (79). Together with lower NCR expression in adaptive NK subsets, these features suggest a decreased functionality. While this seems to be the case for classical tumor targets (78), superior antibody-dependent cellular cytotoxicity (ADCC) responses are emerging as a prominent and distinct characteristic of adaptive NK cells (38, 75, 77, 80–82), augmented by CD2 co-stimulation (67, 83). This functional specialization is accompanied by broad epigenetic modifications, including better accessibility of the *ifn-*γ locus (77, 78, 84). Latent HCMV infection might, therefore, be a worthy trade-off for the host if the interplay of CMV-induced adaptive NK cell populations and antigen-specific humoral immunity *via* ADCC results in elevated resistance to heterologous infection.

Besides deciphering the generation and function of adaptive NK cell subsets, several studies focused on their localization. In mice, a subset of NK cells endowed with antigen-specific memory has been shown to reside specifically in the liver (85, 86). In human liver samples, a subset, phenotypically similar to memory NK cells in the mouse (CD49a$^+$T-bet$^+$Eomes$^-$) that also displayed high NKG2C expression, was identified. Yet, in contrast to peripheral NKG2C$^+$ NK cells, they had an immature phenotype (CD57$^-$, CD16$^-$CD56^{bright}), existed in high numbers in HCMV negative donors, and were incapable of mounting ADCC responses due to lack of CD16 expression (85, 87). Therefore, besides their striking resemblance to murine liver-resident memory NK cells and some shared features with human peripheral blood adaptive NK cells, the function of this unique human liver NK subset needs to be further defined.

Very recently, another liver-resident NK cell population, characterized as Eomes^{hi}, CXCR6$^+$ (88–90), and CD49e$^-$ (91) was described, following up HLA-mismatched human liver-transplants. The authors demonstrated that these Eomes^{hi} NK cells survive for up to 13 years (90). This remarkable longevity makes an involvement of this subset in tissue homeostasis or antiviral responses against chronic infections a plausible scenario, which awaits further investigation.

CONCLUSION

Virus infections, especially with persistent viruses, have a remarkable impact on the NK cell compartment, shape the overall NK cell repertoire, and profoundly affect their effector functions. However, *vice versa*, the different NK subset composition and receptor distribution before infection can also be decisive how well infections can be combated. The diversity of human NK cell subsets is one of the emerging topics in the field. Especially, tissue-resident NK cells and other subsets of helper innate lymphoid cells (ILCs), their development, regulation, antiviral functions, and plasticity in tissues, such as in the liver, are currently an area of intense research. A better understanding of the development and dynamics of ILCs comprising both NK cells and helper ILCs subsets in affected tissues during chronic viral infection might help the design of improved targeted strategies for therapeutic intervention.

AUTHOR CONTRIBUTIONS

JP and AR wrote the manuscript. MH revised the manuscript. AC revised the manuscript and provided conceptual input.

REFERENCES

1. Boldogh I, Albrecht T, Porter DD. Persistent viral infections. In: Baron S, editor. *Medical Microbiology*. Chapter 46, 4th ed. Galveston, TX: University of Texas Medical Branch at Galveston (1996).
2. Harris HE, Eldridge KP, Harbour S, Alexander G, Teo C-G, Ramsay ME. HCV National Register Steering Group*. Does the clinical outcome of hepatitis C infection vary with the infecting hepatitis C virus type? *J Viral Hepat* (2007) 14:213–20. doi:10.1111/j.1365-2893.2006.00795.x
3. Grebely J, Page K, Sacks-Davis R, van der Loeff MS, Rice TM, Bruneau J, et al. The effects of female sex, viral genotype, and IL28B genotype on spontaneous clearance of acute hepatitis C virus infection. *Hepatology* (2014) 59:109–20. doi:10.1002/hep.26639
4. Messina JP, Humphreys I, Flaxman A, Brown A, Cooke GS, Pybus OG, et al. Global distribution and prevalence of hepatitis C virus genotypes. *Hepatology* (2015) 61:77–87. doi:10.1002/hep.27259
5. Lozano R, Naghavi M, Foreman K, Lim S, Shibuya K, Aboyans V, et al. Global and regional mortality from 235 causes of death for 20 age groups in 1990 and 2010: a systematic analysis for the Global Burden of Disease Study 2010. *Lancet* (2012) 380:2095–128. doi:10.1016/S0140-6736(12)61728-0
6. Meylan E, Curran J, Hofmann K, Moradpour D, Binder M, Bartenschlager R, et al. Cardif is an adaptor protein in the RIG-I antiviral pathway and is targeted by hepatitis C virus. *Nature* (2005) 437:1167–72. doi:10.1038/nature04193
7. Petrovic D, Dempsey E, Doherty DG, Kelleher D, Long A. Hepatitis C virus – T-cell responses and viral escape mutations. *Eur J Immunol* (2012) 42:17–26. doi:10.1002/eji.201141593

8. European Association for Study of Liver. EASL clinical practice guidelines: management of hepatitis C virus infection. *J Hepatol* (2014) 60:392–420. doi:10.1016/j.jhep.2013.11.003
9. Shahid I, ALMalki WH, Hafeez MH, Hassan S. Hepatitis C virus infection treatment: an era of game changer direct acting antivirals and novel treatment strategies. *Crit Rev Microbiol* (2016) 42:535–47. doi:10.3109/1040841X.2014.970123
10. Mocarski ES Jr, Pass RF. Human cytomegalovirus: general features. *Encyclopedia of Virology (Third Edition)*. Oxford: Academic Press (2008). p. 474–85.
11. Noriega V, Redmann V, Gardner T, Tortorella D. Diverse immune evasion strategies by human cytomegalovirus. *Immunol Res* (2012) 54:140–51. doi:10.1007/s12026-012-8304-8
12. Rossini G, Cerboni C, Santoni A, Landini MP, Landolfo S, Gatti D, et al. Interplay between human cytomegalovirus and intrinsic/innate host responses: a complex bidirectional relationship. *Mediators Inflamm* (2012) 2012:607276. doi:10.1155/2012/607276
13. Wills MR, Poole E, Lau B, Krishna B, Sinclair JH. The immunology of human cytomegalovirus latency: could latent infection be cleared by novel immunotherapeutic strategies? *Cell Mol Immunol* (2015) 12:128–38. doi:10.1038/cmi.2014.75
14. Rehermann B. Pathogenesis of chronic viral hepatitis: differential roles of T cells and NK cells. *Nat Med* (2013) 19:859–68. doi:10.1038/nm.3251
15. Jost S, Altfeld M. Control of human viral infections by natural killer cells. *Annu Rev Immunol* (2013) 31:163–94. doi:10.1146/annurev-immunol-032712-100001
16. Cerwenka A, Lanier LL. Natural killer cell memory in infection, inflammation and cancer. *Nat Rev Immunol* (2016) 16:112–23. doi:10.1038/nri.2015.9
17. Rölle A, Pollmann J, Cerwenka A. Memory of infections: an emerging role for natural killer cells. *PLoS Pathog* (2013) 9:e1003548. doi:10.1371/journal.ppat.1003548
18. Biron CA, Byron KS, Sullivan JL. Severe herpesvirus infections in an adolescent without natural killer cells. *N Engl J Med* (1989) 320:1731–5. doi:10.1056/NEJM198906293202605
19. Etzioni A, Eidenschenk C, Katz R, Beck R, Casanova JL, Pollack S. Fatal varicella associated with selective natural killer cell deficiency. *J Pediatr* (2005) 146:423–5. doi:10.1016/j.jpeds.2004.11.022
20. Eidenschenk C, Dunne J, Jouanguy E, Fourlinnie C, Gineau L, Bacq D, et al. A novel primary immunodeficiency with specific natural-killer cell deficiency maps to the centromeric region of chromosome 8. *Am J Hum Genet* (2006) 78:721–7. doi:10.1086/503269
21. Pegram HJ, Andrews DM, Smyth MJ, Darcy PK, Kershaw MH. Activating and inhibitory receptors of natural killer cells. *Immunol Cell Biol* (2011) 89:216–24. doi:10.1038/icb.2010.78
22. Cichocki F, Miller JS, Anderson SK. Killer immunoglobulin-like receptor transcriptional regulation: a fascinating dance of multiple promoters. *J Innate Immun* (2011) 3:242–8. doi:10.1159/000323929
23. Rajalingam R. Overview of the killer cell immunoglobulin-like receptor system. *Methods Mol Biol* (2012) 882:391–414. doi:10.1007/978-1-61779-842-9_23
24. Khakoo SI, Thio CL, Martin MP, Brooks CR, Gao X, Astemborski J, et al. HLA and NK cell inhibitory receptor genes in resolving hepatitis C virus infection. *Science* (2004) 305:872–4. doi:10.1126/science.1097670
25. Thoens C, Berger C, Trippler M, Siemann H, Lutterbeck M, Broering R, et al. KIR2DL3+NKG2A natural killer cells are associated with protection from productive hepatitis C virus infection in people who inject drugs. *J Hepatol* (2014) 61(3):475–81. doi:10.1016/j.jhep.2014.04.020
26. Knapp S, Warshow U, Hegazy D, Brackenbury L, Guha IN, Fowell A, et al. Consistent beneficial effects of killer cell immunoglobulin-like receptor 2DL3 and group 1 human leukocyte antigen-C following exposure to hepatitis C virus. *Hepatology* (2010) 51:1168–75. doi:10.1002/hep.23477
27. Montes-Cano MA, Caro-Oleas JL, Romero-Gómez M, Diago M, Andrade R, Carmona I, et al. HLA-C and KIR genes in hepatitis C virus infection. *Hum Immunol* (2005) 66:1106–9. doi:10.1016/j.humimm.2006.02.001
28. Rauch A, Laird R, McKinnon E, Telenti A, Furrer H, Weber R, et al. Influence of inhibitory killer immunoglobulin-like receptors and their HLA-C ligands on resolving hepatitis C virus infection. *Tissue Antigens* (2007) 69(Suppl 1): 237–40. doi:10.1111/j.1399-0039.2006.773_4.x
29. Winter CC, Gumperz JE, Parham P, Long EO, Wagtmann N. Direct binding and functional transfer of NK cell inhibitory receptors reveal novel patterns of HLA-C allotype recognition. *J Immunol* (1998) 161:571–7.
30. Alter G, Heckerman D, Schneidewind A, Fadda L, Kadie CM, Carlson JM, et al. HIV-1 adaptation to NK-cell-mediated immune pressure. *Nature* (2011) 476:96–100. doi:10.1038/nature10237
31. Fadda L, Körner C, Kumar S, van Teijlingen NH, Piechocka-Trocha A, Carrington M, et al. HLA-Cw*0102-restricted HIV-1 p24 epitope variants can modulate the binding of the inhibitory KIR2DL2 receptor and primary NK cell function. *PLoS Pathog* (2012) 8:e1002805. doi:10.1371/journal.ppat.1002805
32. van Teijlingen NH, Hölzemer A, Körner C, Garcia-Beltran WF, Schafer JL, Fadda L, et al. Sequence variations in HIV-1 p24 Gag-derived epitopes can alter binding of KIR2DL2 to HLA-C*03:04 and modulate primary natural killer cell function. *AIDS* (2014) 28:1399–408. doi:10.1097/QAD.0000000000000284
33. Lunemann S, Martrus G, Hölzemer A, Chapel A, Ziegler M, Körner C, et al. Sequence variations in HCV core-derived epitopes alter binding of KIR2DL3 to HLA-C. *J Hepatol* (2016) 65(2):252–8. doi:10.1016/j.jhep.2016.03.016
34. Foley B, Cooley S, Verneris MR, Pitt M, Curtsinger J, Luo X, et al. Cytomegalovirus reactivation after allogeneic transplantation promotes a lasting increase in educated NKG2C+ natural killer cells with potent function. *Blood* (2012) 119:2665–74. doi:10.1182/blood-2011-10-386995
35. Béziat V, Dalgard O, Asselah T, Halfon P, Bedossa P, Boudifa A, et al. CMV drives clonal expansion of NKG2C+ NK cells expressing self-specific KIRs in chronic hepatitis patients. *Eur J Immunol* (2012) 42:447–57. doi:10.1002/eji.201141826
36. Djaoud Z, David G, Bressollette C, Willem C, Rettman P, Gagne K, et al. Amplified NKG2C+ NK cells in cytomegalovirus (CMV) infection preferentially express killer cell Ig-like receptor 2DL: functional impact in controlling CMV-infected dendritic cells. *J Immunol* (2013) 191:2708–16. doi:10.4049/jimmunol.1301138
37. Lopez-Vergès S, Milush JM, Schwartz BS, Pando MJ, Jarjoura J, York VA, et al. Expansion of a unique CD57+ NKG2Chi natural killer cell subset during acute human cytomegalovirus infection. *Proc Natl Acad Sci U S A* (2011) 108: 14725–32. doi:10.1073/pnas.1110900108
38. Béziat V, Liu LL, Malmberg J-A, Ivarsson MA, Sohlberg E, Björklund AT, et al. NK cell responses to cytomegalovirus infection lead to stable imprints in the human KIR repertoire and involve activating KIRs. *Blood* (2013) 121:2678–88. doi:10.1182/blood-2012-10-459545
39. Björkström NK, Ljunggren H-G, Sandberg JK. CD56 negative NK cells: origin, function, and role in chronic viral disease. *Trends Immunol* (2010) 31:401–6. doi:10.1016/j.it.2010.08.003
40. Gonzalez VD, Falconer K, Björkström NK, Blom KG, Weiland O, Ljunggren H-G, et al. Expansion of functionally skewed CD56-negative NK cells in chronic hepatitis C virus infection: correlation with outcome of pegylated IFN-alpha and ribavirin treatment. *J Immunol* (2009) 183:6612–8. doi:10.4049/jimmunol.0901437
41. Conry SJ, Meng Q, Hardy G, Yonkers NL, Sugalski JM, Hirsch A, et al. Genetically associated CD16(+)56(-) natural killer cell interferon (IFN)-αR expression regulates signaling and is implicated in IFN-α-induced hepatitis C virus decline. *J Infect Dis* (2012) 205:1131–41. doi:10.1093/infdis/jis027
42. Alter G, Jost S, Rihn S, Reyor LL, Nolan BE, Ghebremichael M, et al. Reduced frequencies of NKp30+NKp46+, CD161+, and NKG2D+ NK cells in acute HCV infection may predict viral clearance. *J Hepatol* (2011) 55:278–88. doi:10.1016/j.jhep.2010.11.030
43. Gaddy J, Broxmeyer HE. Cord blood CD16+56- cells with low lytic activity are possible precursors of mature natural killer cells. *Cell Immunol* (1997) 180:132–42. doi:10.1006/cimm.1997.1175
44. Anthony DD, Conry SJ, Medvik K, Sandhya Rani MR, Falck-Ytter Y, Blanton RE, et al. Baseline levels of soluble CD14 and CD16+56- natural killer cells are negatively associated with response to interferon/ribavirin therapy during HCV-HIV-1 coinfection. *J Infect Dis* (2012) 206:969–73. doi:10.1093/infdis/jis434
45. Gonzalez VD, Falconer K, Michaëlsson J, Moll M, Reichard O, Alaeus A, et al. Expansion of CD56- NK cells in chronic HCV/HIV-1 co-infection: reversion by antiviral treatment with pegylated IFNalpha and ribavirin. *Clin Immunol* (2008) 128:46–56. doi:10.1016/j.clim.2008.03.521
46. Spaan M, van Oord G, Kreefft K, Hou J, Hansen BE, Janssen HLA, et al. Immunological analysis during interferon-free therapy for chronic hepatitis

C virus infection reveals modulation of the natural killer cell compartment. *J Infect Dis* (2015) 213:216–23. doi:10.1093/infdis/jiv391

47. Serti E, Chepa-Lotrea X, Kim YJ, Keane M, Fryzek N, Liang TJ, et al. Successful interferon-free therapy of chronic hepatitis c virus infection normalizes natural killer cell function. *Gastroenterology* (2015) 149:190–200.e2. doi:10.1053/j.gastro.2015.03.004
48. Burchill MA, Golden-Mason L, Wind-Rotolo M, Rosen HR. Memory re-differentiation and reduced lymphocyte activation in chronic HCV-infected patients receiving direct-acting antivirals. *J Viral Hepat* (2015) 22:983–91. doi:10.1111/jvh.12465
49. Chiesa Della M, Falco M, Podestà M, Locatelli F, Moretta L, Frassoni F, et al. Phenotypic and functional heterogeneity of human NK cells developing after umbilical cord blood transplantation: a role for human cytomegalovirus? *Blood* (2012) 119:399–410. doi:10.1182/blood-2011-08-372003
50. Heeg M, Thimme R. NK cells and hepatitis C: NKp46 expression linked to antiviral and antifibrotic activity. *Hepatology* (2012) 56(4):1197–200. doi:10.1002/hep.25858
51. Yoon JC, Yang CM, Song Y, Lee JM. Natural killer cells in hepatitis C: current progress. *World J Gastroenterol* (2016) 22:1449. doi:10.3748/wjg.v22.i4.1449
52. Krämer B, Körner C, Kebschull M, Glässner A, Eisenhardt M, Nischalke HD, et al. NKp46(High) expression defines a NK cell subset that is potentially involved in control of HCV replication and modulation of liver fibrosis. *Hepatology* (2012) 56(4):1201–13. doi:10.1002/hep.25804
53. Golden-Mason L, Stone AEL, Bambha KM, Cheng L, Rosen HR. Race- and gender-related variation in NKp46 expression associated with differential anti-HCV immunity. *Hepatology* (2012) 56(4):1214–22. doi:10.1002/hep.25771
54. Pembroke T, Christian A, Jones E, Hills RK, Wang ECY, Gallimore AM, et al. The paradox of NKp46 +natural killer cells: drivers of severe hepatitis C virus-induced pathology but in-vivo resistance to interferon α treatment. *Gut* (2014) 63:515–24. doi:10.1136/gutjnl-2013-304472
55. Cosgrove C, Berger CT, Kroy DC, Cheney PC, Ghebremichael M, Aneja J, et al. Chronic HCV infection affects the NK cell phenotype in the blood more than in the liver. *PLoS One* (2014) 9:e105950. doi:10.1371/journal.pone.0105950
56. Gur C, Doron S, Kfir-Erenfeld S, Horwitz E, Abu-Tair L, Safadi R, et al. NKp46-mediated killing of human and mouse hepatic stellate cells attenuates liver fibrosis. *Gut* (2012) 61:885–93. doi:10.1136/gutjnl-2011-301400
57. Bozzano F, Picciotto A, Costa P, Marras F, Fazio V, Hirsch I, et al. Activating NK cell receptor expression/function (NKp30, NKp46, DNAM-1) during chronic viraemic HCV infection is associated with the outcome of combined treatment. *Eur J Immunol* (2011) 41:2905–14. doi:10.1002/eji.201041361
58. Arnon TI, Achdout H, Levi O, Markel G, Saleh N, Katz G, et al. Inhibition of the NKp30 activating receptor by pp65 of human cytomegalovirus. *Nat Immunol* (2005) 6:515–23. doi:10.1038/ni1190
59. Rölle A, Brodin P. Immune adaptation to environmental influence: the case of NK cells and HCMV. *Trends Immunol* (2016) 37:233–43. doi:10.1016/j.it.2016.01.005
60. Gumá M, Angulo A, Vilches C, Gómez-Lozano N, Malats N, López-Botet M. Imprint of human cytomegalovirus infection on the NK cell receptor repertoire. *Blood* (2004) 104:3664–71. doi:10.1182/blood-2004-05-2058
61. Guma M, Budt M, Sáez A, Brckalo T, Hengel H, Angulo A, et al. Expansion of CD94/NKG2C+ NK cells in response to human cytomegalovirus-infected fibroblasts. *Blood* (2006) 107:3624–31. doi:10.1182/blood-2005-09-3682
62. Foley B, Cooley S, Verneris MR, Curtsinger J, Luo X, Waller EK, et al. Human cytomegalovirus (CMV)-induced memory-like NKG2C+ NK cells are transplantable and expand in vivo in response to recipient CMV antigen. *J Immunol* (2012) 189:5082–8. doi:10.4049/jimmunol.1201964
63. Davis ZB, Cooley SA, Cichocki F, Felices M, Wangen R, Luo X, et al. Adaptive natural killer cell and killer cell immunoglobulin-like receptor-expressing T cell responses are induced by cytomegalovirus and are associated with protection against cytomegalovirus reactivation after allogeneic donor hematopoietic cell transplantation. *Biol Blood Marrow Transplant* (2015) 21:1653–62. doi:10.1016/j.bbmt.2015.05.025
64. Rölle A, Pollmann J, Ewen E-M, Le VTK, Halenius A, Hengel H, et al. IL-12-producing monocytes and HLA-E control HCMV-driven NKG2C+ NK cell expansion. *J Clin Invest* (2014) 124:5305–16. doi:10.1172/JCI77440
65. Björkström NK, Lindgren T, Stoltz M, Fauriat C, Braun M, Evander M, et al. Rapid expansion and long-term persistence of elevated NK cell numbers in humans infected with hantavirus. *J Exp Med* (2011) 208:13–21. doi:10.1084/jem.20100762
66. Cichocki F, Cooley S, Davis Z, DeFor TE, Schlums H, Zhang B, et al. CD56dimCD57+NKG2C+ NK cell expansion is associated with reduced leukemia relapse after reduced intensity HCT. *Leukemia* (2016) 30:456–63. doi:10.1038/leu.2015.260
67. Rölle A, Halenius A, Ewen E-M, Cerwenka A, Hengel H, Momburg F. CD2-CD58 interactions are pivotal for the activation and function of adaptive natural killer cells in human cytomegalovirus infection. *Eur J Immunol* (2016) 46:2420–5. doi:10.1002/eji.201646492
68. Gumá M, Cabrera C, Erkizia I, Bofill M, Clotet B, Ruiz L, et al. Human cytomegalovirus infection is associated with increased proportions of NK cells that express the CD94/NKG2C receptor in aviremic HIV-1-positive patients. *J Infect Dis* (2006) 194:38–41. doi:10.1086/504719
69. Goodier MR, Mela CM, Steel A, Gazzard B, Bower M, Gotch F. NKG2C+ NK cells are enriched in AIDS patients with advanced-stage Kaposi's sarcoma. *J Virol* (2007) 81:430–3. doi:10.1128/JVI.01567-06
70. Brunetta E, Fogli M, Varchetta S, Bozzo L, Hudspeth KL, Marcenaro E, et al. Chronic HIV-1 viremia reverses NKG2A/NKG2C ratio on natural killer cells in patients with human cytomegalovirus co-infection. *AIDS* (2010) 24:27–34. doi:10.1097/QAD.0b013e3283328d1f
71. Petitdemange C, Becquart P, Wauquier N, Béziat V, Debré P, Leroy EM, et al. Unconventional repertoire profile is imprinted during acute chikungunya infection for natural killer cells polarization toward cytotoxicity. *PLoS Pathog* (2011) 7:e1002268. doi:10.1371/journal.ppat.1002268
72. Saghafian-Hedengren S, Sohlberg E, Theorell J, Carvalho-Queiroz C, Nagy N, Persson J-O, et al. Epstein-Barr virus coinfection in children boosts cytomegalovirus-induced differentiation of natural killer cells. *J Virol* (2013) 87:13446–55. doi:10.1128/JVI.02382-13
73. Oliviero B, Varchetta S, Paudice E, Michelone G, Zaramella M, Mavilio D, et al. Natural killer cell functional dichotomy in chronic hepatitis B and chronic hepatitis C virus infections. *Gastroenterology* (2009) 137: 1151–60, 1160.e1–7. doi:10.1053/j.gastro.2009.05.047
74. Hwang I, Zhang T, Scott JM, Kim AR, Lee T, Kakarla T, et al. Identification of human NK cells that are deficient for signaling adaptor FcRγ and specialized for antibody-dependent immune functions. *Int Immunol* (2012) 24:793–802. doi:10.1093/intimm/dxs080
75. Zhang T, Scott JM, Hwang I, Kim S. Cutting edge: antibody-dependent memory-like NK cells distinguished by FcRγ deficiency. *J Immunol* (2013) 190:1402–6. doi:10.4049/jimmunol.1203034
76. Oh JS, Ali AK, Kim S, Corsi DJ, Cooper CL, Lee S-H. NK cells lacking FcεRIγ are associated with reduced liver damage in chronic hepatitis C virus infection. *Eur J Immunol* (2016) 46:1020–9. doi:10.1002/eji.201546009
77. Lee J, Zhang T, Hwang I, Kim A, Nitschke L, Kim M, et al. Epigenetic modification and antibody-dependent expansion of memory-like NK cells in human cytomegalovirus-infected individuals. *Immunity* (2015) 42:431–42. doi:10.1016/j.immuni.2015.02.013
78. Schlums H, Cichocki F, Tesi B, Theorell J, Béziat V, Holmes TD, et al. Cytomegalovirus infection drives adaptive epigenetic diversification of NK cells with altered signaling and effector function. *Immunity* (2015) 42:443–56. doi:10.1016/j.immuni.2015.02.008
79. Pesce S, Greppi M, Tabellini G, Rampinelli F, Parolini S, Olive D, et al. Identification of a subset of human natural killer cells expressing high levels of programmed death 1: a phenotypic and functional characterization. *J Allergy Clin Immunol* (2016) 139(1):335–46. doi:10.1016/j.jaci.2016.04.025
80. Wu Z, Sinzger C, Frascaroli G, Reichel J, Bayer C, Wang L, et al. Human cytomegalovirus-induced NKG2Chi CD57hi natural killer cells are effectors dependent on humoral antiviral immunity. *J Virol* (2013) 87:7717–25. doi:10.1128/JVI.01096-13
81. Costa-Garcia M, Vera A, Moraru M, Vilches C, López-Botet M, Muntasell A. Antibody-mediated response of NKG2Cbright NK cells against human cytomegalovirus. *J Immunol* (2015) 194:2715–24. doi:10.4049/jimmunol.1402281
82. Zhou J, Amran FS, Kramski M, Angelovich TA, Elliott J, Hearps AC, et al. An NK cell population lacking FcRγ is expanded in chronically infected HIV patients. *J Immunol* (2015) 194:4688–97. doi:10.4049/jimmunol.1402448
83. Liu LL, Landskron J, Ask EH, Enqvist M, Sohlberg E, Traherne JA, et al. Critical role of CD2 co-stimulation in adaptive natural killer cell responses revealed in NKG2C-deficient humans. *Cell Rep* (2016) 15:1088–99. doi:10.1016/j.celrep.2016.04.005

84. Luetke-Eversloh M, Hammer Q, Durek P, Nordström K, Gasparoni G, Pink M, et al. Human cytomegalovirus drives epigenetic imprinting of the IFNG locus in NKG2Chi natural killer cells. *PLoS Pathog* (2014) 10:e1004441. doi:10.1371/journal.ppat.1004441
85. Marquardt N, Béziat V, Nyström S, Hengst J, Ivarsson MA, Kekäläinen E, et al. Cutting edge: identification and characterization of human intrahepatic CD49a+ NK cells. *J Immunol* (2015) 194:2467–71. doi:10.4049/jimmunol.1402756
86. Paust S, Gill HS, Wang B-Z, Flynn MP, Moseman EA, Senman B, et al. Critical role for the chemokine receptor CXCR6 in NK cell–mediated antigen-specific memory of haptens and viruses. *Nat Immunol* (2010) 11:1127–35. doi:10.1038/ni.1953
87. Hydes T, Abuhilal M, Armstrong T, Primrose J, Takhar A, Khakoo S. Natural killer cell maturation markers in the human liver and expansion of an NKG2C+KIR+ population. *Lancet* (2015) 385(Suppl 1):S45. doi:10.1016/S0140-6736(15)60360-9
88. Stegmann KA, Robertson F, Hansi N, Gill U, Pallant C, Christophides T, et al. CXCR6 marks a novel subset of T-bet(lo)Eomes(hi) natural killer cells residing in human liver. *Sci Rep* (2016) 6:26157. doi:10.1038/srep26157
89. Harmon C, Robinson MW, Fahey R, Whelan S, Houlihan DD, Geoghegan J, et al. Tissue-resident Eomes(hi) T-bet(lo) CD56(bright) NK cells with reduced proinflammatory potential are enriched in the adult human liver. *Eur J Immunol* (2016) 46:2111–20. doi:10.1002/eji.201646559
90. Cuff AO, Robertson FP, Stegmann KA, Pallett LJ, Maini MK, Davidson BR, et al. Eomeshi NK cells in human liver are long-lived and do not recirculate but can be replenished from the circulation. *J Immunol* (2016) 197:4283–91. doi:10.4049/jimmunol.1601424
91. Yeang HX, Piersma SJ, Lin Y, Yang L, Malkova ON, Miner C, et al. Cutting edge: human CD49e– NK cells are tissue resident in the liver. *J Immunol* (2017) 198(4):1417–22. doi:10.4049/jimmunol.1601818

4

Reduction of Relapse after Unrelated Donor Stem Cell Transplantation by KIR-Based Graft Selection

Silke Heidenreich and Nicolaus Kröger*

Department of Stem Cell Transplantation, University Medical Center Hamburg-Eppendorf, Hamburg, Germany

Edited by:
Ulrike Koehl,
Hannover Medical School, Germany

Reviewed by:
Evelyn Ullrich,
Goethe University Frankfurt, Germany
Jan Spanholtz,
Glycostem Therapeutics, Netherlands

***Correspondence:**
Silke Heidenreich
s.heidenreich@uke.de

Besides donor T cells, natural killer (NK) cells are considered to have a major role in preventing relapse after allogeneic hematopoietic stem cell transplantation (HSCT). After T-cell-depleted haploidentical HSCT, a strong NK alloreactivity has been described. These effects have been attributed to killer-cell immunoglobulin-like receptors (KIR). Abundant reports suggest a major role of KIR not only on outcome after haploidentical HSCT but also in the unrelated donor setting. In this review, we give a brief overview of the mechanism of NK cell activation, nomenclature of KIR haplotypes, human leukocyte antigen (HLA) groups, and distinct models for prediction of NK cell alloreactivity. It can be concluded that KIR-ligand mismatch seems to provoke adverse effects in unrelated donor HSCT with reduced overall survival and increased risk for high-grade acute graft-versus-host disease. The presence of activating KIR, as seen in KIR haplotype B, as well as the patient's HLA C1/x haplotype might reduce relapse in myeloid malignancies.

Keywords: NK-cell, killer-cell immunoglobulin-like receptor, unrelated, stem cell, transplantation, HSCT, haplotype

INTRODUCTION

Natural killer (NK) cells are considered to contribute important immune effects against leukemia [graft-versus-leukemia (GVL) effect] after allogeneic hematopoietic stem cell transplantation (HSCT). Alloreactive NK cells are considered rather save concerning the development of graft-versus-host disease (GVHD) (1–5), although a high number of activating killer-cell immunoglobulin-like receptors (KIR) (6) or extensive NK cell stimulation (7) might promote GVHD, maybe due to remaining T-cells in the graft. Shah et al. (7) found an association between infusion of activated NK cells and occurrence of acute GVHD (aGVHD): Children with ultrahigh-risk sarcoma received T-cell-repleted grafts from matched unrelated donors (URDs) or matched sibling donors with subsequent infusion of IL-15 and 4-1BBL preactivated NK cells. Five of nine patients developed aGVHD. Those effects were attributed to NK cell-mediated T-cell activation (7). The biology of NK cells is complex, but activation by human leukocyte antigen (HLA) *via* the group of KIR is considered to be a relevant mechanism of activation. Within this review, we will provide a summary of concepts of KIR-mediated NK cell activation and an overview of GVL effects in haploidentical (haplo), but especially in URD HSCT.

Biology and Activation of NK Cells

Natural killer cells were named after their ability to kill infected or tumor cells without the need for prior antigen contact (8–10). They are defined by surface expression of CD56 and lack of CD3 (11). Unlike T cells, NK-cell receptors do not undergo rearrangement. In a process called licensing, NK cells with inhibitory receptors for present HLA class I (HLA-I) molecules (indicating "self") are positively selected and stimulated for proliferation, leading to a licensed and self-tolerant subset. Missing inhibitory receptors against HLA-I do not lead to depletion but to a second subset of unlicensed but self-tolerant NK cells (12). Activation of NK cells might be initiated by antigen contact, but it is executed only after integration of abundant activating and inhibitory signals (13, 14). Today, several NK-cell receptors are known. Besides KIR, other NK-cell receptors that have been shown to have the potential to positively influence outcome after allogeneic HSCT are natural cytotoxicity receptors (15–17) as well as activating NKG2D (18) and DNAM-1 (19, 20) that bind to MICA/B and ULBPs or CD112/CD155, respectively. Both can be induced by DNA damage (21) and seem to play a role in negative regulation of T-cell responses (22) and acute myeloid leukemia (AML)/myelodysplastic syndrome immune evasion (15, 23).

KIR and HLA

Killer-cell immunoglobulin-like receptors belong to type-I transmembrane proteins of the immunoglobulin-like receptor superfamily and recognize classical HLA-I molecules (14). The 15 KIR genes and 2 pseudogenes are located on chromosome 19q13.4. According to the number of extracellular immunoglobulin-like domains (D), the receptors are named KIR2D and KIR3D (24, 25). On the cytoplasmic side, they have either long (L) inhibitory or short (S) activating domains (14). Inhibitory KIR bind to the highly polymorphic regions of HLA-I molecules: HLA-A, B, and C (26), while the ligands for activating KIR are poorly defined (14, 27).

To facilitate description of KIR-ligands, HLA-C phenotypes can be grouped into HLA-C group 1 and 2 according to their respective KIR-binding motif. HLA-C group 1 contains all ligands with serine at residue 77 and asparagine at residue 80 of the α1 helix (HLA-C^{asn80}), binding KIR2DL**2**/**3** and 2DS2. Members of this group are HLA-C*01/*03/*07/*08/*12/*14/*16. HLA-C group 2 (HLA-C^{lys80}) has asparagine at residue 77 and lysine at residue 80 and contains HLA-C*02/*04/*05/*06/*15/*17/*18. They are ligands for KIR2DL1 and KIR2DS1 (28–31).

KIR3DL1 binds HLA-Bw4, and KIR3DL2 and 2DS2 bind HLA-A3 and A11 (14, 18, 32–38). Despite its structure, KIR2DL4 exhibits activating capacities and might bind soluble HLA-G (39–45). The KIR phenotype of an individual is his or her distinct set of inhibitory or activating KIR with an underlying distinct genotype (27, 46, 47). All genotypes can be summarized to a set of distinct haplotypes, which again result in the superordinated KIR haplotypes A or B (27, 46). KIR haplotype B is defined as the presence of KIR2DL5, 2DS1/2/3/5, or 3DS1, which have to be absent in KIR haplotype A (48). KIR2DS4 is the only activating KIR in haplotype A (46). KIR haplotype B/x (B/B or B/A) is found in about 30% of the Caucasian population (49). A more detailed analysis includes the information, whether the individual KIR is coded in the centromeric (Cen) or telomeric (Tel) gene motif of the KIR locus, resulting in Cen-A/A, Cen-B/x, and the respective Tel haplotypes (49–52). Thus, each individual expresses a certain KIR haplotype and a distinct HLA-C haplotype (C1/C1, C1/C2, or C2/C2). For prediction of alloreactive NK cell effects, the presence of HLA-C1, C2, and Bw4, as well as their respective KIR, are investigated (53). KIR2DL4 stimulation by HLA-G is considered to induce tolerance at the maternal–fetal barrier as well as IFN-gamma release of NK cells but not cytotoxicity (39, 43). KIR3DL2 and 2DS2 stimulation by HLA-A3 and A11 is also not in the primary focus of altering NK cell alloreactivity. KIR3DL2 has been identified as a surface marker in cutaneous T-cell lymphoma (54–56). For KIR2DS2, a reduced survival after URD-HSCT is suspected due to higher incidence of GVHD (57).

Model Situations Predicting NK Cell Alloreactivity

Different definitions of a mismatch between the donor's NK cells and the recipient's HLA exist, depending on the method that was chosen for KIR and HLA (HLA-C1, C2, and Bw4) evaluation (**Figure 1**).

Missing-Self/KIR-Ligand Mismatch (Figure 1A)

Alloreactivity was initially thought to be only dependent on lack of inhibitory HLA-I molecules in the recipient that are present in the donor ("missing-self" or "KIR-ligand mismatch" or "ligand-incompability model") (53, 58–60). For evaluation of KIR-ligand mismatch, donor and recipient are screened for expression of HLA: NK cells from a HLA C1/C1 donor will be alloreactive against a C2/C2 recipient. If a recipient expresses HLA-C1, C2, and Bw4, he will be resistant toward NK cell killing, as seen in one-third of the population (61). It is assumed, but not verified that the respective KIR, necessary for alloreactivity, is present in the donor.

Receptor–Ligand Mismatch (Figure 1B)

The receptor–ligand model states that donor NK cells become activated in the graft-versus-host direction; if they have inhibitory KIR, for which the HLA ligands in the recipient are missing, the NK cells become "uninhibited" (4). Thus, in addition to the HLA status of the recipient, confirmatory KIR genotyping of the patient is required. Other than in the first model, KIR on donor cells and HLA on recipient cells are investigated, not "assumed." This model can be considered as an improvement of the "missing ligand model."

Missing Ligand Model (Figure 1C)

Here, only the recipients' HLA are genotyped, and missing HLA-C1, C2, or Bw4 for inhibitory KIR predict an alloreactivity of the graft; the presence of the respective KIR that would bind to the missing HLA is only assumed (53).

Presence of Activating KIR (Figure 1D)

Some theories emphasize that to achieve NK cell alloreactivity, "un-inhibiton" of NK cells by missing inhibitory HLA ligands might not be sufficient. Activation requires additional stimulation of activating KIR in the graft (62). In this model, alloreactivity can be predicted by measurement of activating KIR on donor cells. Some studies increase the predictive validity by detection of the respective activating ligands on donor cells.

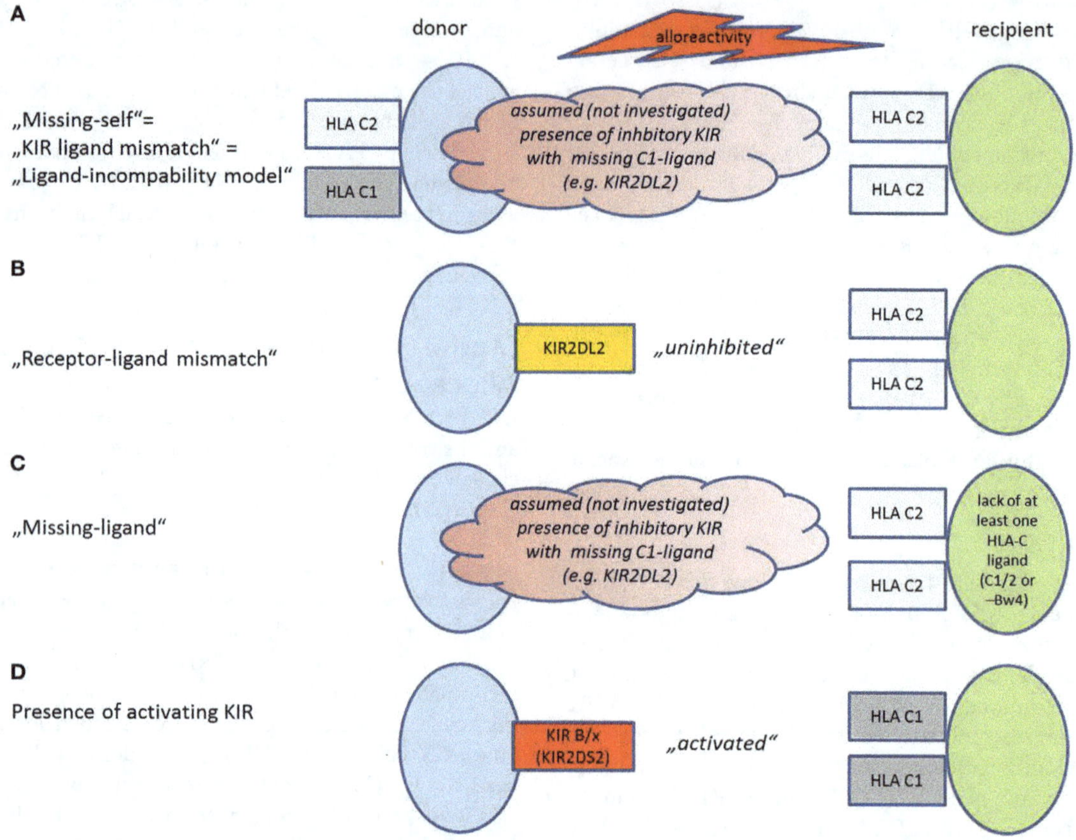

FIGURE 1 | Model situations that provoke natural killer (NK) cell alloreactivity. Models are depicted as used in the present review, adopted and modified from Symons and Fuchs (53). Details concerning the activation mechanism are provided in the text. **(A)** Missing-self model, also described as "killer-cell immunoglobulin-like receptors (KIR)-ligand mismatch" or "ligand-incompability model": Potential alloreactivity in the graft-versus-host direction is predicted by investigation of human leukocyte antigen (HLA) on donor and recipient. An HLA for inhibitory KIR that is present in the donor lacks in the recipient. The presence of the respective inhibitory KIR in the donor is assumed but not verified. **(B)** Receptor–ligand mismatch: NK cells become activated in the graft-versus-host direction, if they have an inhibitory KIR, for which the HLA ligand in the recipient is missing. Thus, the NK cells are "uninhibited." Other than in **(A)**, KIR on donor cells and HLA on recipient cells are investigated, not "assumed." **(C)** Missing ligand: If the presence of the respective inhibitory KIR is not evaluated, but assumed in a model where at least one HLA-ligand is missing (HLA-C1/2 or Bw4). Other than in **(A)**, only HLA on recipient cells but not on donor cells are evaluated. **(D)** The presence of activating KIR predicts alloreactivity in the presence of the respective activating ligand. KIR haplotype B/x contains more activating KIR than KIR haplotype A/A.

Therefore, we determine mismatches on the donor and recipient side between ligand–ligand, receptor–ligand, and receptor–receptor or identify activating KIR in the donor (49, 53). HLA and KIR can be investigated by genotyping, phenotyping, or functional NK cell assays to predict alloreactivity. We would suggest to follow the well-described methods of Ruggeri et al. for genotyping and phenotyping (61).

All approaches were initially tested in the haploidentical setting: The Perugia group suggests donor and recipient HLA typing to identify mismatch (2), followed by confirmatory donor KIR typing to verify a mismatch between donor KIR and recipient HLA ligand ("KIR-ligand mismatch" combined with a "receptor–ligand" concept) (61). They found ligand incompatibility between donor and recipient in haploidentical HSCT to be associated with increased GVL effects and lower relapse in acute leukemia (2, 4). KIR-ligand mismatch can be prevalent either in the graft-versus-host direction when the donor's KIR ligand is not shared by the recipient or in the host-versus-graft direction when the recipient's KIR ligand is not present in the donor. The St. Jude group rather focuses on receptor–ligand mismatch in the haplo setting (63), while the researchers from Minnesota implemented their strategy for the URD setting by selecting KIR B/x donors for HLA-C1-positive recipients for improved alloreactivity (50, 64, 65).

EVIDENCE OF NK CELL-MEDIATED GVL EFFECTS

Lessons Learned from the Haploidentical HSCT Setting

Much knowledge concerning NK cell-mediated alloreactivity has been collected due to the implementation of haploidentical HSCT. To reduce the risk of GVHD, T-cell depletion was performed before graft infusion at the cost of graft rejection (66). These effects could be partially overcome by infusion of high numbers of stem cells (67). Ruggeri et al. were the first to show NK cell-mediated alloreactivity in the T-cell-depleted haploidentical graft (5). Facilitated engraftment as well as tumor lysis by NK cells occurred by donor grafts that were KIR-ligand incompatible in the graft-versus-host direction without occurrence of GVHD. Since then, many other groups have investigated the beneficial effect of alloreactive NK cells in the haploidentical HSCT (2, 68) and have refined criteria for potential donor choice (61). The results are promising for AML (4, 68, 69), while lymphoid malignancies have been shown to be resistant in some (2, 69) but not all cases (63, 70) for KIR-mediated NK cell effects. The present status of NK cell-mediated effects in haploidentical HSCT has been reviewed elsewhere (53, 71, 72).

Results in the Unrelated-Donor HSCT Setting

After the identification of beneficial NK cell-mediated alloreactivity in haploidentical HSCT, efforts were made to adopt the findings for transplantations with URD (**Table 1**). Even though many patients already have the opportunity to receive a graft from an HLA-matched donor, donor choice by KIR repertoire is useful. Since HLA and KIR are inherited separately, approximately 75% of HLA-identical sibling donors and almost 100% of matched URDs will show KIR disparities and might therefore be a potential source for alloreactive NK cells (73, 74).

KIR-Ligand Mismatch Seems to Induce Adverse Effects in URD HSCT

Davies et al. (75) were the first to perform a retrospective analysis of patients with HLA mismatched URD HSCT, comparing KIR-ligand mismatch. In the analysis, no difference in any of the primary endpoints was achieved. Concerning the subgroup of myeloid malignancies, KIR-ligand mismatch resulted in worse OS at 5 years [13 versus 38%, $P < 0.01$, no use of antithymocyte globulin (ATG)], which was even more surprising. Others confirmed worse outcome for KIR-ligand mismatch in URD HSCT after conditioning with ATG (76–79) or without ATG (80), accompanied with higher infections in the early posttransplant period (78) or increased graft rejection, TRM, and GVHD (80). A recent study confirmed higher mortality and higher TRM without difference in relapse in 3–5/8 HLA-mismatched KIR-ligand mismatched (in the host-versus-graft direction) unrelated cord blood transplantations for AML and acute lymphoid leukemia (ALL) compared to KIR-matched cord blood, while no difference was found for mismatch in the graft-versus-host direction or in a higher HLA-matched subgroup or the complete patient cohort (81). The authors did suggest to not using KIR-ligand mismatch as a criterion for cord blood selection. An earlier Eurocord study (82) detected favorable outcome for KIR-ligand mismatched transplantations in AML and ALL but used lower HLA-resolution techniques.

No difference in mortality after either KIR-ligand mismatched or HLA-mismatched but KIR-ligand matched donor–recipient pairs was detected by a comprehensive study of CIBMT, EBMT, and the Dutch transplant registry (83), investigating the results of 1,571 patients with myeloid malignancies with or without T-cell depletion. KIR-ligand mismatch was associated with significantly higher high-grade aGVHD, just as HLA mismatch at HLA-C and/or B. No predictive effects of KIR-ligand mismatch on outcome after T-cell-repleted unrelated HSCT were detected in a retrospective multicenter study in France (84). Here, different models of NK cell alloreactivity were compared in a very heterogeneous cohort of patients. These investigations were partially designed as a response to the positive results in haploidentical HSCT and in a previous study by Giebel et al. (85) with different results: KIR-ligand mismatch in patients with myeloid malignancies achieved significant higher OS and RFS as well as lower TRM and relapse compared to HLA mismatch with KIR ligand match or compared to matched URD HSCT with the use of pretransplant ATG. The differing results could be only partially attributed to the use or sparing of ATG (85): Although toxic (86) or immunosuppressive (87, 88) on NK cells, ATG has been shown to accelerate NK-cell and B-cell reconstitution in some (89) but not all investigations (90, 91). It has also been shown to decelerate the recovery of CD4+ and CD8+ T cells (89, 91) while sparing effector-memory T cells and T-regulatory cells (91). The results indicated that knowledge from haploidentical cannot be transferred to unrelated HSCT without further adaptations (75). Grafts for haploidentical HSCT were mainly highly T-cell depleted and performed with high stem-cell doses as well as no or low immunosuppression, resulting in fast NK cell but slow T-cell reconstitution with low T-cell numbers and eradication of antigen-presenting cells by alloreactive NK clones (2, 67, 92). Therefore, the immunological environment during engraftment in haploidentical HSCT is much different from URD-HSCT.

Missing-Ligand Model and Presence of Activating KIR Are Predictive for Outcome

Later, Hsu et al. (60) identified not only KIR-ligand mismatch but also missing KIR ligands as protective against relapse in HLA mismatched but not in matched URD HSCT. These effects were seen in myeloid and lymphoid malignancies and supported by later investigations by other authors (93). In the study by Hsu et al. (60), the absence of HLA-C2 or HLA-Bw4 KIR ligands was associated with lower relapse. Other authors confirmed the impact of HLA-C2: Absence of HLA-C2 in recipients of KIR2DL1-positive grafts resulted in higher incidence of aGVHD after myeloablative (94) as well as reduced intensity (95) conditioning. The absence of C1 epitopes, as seen for C2/C2 recipients, has been claimed responsible for poorer outcome (57, 65, 96). In search for favorable KIR in URD HSCT, Sun et al. (97) prospectively analyzed outcome of URD AML patients without *in vivo* T-cell depletion by ATG. According to the presence or absence of activating or inhibitory KIR in donor and recipient, they calculated a new predictive algorithm

TABLE 1 | Studies on NK cell alloreactivity for unrelated donors.

Reference	*N*	Median age (years)[a]	Disease (*n*)	Tx (*n*)	Model	Conditioning and graft source	Immunosuppression	Main results
Davies et al. (75)	175	17	CML, AML, ALL, MDS, others	MMUD (175)	KIR-L MM[b]	Myeloablative • BM	TCD or CSA ± MTX	*Adverse KIR-L MM in myeloid malignancies*: Lower OS at 1 and 5 years ($P < 0.01$). No difference between KIR-L M/MM in any endpoint for total cohort
Schaffer et al. (76)	104	29	Diverse	MUD (62)/ MMUD (42)	KIR-L MM	Myeloablative • BM (80) • PB (24)	MTX + CSA, ATG	*Adverse KIR-L MM*: Reduced OS and RFS
Giebel et al. (85)	130	18–20.5	Diverse	MUD (61)/ MMUD (49)	KIR-L MM	Myeloablative • BM (125) • PB (5)	CSA, MTX, ATG	*Beneficial KIR-L MM*: Higher OS and RFS ($P = 0.0007$; 4.5 years). No influence of HLA-MM in the patients without KIR-L MM
Bornhauser et al. (77)	118	42–44	AML, CML, MDS	MUD (54)/ MMUD (64)	KIR-L MM	Myeloablative • BM (54) • PB (64)	ATG (118)	*Adverse KIR-L MM*: Higher relapse for KIR-L MM ($P = 0.02$), but no difference in survival after KIR-L MM, MUD, and MMUD transplantation
Schaffer et al. (78)	190	35–39	Diverse	MUD (94)/ MMUD (96)	KIR-L MM	Myeloablative (168) RIC (22) • BM (118) • PB (72)	CSA based (179) or TCD (11) plus ATG (all)	*Adverse KIR-L MM*: Higher infections, leading to increased TRM and reduced OS ($P = 0.01$), but no increase of relapse or GVHD
Venstrom et al. (96)	1,277	40.5–41.7	AML	MUD (664)/ MMUD (613)	Missing ligand Receptor–ligand KIR genes	Myeloablative (1,069) RIC/NMA (189) • BM (689) • PB (588)	Diverse, no ATG	*Adverse absence of C1*: HLAC2/C2 recipients have higher relapse than HLAC1/x recipients ($P = 0.05$) *Beneficial KIR2DS1 from C1/x donor* associated with lower relapse compared to absence of KIR2DS1 ($P = 0.003$) and lower mortality ($P = 0.04$) w/o higher high-grade aGVHD or TRM *Beneficial KIR3DS1* associated with lower mortality ($P = 0.01$) by lower TRM and aGVHD No predictive effects in ALL patients (separate cohort)
De Santis et al. (80)	104	24	Diverse	MMUD (104)	KIR-L MM	Myeloablative • BM (65) • PB (39)	No ATG BM: CSA, MTX (59), T-cell depletion (9) PB: No CSA (39)	*Adverse: KIR-L MM* (HVG): Increased graft rejection Adverse KIR-L MM (GVH): Increased aGVHD grade 3–4 Adverse KIR-L MM (GVH or HVG): Increased TRM, decrease RFS *Beneficial high number of donor KIR*: Lower GVHD and improved survival
Giebel et al. (57)	111	18.5–21	Diverse	MUD (90)/ MMUD (21)	Missing ligand	Myloablative • BM (96) • PB (15)	CSA, MTX, ATG	*Adverse absence of C1*: C2/C patients have lower OS and DFS, due to higher relapse
Sun et al. (97)	65	45–46	AML	MUD (39)/ MMUD (26)	Receptor–receptor	Diverse	CSA + MTX (65) No ATG or TCD	*Prediction of incidence of aGVHD possible*: Activating KIR in the donor that lack in recipient and the lack of inhibitory KIR in the donor that are present in the recipient predict increased aGVHD Indifferent results for KIR-L MM, missing ligand, number of activating KIR

(Continued)

TABLE 1 | **Continued**

Reference	*N*	Median age (years)[a]	Disease (*n*)	Tx (*n*)	Model	Conditioning and graft source	Immunosuppression	Main results
Giebel et al. (98)	25	27	ALL, AML, MDS, CML, NHL	MUD (23)/ MMUD (2)	KIR genes	Myeloablative • BM (20) • PB (7)	CSA, MTX, ATG	*Adverse presence of KIR2DS1*: Reduced OS and DFS due to increased GVHD and relapse *Indifferent presence of KIR2DS1*
Kröger et al. (79)	142	33	AML, MDS, CMML, CML, ALL	MUD (103)/ MMUD (39)	KIR haplotype KIR-L MM	Myeloablative • BM (67) • PB (75)	ATG, CSA, MTX	*Adverse KIR B/x*: Higher relapse than KIR A/A ($P = 0.03$), but only in AML/MDS/CML/CMML, not ALL, resulting in lower OS *Adverse KIR-L MM*: Higher TRM, lower OS, no increase of GVHD *Adverse KIR3DS1, 2DS1, 2DS5* in UVA, only 2DS5 in MVA, all resulting in higher relapse
Farag et al. (83)	1,571	59–68	AML, MDS, CML	MMUD KIR-L MM GVH (137) MMUD KIR-L MM HVG (170) MMUD KIR-L M (260) MUD (1,004)	KIR-L MM	Myeloablative • BM	± T-cell depletion	*Indifferent KIR-L MM*: For KIR-L MM (GVH/HVG) as well as KIR-L M but HLA MM at HLA B ± C versus HLA- and KIR-L M grafts: Same rates of increased aGVHD grade 3–4, TRM, treatment failure, and overall mortality compared to HLA- and KIR-L matched grafts
Hsu et al. (60)	1,770	34.5–35	AML, MDS, CML, ALL	MMUD (1,190)/ MUD (580)	Missing ligand KIR-L MM	Myeloablative • BM or PB	T-cell replete grafts	*Beneficial: missing ligand in MMUD* (defined as homozygosity of recipient HLA-B or C epitopes) resulting in lower relapse ($P = 0.004$), but not for MUD *Absence of HLA-C2 or Bw4 associated with reduced relapse*, no survival benefit *Indifferent KIR-L MM* model in subgroup of 428 patients: no difference in relapse (but also not with applied missing-ligand model in same subgroup $P = 0.07$)
Miller et al. (93)	2,062	–	AML, CML, MDS	MMUD/ MUD	Missing ligand	-	± ATG or TCD	*Beneficial absence of one ligand in early stage AML or MDS*: reduced relapse, independent from HLA match (C1/C2/Bw4) *Adverse absence of ≥1 ligand in CML*: Increased late-onset high-grade acute GVHD
Willemze et al. (82)	218	12.8–15	AML, ALL	MUD (42)/ MMUD (176)	KIR-L MM	RIC (202) Myeloablative (6) • CB (single)	CSA based (174) Other (44)± ATG (196)	*Beneficial KIR-L MM*: Improved DFS, OS, and decrease relapse
Gagne et al. (84)	264	24.5	Diverse	MUD (164)/ MMUD (100)	KIR-L MM Missing ligand Receptor–ligand Receptor–receptor	Myeloablative • BM	Unmanipulated BM	*Indifferent KIR-L MM* *Adverse missing-ligand*: Decreased survival but only in C1-deficient recipients, in myeloid malignancies *Adverse receptor–ligand mismatch: KIR3DL1* as well as KIR3DL1/3DS1 mismatch (GVH: D+ R–, absence of recipient HLA-Bw4) from a HLA-Bw4-negative donor is correlated with low OS in HLA-identical and high relapse in MMUD HSCT

(*Continued*)

TABLE 1 | **Continued**

Reference	*N*	Median age (years)[a]	Disease (*n*)	Tx (*n*)	Model	Conditioning and graft source	Immunosuppression	Main results
Ludajic et al. (94)	124	42	Diverse	MUD	Missing ligand	Myeloablative (90) RIC (34) • BM (54) • PB (70)	CSA-based (124) ± ATG (30)	*Adverse absence of HLA-C2* in recipients of KIR2DL1-positive grafts or KIR A/A grafts: Increased aGVHD *Beneficial absence of HLA-C2* in recipients of KIR2DS2-positive grafts: Decreased aGVHD
Cooley et al. (64)	448	33–34	AML	MUD (209)/ MMUD (239)	KIR haplotype KIR-L MM	Myeloablative • BM (397) • PB (51)	T-cell replete MMUD grafts	*Beneficial KIR B/x in KIR-L M HSCT*: Compared to KIR A/A higher RFS in KIR-L M (MUD and MMUD) but not in KIR-L MM (MMUD) *Beneficial survival rates for KIR2DL2 and 2DS2 positive grafts*
Cooley et al. (50)	1,409	19/39	ALL, AML	MUD (687) MMUD (722)	KIR haplotype	Myeloablative • BM (942) • PB (467)	T-cell replete MMUD grafts	*Beneficial KIR B/x*: Higher RFS in AML but not ALL Cen-B motifs improve outcome without increased aGVHD/cGVHD or TRM
Venstrom et al. (99)	1,087	35.3–37.5	AML, MDS, CML, ALL	MUD (670)/ MMUD (417)	KIR genes KIR haplotype	Myloablative • BM (1,050) • PB (37)	CSA (751) No CSA (120) TCD (216)	*Beneficial presence of KIR3DS1*: Same rate of relapse but reduced TRM and aGVHD, resulting in lower mortality in AML and MDS. Beneficial effects increase with copy numbers of donor KIR3DS1 *Beneficial effect of KIR B/x (including KIR3DS1) similar but weaker*
Kröger et al. (100)	118	51	MM	Unrelated (81) Related (37)	KIR haplotype	Myeloablative (12) RIC (106) • BM (13) • PB (105)	ATG (110)	*Beneficial KIR B/x B in MUD*: MUD but not MMUD haplotype B/x reaches lower 1-year relapse than haplotype AA ($P = 0.005$), resulting in higher 5-year DFS ($P = 0.009$).
Venstrom et al. (96)	1,277	40.5–41.7	AML	MUD (664)/ MMUD (613)	Missing ligand Receptor–ligand KIR genes	Myloablative (1,069) • BM (689) • PB (588)	CSA (346) Tac (428) TCD (348)	*Adverse absence of C1 and beneficial KIRSDS1*: Reduced risk of relapse, if the allograft was derived from an HLA-C1/x donor *Beneficial presence of KIR3DS1*: Not lower relapse but reduced TRM and aGVHD, resulting in lower mortality in AML
Cooley et al. (65)	1,532	Adults and children	AML	MUD (856)/ MMUD (676)	KIR haplotype KIR gene content Missing-ligand	Myeloablative	T-cell replete MMUD grafts	*Beneficial KIR B/x, adverse absence of C1*: Relapse protection improved by high KIR-B content in recipients HLA-C1/x but not C2/C2 (significant only in MMUD, not MUD). No effect of donor HLA
Sobecks et al. (95)	909	56–57	AML, MDS	MUD (712)/ MMUD (197)	Missing ligand	RIC • BM (169) • PB (740)	Diverse ± ATG (317)	*Adverse KIR2DS1 educated in a C2/C2 donor*: Higher GVHD and TRM without reduced relapse (AML) *Adverse ≥1 missing ligand or absence of HLA-C2*: Higher aGVHD (AML) *Indifferent* KIR centromeric gene content or donor activating KIR
Faridi et al. (49)	281	50	AML, ALL	MSD (153)/ MUD (128)	Comparison of different models	Myeloablative • BM (10) • PB (271)	ATG, CSA, MTX	*Adverse KIR-KIR mismatch*: Increased cGVHD in HLA C1/x recipients *Beneficial ≥1 missing ligand*: Reduced relapse without improved OS *Indifferent* results for KIR B

(*Continued*)

TABLE 1 | Continued

Reference	N	Median age (years)[a]	Disease (n)	Tx (n)	Model	Conditioning and graft source	Immunosuppression	Main results
Bachanova et al. (101)	614	48–52	NHL	MUD (396)/ MMUD (218)	KIR haplotype	Myeloablative (253) RIC (361) • BM (227) • PB (387)	Diverse	*Beneficial KIR B/x in MUD HSCT*: Lower relapse after 5 years compared to KIR A/A donors ($P = 0.5$) with improved progression-free survival ($P = 0.007$)
Rocha et al. (81)	461	Adults and children	AML	MMUD (461) 3–5/8 HLA-MM (212)	KIR-L MM	Myeloablative • CB (single)	With (145) or w/o (35) *in vivo* T-cell depletion in 3–5/8 MM cohort	*Adverse: KIR-L MM (HVG)*: At 3–5/8 HLA-MM level: Higher mortality ($P = 0.008$) and NRM ($P = 0.008$), no difference in relapse or GVHD in KIR-L MM versus KIR-L M (HVG) for AML and ALL *Indifferent KIR-L MM (GVH)*[b]: No differences in GVH direction or higher HLA-matched subgroup or complete patient cohort

ALL, acute lymphoid leukemia; AML, acute myeloid leukemia; ATG, antithymocyte globulin (rabbit ATG, Campath or OTK3); BM, bone marrow; Cen, centromeric; CML, chronic myeloid leukemia; CMML, chronic myelomonocytic leukemia; CSA, cyclosporine A; GVH, graft-versus-host direction; GVHD, graft-versus-host disease (a, acute; c, chronic); HD, Hodgkin's disease; HLA, human leukocyte antigen; HVG, host-versus-graft direction; KIR, killer-cell immunoglobulin-like receptors; KIR-L M, KIR-ligand match; KIR-L MM, KIR-ligand mismatch; M, matched; MDS, myelodysplastic syndrome; MM, multiple myeloma or mismatch; MSD, matched sibling donor; MMUD, mismatched unrelated donor, defined as HLA 10/10 – 8/8 or not further specified; MTX, methotrexate; MUD, matched unrelated donor; MVA, multivariate analysis; NMA, non-myeloablative; n.s., not significant; Tel, telomeric; PB, peripheral blood stem cells; RIC, reduced intensity conditioning; TCD, T-cell depletion; Tx, transplantation details; UVA, univariate analysis.

[a]Median age was not always declared for the total cohort but only the investigated subgroups. In this case, age-age does not describe a range but the youngest and oldest median age stated.

[b]KIR-L MM in the GVH direction unless stated otherwise.

for GVHD, in which an inhibitory KIR in the donor that lacks in the patients has a negative value *vice versa* a positive value. On the other hand, they could not find other models such as KIR-ligand, missing-ligand, or high numbers of activating KIR to be predictive for aGVHD (97). In general, among the activating receptors, the presence of KIR2DS2 has been shown to be associated with lower OS and DFS as well as higher incidence of GVHD, resulting in high TRM (98). The alloreactivity of KIR2DS1 educated in a C2/C2 donor results in higher GVHD and TRM without reduced relapse (95). KIRSDS1 has been claimed responsible for reduced risk of relapse, if the allograft was derived from an HLA-C1/x donor (96), but did not show any beneficial effects in other investigations (98). The presence of KIR3DS1 was not associated with lower relapse but reduced TRM and aGVHD, resulting in lower mortality in AML patients (96, 99). KIR3DL1 and KIR3DL1/3DS1 mismatch in the GVH direction (donor positive, recipient negative, absence of recipient HLA-Bw4) from a HLA-Bw4-negative donor is correlated with low OS in HLA-identical and high relapse in HLA-mismatched URD HSCT (84). There are several other investigations apart from the environment of URD HSCT, which might be even more conflicting and difficult to transfer. Our early investigations showed the low-alloreactive KIR haplotype A to be associated with lower relapse after HSCT for leukemia (79), while in a later analysis, KIR haplotype B was associated with improved PFS and OS in patients with multiple myeloma (100). Cooley et al. (50, 64, 65) systematically investigated the influence of the KIR haplotype B. In summary, a high number of KIR haplotype B defining receptors, especially of those coded in the centromeric regions, showed beneficial effects on survival of HLA C1/x AML recipients after ATG-free HSCT without increased GVHD and without benefit of KIR-ligand mismatch. No positive influence of haplotype B was seen in recent investigations for leukemia (49) but in HLA-matched URD-HSCT of non-Hogdkin lymphoma patients, where KIR B/x grafts led to significant lower relapse after 5 years compared to KIR A/A donors ($P = 0.5$) (101). The role of KIR genotypes in matched unrelated and sibling HSCT has recently also been investigated by Faridi et al. (49). Their aim was to compare the predictive value of KIR-ligand mismatch (61) versus the "missing-ligand" hypothesis (63) or the advantage of a specific KIR haplotype (50, 64, 65). They found KIR–KIR match to be associated with lower cGVHD for HLA C1/x recipients as well as lower RFS. One or more missing ligand in the unrelated recipient for donor KIR resulted in reduced relapse (21.6 versus 63.6%, $P = 0.001$) and higher RFS without improved OS. None of the tested hypotheses had influence on OS, and no effect of donor KIR haplotype was detected.

FUTURE DIRECTIONS

During the past years, improvement in understanding NK cell alloreactivity has been made by wisely modeled analyzes (49, 60, 84). Despite clinical relevance (102–105), we still know too little about the NK cell education after HSCT (95). The interplay of NK cells and T-cells after HSCT is still subject of further investigation (105), and as we now know about KIR expression on

T-cells (106), we need to be precise in our technical methods. To overcome the problem of heterogeneity, we would suggest beginning with a simple multicenter prospective trial in adult patients with AML in first molecular complete remission, testing the hypothesis that the number of activating KIR in the unmanipulated graft improves overall survival without increasing GVHD. KIR and HLA of donor and recipient should be measured by high-resolution genotyping and phenotyping. Every patient should receive the same conditioning and first-line immune suppression.

CONCLUSION

Due to heterogeneity of the conducted studies, a general recommendation cannot be made. In matched URD-HSCT, a donor with high numbers of activating KIR can be chosen to optimize patient's chances for survival.

AUTHOR CONTRIBUTIONS

SH and NK contributed equally to the manuscript writing.

REFERENCES

1. Asai O, Longo DL, Tian ZG, Hornung RL, Taub DD, Ruscetti FW, et al. Suppression of graft-versus-host disease and amplification of graft-versus-tumor effects by activated natural killer cells after allogeneic bone marrow transplantation. *J Clin Invest* (1998) 101(9):1835–42. doi:10.1172/JCI1268
2. Ruggeri L, Capanni M, Urbani E, Perruccio K, Shlomchik WD, Tosti A, et al. Effectiveness of donor natural killer cell alloreactivity in mismatched hematopoietic transplants. *Science* (2002) 295(5562):2097–100. doi:10.1126/science.1068440
3. Olson JA, Leveson-Gower DB, Gill S, Baker J, Beilhack A, Negrin RS. NK cells mediate reduction of GVHD by inhibiting activated, alloreactive T cells while retaining GVT effects. *Blood* (2010) 115(21):4293–301. doi:10.1182/blood-2009-05-222190
4. Ruggeri L, Mancusi A, Capanni M, Urbani E, Carotti A, Aloisi T, et al. Donor natural killer cell allorecognition of missing self in haploidentical hematopoietic transplantation for acute myeloid leukemia: challenging its predictive value. *Blood* (2007) 110(1):433–40. doi:10.1182/blood-2006-07-038687
5. Ruggeri L, Capanni M, Casucci M, Volpi I, Tosti A, Perruccio K, et al. Role of natural killer cell alloreactivity in HLA-mismatched hematopoietic stem cell transplantation. *Blood* (1999) 94(1):333–9.
6. Bishara A, De Santis D, Witt CC, Brautbar C, Christiansen FT, Or R, et al. The beneficial role of inhibitory KIR genes of HLA class I NK epitopes in haploidentically mismatched stem cell allografts may be masked by residual donor-alloreactive T cells causing GVHD. *Tissue Antigens* (2004) 63(3):204–11. doi:10.1111/j.0001-2815.2004.00182.x
7. Shah NN, Baird K, Delbrook CP, Fleisher TA, Kohler ME, Rampertaap S, et al. Acute GVHD in patients receiving IL-15/4-1BBL activated NK cells following T-cell-depleted stem cell transplantation. *Blood* (2015) 125(5):784–92. doi:10.1182/blood-2014-07-592881
8. Herberman RB, Nunn ME, Holden HT, Lavrin DH. Natural cytotoxic reactivity of mouse lymphoid cells against syngeneic and allogeneic tumors. II. Characterization of effector cells. *Int J Cancer* (1975) 16(2):230–9. doi:10.1002/ijc.2910160205
9. Garcia-Penarrubia P, Koster FT, Kelley RO, McDowell TD, Bankhurst AD. Antibacterial activity of human natural killer cells. *J Exp Med* (1989) 169(1):99–113. doi:10.1084/jem.169.1.99
10. Biron CA, Brossay L. NK cells and NKT cells in innate defense against viral infections. *Curr Opin Immunol* (2001) 13(4):458–64. doi:10.1016/S0952-7915(00)00241-7
11. Robertson MJ, Ritz J. Biology and clinical relevance of human natural killer cells. *Blood* (1990) 76(12):2421–38.
12. Kim S, Poursine-Laurent J, Truscott SM, Lybarger L, Song YJ, Yang L, et al. Licensing of natural killer cells by host major histocompatibility complex class I molecules. *Nature* (2005) 436(7051):709–13. doi:10.1038/nature03847
13. Leiden JM, Gottesdiener KM, Quertermous T, Coury L, Bray RA, Gottschalk L, et al. T-cell receptor gene rearrangement and expression in human natural killer cells: natural killer activity is not dependent on the rearrangement and expression of T-cell receptor alpha, beta, or gamma genes. *Immunogenetics* (1988) 27(4):231–8. doi:10.1007/BF00376117
14. Lanier LL. NK cell recognition. *Annu Rev Immunol* (2005) 23:225–74. doi:10.1146/annurev.immunol.23.021704.115526
15. Sanchez-Correa B, Gayoso I, Bergua JM, Casado JG, Morgado S, Solana R, et al. Decreased expression of DNAM-1 on NK cells from acute myeloid leukemia patients. *Immunol Cell Biol* (2012) 90(1):109–15. doi:10.1038/icb.2011.15
16. Sivori S, Carlomagno S, Pesce S, Moretta A, Vitale M, Marcenaro E. TLR/NCR/KIR: which one to use and when? *Front Immunol* (2014) 5:105. doi:10.3389/fimmu.2014.00105
17. Ghasemzadeh M, Hosseini E, Schwarer AP, Pourfathollah AA. NK cell maturation to CD56(dim) subset associated with high levels of NCRs overrides the inhibitory effect of NKG2A and recovers impaired NK cell cytolytic potential after allogeneic hematopoietic stem cell transplantation. *Leuk Res* (2016) 43:58–65. doi:10.1016/j.leukres.2015.12.002
18. Farag SS, Caligiuri MA. Human natural killer cell development and biology. *Blood Rev* (2006) 20(3):123–37. doi:10.1016/j.blre.2005.10.001
19. Kearney CJ, Ramsbottom KM, Voskoboinik I, Darcy PK, Oliaro J. Loss of DNAM-1 ligand expression by acute myeloid leukemia cells renders them resistant to NK cell killing. *Oncoimmunology* (2016) 5(8):e1196308. doi:10.1080/2162402X.2016.1196308
20. Kruse V, Hamann C, Monecke S, Cyganek L, Elsner L, Hubscher D, et al. Human induced pluripotent stem cells are targets for allogeneic and autologous natural killer (NK) cells and killing is partly mediated by the activating NK receptor DNAM-1. *PLoS One* (2015) 10(5):e0125544. doi:10.1371/journal.pone.0125544
21. Cerboni C, Fionda C, Soriani A, Zingoni A, Doria M, Cippitelli M, et al. The DNA damage response: a common pathway in the regulation of NKG2D and DNAM-1 ligand expression in normal, infected, and cancer cells. *Front Immunol* (2014) 4:508. doi:10.3389/fimmu.2013.00508
22. Zingoni A, Ardolino M, Santoni A, Cerboni C. NKG2D and DNAM-1 activating receptors and their ligands in NK-T cell interactions: role in the NK cell-mediated negative regulation of T cell responses. *Front Immunol* (2012) 3:408. doi:10.3389/fimmu.2012.00408
23. Carlsten M, Baumann BC, Simonsson M, Jadersten M, Forsblom AM, Hammarstedt C, et al. Reduced DNAM-1 expression on bone marrow NK cells associated with impaired killing of CD34+ blasts in myelodysplastic syndrome. *Leukemia* (2010) 24(9):1607–16. doi:10.1038/leu.2010.149
24. Colonna M, Samaridis J. Cloning of immunoglobulin-superfamily members associated with HLA-C and HLA-B recognition by human natural killer cells. *Science* (1995) 268(5209):405–8. doi:10.1126/science.7716543
25. Wagtmann N, Biassoni R, Cantoni C, Verdiani S, Malnati MS, Vitale M, et al. Molecular clones of the p58 NK cell receptor reveal immunoglobulin-related molecules with diversity in both the extra- and intracellular domains. *Immunity* (1995) 2(5):439–49. doi:10.1016/1074-7613(95)90025-X
26. Borges L, Cosman D. LIRs/ILTs/MIRs, inhibitory and stimulatory Ig-superfamily receptors expressed in myeloid and lymphoid cells. *Cytokine Growth Factor Rev* (2000) 11(3):209–17. doi:10.1016/S1359-6101(00)00007-1
27. Hsu KC, Chida S, Geraghty DE, Dupont B. The killer cell immunoglobulin-like receptor (KIR) genomic region: gene-order, haplotypes and allelic polymorphism. *Immunol Rev* (2002) 190:40–52. doi:10.1034/j.1600-065X.2002.19004.x
28. Mandelboim O, Reyburn HT, Vales-Gomez M, Pazmany L, Colonna M, Borsellino G, et al. Protection from lysis by natural killer cells of group 1 and 2 specificity is mediated by residue 80 in human histocompatibility leukocyte antigen C alleles and also occurs with empty major histocompatibility complex molecules. *J Exp Med* (1996) 184(3):913–22. doi:10.1084/jem.184.3.913
29. Boyington JC, Sun PD. A structural perspective on MHC class I recognition by killer cell immunoglobulin-like receptors. *Mol Immunol* (2002) 38(14):1007–21. doi:10.1016/S0161-5890(02)00030-5

30. Igarashi T, Wynberg J, Srinivasan R, Becknell B, McCoy JP Jr, Takahashi Y, et al. Enhanced cytotoxicity of allogeneic NK cells with killer immunoglobulin-like receptor ligand incompatibility against melanoma and renal cell carcinoma cells. *Blood* (2004) 104(1):170–7. doi:10.1182/blood-2003-12-4438
31. Faridi RM, Agrawal S. Killer immunoglobulin-like receptors (KIRs) and HLA-C allorecognition patterns implicative of dominant activation of natural killer cells contribute to recurrent miscarriages. *Hum Reprod* (2011) 26(2):491–7. doi:10.1093/humrep/deq341
32. Colonna M, Navarro F, Bellon T, Llano M, Garcia P, Samaridis J, et al. A common inhibitory receptor for major histocompatibility complex class I molecules on human lymphoid and myelomonocytic cells. *J Exp Med* (1997) 186(11):1809–18. doi:10.1084/jem.186.11.1809
33. Samaridis J, Colonna M. Cloning of novel immunoglobulin superfamily receptors expressed on human myeloid and lymphoid cells: structural evidence for new stimulatory and inhibitory pathways. *Eur J Immunol* (1997) 27(3):660–5. doi:10.1002/eji.1830270313
34. Fehniger TA, Shah MH, Turner MJ, VanDeusen JB, Whitman SP, Cooper MA, et al. Differential cytokine and chemokine gene expression by human NK cells following activation with IL-18 or IL-15 in combination with IL-12: implications for the innate immune response. *J Immunol* (1999) 162(8):4511–20.
35. Vivier E, Anfossi N. Inhibitory NK-cell receptors on T cells: witness of the past, actors of the future. *Nat Rev Immunol* (2004) 4(3):190–8. doi:10.1038/nri1306
36. Fuchs A, Colonna M. The role of NK cell recognition of nectin and nectin-like proteins in tumor immunosurveillance. *Semin Cancer Biol* (2006) 16(5):359–66. doi:10.1016/j.semcancer.2006.07.002
37. Guo H, Qian X. Clinical applications of adoptive natural killer cell immunotherapy for cancer: current status and future prospects. *Onkologie* (2010) 33(7):389–95. doi:10.1159/000315698
38. Orr MT, Lanier LL. Natural killer cell education and tolerance. *Cell* (2010) 142(6):847–56. doi:10.1016/j.cell.2010.08.031
39. Rajagopalan S, Fu J, Long EO. Cutting edge: induction of IFN-gamma production but not cytotoxicity by the killer cell Ig-like receptor KIR2DL4 (CD158d) in resting NK cells. *J Immunol* (2001) 167(4):1877–81. doi:10.4049/jimmunol.167.4.1877
40. Kikuchi-Maki A, Catina TL, Campbell KS. Cutting edge: KIR2DL4 transduces signals into human NK cells through association with the Fc receptor gamma protein. *J Immunol* (2005) 174(7):3859–63. doi:10.4049/jimmunol.174.7.3859
41. Cantoni C, Verdiani S, Falco M, Pessino A, Cilli M, Conte R, et al. p49, a putative HLA class I-specific inhibitory NK receptor belonging to the immunoglobulin superfamily. *Eur J Immunol* (1998) 28(6):1980–90. doi:10.1002/(SICI)1521-4141(199806)28:06<1980::AID-IMMU1980>3.0.CO;2-F
42. Allan DS, Colonna M, Lanier LL, Churakova TD, Abrams JS, Ellis SA, et al. Tetrameric complexes of human histocompatibility leukocyte antigen (HLA)-G bind to peripheral blood myelomonocytic cells. *J Exp Med* (1999) 189(7):1149–56. doi:10.1084/jem.189.7.1149
43. Ponte M, Cantoni C, Biassoni R, Tradori-Cappai A, Bentivoglio G, Vitale C, et al. Inhibitory receptors sensing HLA-G1 molecules in pregnancy: decidua-associated natural killer cells express LIR-1 and CD94/NKG2A and acquire p49, an HLA-G1-specific receptor. *Proc Natl Acad Sci U S A* (1999) 96(10):5674–9. doi:10.1073/pnas.96.10.5674
44. Rajagopalan S, Long EO. A human histocompatibility leukocyte antigen (HLA)-G-specific receptor expressed on all natural killer cells. *J Exp Med* (1999) 189(7):1093–100. doi:10.1084/jem.189.7.1093
45. Boyson JE, Erskine R, Whitman MC, Chiu M, Lau JM, Koopman LA, et al. Disulfide bond-mediated dimerization of HLA-G on the cell surface. *Proc Natl Acad Sci U S A* (2002) 99(25):16180–5. doi:10.1073/pnas.212643199
46. Uhrberg M, Valiante NM, Shum BP, Shilling HG, Lienert-Weidenbach K, Corliss B, et al. Human diversity in killer cell inhibitory receptor genes. *Immunity* (1997) 7(6):753–63. doi:10.1016/S1074-7613(00)80394-5
47. Shilling HG, Guethlein LA, Cheng NW, Gardiner CM, Rodriguez R, Tyan D, et al. Allelic polymorphism synergizes with variable gene content to individualize human KIR genotype. *J Immunol* (2002) 168(5):2307–15. doi:10.4049/jimmunol.168.5.2307
48. Marsh SG, Parham P, Dupont B, Geraghty DE, Trowsdale J, Middleton D, et al. Killer-cell immunoglobulin-like receptor (KIR) nomenclature report, 2002. *Tissue Antigens* (2003) 62(1):79–86. doi:10.1034/j.1399-0039.2003.00072.x
49. Faridi RM, Kemp TJ, Dharmani-Khan P, Lewis V, Tripathi G, Rajalingam R, et al. Donor-recipient matching for KIR genotypes reduces chronic GVHD and missing inhibitory KIR ligands protect against relapse after myeloablative, HLA matched hematopoietic cell transplantation. *PLoS One* (2016) 11(6):e0158242. doi:10.1371/journal.pone.0158242
50. Cooley S, Weisdorf DJ, Guethlein LA, Klein JP, Wang T, Le CT, et al. Donor selection for natural killer cell receptor genes leads to superior survival after unrelated transplantation for acute myelogenous leukemia. *Blood* (2010) 116(14):2411–9. doi:10.1182/blood-2010-05-283051
51. Yawata M, Yawata N, Abi-Rached L, Parham P. Variation within the human killer cell immunoglobulin-like receptor (KIR) gene family. *Crit Rev Immunol* (2002) 22(5–6):463–82. doi:10.1615/CritRevImmunol.v22.i5-6.70
52. Martin AM, Kulski JK, Gaudieri S, Witt CS, Freitas EM, Trowsdale J, et al. Comparative genomic analysis, diversity and evolution of two KIR haplotypes A and B. *Gene* (2004) 335:121–31. doi:10.1016/j.gene.2004.03.018
53. Symons HJ, Fuchs EJ. Hematopoietic SCT from partially HLA-mismatched (HLA-haploidentical) related donors. *Bone Marrow Transplant* (2008) 42(6):365–77. doi:10.1038/bmt.2008.215
54. Musette P, Michel L, Jean-Louis F, Bagot M, Bensussan A. Polymorphic expression of CD158k/p140/KIR3DL2 in Sezary patients. *Blood* (2003) 101(3):1203. doi:10.1182/blood-2002-09-2915
55. Bagot M, Moretta A, Sivori S, Biassoni R, Cantoni C, Bottino C, et al. CD4(+) cutaneous T-cell lymphoma cells express the p140-killer cell immunoglobulin-like receptor. *Blood* (2001) 97(5):1388–91. doi:10.1182/blood.V97.5.1388
56. Battistella M, Janin A, Jean-Louis F, Collomb C, Leboeuf C, Sicard H, et al. KIR3DL2 (CD158k) is a potential therapeutic target in primary cutaneous anaplastic large-cell lymphoma. *Br J Dermatol* (2016) 175(2):325–33. doi:10.1111/bjd.14626
57. Giebel S, Locatelli F, Wojnar J, Velardi A, Mina T, Giorgiani G, et al. Homozygosity for human leucocyte antigen-C ligands of KIR2DL1 is associated with increased risk of relapse after human leucocyte antigen-C-matched unrelated donor haematopoietic stem cell transplantation. *Br J Haematol* (2005) 131(4):483–6. doi:10.1111/j.1365-2141.2005.05797.x
58. Karre K, Ljunggren HG, Piontek G, Kiessling R. Selective rejection of H-2-deficient lymphoma variants suggests alternative immune defence strategy. *Nature* (1986) 319(6055):675–8. doi:10.1038/319675a0
59. Ljunggren HG, Karre K. In search of the 'missing self': MHC molecules and NK cell recognition. *Immunol Today* (1990) 11(7):237–44. doi:10.1016/0167-5699(90)90097-S
60. Hsu KC, Gooley T, Malkki M, Pinto-Agnello C, Dupont B, Bignon JD, et al. KIR ligands and prediction of relapse after unrelated donor hematopoietic cell transplantation for hematologic malignancy. *Biol Blood Marrow Transplant* (2006) 12(8):828–36. doi:10.1016/j.bbmt.2006.04.008
61. Ruggeri L, Mancusi A, Urbani E, Velardi A. Identifying NK alloreactive donors for haploidentical hematopoietic stem cell transplantation. *Methods Mol Biol* (2016) 1393:141–5. doi:10.1007/978-1-4939-3338-9_14
62. Chewning JH, Gudme CN, Hsu KC, Selvakumar A, Dupont B. KIR2DS1-positive NK cells mediate alloresponse against the C2 HLA-KIR ligand group in vitro. *J Immunol* (2007) 179(2):854–68. doi:10.4049/jimmunol.179.2.854
63. Leung W, Iyengar R, Turner V, Lang P, Bader P, Conn P, et al. Determinants of antileukemia effects of allogeneic NK cells. *J Immunol* (2004) 172(1):644–50. doi:10.4049/jimmunol.172.1.644
64. Cooley S, Trachtenberg E, Bergemann TL, Saeteurn K, Klein J, Le CT, et al. Donors with group B KIR haplotypes improve relapse-free survival after unrelated hematopoietic cell transplantation for acute myelogenous leukemia. *Blood* (2009) 113(3):726–32. doi:10.1182/blood-2008-07-171926
65. Cooley S, Weisdorf DJ, Guethlein LA, Klein JP, Wang T, Marsh SG, et al. Donor killer cell Ig-like receptor B haplotypes, recipient HLA-C1, and HLA-C mismatch enhance the clinical benefit of unrelated transplantation for acute myelogenous leukemia. *J Immunol* (2014) 192(10):4592–600. doi:10.4049/jimmunol.1302517
66. Reisner Y, Kapoor N, Kirkpatrick D, Pollack MS, Cunningham-Rundles S, Dupont B, et al. Transplantation for severe combined immunodeficiency with HLA-A,B,D,DR incompatible parental marrow cells fractionated by soybean agglutinin and sheep red blood cells. *Blood* (1983) 61(2):341–8.
67. Aversa F, Tabilio A, Terenzi A, Velardi A, Falzetti F, Giannoni C, et al. Successful engraftment of T-cell-depleted haploidentical "three-loci" incompatible transplants in leukemia patients by addition of recombinant

human granulocyte colony-stimulating factor-mobilized peripheral blood progenitor cells to bone marrow inoculum. *Blood* (1994) 84(11):3948–55.

68. Bachanova V, Cooley S, Defor TE, Verneris MR, Zhang B, McKenna DH, et al. Clearance of acute myeloid leukemia by haploidentical natural killer cells is improved using IL-2 diphtheria toxin fusion protein. *Blood* (2014) 123(25):3855–63. doi:10.1182/blood-2013-10-532531
69. Stringaris K, Adams S, Uribe M, Eniafe R, Wu CO, Savani BN, et al. Donor KIR Genes 2DL5A, 2DS1 and 3DS1 are associated with a reduced rate of leukemia relapse after HLA-identical sibling stem cell transplantation for acute myeloid leukemia but not other hematologic malignancies. *Biol Blood Marrow Transplant* (2010) 16(9):1257–64. doi:10.1016/j.bbmt.2010.03.004
70. Pende D, Marcenaro S, Falco M, Martini S, Bernardo ME, Montagna D, et al. Anti-leukemia activity of alloreactive NK cells in KIR ligand-mismatched haploidentical HSCT for pediatric patients: evaluation of the functional role of activating KIR and redefinition of inhibitory KIR specificity. *Blood* (2009) 113(13):3119–29. doi:10.1182/blood-2008-06-164103
71. Moretta A, Pende D, Locatelli F, Moretta L. Activating and inhibitory killer immunoglobulin-like receptors (KIR) in haploidentical haemopoietic stem cell transplantation to cure high-risk leukaemias. *Clin Exp Immunol* (2009) 157(3):325–31. doi:10.1111/j.1365-2249.2009.03983.x
72. Rezvani K, Rouce RH. The application of natural killer cell immunotherapy for the treatment of cancer. *Front Immunol* (2015) 6:578. doi:10.3389/fimmu.2015.00578
73. Shilling HG, Young N, Guethlein LA, Cheng NW, Gardiner CM, Tyan D, et al. Genetic control of human NK cell repertoire. *J Immunol* (2002) 169(1):239–47. doi:10.4049/jimmunol.169.1.239
74. Shilling HG, McQueen KL, Cheng NW, Shizuru JA, Negrin RS, Parham P. Reconstitution of NK cell receptor repertoire following HLA-matched hematopoietic cell transplantation. *Blood* (2003) 101(9):3730–40. doi:10.1182/blood-2002-08-2568
75. Davies SM, Ruggieri L, DeFor T, Wagner JE, Weisdorf DJ, Miller JS, et al. Evaluation of KIR ligand incompatibility in mismatched unrelated donor hematopoietic transplants. Killer immunoglobulin-like receptor. *Blood* (2002) 100(10):3825–7. doi:10.1182/blood-2002-04-1197
76. Schaffer M, Aldener-Cannava A, Remberger M, Ringden O, Olerup O. Roles of HLA-B, HLA-C and HLA-DPA1 incompatibilities in the outcome of unrelated stem-cell transplantation. *Tissue Antigens* (2003) 62(3):243–50. doi:10.1034/j.1399-0039.2003.00089.x
77. Bornhauser M, Schwerdtfeger R, Martin H, Frank KH, Theuser C, Ehninger G. Role of KIR ligand incompatibility in hematopoietic stem cell transplantation using unrelated donors. *Blood* (2004) 103(7):2860–1; author reply 2. doi:10.1182/blood-2003-11-3893
78. Schaffer M, Malmberg KJ, Ringden O, Ljunggren HG, Remberger M. Increased infection-related mortality in KIR-ligand-mismatched unrelated allogeneic hematopoietic stem-cell transplantation. *Transplantation* (2004) 78(7):1081–5. doi:10.1097/01.TP.0000137103.19717.86
79. Kröger N, Binder T, Zabelina T, Wolschke C, Schieder H, Renges H, et al. Low number of donor activating killer immunoglobulin-like receptors (KIR) genes but not KIR-ligand mismatch prevents relapse and improves disease-free survival in leukemia patients after in vivo T-cell depleted unrelated stem cell transplantation. *Transplantation* (2006) 82(8):1024–30. doi:10.1097/01.tp.0000235859.24513.43
80. De Santis D, Bishara A, Witt CS, Nagler A, Brautbar C, Slavin S, et al. Natural killer cell HLA-C epitopes and killer cell immunoglobulin-like receptors both influence outcome of mismatched unrelated donor bone marrow transplants. *Tissue Antigens* (2005) 65(6):519–28. doi:10.1111/j.1399-0039.2005.00396.x
81. Rocha V, Ruggeri A, Spellman S, Wang T, Sobecks R, Locatelli F, et al. Killer cell immunoglobulin-like receptor-ligand matching and outcomes after unrelated cord blood transplantation in acute myeloid leukemia. *Biol Blood Marrow Transplant* (2016) 22(7):1284–9. doi:10.1016/j.bbmt.2016.04.007
82. Willemze R, Rodrigues CA, Labopin M, Sanz G, Michel G, Socie G, et al. KIR-ligand incompatibility in the graft-versus-host direction improves outcomes after umbilical cord blood transplantation for acute leukemia. *Leukemia* (2009) 23(3):492–500. doi:10.1038/leu.2008.365
83. Farag SS, Bacigalupo A, Eapen M, Hurley C, Dupont B, Caligiuri MA, et al. The effect of KIR ligand incompatibility on the outcome of unrelated donor transplantation: a report from the center for international blood and marrow transplant research, the European blood and marrow transplant registry, and the Dutch registry. *Biol Blood Marrow Transplant* (2006) 12(8):876–84. doi:10.1016/j.bbmt.2006.05.007
84. Gagne K, Busson M, Bignon JD, Balere-Appert ML, Loiseau P, Dormoy A, et al. Donor KIR3DL1/3DS1 gene and recipient Bw4 KIR ligand as prognostic markers for outcome in unrelated hematopoietic stem cell transplantation. *Biol Blood Marrow Transplant* (2009) 15(11):1366–75. doi:10.1016/j.bbmt.2009.06.015
85. Giebel S, Locatelli F, Lamparelli T, Velardi A, Davies S, Frumento G, et al. Survival advantage with KIR ligand incompatibility in hematopoietic stem cell transplantation from unrelated donors. *Blood* (2003) 102(3):814–9. doi:10.1182/blood-2003-01-0091
86. Penack O, Fischer L, Gentilini C, Nogai A, Muessig A, Rieger K, et al. The type of ATG matters – natural killer cells are influenced differentially by Thymoglobulin, Lymphoglobulin and ATG-Fresenius. *Transpl Immunol* (2007) 18(2):85–7. doi:10.1016/j.trim.2007.05.001
87. Neudorf S, Jones M. The effects of antithymocyte globulin on natural killer cells. *Exp Hematol* (1988) 16(10):831–5.
88. Stauch D, Dernier A, Sarmiento Marchese E, Kunert K, Volk HD, Pratschke J, et al. Targeting of natural killer cells by rabbit antithymocyte globulin and campath-1H: similar effects independent of specificity. *PLoS One* (2009) 4(3):e4709. doi:10.1371/journal.pone.0004709
89. Bosch M, Dhadda M, Hoegh-Petersen M, Liu Y, Hagel LM, Podgorny P, et al. Immune reconstitution after anti-thymocyte globulin-conditioned hematopoietic cell transplantation. *Cytotherapy* (2012) 14(10):1258–75. doi:10.3109/14653249.2012.715243
90. Giebel S, Dziaczkowska J, Wojnar J, Krawczyk-Kulis M, Markiewicz M, Kruzel T, et al. The impact of immunosuppressive therapy on an early quantitative NK cell reconstitution after allogeneic haematopoietic cell transplantation. *Ann Transplant* (2005) 10(2):29–33.
91. Servais S, Menten-Dedoyart C, Beguin Y, Seidel L, Gothot A, Daulne C, et al. Impact of pre-transplant anti-t cell globulin (ATG) on immune recovery after myeloablative allogeneic peripheral blood stem cell transplantation. *PLoS One* (2015) 10(6):e0130026. doi:10.1371/journal.pone.0130026
92. Aversa F, Tabilio A, Velardi A, Cunningham I, Terenzi A, Falzetti F, et al. Treatment of high-risk acute leukemia with T-cell-depleted stem cells from related donors with one fully mismatched HLA haplotype. *N Engl J Med* (1998) 339(17):1186–93. doi:10.1056/NEJM199810223391702
93. Miller JS, Cooley S, Parham P, Farag SS, Verneris MR, McQueen KL, et al. Missing KIR ligands are associated with less relapse and increased graft-versus-host disease (GVHD) following unrelated donor allogeneic HCT. *Blood* (2007) 109(11):5058–61. doi:10.1182/blood-2007-01-065383
94. Ludajic K, Balavarca Y, Bickeboller H, Rosenmayr A, Fae I, Fischer GF, et al. KIR genes and KIR ligands affect occurrence of acute GVHD after unrelated, 12/12 HLA matched, hematopoietic stem cell transplantation. *Bone Marrow Transplant* (2009) 44(2):97–103. doi:10.1038/bmt.2008.432
95. Sobecks RM, Wang T, Askar M, Gallagher MM, Haagenson M, Spellman S, et al. Impact of KIR and HLA genotypes on outcomes after reduced-intensity conditioning hematopoietic cell transplantation. *Biol Blood Marrow Transplant* (2015) 21(9):1589–96. doi:10.1016/j.bbmt.2015.05.002
96. Venstrom JM, Pittari G, Gooley TA, Chewning JH, Spellman S, Haagenson M, et al. HLA-C-dependent prevention of leukemia relapse by donor activating KIR2DS1. *N Engl J Med* (2012) 367(9):805–16. doi:10.1056/NEJMoa1200503
97. Sun JY, Gaidulis L, Dagis A, Palmer J, Rodriguez R, Miller MM, et al. Killer Ig-like receptor (KIR) compatibility plays a role in the prevalence of acute GVHD in unrelated hematopoietic cell transplants for AML. *Bone Marrow Transplant* (2005) 36(6):525–30. doi:10.1038/sj.bmt.1705089
98. Giebel S, Nowak I, Wojnar J, Markiewicz M, Dziaczkowska J, Wylezol I, et al. Impact of activating killer immunoglobulin-like receptor genotype on outcome of unrelated donor-hematopoietic cell transplantation. *Transplant Proc* (2006) 38(1):287–91. doi:10.1016/j.transproceed.2005.11.091
99. Venstrom JM, Gooley TA, Spellman S, Pring J, Malkki M, Dupont B, et al. Donor activating KIR3DS1 is associated with decreased acute GVHD in unrelated allogeneic hematopoietic stem cell transplantation. *Blood* (2010) 115(15):3162–5. doi:10.1182/blood-2009-08-236943
100. Kröger N, Zabelina T, Berger J, Duske H, Klyuchnikov E, Binder T, et al. Donor KIR haplotype B improves progression-free and overall survival after

allogeneic hematopoietic stem cell transplantation for multiple myeloma. *Leukemia* (2011) 25(10):1657–61. doi:10.1038/leu.2011.138

101. Bachanova V, Weisdorf DJ, Wang T, Marsh SG, Trachtenberg E, Haagenson MD, et al. Donor KIR B genotype improves progression-free survival of non-Hodgkin lymphoma patients receiving unrelated donor transplantation. *Biol Blood Marrow Transplant* (2016) 22(9):1602–7. doi:10.1016/j.bbmt.2016.05.016
102. Nowak J, Koscinska K, Mika-Witkowska R, Rogatko-Koros M, Mizia S, Jaskula E, et al. Donor NK cell licensing in control of malignancy in hematopoietic stem cell transplant recipients. *Am J Hematol* (2014) 89(10):E176–83. doi:10.1002/ajh.23802
103. Foley B, Felices M, Cichocki F, Cooley S, Verneris MR, Miller JS. The biology of NK cells and their receptors affects clinical outcomes after hematopoietic cell transplantation (HCT). *Immunol Rev* (2014) 258(1):45–63. doi:10.1111/imr.12157
104. Nowak J, Koscinska K, Mika-Witkowska R, Rogatko-Koros M, Mizia S, Jaskula E, et al. Role of donor activating KIR-HLA ligand-mediated NK cell education status in control of malignancy in hematopoietic cell transplant recipients. *Biol Blood Marrow Transplant* (2015) 21(5):829–39. doi:10.1016/j.bbmt.2015.01.018
105. Foley B, Cooley S, Verneris MR, Curtsinger J, Luo X, Waller EK, et al. NK cell education after allogeneic transplantation: dissociation between recovery of cytokine-producing and cytotoxic functions. *Blood* (2011) 118(10):2784–92. doi:10.1182/blood-2011-04-347070
106. Horowitz A, Guethlein LA, Nemat-Gorgani N, Norman PJ, Cooley S, Miller JS, et al. Regulation of adaptive NK cells and CD8 T cells by HLA-C correlates with allogeneic hematopoietic cell transplantation and with cytomegalovirus reactivation. *J Immunol* (2015) 195(9):4524–36. doi:10.4049/jimmunol.1401990

5

HLA-Bw4-I-80 Isoform Differentially Influences Clinical Outcome as Compared to HLA-Bw4-T-80 and HLA-A-Bw4 Isoforms in Rituximab or Dinutuximab-Based Cancer Immunotherapy

Amy K. Erbe [1†], *Wei Wang* [1†], *Patrick K. Reville* [1], *Lakeesha Carmichael* [2], *KyungMann Kim* [2], *Eneida A. Mendonca* [2], *Yiqiang Song* [2], *Jacquelyn A. Hank* [1], *Wendy B. London* [3], *Arlene Naranjo* [4], *Fangxin Hong* [5], *Michael D. Hogarty* [6], *John M. Maris* [7], *Julie R. Park* [8,9], *M. F. Ozkaynak* [10], *Jeffrey S. Miller* [11], *Andrew L. Gilman* [12], *Brad Kahl* [13], *Alice L. Yu* [14,15] and *Paul M. Sondel* [1,16*]

[1] Department of Human Oncology, University of Wisconsin, Madison, WI, United States, [2] Department of Biostatistics and Medical Informatics, University of Wisconsin, Madison, WI, United States, [3] Dana-Farber Cancer Institute/Boston Children's Cancer and Blood Disorder Center, Harvard Medical School, Boston, MA, United States, [4] COG Statistics and Data Center, Department of Biostatistics, University of Florida, Gainesville, FL, United States, [5] Department of Biostatistics, Harvard University, Dana Farber Cancer Institute, Boston, MA, United States, [6] Children's Hospital of Philadelphia, University of Pennsylvania School of Medicine, Philadelphia, PA, United States, [7] Provenance Biopharmaceuticals, Carlisle, MA, United States, [8] Seattle Children's Hospital/University, Seattle, WA, United States, [9] University of Washington, Seattle, WA, United States, [10] New York Medical College, Valhalla, NY, United States, [11] Department of Medicine, University of Minnesota, Minneapolis, MN, United States, [12] Levine Children's Hospital, Charlotte, NC, United States, [13] Department of Medicine, Washington University, St. Louis, MO, United States, [14] Department of Pediatrics, Hematology/Oncology, Moores Cancer Center, University of California San Diego, San Diego, CA, United States, [15] Institute of Stem Cell and Translational Cancer Research, Chang Gung Memorial Hospital, Taoyuan, Taiwan, [16] Department of Pediatrics, University of Wisconsin-Madison, Madison, WI, United States

Edited by:
Gianfranco Pittari, Hamad Medical Corporation, Qatar

Reviewed by:
Daniel Olive, Institut national de la santé et de la recherche médicale (INSERM), France
Raquel Tarazona, University of Extremadura, Spain
Alessandro Moretta, Università di Genova, Italy

***Correspondence:**
Paul M. Sondel
pmsondel@humonc.wisc.edu

†Co-first authors.

Killer-cell immunoglobulin-like receptors (KIRs) are a family of glycoproteins expressed primarily on natural killer cells that can regulate their function. Inhibitory KIRs recognize MHC class I molecules (KIR-ligands) as ligands. We have reported associations of KIRs and KIR-ligands for patients in two monoclonal antibody (mAb)-based trials: (1) A Children's Oncology Group (COG) trial for children with high-risk neuroblastoma randomized to immunotherapy treatment with dinutuximab (anti-GD2 mAb) + GM-CSF + IL-2 + isotretinion or to treatment with isotretinoin alone and (2) An Eastern Cooperative Oncology Group (ECOG) trial for adults with low-tumor burden follicular lymphoma responding to an induction course of rituximab (anti-CD20 mAb) and randomized to treatment with maintenance rituximab or no-maintenance rituximab. In each trial, certain KIR/KIR-ligand genotypes were associated with clinical benefit for patients randomized to immunotherapy treatment (immunotherapy in COG; maintenance rituximab in ECOG) as compared to patients that did not receive the immunotherapy [isotretinoin alone (COG); no-maintenance (ECOG)]. Namely, patients with both KIR3DL1 and its HLA-Bw4 ligand (KIR3DL1+/HLA-Bw4+ genotype) had improved clinical outcomes if randomized to immunotherapy regimens, as compared to patients with the KIR3DL1+/

HLA-Bw4+ genotype randomized to the non-immunotherapy regimen. Conversely, patients that did not have the KIR3DL1+/HLA-Bw4+ genotype showed no evidence of a difference in outcome if receiving the immunotherapy vs. no-immunotherapy. For each trial, HLA-Bw4 status was determined by assessing the genotypes of three separate isoforms of HLA-Bw4: (1) HLA-B-Bw4 with threonine at amino acid 80 (B-Bw4-T80); (2) HLA-B-Bw4 with isoleucine at amino acid 80 (HLA-B-Bw4-I80); and (3) HLA-A with a Bw4 epitope (HLA-A-Bw4). Here, we report on associations with clinical outcome for patients with KIR3DL1 and these separate isoforms of HLA-Bw4. Patients randomized to immunotherapy with KIR3DL1+/A-Bw4+ or with KIR3DL1+/B-Bw4-T80+ had better outcome vs. those randomized to no-immunotherapy, whereas for those with KIR3DL1+/B-Bw4-I80+ there was no evidence of a difference based on immunotherapy vs. no-immunotherapy. Additionally, we observed differences within treatment types (either within immunotherapy or no-immunotherapy) that were associated with the genotype status for the different KIR3DL1/HLA-Bw4-isoforms. These studies suggest that specific HLA-Bw4 isoforms may differentially influence response to these mAb-based immunotherapy, further confirming the involvement of KIR-bearing cells in tumor-reactive mAb-based cancer immunotherapy.

Keywords: KIR, KIR-ligand, HLA-Bw4, HLA, MHC class I, natural killer cells, cancer immunotherapy

INTRODUCTION

One modality of cancer immunotherapy utilizes tumor-reactive monoclonal antibodies (mAbs) to elicit a tumor-targeted immune response. Two recently completed clinical trials, in separate disease settings, utilized tumor-reactive mAbs to successfully target and treat the tumors: (1) the combination of dinutuximab with IL-2, GM-CSF, and isotretinoin for patients with high-risk neuroblastoma (1) and (2) rituximab for the treatment of patients with low-tumor burden follicular lymphoma (FL) (2).

Natural killer (NK) cells can contribute to the response to tumor-reactive mAb-based immunotherapeutics through antibody-dependent cellular cytotoxicity (ADCC). The ability of NK cells to elicit ADCC is regulated by activating and inhibiting signaling. Killer-cell immunoglobulin-like receptors (KIRs) are a class of receptors expressed on NK cells that influence such signaling (3,4). Most inhibitory KIRs interact with HLA class I molecules as their ligands (KIR-ligand) (5). Specifically, KIR2DL1 binds to HLA-C2, KIR2DL2 and KIR2DL3 bind to HLA-C1, and KIR3DL1 recognizes the Bw4 epitope of HLA-A and HLA-B (6, 7). The independent segregation and inheritance of KIRs and KIR-ligands help to shape NK cell function and response to immunotherapeutic agents (8–11). When inhibitory KIRs interact with class I HLA molecules on target cells, NK cell-mediated lysis and ADCC are inhibited. During development, KIR/KIR-ligand interactions lead to self tolerance and NK cells become "licensed NK cells" (12–14). Licensed NK cells have augmented cytotoxicity against class I negative tumors compared to unlicensed NK cells (15, 16).

Killer-cell immunoglobulin-like receptors and KIR-ligands segregate independently: KIR genes are located on chromosome 19; HLA genes (KIR-ligands) are located on chromosome 6. Several studies have shown that genotypic differences of KIR and KIR-ligands can influence clinical outcome of certain cancer immunotherapies (8, 11, 17–19). We recently showed in two clinical trials that KIR3DL1 and its KIR-ligand, HLA-Bw4, appear to influence clinical outcome.

In a phase III trial (ANBL0032) of high-risk neuroblastoma patients, conducted by the Children's Oncology Group (COG) (1), patients who inherited the KIR3DL1 gene and the gene for its HLA-Bw4 ligand (KIR3DL1+/Bw4+ genotype) and were treated with an immunotherapy regimen [dinutuximab (anti-GD2), IL-2, GM-CSF, and isotretinoin] had improved event-free survival (EFS) and overall survival as compared to those treated with isotretinoin alone (20, 21). In a separate Eastern Cooperative Oncology Group (ECOG) Phase III clinical trial of low-tumor burden FL (2), patients who were KIR3DL1+/HLA-Bw4+ and treated with a continuous regimen of maintenance rituximab had improved duration of response and % tumor shrinkage compared to KIR3DL1+/HLA-Bw4+ patients who were randomized to not receive maintenance rituximab (22, 23). Conversely, we did not observe improved outcome for patients that were *not* KIR3DL1+/HLA-Bw4+ when randomized to immunotherapy, in either study (22, 23). Furthermore, in both the COG and ECOG studies, patients who were randomized to the immunotherapy regimen that were KIR3DL1+/HLA-Bw4+ had better outcome compared to patients who were *not* KIR3DL1+/HLA-Bw4+.

Given these similar associations with outcome for the KIR3DL1/HLA-Bw4 interaction in these two clinical trials, we chose to evaluate these more deeply by evaluating the potential influence of distinct HLA-Bw4 isoforms. Polymorphisms in the α1 helix (positions 77–83) of HLA class I correspond to the sequence site of the Bw4 epitope that is recognized by KIR3DL1 (24). In KIR/KIR-ligand associations, we analyzed in these COG and ECOG trials, individuals were considered positive for HLA-Bw4 if they were found to have at least one of the three isoforms of HLA-Bw4: (1) HLA-B allele with a threonine at amino acid position 80 (B-Bw4-T80), (2) HLA-B allele with an isoleucine at amino acid position 80 (B-Bw4-I80), or (3) HLA-A with a Bw4 epitope (A-Bw4). Patients were negative for HLA-Bw4 if they did not have any of these three isoforms. These polymorphisms of this Bw4 epitope can impact KIR3DL1 recognition (25–29). As such, we describe the impact of the genotype status of B-Bw4-T80, B-Bw4-I80, and A-Bw4, together with the genotype status of KIR3DL1, on the clinical outcome, based on a clinical outcome parameter that measured the duration of response to the treatment regimen (EFS in COG; duration of response in ECOG).

MATERIALS AND METHODS

Patients

COG ANBL0032 Patients

The phase III neuroblastoma clinical trial (ANBL0032; Clinicaltrials.gov # NCT00026312) evaluated the efficacy of isotretinoin alone as compared to an immunotherapeutic regimen consisting of dinutuximab (anti-GD2), aldesleukin (IL-2), sargramostim (GM-CSF), and isotretinoin (1). Of the 226 patients randomized, 174 patients (immunotherapy: n = 88; isotretinoin: n = 86) had DNA available, allowing evaluation of KIR/KIR-ligand genotype association with updated clinical outcome (>5-year follow-up if no event). All analyses in this study were conducted utilizing an intent-to-treat approach. All patients signed IRB approved consent forms enabling lab-based immune correlative analyses, and the genotyping done at UW-Madison was approved by the UW-IRB.

ECOG E4402 Patients

The Phase III ECOG clinical trial (E4402; ClinicalTrials.gov #NCT00075946) evaluated the efficacy of single agent, rituximab therapy for adults with low-tumor burden FL. Clinical results from this study have been reported elsewhere (2). A total of 408 patients with FL were entered, with 289 patients responding and randomized to no-maintenance or maintenance rituximab regimens. Disease measurements were obtained every 13 weeks (2). Of the 289 randomized patients from this trial, 213 patients had evaluable DNA and clinical data for this study, and 159 of them were randomized to no-maintenance (n = 80) or maintenance rituximab (n = 79) treatment. Of these 79 patients treated with maintenance rituximab, 75 patients had clinical data available for duration of response. All patients signed IRB approved consent forms enabling lab-based immune correlative analyses, and the genotyping done at UW-Madison was approved by the UW-IRB.

Genotyping

KIR3DL1 gene status was determined by a SYBR green real-time PCR reaction (30, 31). The genotype for HLA-Bw4, which includes three known HLA-Bw4 epitopes (B-Bw4-T80, B-Bw4-I80, and A-Bw4) were determined by PCR-SSP reactions using the KIR HLA Ligand SSP typing kit (product number 104.201-12u from Olerup, West Chester, PA, USA) with GoTaq DNA polymerase (M8295, Promega, WI, USA). All genotyping was conducted in a blinded manner, whereby individuals who determined the genotype of the patients did not have access to the clinical outcome data.

Statistical Methods

The goal of these analyses was to evaluate the association of KIR3DL1 in combination with each HLA-Bw4 isoform (B-Bw4-T80, B-Bw4-I80, and A-Bw4) on response to therapy (EFS or duration of response). For the COG trial, EFS time was defined as the time from study enrollment until the first occurrence of relapse, progressive disease, secondary cancer, or death or until the last contact with the patient if none of these events occurred (censored). For the ECOG trial, the duration of response was defined as the time from randomization (following an initial response to the induction rituximab treatment) to documented disease progression (2).

Cox proportional hazards regression models and log-rank tests were used to compare EFS/duration of response curves by treatment and genotype combinations. The proportional hazards assumption was tested, and when the assumption was not met, adjustments were made by incorporating time-dependent covariates into the model. For both trials, only randomized patients were included in the analyses. Statistical analyses were performed using SAS v9.4 (SAS Institute, Cary, NC, USA).

RESULTS

HLA-Bw4 Isoforms, Together with KIR3DL1, Differentially Influence the Impact of mAb-Based Immunotherapy on Clinical Outcome of Neuroblastoma Patients

In our analyses of associations of KIR/KIR-ligand genotypic influence on clinical response in the neuroblastoma study (ANBL0032), we reported on differences in clinical outcome for those KIR3DL1+/Bw4+ (immunotherapy n = 58; isotretinoin n = 61) and those *not* KIR3DL1+/Bw4+ (immunotherapy n = 30; isotretinoin n = 25), and differences in response were observed dependent upon treatment type (20, 21). Since not all of the isoforms of HLA-Bw4 may interact with KIR3DL1 to the same degree, we further assessed patients with different HLA-Bw4 isoforms in this setting.

To better understand the KIR/KIR-ligand genotypic influence on clinical outcome, we evaluated the effect of Bw4 epitope on either an HLA-A or HLA-B allele. In this study, patients who were KIR3DL1+/A-Bw4+ had a trend toward improved EFS if they were treated with immunotherapy as compared to those treated with isotretinoin alone (p = 0.06; **Figure 1A**) (Table S1 in Supplementary Material). In contrast, we did not find a significant difference in EFS for patients receiving the immunotherapy vs. those randomized to not receive the immunotherapy (i.e., isotretinoin alone) in the patients that were *not* KIR3DL1+/A-Bw4+ (p = 0.35; **Figure 1A**).

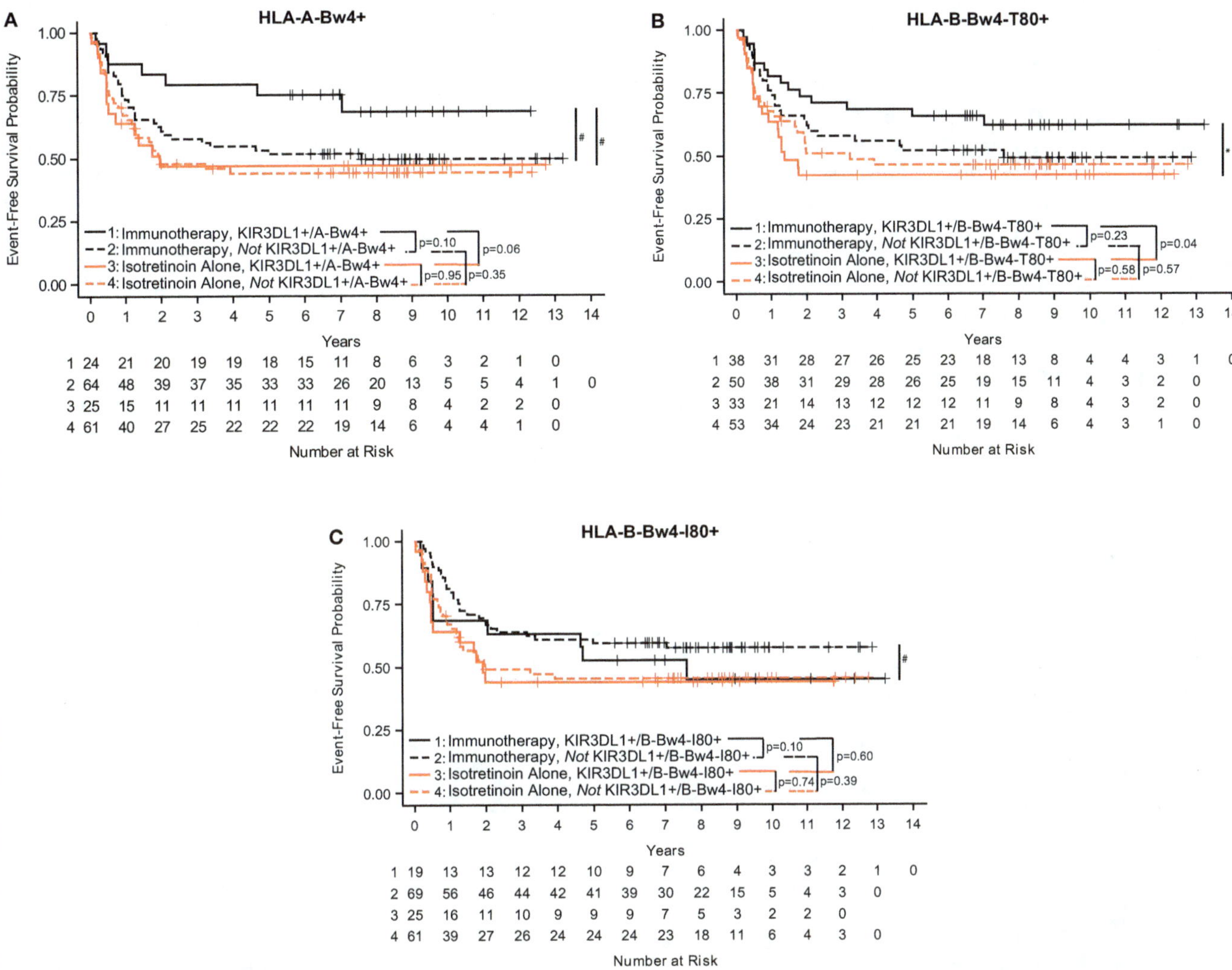

FIGURE 1 | KIR3DL1/HLA-Bw4 isoforms differentially influence event-free survival (EFS) in neuroblastoma patients. In neuroblastoma patients treated with immunotherapy (black lines) or isotretinoin alone (red lines), genotype statuses of KIR3DL1/A-Bw4 **(A)**, KIR3DL1/B-Bw4-T80 **(B)**, and KIR3DL1/B-Bw4-I80 **(C)** had differential effects on EFS. Those with the genes present are represented by solid lines, and those without the genes present are represented by dotted lines. The number of patients at risk for EFS at a given time point are provided below the *x*-axis. $^{*}p < 0.5$ and $\#p < 0.10$.

We found that B-Bw4-T80 and B-Bw4-I80 differentially influenced EFS in these neuroblastoma patients (Table S1 in Supplementary Material). Similar to results in **Figure 1A**, patients who were KIR3DL1+/B-Bw4-T80+ showed significantly improved EFS if they received immunotherapy compared with isotretinoin alone ($p = 0.04$; **Figure 1B**), whereas those that were *not* KIR3DL1+/B-Bw4-T80+ showed no difference in EFS for patients receiving the immunotherapy vs. those randomized to receive isotretinoin alone ($p = 0.57$; **Figure 1B**). However, for B-Bw4-I80+, the results were converse. Patients who were KIR3DL1+/B-Bw4-I80+ showed no sign of improved EFS if they received immunotherapy compared with isotretinoin alone ($p = 0.60$; **Figure 1C**). Furthermore, and in contrast to results in **Figures 1A,B**, while not significant, there appears to be improved EFS for patients receiving the immunotherapy vs. isotretinoin alone in the patients who were *not* KIR3DL1+/B-Bw4-I80+ ($p = 0.10$; **Figure 1C**).

These findings suggest that the different isoforms of HLA-Bw4 differentially influence the impact of anti-GD2-based immunotherapy on EFS for high-risk neuroblastoma patients.

HLA-Bw4 Isoforms, Together with KIR3DL1, Differentially Influence the Impact of mAb-Based Immunotherapy on Clinical Outcome of FL Patients

The ECOG E4402 Phase III clinical trial sought to optimize the rituximab treatment regimen for low-tumor burden FL patients (2). As such, different from the design of the neuroblastoma COG trial described above where one treatment arm was treated with immunotherapy and the other was not, in E4402 *all* patients were initially treated with rituximab. In E4402, all FL patients received induction rituximab, consisting of four weekly rituximab treatments. After 13 weeks, those patients who achieved ≥50% tumor shrinkage were randomized to two separate treatment regimens: (1) "maintenance" rituximab was given every 13 weeks or (2) "no-maintenance" where rituximab was given only upon disease progression (2). Thus, for the parameter of disease progression, the no-maintenance group received no rituximab between randomization and disease progression. Similar to the COG findings regarding the genotype status of KIR3DL1/Bw4, in this ECOG study, we also found that those KIR3DL1+/Bw4+ (maintenance $n = 49$; no-maintenance $n = 53$) had different clinical outcome than those not KIR3DL1+/Bw4+ (maintenance $n = 27$; no-maintenance $n = 26$), which was also influenced by the treatment arm.

Analyses of the three separate HLA-Bw4 isoforms suggest that the isoforms of HLA-Bw4 differently influenced the impact of maintenance rituximab. FL patients who were KIR3DL1+/A-Bw4+ that were treated with maintenance rituximab had a longer duration of response (0 of 23 progressed, **Figure 2A**) as compared to patients who were *not* KIR3DL1+/A-Bw4+ [13 out of 53 progressed ($p = 0.008$, **Figure 2A**) (Table S1 in Supplementary Material)]. Separately, patients who were KIR3DL1+/B-Bw4-T80+ also showed significantly prolonged duration of response if they received maintenance as compared with no-maintenance rituximab ($p = 0.007$; **Figure 2B**). In addition, those patients whowere *not* KIR3DL1+/B-Bw4-T80+ had a trend toward improved duration of response if treated with maintenance as compared with no-maintenance rituximab ($p = 0.07$; **Figure 2B**) (Table S1 in Supplementary Material). However, patients who were KIR3DL1+/B-Bw4-I80+ did not show prolonged duration of response if they received maintenance as compared with no-maintenance rituximab ($p = 0.40$; **Figure 2C**). Similar to the trends for improved EFS observed in neuroblastoma patients treated

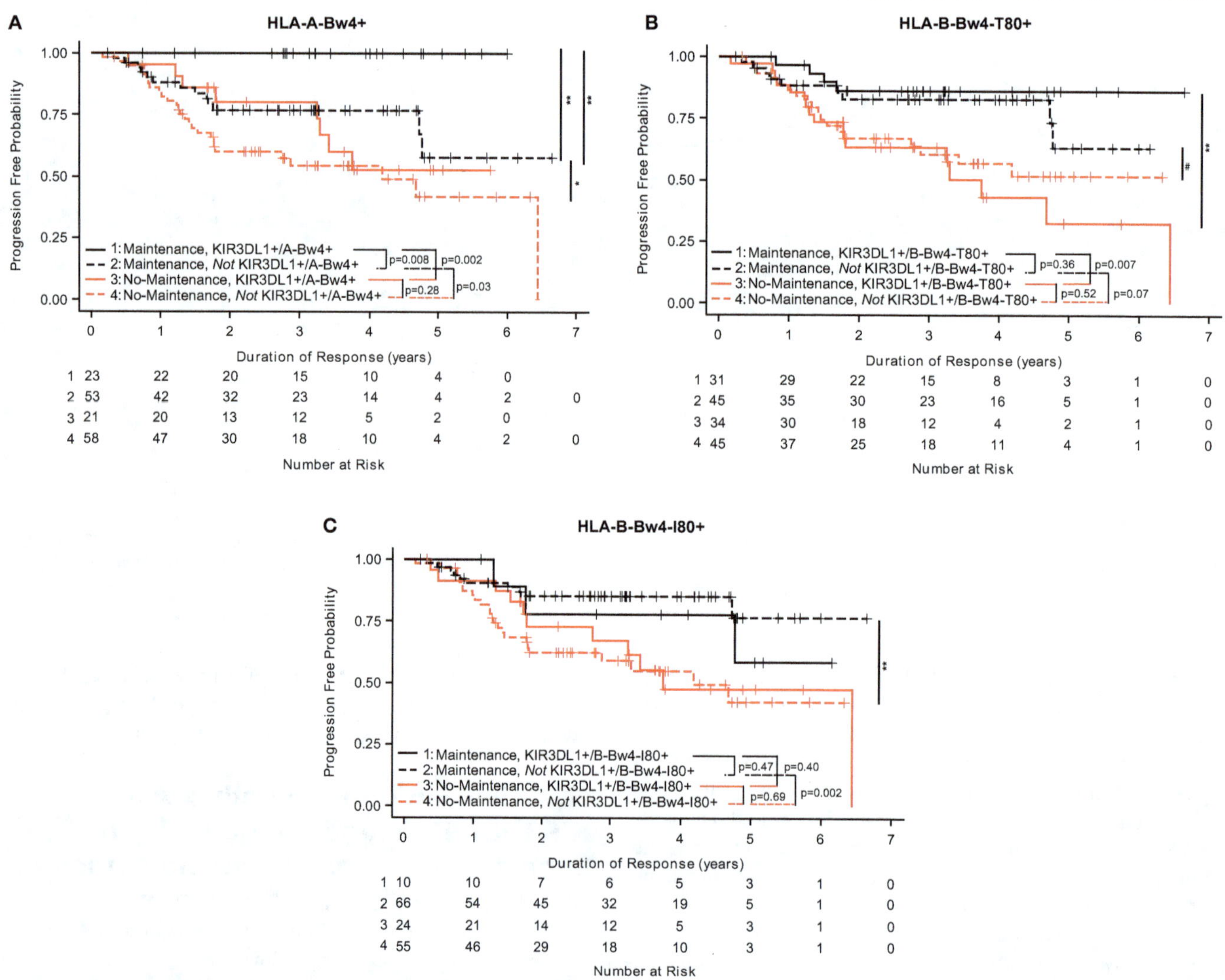

FIGURE 2 | KIR3DL1/HLA-Bw4 isoforms differentially influence duration of response in follicular lymphoma (FL) patients. In FL patients treated with maintenance rituximab (red lines) or no-maintenance rituximab (black lines), genotype statuses of KIR3DL1/A-Bw4 **(A)**, KIR3DL1/B-Bw4-T80 **(B)**, and KIR3DL1/B-Bw4-I80 **(C)** had differential effects on duration of response. Those with the genes present are represented by solid lines, and those without the genes present are represented by dotted lines. The number of patients at risk for duration of response at a given time point are provided below the *x*-axis. $^{**}p < 0.01$, $^{*}p < 0.5$, and $\#p < 0.10$.

with immunotherapy (**Figure 1C**), those FL patients who were *not* KIR3DL1+/B-Bw4-I80+ had improved duration of response if treated with maintenance rituximab as compared to no-maintenance ($p = 0.002$; **Figure 2C**) (Table S1 in Supplementary Material).

These findings suggest that the different isoforms of HLA-Bw4 differentially influence the impact of rituximab maintenance treatment for these low-tumor burden FL.

DISCUSSION

In both of these clinical trials, in separate disease settings, tumor-reactive mAbs were used to treat the tumors. In the analysis of KIR/KIR-ligand genotypes in each of these studies, we found similar associations with outcomes based upon the influence of KIR3DL1/HLA-Bw4. Specifically, those patients who had both KIR3DL1 and HLA-Bw4 had improved clinical outcomes if they were treated with either the COG immunotherapy regimen or the maintenance rituximab regimen in ECOG as compared to those who did not receive these same immunotherapeutic regimens (20, 21). Here, we report on the analyses of the specific HLA-Bw4 isoforms in both trials. In the ECOG trial of FL patients, patients with a KIR3DL1+/A-Bw4+ genotype or a KIR3DL1+/B-Bw4-T80+ genotype showed improved outcome when randomized to the maintenance regimen rather than to the no-maintenance regimen. In contrast, patients with a KIR3DL1+/B-Bw4-I80+ genotype showed no evidence of improved outcome when randomized to the maintenance treatment vs. no-maintenance regimen. We also observed similar trends for these same analyses in the COG trial of neuroblastoma patients.

Although other mechanisms, such as antibody-dependent cellular phagocytosis and complement-dependent cellular cytotoxicity (32, 33), could also contribute to the anti-tumor efficacy of tumor antigen-specific monoclonal antibodies, we hypothesize that the anti-tumor effect of rituximab and dinutuximab in these FL and neuroblastoma patients, respectively, is primarily through ADCC. NK cells are major contributors to ADCC, and their activity is regulated *via* the interactions between KIRs/KIR-ligands (34). As such, we hypothesize that the KIR/KIR-ligand genotypes could influence the degree that patients respond to antibody-based immunotherapies. Besides NK cells, KIRs are also expressed by a subset of T cells as well as NKT cells (35, 36). Therefore, it is possible that these other cell types may also be influenced by KIR/KIR-ligand genotypes.

Besides inherited genetic differences in KIR and KIR-ligand genotypes, other individual genetic differences, such as polymorphisms in Fc gamma receptors (FCGRs), may influence patient outcome to immunotherapy. FCGR polymorphisms can alter the affinity of FCGRs for the Fc portion of antibodies (mAbs or endogenous antibodies) (37). For example, in a separate study of patients with metastatic renal cell carcinoma treated with high-dose IL-2, we found that patients with a "higher affinity" FCGR genotype had improved clinical outcome as compared to those patients with a "lower affinity" FCGR genotype (38). In our analysis of those same metastatic renal cell carcinoma patients for KIR/KIR-ligand genotype influence on outcome, we did not observe differences in clinical outcome associated with KIR3DL1 and HLA-Bw4 genotype status (39). The influence of FCGR polymorphisms on clinical outcome to rituximab is variable (40–42). For the FL patients analyzed here from this ECOG study, Kenkre and colleagues reported no association of FCGR genotype polymorphisms with patient outcome (43). In addition, some groups have found associations of FCGR genotype with clinical outcome for patients treated with anti-GD2 immunotherapy (8, 44, 45). For the neuroblastoma patients from this COG trial, FCGR genotype associations with clinical outcome are still under investigation. In addition, it has been reported that the influence from KIR/KIR-ligand interactions on NK cells may be affected by the affinity of the Fc portion of different therapeutic mAb used (46), the rituximab used in this ECOG trial and the dinutuximab used in this COG trial have similar human IgG1 Fc components, which may also help account for why we observed similar influences from HLA-Bw4 epitopes in these two separate studies where two different therapeutic mAbs were used.

These clinical data are consistent with the B-Bw4-I80 isoform functioning somewhat differently than the B-Bw4-T80 or A-Bw4 isoforms, and potentially making the tumor cells less responsive to the potential benefit of the anti-GD2 or anti-CD20 mAb-based immunotherapy. *In vitro* analyses have shown that a subset of HLA-Bw4 alleles (those with an B-Bw4-I80 isoform) show relative protection from lysis by NK cells (47, 48). The data presented here are consistent with these *in vitro* results; mAb-based immunotherapy may provide more benefit for patients with weaker NK cell inhibition from B-Bw4-T80 or A-Bw4, than for patients with stronger NK inhibition from B-Bw4-I80.

Given that patients assessed in either trial could be positive for more than one of the HLA-Bw4 epitopes, we did consider whether the HLA-Bw4 epitopes were in linkage disequilibrium. We found that A-Bw4 was not in linkage disequilibrium with either B-Bw4-I80 or B-Bw4-T80 (Table S2 in Supplementary Material). Thus, the influence that each of these HLA-Bw4 epitopes had on the length of patient response in either trial is presumably not due to linkage disequilibrium with each other.

We also considered whether the interaction of KIR3DL1 with these three different HLA-Bw4 isoforms showed any association of outcome among patients randomized to receive the immunotherapy regimens. Within the COG study, we observed a trend for improved outcome for those KIR3DL1+/HLA-A-Bw4+ vs. those not KIR3DL1+/HLA-A-Bw4+ (**Figure 1A**), and we also observed a trend in the opposite direction for HLA-Bw4-I80, namely, there was a trend for improved outcome for those not KIR3DL1+/HLA-B-Bw4-I80+ vs. those who were KIR3DL1+/HLA-B-Bw4-I80+ (**Figure 1C**). Although only a trend, this difference in **Figure 1A** and **Figure 1C** is consistent with differential function of HLA-A-Bw4 and HLA-B-Bw4-I80. No significant differences or trends were noted when we evaluated among the FL patients randomized to receive the maintenance rituximab regimen (**Figures 2A–C**).

The interaction of KIR3DL1 with the Bw4 epitope is dependent not only on the architecture of Bw4 but also on the sequence of the bound peptide (25, 28, 49–51). Additionally, the differences we observed between A-Bw4, B-Bw4-T80, and B-Bw4-I80 may be due to the different inhibition strength for KIR3DL1 from these isoforms. For instance, HLA-A*32:01, HLA-B*51:01, and HLA-B*58:01 strongly inhibit target cells from lysis by KIR3DL1+ NK cells, yet HLA-B*15:13 and HLA-B*27:05 have weaker inhibitory effects, despite all being HLA-Bw4 alleles (6, 25, 26, 48, 52–54). In addition, depending on the KIR3DL1 allele, expression of KIR3DL1 can vary; different HLA-A-Bw4 alleles have differential affinity for KIR3DL1 that is attributed to high vs. low expression of KIR3DL1 (55). Furthermore, the specific Bw4 allele, as well as the KIR3DL1 allele, the strength of KIR3DL1/HLA-Bw4 interaction and the binding avidity can vary (29). For example, Saunders et al. recently showed that HLA-A*24:02 acts as a poor ligand for KIR3DL1, and the strength of its interaction with KIR3DL1 differed depending on the allele of KIR3DL1 (29). The genotyping methodology employed for analyzing the many patients in these two clinical trials reported here was not able to address these more subtle allele-specific or peptide-related issues.

Another possible cause of the differences observed in these HLA-Bw4 isoforms may be due to genetic polymorphisms of KIR3DL1 (26, 28, 29, 56–60). More than 100 alleles of KIR3DL1 have been described. Phylogenetically, these alleles span three lineages based on the polymorphism of the three extracellular domains (D0–D1–D2) (53, 61). In both of these clinical studies analyzed, we did not determine the allelic differences of the KIR genes, but rather we determined their presence or absence. Thus, we cannot assess how different KIR3DL1 alleles may affect the interactions between different isoforms of HLA-Bw4. We did, however, assess if KIR3DL1 allelic status could influence the interactions of KIR3DL1 with HLA-Bw4 and with the separate HLA-Bw4 isoforms. KIR3DL1 and KIR3DS1 are alleles, thus individuals can have 2, 1, or 0 copies of KIR3DL1 (2 copies: KIR3DL1/KIR3DL1, 1 copy: KIR3DL1/KIR3DS1, or 0 copies: KIR3DS1/KIRDS1). Although KIR3DS1 has not been shown to utilize HLA-Bw4 as a ligand *in vitro*, whether KIR3DS1 may still interact with HLA-Bw4 *in vivo* is controversial (62–65). We assessed whether the allelic status of KIR3DL1/KIR3DS1 together with HLA-Bw4 (and HLA-Bw4 isoforms) influenced patient response. We found that there was no evidence of an association with outcome in either the COG or the ECOG study that could be linked to the allelic status of KIR3DL1/KIR3DLS1 (data not shown), nor was there evidence of an association of clinical outcome linked to KIR3DL1/KIR3DS1 status together with the HLA-Bw4 ligand isoforms (data not shown). Rather, the mere presence of KIR3DL1 together with its ligand, HLA-Bw4, seemed to influence patients' response to immunotherapy in both clinical trials. These observations will require validation in a separate study.

In conclusion, this work sheds further light on the role of KIR receptors on NK cells in the antitumor response to immunotherapeutic mAbs. We demonstrate that the KIR3DL1/HLA-Bw4 axis influences response to tumor-targeted mAbs in two separate clinical trials and that the presence of the B-Bw4-T80 isoform or the A-Bw4 isoform is associated with improved response to mAb-based immunotherapy, while the presence of the B-Bw4-I80 isoform is not.

ETHICS STATEMENT

This study was carried out in accordance with the recommendations of University of Wisconsin Health Sciences Institutional Review Board with written informed consent from all subjects. All subjects gave written informed consent in accordance with the Declaration of Helsinki. The protocol was approved by the University of Wisconsin Health Sciences Institutional Review Board.

AUTHOR CONTRIBUTIONS

Each author made substantial contributions to the conception and/or design of this research, including the acquisition, analysis, and/or interpretation of data for the work; the drafting of this manuscript, including critical revisions important intellectual content, were shared duties by all authors; each author submitted final approval of this manuscript as submitted to be published; each author is in agreement to be accountable for all aspects of the work in ensuring that questions related to the accuracy or integrity of any part of the work are appropriately investigated and resolved (AE, WW, PR, LC, KK, EM, YS, JH, WL, AN, FH, MH, JM, JP, MO, JM, AG, BK, AY, and PS).

ACKNOWLEDGMENTS

We would like to thank the COG and ECOG for allowing us to assess the role of the genotypic influences of KIR and KIR-ligands on the clinical outcome from the treatments administered in each of these trials. In addition, we thank the medical and nursing staff who participated in the care of patients in each study, and especially all patients participating in each study.

REFERENCES

1. Yu AL, Gilman AL, Ozkaynak MF, London WB, Kreissman SG, Chen HX, et al. Anti-GD2 antibody with GM-CSF, interleukin-2, and isotretinoin for neuroblastoma. *N Engl J Med* (2010) 363(14):1324–34. doi:10.1056/NEJMoa0911123
2. Kahl BS, Hong F, Williams ME, Gascoyne RD, Wagner LI, Krauss JC, et al. Rituximab extended schedule or re-treatment trial for low-tumor burden follicular lymphoma: Eastern Cooperative Oncology Group Protocol e4402. *J Clin Oncol* (2014) 32(28):3096–102. doi:10.1200/JCO.2014.56.5853
3. Boyington JC, Sun PD. A structural perspective on MHC class I recognition by killer cell immunoglobulin-like receptors. *Mol Immunol* (2002) 38(14):1007–21. doi:10.1016/S0161-5890(02)00030-5
4. Vilches C, Parham P. KIR: diverse, rapidly evolving receptors of innate and adaptive immunity. *Annu Rev Immunol* (2002) 20:217–51. doi:10.1146/annurev.immunol.20.092501.134942
5. Horton R, Wilming L, Rand V, Lovering RC, Bruford EA, Khodiyar VK, et al. Gene map of the extended human MHC. *Nat Rev Genet* (2004) 5(12):889–99. doi:10.1038/nrg1489

3. Boyington JC, Sun PD. A structural perspective on MHC class I recognition by killer cell immunoglobulin-like receptors. *Mol Immunol* (2002) 38(14):1007–21. doi:10.1016/S0161-5890(02)00030-5
4. Vilches C, Parham P. KIR: diverse, rapidly evolving receptors of innate and adaptive immunity. *Annu Rev Immunol* (2002) 20:217–51. doi:10.1146/annurev.immunol.20.092501.134942
5. Horton R, Wilming L, Rand V, Lovering RC, Bruford EA, Khodiyar VK, et al. Gene map of the extended human MHC. *Nat Rev Genet* (2004) 5(12):889–99. doi:10.1038/nrg1489
6. Litwin V, Gumperz J, Parham P, Phillips JH, Lanier LL. NKB1: a natural killer cell receptor involved in the recognition of polymorphic HLA-B molecules. *J Exp Med* (1994) 180(2):537–43. doi:10.1084/jem.180.2.537
7. Pende D, Biassoni R, Cantoni C, Verdiani S, Falco M, di Donato C, et al. The natural killer cell receptor specific for HLA-A allotypes: a novel member of the p58/p70 family of inhibitory receptors that is characterized by three immunoglobulin-like domains and is expressed as a 140-kD disulphide-linked dimer. *J Exp Med* (1996) 184(2):505–18. doi:10.1084/jem.184.2.505
8. Delgado DC, Hank JA, Kolesar J, Lorentzen D, Gan J, Seo S, et al. Genotypes of NK cell KIR receptors, their ligands, and Fcgamma receptors in the response of neuroblastoma patients to Hu14.18-IL2 immunotherapy. *Cancer Res* (2010) 70(23):9554–61. doi:10.1158/0008-5472.CAN-10-2211
9. Du J, Lopez-Verges S, Pitcher BN, Johnson J, Jung SH, Zhou L, et al. CALGB 150905 (alliance): rituximab broadens the anti-lymphoma response by activating unlicensed NK cells. *Cancer Immunol Res* (2014) 2(9):878–89. doi:10.1158/2326-6066.CIR-13-0158
10. Moesta AK, Norman PJ, Yawata M, Yawata N, Gleimer M, Parham P. Synergistic polymorphism at two positions distal to the ligand-binding site makes KIR2DL2 a stronger receptor for HLA-C than KIR2DL3. *J Immunol* (2008) 180(6):3969–79. doi:10.4049/jimmunol.180.6.3969
11. Tarek N, Le Luduec JB, Gallagher MM, Zheng J, Venstrom JM, Chamberlain E, et al. Unlicensed NK cells target neuroblastoma following anti-GD2 antibody treatment. *J Clin Invest* (2012) 122(9):3260–70. doi:10.1172/JCI62749
12. Jonsson AH, Yokoyama WM. Natural killer cell tolerance licensing and other mechanisms. *Adv Immunol* (2009) 101:27–79. doi:10.1016/S0065-2776(08)01002-X
13. Kim S, Poursine-Laurent J, Truscott SM, Lybarger L, Song YJ, Yang L, et al. Licensing of natural killer cells by host major histocompatibility complex class I molecules. *Nature* (2005) 436(7051):709–13. doi:10.1038/nature03847
14. Yokoyama WM, Kim S. Licensing of natural killer cells by self-major histocompatibility complex class I. *Immunol Rev* (2006) 214:143–54. doi:10.1111/j.1600-065X.2006.00458.x
15. Liao NS, Bix M, Zijlstra M, Jaenisch R, Raulet D. MHC class I deficiency: susceptibility to natural killer (NK) cells and impaired NK activity. *Science* (1991) 253(5016):199–202. doi:10.1126/science.1853205
16. Orr MT, Lanier LL. Natural killer cell education and tolerance. *Cell* (2010) 142(6):847–56. doi:10.1016/j.cell.2010.08.031
17. Cheung NK, Cheung IY, Kushner BH, Ostrovnaya I, Chamberlain E, Kramer K, et al. Murine anti-GD2 monoclonal antibody 3F8 combined with granulocyte-macrophage colony-stimulating factor and 13-cis-retinoic acid in high-risk patients with stage 4 neuroblastoma in first remission. *J Clin Oncol* (2012) 30(26):3264–70. doi:10.1200/JCO.2011.41.3807
18. Forlenza CJ, Boudreau JE, Zheng J, Le Luduec JB, Chamberlain E, Heller G, et al. KIR3DL1 allelic polymorphism and HLA-B epitopes modulate response to anti-GD2 monoclonal antibody in patients with neuroblastoma. *J Clin Oncol* (2016) 34(21):2443–51. doi:10.1200/JCO.2015.64.9558
19. Venstrom JM, Zheng J, Noor N, Danis KE, Yeh AW, Cheung IY, et al. KIR and HLA genotypes are associated with disease progression and survival following autologous hematopoietic stem cell transplantation for high-risk neuroblastoma. *Clin Cancer Res* (2009) 15(23):7330–4. doi:10.1158/1078-0432.CCR-09-1720
20. Erbe A, Wang W, Carmichael L, Kim K, Reville P, London W, et al. Impact of KIR/KIR ligand genotype for neuroblastoma patients in a phase 3 COG immunotherapy trial. *AACR Annual Meeting (Abstract)*. Washington, DC: Cancer Research (2017).
21. Reville P, Erbe A, Wang W, Carmichael L, Kim K, London W, et al. KIR/KIR-ligand genotypes influence clinical response to dinutuximab-based immunotherapy in high-risk neuroblastoma patients. *Frontiers in Cancer Immunotherapy Conference (Abstract)*. New York, NY: The New York Academy of Sciences (2017).
22. Erbe A, Wang W, Grzywacz B, Ranheim E, Hank J, Kim K, et al. Rituximab response in follicular lymphoma: contributions from KIR 2DS1 and HLA-C. *Society for Immunotherapy of Cancer Annual Conference (abstract): Journal for ImmunoTherapy of Cancer*. Maryland: National Harbor (2013).
23. Grzywacz B, Erbe A, Wang W, Ranheim E, Hank J, Kim K, et al. Specific KIR and HLA genotypes affect outcomes of single-agent anti-CD20 immunotherapy of follicular lymphoma. *ASH Annual Meeting (Abstract): Blood*. New Orleans, LA (2013).
24. Natarajan K, Dimasi N, Wang J, Mariuzza RA, Margulies DH. Structure and function of natural killer cell receptors: multiple molecular solutions to self, nonself discrimination. *Annu Rev Immunol* (2002) 20:853–85. doi:10.1146/annurev.immunol.20.100301.064812
25. Gumperz JE, Barber LD, Valiante NM, Percival L, Phillips JH, Lanier LL, et al. Conserved and variable residues within the Bw4 motif of HLA-B make separable contributions to recognition by the NKB1 killer cell-inhibitory receptor. *J Immunol* (1997) 158(11):5237–41.
26. O'Connor GM, Vivian JP, Widjaja JM, Bridgeman JS, Gostick E, Lafont BA, et al. Mutational and structural analysis of KIR3DL1 reveals a lineage-defining allotypic dimorphism that impacts both HLA and peptide sensitivity. *J Immunol* (2014) 192(6):2875–84. doi:10.4049/jimmunol.1303142
27. Sanjanwala B, Draghi M, Norman PJ, Guethlein LA, Parham P. Polymorphic sites away from the Bw4 epitope that affect interaction of Bw4+ HLA-B with KIR3DL1. *J Immunol* (2008) 181(9):6293–300. doi:10.4049/jimmunol.181.9.6293
28. Thananchai H, Gillespie G, Martin MP, Bashirova A, Yawata N, Yawata M, et al. Cutting edge: allele-specific and peptide-dependent interactions between KIR3DL1 and HLA-A and HLA-B. *J Immunol* (2007) 178(1):33–7. doi:10.4049/jimmunol.178.1.33
29. Saunders PM, Pymm P, Pietra G, Hughes VA, Hitchen C, O'Connor GM, et al. Killer cell immunoglobulin-like receptor 3DL1 polymorphism defines distinct hierarchies of HLA class I recognition. *J Exp Med* (2016) 213(5):791–807. doi:10.1084/jem.20152023
30. Alves LG, Rajalingam R, Canavez F. A novel real-time PCR method for KIR genotyping. *Tissue Antigens* (2009) 73(2):188–91. doi:10.1111/j.1399-0039.2008.01184.x
31. Vilches C, Castano J, Gomez-Lozano N, Estefania E. Facilitation of KIR genotyping by a PCR-SSP method that amplifies short DNA fragments. *Tissue Antigens* (2007) 70(5):415–22. doi:10.1111/j.1399-0039.2007.00923.x
32. Sondel PM, Hank JA. Antibody-directed, effector cell-mediated tumor destruction. *Hematol Oncol Clin North Am* (2001) 15(4):703–21. doi:10.1016/S0889-8588(05)70243-4
33. Weiskopf K, Weissman IL. Macrophages are critical effectors of antibody therapies for cancer. *MAbs* (2015) 7(2):303–10. doi:10.1080/19420862.2015.1011450
34. Wang W, Erbe AK, Hank JA, Morris ZS, Sondel PM. NK cell-mediated antibody-dependent cellular cytotoxicity in cancer immunotherapy. *Front Immunol* (2015) 6:368. doi:10.3389/fimmu.2015.00368
35. Vivier E, Anfossi N. Inhibitory NK-cell receptors on T cells: witness of the past, actors of the future. *Nat Rev Immunol* (2004) 4(3):190–8. doi:10.1038/nri1306
36. Patterson S, Chaidos A, Neville DC, Poggi A, Butters TD, Roberts IA, et al. Human invariant NKT cells display alloreactivity instructed by invariant TCR-CD1d interaction and killer Ig receptors. *J Immunol* (2008) 181(5):3268–76. doi:10.4049/jimmunol.181.5.3268
37. Mellor JD, Brown MP, Irving HR, Zalcberg JR, Dobrovic A. A critical review of the role of Fc gamma receptor polymorphisms in the response to monoclonal antibodies in cancer. *J Hematol Oncol* (2013) 6:1. doi:10.1186/1756-8722-6-1
38. Erbe AK, Wang W, Goldberg J, Gallenberger M, Kim K, Carmichael L, et al. FCGR polymorphisms influence response to IL2 in metastatic renal cell carcinoma. *Clin Cancer Res* (2017) 23(9):2159–68. doi:10.1158/1078-0432.CCR-16-1874
39. Wang W, Erbe AK, Gallenberger M, Kim K, Carmichael L, Hess D, et al. Killer immunoglobulin-like receptor (KIR) and KIR-ligand genotype do not correlate with clinical outcome of renal cell carcinoma patients receiving high-dose IL2. *Cancer Immunol Immunother* (2016) 65(12):1523–32. doi:10.1007/s00262-016-1904-8

40. Carlotti E, Palumbo GA, Oldani E, Tibullo D, Salmoiraghi S, Rossi A, et al. FcgammaRIIIA and FcgammaRIIA polymorphisms do not predict clinical outcome of follicular non-Hodgkin's lymphoma patients treated with sequential CHOP and rituximab. *Haematologica* (2007) 92(8):1127–30. doi:10.3324/haematol.11288
41. Cartron G, Dacheux L, Salles G, Solal-Celigny P, Bardos P, Colombat P, et al. Therapeutic activity of humanized anti-CD20 monoclonal antibody and polymorphism in IgG Fc receptor FcgammaRIIIa gene. *Blood* (2002) 99(3):754–8. doi:10.1182/blood.V99.3.754
42. Weng WK, Weng WK, Levy R. Immunoglobulin G Fc receptor polymorphisms do not correlate with response to chemotherapy or clinical course in patients with follicular lymphoma. *Leuk Lymphoma* (2009) 50(9):1494–500. doi:10.1080/10428190903128660
43. Kenkre VP, Hong F, Cerhan JR, Lewis M, Sullivan L, Williams ME, et al. Fc gamma receptor 3A and 2A polymorphisms do not predict response to rituximab in follicular lymphoma. *Clin Cancer Res* (2016) 22(4):821–6. doi:10.1158/1078-0432.CCR-15-1848
44. Cheung NK, Sowers R, Vickers AJ, Cheung IY, Kushner BH, Gorlick R. FCGR2A polymorphism is correlated with clinical outcome after immunotherapy of neuroblastoma with anti-GD2 antibody and granulocyte macrophage colony-stimulating factor. *J Clin Oncol* (2006) 24(18):2885–90. doi:10.1200/JCO.2005.04.6011
45. Lode HN, Troschke-Meurer S, Valteau-Couanet D, Garaventa A, Gray J, Castel V, et al. Correlation of killer-cell Ig like receptor (KIR) haplotypes and Fcγ-receptor polymorphisms with survival of high-risk relapsed/refractory neuroblastoma patients treated by long-term infusion of anti-GD2 antibody ch14.18/CHO. *ASCO Annual Meeting (abstract): Journal of Clinical Oncology*. Chicago, IL (2016).
46. Terszowski G, Klein C, Stern M. KIR/HLA interactions negatively affect rituximab- but not GA101 (obinutuzumab)-induced antibody-dependent cellular cytotoxicity. *J Immunol* (2014) 192(12):5618–24. doi:10.4049/jimmunol.1400288
47. Martin MP, Qi Y, Gao X, Yamada E, Martin JN, Pereyra F, et al. Innate partnership of HLA-B and KIR3DL1 subtypes against HIV-1. *Nat Genet* (2007) 39(6):733–40. doi:10.1038/ng2035
48. Cella M, Longo A, Ferrara GB, Strominger JL, Colonna M. NK3-specific natural killer cells are selectively inhibited by Bw4-positive HLA alleles with isoleucine 80. *J Exp Med* (1994) 180(4):1235–42. doi:10.1084/jem.180.4.1235
49. Gumperz JE, Litwin V, Phillips JH, Lanier LL, Parham P. The Bw4 public epitope of HLA-B molecules confers reactivity with natural killer cell clones that express NKB1, a putative HLA receptor. *J Exp Med* (1995) 181(3):1133–44. doi:10.1084/jem.181.3.1133
50. Saunders PM, Vivian JP, Baschuk N, Beddoe T, Widjaja J, O'Connor GM, et al. The interaction of KIR3DL1*001 with HLA class I molecules is dependent upon molecular microarchitecture within the Bw4 epitope. *J Immunol* (2015) 194(2):781–9. doi:10.4049/jimmunol.1402542
51. Vivian JP, Duncan RC, Berry R, O'Connor GM, Reid HH, Beddoe T, et al. Killer cell immunoglobulin-like receptor 3DL1-mediated recognition of human leukocyte antigen B. *Nature* (2011) 479(7373):401–5. doi:10.1038/nature10517
52. Luque I, Solana R, Galiani MD, Gonzalez R, Garcia F, Lopez de Castro JA, et al. Threonine 80 on HLA-B27 confers protection against lysis by a group of natural killer clones. *Eur J Immunol* (1996) 26(8):1974–7. doi:10.1002/eji.1830260845
53. Norman PJ, Abi-Rached L, Gendzekhadze K, Hammond JA, Moesta AK, Sharma D, et al. Meiotic recombination generates rich diversity in NK cell receptor genes, alleles, and haplotypes. *Genome Res* (2009) 19(5):757–69. doi:10.1101/gr.085738.108
54. Rojo S, Wagtmann N, Long EO. Binding of a soluble p70 killer cell inhibitory receptor to HLA-B*5101: requirement for all three p70 immunoglobulin domains. *Eur J Immunol* (1997) 27(2):568–71. doi:10.1002/eji.1830270231
55. Foley BA, De Santis D, Van Beelen E, Lathbury LJ, Christiansen FT, Witt CS. The reactivity of Bw4+ HLA-B and HLA-A alleles with KIR3DL1: implications for patient and donor suitability for haploidentical stem cell transplantations. *Blood* (2008) 112(2):435–43. doi:10.1182/blood-2008-01-132902
56. Carr WH, Pando MJ, Parham P. KIR3DL1 polymorphisms that affect NK cell inhibition by HLA-Bw4 ligand. *J Immunol* (2005) 175(8):5222–9. doi:10.4049/jimmunol.175.8.5222
57. Yawata M, Yawata N, Draghi M, Little AM, Partheniou F, Parham P. Roles for HLA and KIR polymorphisms in natural killer cell repertoire selection and modulation of effector function. *J Exp Med* (2006) 203(3):633–45. doi:10.1084/jem.20051884
58. Hughes AL. Natural selection and the diversification of vertebrate immune effectors. *Immunol Rev* (2002) 190:161–8. doi:10.1034/j.1600-065X.2002.19012.x
59. Selvakumar A, Steffens U, Dupont B. Polymorphism and domain variability of human killer cell inhibitory receptors. *Immunol Rev* (1997) 155:183–96. doi:10.1111/j.1600-065X.1997.tb00951.x
60. Shilling HG, Guethlein LA, Cheng NW, Gardiner CM, Rodriguez R, Tyan D, et al. Allelic polymorphism synergizes with variable gene content to individualize human KIR genotype. *J Immunol* (2002) 168(5):2307–15. doi:10.4049/jimmunol.168.5.2307
61. Robinson J, Halliwell JA, Hayhurst JD, Flicek P, Parham P, Marsh SG. The IPD and IMGT/HLA database: allele variant databases. *Nucleic Acids Res* (2015) 43(Database issue):D423–31. doi:10.1093/nar/gku1161
62. Morvan M, Willem C, Gagne K, Kerdudou N, David G, Sebille V, et al. Phenotypic and functional analyses of KIR3DL1+ and KIR3DS1+ NK cell subsets demonstrate differential regulation by Bw4 molecules and induced KIR3DS1 expression on stimulated NK cells. *J Immunol* (2009) 182(11): 6727–35. doi:10.4049/jimmunol.0900212
63. Jiang Y, Chen O, Cui C, Zhao B, Han X, Zhang Z, et al. KIR3DS1/L1 and HLA-Bw4-80I are associated with HIV disease progression among HIV typical progressors and long-term nonprogressors. *BMC Infect Dis* (2013) 13:405. doi:10.1186/1471-2334-13-405
64. Martin MP, Gao X, Lee JH, Nelson GW, Detels R, Goedert JJ, et al. Epistatic interaction between KIR3DS1 and HLA-B delays the progression to AIDS. *Nat Genet* (2002) 31(4):429–34. doi:10.1038/ng934
65. Mulrooney TJ, Zhang AC, Goldgur Y, Boudreau JE, Hsu KC. KIR3DS1-specific D0 domain polymorphisms disrupt KIR3DL1 surface expression and HLA binding. *J Immunol* (2015) 195(3):1242–50. doi:10.4049/jimmunol.1500243

6

The Transcription Factor ZNF683/HOBIT Regulates Human NK-Cell Development

Mirte Post[1‡], Angelica Cuapio[1‡], Markus Osl[1], Dorit Lehmann[1], Ulrike Resch[1], David M. Davies[2†], Martin Bilban[3], Bernhard Schlechta[4], Wolfgang Eppel[4], Amit Nathwani[2], Dagmar Stoiber[5,6], Jan Spanholtz[7], Emilio Casanova[5,8] and Erhard Hofer[1]*

[1] Institute of Vascular Biology and Thrombosis Research, Center of Physiology and Pharmacology, Medical University of Vienna, Vienna, Austria, [2] Department of Oncology, University College London Cancer Institute, London, UK, [3] Department of Laboratory Medicine, Medical University of Vienna, Vienna, Austria, [4] Department of Obstetrics and Gynecology, Medical University of Vienna, Vienna, Austria, [5] Ludwig Boltzmann Institute of Cancer Research, Vienna, Austria, [6] Institute of Pharmacology, Center of Physiology and Pharmacology, Medical University of Vienna, Vienna, Austria, [7] Glycostem Therapeutics, Oss, Netherlands, [8] Institute of Physiology, Center of Physiology and Pharmacology, Comprehensive Cancer Center, Medical University of Vienna, Vienna, Austria

Edited by:
Gianfranco Pittari,
Hamad Medical Corporation, Qatar

Reviewed by:
Evren Alici,
Karolinska Institutet, Sweden
Ralf Dressel,
Universitätsmedizin Göttingen, Germany

***Correspondence:**
Erhard Hofer
erhard.hofer@meduniwien.ac.at,
erhard.hofer@gmx.at

We identified *ZNF683/HOBIT* as the most highly upregulated transcription factor gene during *ex vivo* differentiation of human CD34⁺ cord blood progenitor cells to CD56⁺ natural killer (NK) cells. ZNF683/HOBIT mRNA was preferentially expressed in NK cells compared to other human peripheral blood lymphocytes and monocytes. During *ex vivo* differentiation, ZNF683/HOBIT mRNA started to increase shortly after addition of IL-15 and further accumulated in parallel to the generation of CD56⁺ NK cells. shRNA-mediated knockdown of ZNF683/HOBIT resulted in a substantial reduction of CD56⁻CD14⁻ NK-cell progenitors and the following generation of CD56⁺ NK cells was largely abrogated. The few CD56⁺ NK cells, which escaped the developmental inhibition in the ZNF683/HOBIT knockdown cultures, displayed normal levels of NKG2A and KIR receptors. Functional analyses of these cells showed no differences in degranulation capacity from control cultures. However, the proportion of IFN-γ-producing cells appeared to be increased upon ZNF683/HOBIT knockdown. These results indicate a key role of ZNF683/HOBIT for the differentiation of the human NK-cell lineage and further suggest a potential negative control on IFN-γ production in more mature human NK cells.

Keywords: ZNF683/HOBIT, natural killer cells, CD56, *ex vivo* differentiation, NK-cell development

INTRODUCTION

Natural killer (NK) cells are the third largest group of lymphocytes in peripheral blood and an important component of the first line of immune defense. They act against a wide spectrum of virally infected and neoplastic cells by direct killing of these cells or production of cytokines, such as IFN-γ. As components of the innate part of the immune system, they display immediate reactivity and do not require prior sensitization (1, 2). This traditional characterization of NK cells has been expanded over the recent years as they have been described to be able to incorporate features previously thought to be restricted to the adaptive immune system, such as interaction with dendritic cells and immunological memory (3–5). In contrast to the adaptive T and B lymphocytes,

NK cells lack somatically recombined and clonally distributed antigen receptors, and their activity is controlled by a varied repertoire of germline-encoded inhibitory and activating receptors (6). Recently, additional tissue-resident subsets of innate lymphoid cells (ILCs), distinct from NK cells, became apparent and significantly expanded the complexity of innate lymphoid lineages. Whereas conventional NK cells resemble cytotoxic T lymphocytes in many aspects, ILCs 1–3 rather mirror T helper-like cells (7).

Based on currently available data, the relationship between the different innate and adaptive lymphoid lineages is reflected by their initial common differentiation from the hematopoietic stem cell (HSC) and in similarities of their transcriptional networks. According to the current hypothesis, the HSC develops through a multipotent progenitor to a common lymphoid progenitor (CLP) (8). This CLP can further differentiate into adaptive lymphocytes under the control of E-proteins, whereas the development of innate lymphocytes requires antagonism of E-proteins and likely proceeds through a common innate lymphoid progenitor (9, 10). The following pre-NK progenitor (NKP) stage restricts the differentiating cells to the NK lineage and develops into NKP and subsequently into immature NK (iNK) cells. The final step involves maturation from the iNK cells to mature NK (mNK) cells, both stages expressing the NK marker CD56 (11, 12).

Substantial data have been obtained to identify key transcription factors essential for the differentiation of adaptive and innate lymphocytes. A common theme appears to be the mutual inhibition of factors determining different lineages. For example, EBF strongly inhibits ID2 expression, thereby allowing E2A to function during B-cell development. In addition, EBF and PAX5 support B-cell differentiation by repressing additional critical regulators of T-cell and ILC differentiation, such as NOTCH1, GATA-3, and TCF-1 (8, 9). Conversely, all ILCs including NK cells are dependent for their differentiation on ID2 that heterodimerizes with E proteins and neutralizes their activity (10). Subsequently, a complex network of transcription factors guides the cells through the distinct steps of NK-cell differentiation (13). The most important transcription factors for the early stages of murine NK-cell development include STAT5, two ETS family members (PU.1 and ETS-1), and NFIL3 (also known as E4BP4) (14–17). The maturation stage from iNK to mNK cells and NK cells' function are coordinated by BLIMP-1, T-BET, EOMES, and MEF among others (18–20). Compared to the data obtained from the murine system, experimental evidence on transcription factors governing human NK-cell development is far less available. This has been partially caused by the lack of an easy and robust system to mimic human NK-cell differentiation from HSC *ex vivo*. Of the few transcription factors described so far, TOX1 and 2 are important in the early stages, T-BET and BLIMP-1 both play critical roles in the later phases and effector functions of human NK cells (21–24). From the currently available data, it appears that the precise function and sequential order of transcription factors directing NK-cell development may to some extent differ between mice and humans.

Due to their important role in immunosurveillance, NK cells and their modulation are currently being explored as a therapeutic approach in a wide variety of cancers, autoimmune diseases, allergies, and transplantation (1, 25). These attempts have led to the development of methods for the specific modulation of endogenous NK cells as well as for *ex vivo* amplification of NK cells from patients or allogeneic donors for NK cell-infusion therapies. The obtained results show clear benefits of NK cell-based therapies, in comparison to T lymphocyte-based, including a good tolerance of allogeneic NK cells by the patients and the lack of a graft-versus-host disease (1, 25). Different strategies are exploited to obtain a sufficient number of NK cells for infusion therapies, including cytokine- and/or feeder cell-mediated expansion of peripheral blood NK cells as well as *ex vivo* differentiation from cord or peripheral blood-derived HSC (26, 27). In this regard, we have previously analyzed a feeder cell-free *ex vivo* system to generate large-scale therapeutic NK cells from cord blood stem cells that faithfully reproduces different steps of human NK-cell differentiation (28). This system is, therefore, ideally suited to investigate human NK-cell differentiation in addition to being a reliable method to generate NK cells for therapy (29, 30).

Here, we used this *ex vivo* differentiation system for a complete transcriptomic profiling of cells in different stages of NK-cell development. We identified the 20 most differentially expressed transcription factor genes and confirmed ZNF683/HOBIT mRNA as the highest upregulated transcription factor mRNA. The analysis of different human peripheral blood cell types showed preferential expression of ZNF683/HOBIT mRNA in NK cells. Then, we analyzed the role of ZNF683/HOBIT during NK-cell differentiation in more detail. In the *ex vivo* differentiation system, shRNA-mediated knockdown of ZNF683/HOBIT significantly reduced $CD56^{-}CD14^{-}$ NKPs up to day 21 of culture and nearly abrogated the following generation of $CD56^{+}$ NK cells. The few $CD56^{+}$ cells that continued to mature displayed normal levels of NKG2A and KIR as well as degranulation capacities similar to control cells. However, the number of IFN-γ-producing cells significantly increased upon ZNF683/HOBIT knockdown. Taken together, these results support that ZNF683/HOBIT is a key regulator of early stages of human NK-cell differentiation and, in later stages, may function to repress IFN-γ production.

MATERIALS AND METHODS

Ex Vivo Differentiation of $CD34^{+}$ Stem Cells into NK Cells

Human umbilical cord blood samples were obtained at birth after full-term delivery from the Department of Obstetrics and Gynecology of the University Hospital of Vienna, Austria. Cord blood mononuclear cells (CBMCs) were collected by Ficoll density gradient centrifugation (Lympholyte Cell Separation Media, human, Cedarlane®, Burlington, ON, Canada). Stem cells were isolated from CBMCs according to manufacturer's protocol, using a magnetic bead-based $CD34^{+}$ isolation kit (CD34 MicroBead Kit, human; Miltenyi Biotec, Bergisch Gladbach, Germany). The purity of the stem cells was evaluated by flow cytometry ($CD34^{+}$ reached at least 95% purity) after which the cells were cultured as previously described (28, 30). In short, stem cells at an initial density of 10^{5} cells/ml were seeded into 6-well plates (Corning Incorporated, Corning, NY, USA) for 10 days in basal expansion medium

(GBGM©; Glycostem Therapeutics, Oss, The Netherlands) supplemented with stem cell factor (SCF), interleukin-7 (IL-7), thrombopoietin (TPO), and FMS-like tyrosine kinase 3 ligand (Flt3L); all factors at a concentration of 25 ng/ml (CellGro®, CellGenix GmbH, Freiburg, Germany) and granulocyte colony-stimulating factor (G-CSF, 250 pg/ml; Stemcell Technologies, Vancouver, BC, Canada), granulocyte-macrophage colony-stimulating factor (GM-CSF, 10 pg/ml; Stemcell Technologies), and IL-6 (50 pg/ml; CellGenix GmbH). At day 10, TPO was replaced with IL-15 (20 ng/ml; CellGro®, CellGenix GmbH), and at day 14, Flt3L with IL-2 (1,000 U/mL; Chiron Corporation, Emeryville, CA, USA) by refreshing half of the medium. After a total culture period of 35 days, a regular culture consisted of >95% $CD56^+CD3^-$ NK cells as evaluated by flow cytometry.

Microarray

Total RNA was extracted from cells at different time points of the *ex vivo* culture and transcribed into cDNA using the GeneChip® Whole Transcript Sense Target Labeling Kit (Affymetrix, High Wycombe, UK). The labeled cDNA was hybridized to GeneChip Human Gene 1.0ST Arrays, and the arrays were scanned and analyzed according to protocols of the manufacturer[1] as described in Ref. (31). Robust multiarray average signal extraction and normalization were performed, as detailed at the Bioconductor website[2] (32). The microarray data have been submitted to the GEO database under the accession number GSE95018.

RNA Sampling, cDNA Synthesis, and Real-time RT-PCR

Cultured cells (5×10^5) were lysed in Trizol (QIAzol Lysis Reagent, Qiagen Biosciences, MD, USA) and stored at −80°C. Total RNA was extracted following ThermoFishers manual[3], and 1 µg RNA was used for cDNA synthesis according to manufacturer's protocol (RevertAid H Minus First Strand cDNA Synthesis Kit, ThermoFisher Scientific, MA, USA). Two hundred nanograms of transcribed cDNA were analyzed by real-time PCR using the KAPA SYBR FAST UNIVERSAL kit (Kapa Biosystems, Inc., Wilmington, MA, USA) and primer sequences as shown in Table S3 in Supplementary Material. As internal controls, primers for either hypoxanthine-guanine phosphoribosyltransferase or β-actin were used. Samples were measured and analyzed with QIAGEN's real-time PCR cycler and corresponding software (Rotor-Gene Q, Qiagen, Hilden, Germany).

Western Blot

For protein-expression analysis, cell pellets (5×10^5 cells/sample) were resuspended in 2× Laemmli buffer, boiled for 10 min at 95°C and subsequently separated by SDS-polyacrylamide gel electrophoresis (10% Bis-Tris gels, Acrylamide:Bis 37.5:1) and transferred to a nitrocellulose membrane (Amersham Protran Supported 0.45 µm NC, GE Healthcare Europe GmbH, Eindhoven, The Netherlands) by semi-dry electrophoretic blotting in Towbin buffer with 20% methanol (PerfectBlue, "Semi-Dry" Electroblotter Sedec, Peqlab, Southampton, UK). The membrane was blocked with 5% non-fat dry milk in TBS-T (0.2% Tween in TBS) followed by overnight incubation at 4°C with primary anti-ZNF683 antibodies (goat-anti-ZNF683 antibody C-12, 1:500, Santa Cruz Biotechnology, Heidelberg, Germany) or (mouse-anti-ZNF683 antibody, 1:500, Sigma-Aldrich, Saint Louis, MO, USA) and as control anti-GAPDH (mouse-anti-GAPDH, 1:10,000, Merck KGaA, Darmstadt, Germany). This was followed by 2 h incubation at room temperature (RT) with secondary antibodies, all 1:5,000 (donkey-anti-goat, Santa Cruz Biotechnology or goat-anti-mouse, Thermo Fisher Scientific) and corresponding washing steps with TBS-T.

Lentiviral Vectors

To generate a lentiviral construct for knockdown studies, a sense oligo, 5′-TGGAAACACATGGGCTATGACATTTCAAGAGAATGTCATAGCCCATGTGTTTCTTTTTTC-3′, corresponding with the sequence "GAAACACATGGGCTATGACAT" to position 1,402 to 1,412 from the start ATG of *ZNF683/HOBIT* cDNA (NCBI Reference Sequence: NM_001114759.2) and a corresponding antisense oligo, 5′-TCGAGAAAAAAGAAACACATGGGCTATGACATTCTCTTGAAATGTCATAGCCCATGTGTTTCCA-3′ were synthesized (Integrated DNA Technologies Inc., Coralville, IA, USA) as complementary overlapping oligos, with a *XhoI* overhang at the 5′end of the antisense oligo, as detailed.[4] The complementary oligos were annealed and subsequently ligated into the LeGO-G/BSD lentiviral vector (LeGO-G/BSD was a gift from Boris Fehse, Addgene plasmid #27354) that had been digested with *Hpa*I and *Xho*I (New England Biolabs, Ipswich, MA, USA). The ligated product was transformed into the *E. coli* strain Stbl3 (New England Biolabs, Ipswich, MA, USA), recombinant colonies were selected, and plasmid-DNA isolated and verified by restriction digestion and Sanger sequencing. As a control, oligos containing a scrambled shRNA were synthesized and cloned into the LeGO-G/BSD vector using an identical strategy. The control shRNA sequence was obtained from the "Open Biosystems pGIPZ shRNAmir library" situated at University College London.[5] To confirm the functioning of the shRNA, shRNA expression plasmids were co-transfected into HEK293 cells in the absence or presence of a commercially available ZNF683/HOBIT expression plasmid (ORF of ZNF683, transcript variant 1, in pEnter, with C-terminal Flag and His tag; Vigene Biosciences, Rockville, MD, USA) employing the $CaPO_4$ method using 4 µg of total DNA/6-well. Six hours after, transfection medium was changed, and cells were harvested 96 h after transfection by either direct lysis in Trizol (Qiagen) for RNA isolation or 1× Laemmli sample buffer for protein-expression analysis.

Virus Production

Virus production and all experiments involving the generated recombinant viruses were performed in a separate room under biosafety level (BSL)-2 conditions following the instructions

[1] Affymetrix support site; www.affymetrix.com/support/index.affx.
[2] www.bioconductor.org/.
[3] https://www.thermofisher.com/order/catalog/product/15596026?ICID=cvc-rna-cultured-cells-c1t1.
[4] http://www.sciencegateway.org/protocols/lentivirus/cloning.htm.
[5] https://www.ucl.ac.uk/cancer/research/scientific-facilities-and-services/cancer-genomics-engineering-facility/rnai-library/open-biosystems-pgipz-rna-library.

given by the corresponding directives of the European Union (Council Directive 90/679/EEC) and the Austrian government (GTG-BGBl. Nr. 510/1994 and 114/2012). The specific work with the replication-defective human lentiviruses was registered at the Austrian Ministry of Science, Research and Economy (BMWFW-5.011/009-WF/V/3b/2015).

To produce lentiviruses, HEK293T cells (ATCC # CRL-11268) were seeded at 10^7 cells per 15 cm^2 tissue culture plate in 15 ml DMEM (DMEM/high glucose, GE Healthcare) supplemented with 10% FBS and glutamine (200 mM, Sigma-Aldrich). After 24 h, 5 ml DMEM was added, and the 60–70% dense cultures were transfected with the LeGO-G/BSD-*ZNF683/HOBIT*-shRNA expression plasmid plus three complementing plasmids providing the essential *Gag*, *Pol*, *Rev*, and *Tat* gene products missing in the replication-defective virus. To this end, the following 2 transfection solutions were prepared: (1) DMEM with 10% polyethyleneimine (PEI, Polysciences, Warrington, PA, USA) (1.5 ml/15 cm^2 plate) and (2) DMEM with 2.3 µg pCAG-KGP3R, 1.2 µg pCAG-RTR2, 2.5 µg pCAG-VSVg, and 24 µg vector of interest (in total 30 µg plasmids in 1.55 ml/15 cm^2 plate). After 15 min incubation at RT, solution 2 was sterile filtered and added to solution 1. The combined solutions were added dropwise to HEK293T cells at a density of 70%. After 12 h, the transfection mix was replaced with fresh medium (12 ml DMEM/15 cm^2 plate). The supernatants were harvested after 48 and 72 h post-transfection. To remove cell debris, supernatants were centrifuged (1,500 rpm, 5 min, 4°C), filtered (Puradisc FP30mm cellulose acetate syringe filter, 0.45 µm, sterile, GE Healthcare Life Sciences), and finally the virus particles were concentrated by ultracentrifugation for 90 min at 4°C at 25,000 rpm/76,000 × *g* in a SW-32 rotor equipped with 32 ml open-top thickwall polycarbonate tubes in a XL-70 ultracentrifuge (Beckman Coulter, Mystic, CT, USA). After ultracentrifugation, the supernatant was discarded leaving a 100 µl drop at the bottom of the tube. The virus particles were gently resuspended after keeping the tubes on ice for 1 h, aliquoted, and stored at −80°C until use.

Virus Transduction of $CD34^+$ Stem Cells

After culturing freshly isolated cord blood $CD34^+$ stem cells for 5 days in basal expansion medium supplemented with SCF, IL-7, TPO, and Flt3L as described under "*Ex vivo differentiation of $CD34^+$ stem cells into NK cells*," the expanded cells were transduced with lentiviral particles. To this end, 24-well plates (multi-well plate for suspension culture, Greiner Bio-one GmbH, Frickenhausen, Germany) were coated with 8 µg retronectin/500 µl/well (recombinant human fibronectin fragment, Takara Bio Inc., Shiga, Japan) 24 h prior virus transduction and stored at 4°C until use. The retronectin solution was removed, 250 µl GBGM containing virus particles (40 Transduction Units/cell) were added per well, and the plate was centrifuged at 4,000 rpm/1,900 × *g* for 2 h at 4°C. Immediately after centrifugation, the 5 days cultivated $CD34^+$ stem cells were added (5×10^4 cells in 250 µl basal expansion medium/well). The following day, 500 µl of basal expansion medium was added. Two days after transduction the cells were centrifuged (1,500 rpm for 5 min), suspended in fresh basal medium, and transferred to new uncoated 24-well cell culture plates (Corning® Costar®, Sigma-Aldrich). Transduction efficiency, determined as GFP^+ cells, was measured by flow cytometry 3 days after transduction, and cultures were continued as described in the first section (for *ex vivo* differentiation). Further cell expansion was calculated as follows. Cell numbers were obtained at the various differentiation stages and divided by the cell numbers for the corresponding GFP^+ and GFP^- fractions measured 3 days after transduction, to correct for different transduction efficiencies.

Flow Cytometry

Cell surface expression of NK and monocytic markers was monitored using anti-CD56-PeCy7, anti-CD14-PerCPCy5.5 (all BD Biosciences, San Jose, CA, USA), and anti-KIR-PE antibodies (R&D Systems, Vienna, Austria) on a FACS Canto II (BD Biosciences) and data were analyzed using both FACS DIVA software v6.0 (BD Biosciences) and Flowjo v10.0.8 (Tree Star, Yorba Linda, CA, USA). Details on antibodies are given in Table S4 in Supplementary Material.

Cytotoxicity and IFN-γ Assay

Target K562 and effector NK cells were cocultured at a 1:1 ratio (15×10^4 cells of each cell type) in 200 µl RPMI medium (RPMI 1640 medium, Life Technologies, Carlsbad, CA, USA) in an U-bottom 96-well plate (Greiner Cellstar® 96-well plates, Sigma-Aldrich) in the presence of anti-CD107a-APC (BD Biosciences). Brefeldin A and Monensin (BD Golgiplug and Golgistop, BD Biosciences) were added after 1 h of culture. After an additional 5 h of culture, cells were collected, stained for surface CD56 (anti-CD56-PeCy7, BD Biosciences), subsequently prepared for intracellular staining with IFN-γ (anti-IFN-γ-PE, BD Biosciences) using a fixation/permeabilization solution kit (BD Biosciences), and finally measured on a FACS Canto II.

Statistical Analysis

Statistical analysis was performed with Prism 6 software (GraphPad, San Diego, CA, USA) using Student's *t*-test or a two-way ANOVA as indicated in Figure Legends. A *p*-value of 0.05 was considered as statistically significant.

RESULTS

ZNF683/HOBIT Is the Most Highly Upregulated Transcription Factor Gene During *Ex Vivo* Differentiation of Human NK Cells

Initially, we were interested to identify novel transcription factors potentially contributing to human NK-cell differentiation that have not been described in this function before. For this purpose, we employed a recently developed *ex vivo* differentiation system (29). In this system, cord blood $CD34^+$ stem cells are initially expanded for 10 days, prior to addition of IL-15. Following further addition of IL-2 from day 14, the differentiating cells are cultured for a total period of 35–42 days, by which time a regular culture comprises over 95% NK cells. To determine the repertoire of transcription factors differentially expressed during NK-cell development, we first performed a transcriptomic profiling study comparing samples from different time points up to day 35 with cultures at day 10 (start of the NK-cell differentiation). The 20 most highly upregulated transcription factor genes detected are shown in Table S1 in Supplementary Material. When real-time RT-PCR was performed for the corresponding transcripts, ZNF683/HOBIT mRNA was found to be by far the most highly upregulated mRNA at day 35 in relative terms (about 9,000-fold; **Figure 1A**). Among the most differentially expressed genes are many that have previously been reported to be important in NK-cell development and/or maturation such as *GATA3* (33), TOX (34), *ID2* (10), and *ETS1* (16) (**Figure 1A**). Based on its near absence in the stem cell cultures before addition of IL-15 and its high upregulation during differentiation, we decided to focus our further studies on ZNF683/HOBIT, especially as the role of this transcription factor in NK-cell development has not previously been reported.

Human Peripheral Blood NK Cells Preferentially Express ZNF683/HOBIT mRNA

To determine if *ZNF683/HOBIT* is also highly expressed in adult peripheral blood NK cells, we performed real-time RT-PCR on different peripheral blood leukocyte subsets of multiple donors. Indeed, compared to T and B lymphocytes as well as monocytes, ZNF683/HOBIT mRNA is predominantly expressed in NK cells (**Figure 1B**). CD19^{+} B lymphocytes display lower levels of ZNF683/HOBIT mRNA (about 7-fold less as compared to CD56^{dim} NK cells) and the levels in T lymphocytes and monocytes are even much lower or at detection limit. We were furthermore interested to compare the ZNF683/HOBIT mRNA levels between the more immature CD56^{bright} and the rather mature CD56^{dim} subpopulations of NK cells. The data show that CD56^{bright} cells express the highest levels of ZNF683/HOBIT mRNA, whereas CD56^{dim} cells display about 20% lower levels (**Figure 1B**).

ZNF683/HOBIT mRNA Is Upregulated in Parallel to the Generation of NK Cells

To define the potential role(s) of ZNF683/HOBIT throughout NK-cell development, we evaluated the kinetics of accumulation of ZNF683/HOBIT mRNA during *ex vivo* differentiation. During this process, three different cell types can be distinguished based upon the monocytic and NK-cell markers, CD14 and CD56, respectively (28). One of the subpopulations arising from the amplifying CD34^{+} stem cells consists of monocytic CD14^{+} cells that accumulate until day 14 after which they gradually disappear (**Figure 2A**). The second population consists of CD56^{-}CD14^{-} cells, which peak at day 18. These CD56^{-}CD14^{-} cells are presumed to comprise at least in part the NK-cell progenitors as indicated by their rapid decrease from day 18 onward, along with the concomitant rapid generation of CD56^{+} NK cells (**Figure 2A**). The CD56^{+} NK cells become the predominant population after day 25 and from day 35 onward the culture consists of over 95% CD56^{+} NK cells (**Figure 2A**).

We obtained samples at different time points during the *ex vivo* cultures and separated CD14^{-} and CD14^{+} cells by flow

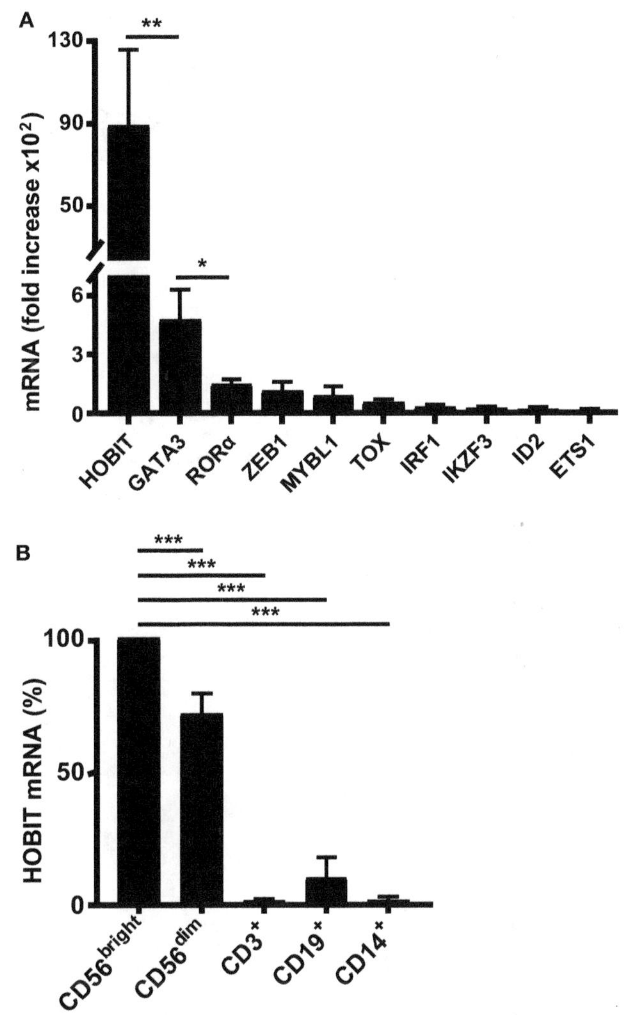

FIGURE 1 | High upregulation of ZNF683/HOBIT mRNA during *ex vivo* differentiation of human natural killer (NK) cells correlates with preferential expression in peripheral blood NK cells. (A) Ten most highly upregulated transcription factor mRNAs during *ex vivo* NK cell differentiation: CD34^{+} stem cells from cord blood were expanded and *ex vivo* differentiated into NK cells. Cells were sampled at day 10, just before the differentiation into NK cells was initiated by the addition of IL-15, and at day 35, after 25 days of differentiation. RNA was isolated and subjected to real-time RT-PCR analysis. β-actin was used as internal control. Fold upregulation of specific mRNAs at day 35 compared to day 10 is shown. Results were calculated from three series of experiments performed in triplicates with cells from different donors and are displayed as mean ± SEM. **(B)** High expression of ZNF683/HOBIT mRNA in peripheral NK cells: mononuclear cells were isolated from human peripheral blood and one half of the cells used for isolation of NK cells by negative magnetic sorting. The NK cell fraction was further separated by flow cytometry into CD56^{bright} and CD56^{dim} NK cells. The second half of the mononuclear fraction was used to isolate CD3^{+} T lymphocytes, CD19^{+} B lymphocytes, and CD14^{+} monocytes by flow cytometry. RNA was isolated from the different cell samples and subjected to real-time RT-PCR analysis using β-actin as internal control. ZNF683/HOBIT mRNA levels within the different cell types are compared to the levels in the CD56^{bright} NK cells set to 100%. Results were obtained from three independent experiments using three different donors and are displayed as mean ± SEM ($^{*}p < 0.05$, $^{**}p < 0.01$, $^{***}p < 0.001$).

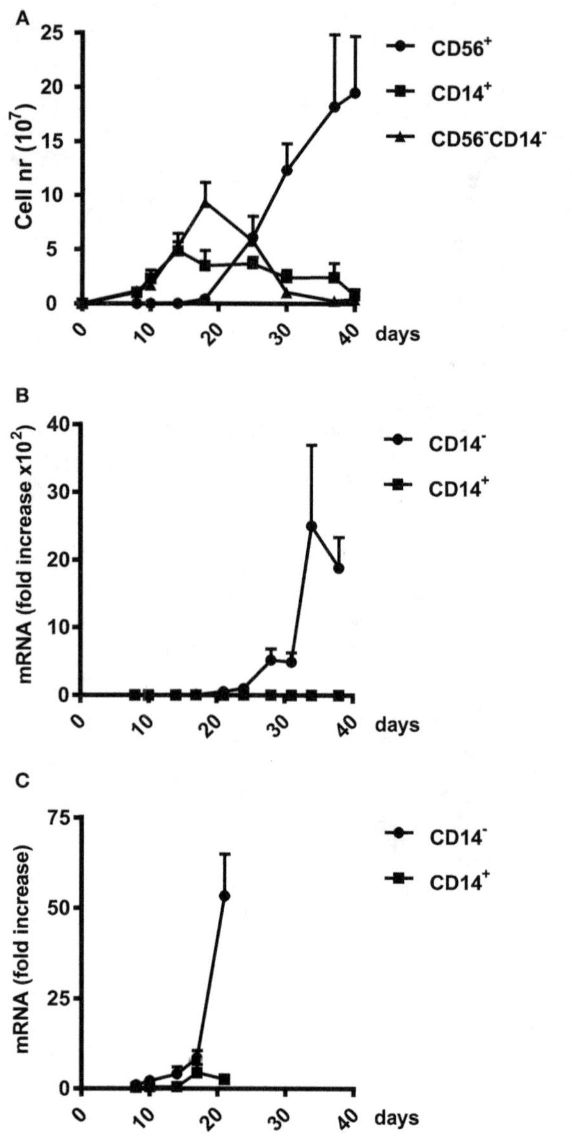

FIGURE 2 | Upregulation of ZNF683/HOBIT mRNA during natural killer (NK) cell differentiation. (A) Amplification of three different cell populations in the *ex vivo* NK-cell differentiation cultures. Cord blood CD34+ cells were differentiated into NK cells over a culture period of 40 days. In regular intervals, cells were analyzed by flow cytometry for expression of the monocytic marker CD14 and the NK cell marker CD56. The numbers of CD56−CD14−, CD56+CD14−, and CD56−CD14+ cells were plotted. Results are calculated from 10 independent experiments using cells of different donors and are displayed as mean ± SEM. **(B,C)** Upregulation of ZNF683/HOBIT mRNA levels. Cell samples were taken at the indicated time points and CD14+ cells separated from the CD14− population using magnetic sorting. RNA was isolated and subjected to real-time RT-PCR analysis with β-actin as internal control. Results are calculated from three independent series of experiments performed in triplicates using different donors. Fold upregulation in comparison to the values obtained for day 8 cells is shown as mean ± SEM **(B)**. To display early ZNF683/HOBIT mRNA upregulation the period until day 21 is shown at a larger scale **(C)**.

cytometry. We detected that ZNF683/HOBIT mRNA is nearly absent after the initial expansion phase of 10 days. A tremendous accumulation of ZNF683/HOBIT mRNA is observed from day 24 onward in CD14− cells comprising NKPs and CD56+ NK cells (**Figure 2B**). The magnitude of ZNF683/HOBIT mRNA accumulation actually blurs the onset of expression in the display of **Figure 2B**. When plotted at a higher magnification, it is apparent that ZNF683/HOBIT mRNA increased in CD14− cells already until day 17 followed by a first exponential accumulation up to day 21 (**Figure 2C**). Although we observed low expression of ZNF683/HOBIT mRNA in the CD14+ monocytes at day 17, the following first accumulation is absent (**Figure 2C**) as well as the exponential increase later during differentiation (**Figure 2B**).

ZNF683/HOBIT Downmodulation Substantially Reduces Expansion of NK Progenitors

After observing that ZNF683/HOBIT mRNA is mainly present in NK cells and their progenitors, we set out to elucidate the potential role(s) of ZNF683/HOBIT during NK-cell development and maturation in more detail. For this purpose, we performed loss-of-function studies using a lentiviral-based shRNA-mediated knockdown of ZNF683/HOBIT. The appropriate functioning of the shRNA was confirmed by shRNA-mediated knockdown of ectopically expressed ZNF683/HOBIT in HEK293T cells. As shown in **Figure 3A**, expression of ZNF683/HOBIT mRNA and protein was strongly reduced (>90%) upon co-expression of the shRNA, thus confirming the proper functioning of the shRNA (**Figure 3A**).

We then continued to investigate the effects of ZNF683/HOBIT knockdown on *ex vivo* differentiation cultures. These cultures were transduced with lentiviruses expressing either ZNF683/HOBIT shRNA (shHOBIT) or a scrambled control shRNA (shControl). Cultures, mock-treated for the transduction procedure but without addition of viruses, were performed in parallel. Based on initial experiments that evaluated best transduction rates and survival in relation to days in culture, cells were transduced during the stem cell expansion phase 5 days after isolation. The presence of a GFP cassette driven by a spleen focus-forming virus (SFFV)-promoter enabled us to discriminate transduced GFP+ cells expressing the shRNA from non-transduced GFP− cells within the same culture in parallel (35). Regularly, transduction rates between 30 and 50% were achieved. Since these varied to some extent, we normalized obtained expansion rates for the analyzed fractions to the different transduction rates measured 3 days after transduction (8 days after isolation).

Based on the fact that the expression of the NK-cell marker CD56 becomes significant only from day 24 onward, we decided to split the analysis in two parts. In the first part, we analyzed the NKP stage until day 21, in the second part the generation and maturation of CD56+ NK cells. At day 21, we observed a fourfold reduced expansion of the shHOBIT GFP+ cells compared to the shControl GFP+ cells (**Figure 3C**). Exemplary dot plots of the GFP staining are shown in **Figure 3B**. The GFP− cells of the shHOBIT culture also seemed to be somewhat reduced in comparison to the scrambled control and regular mock cultures, but the difference did not reach statistical significance.

Since at day 21 the culture comprises mainly two different cell types, CD56−CD14− progenitor cells and CD14+ monocytic cells (**Figure 2A**), we continued to analyze these populations separately. Exemplary dot plots of this analysis are displayed in **Figure 4A**. The data showed that the effect can be mainly traced to a fourfold reduced expansion of the CD56$^-$CD14$^-$ progenitors in the shHOBIT GFP$^+$ fraction. Although smaller inhibitory effects were also visible for CD14$^+$ cells of the shHOBIT GFP$^+$ pool and the GFP$^-$ cells of the shHOBIT culture, both did not reach statistical significance (**Figure 4B**). However, this may

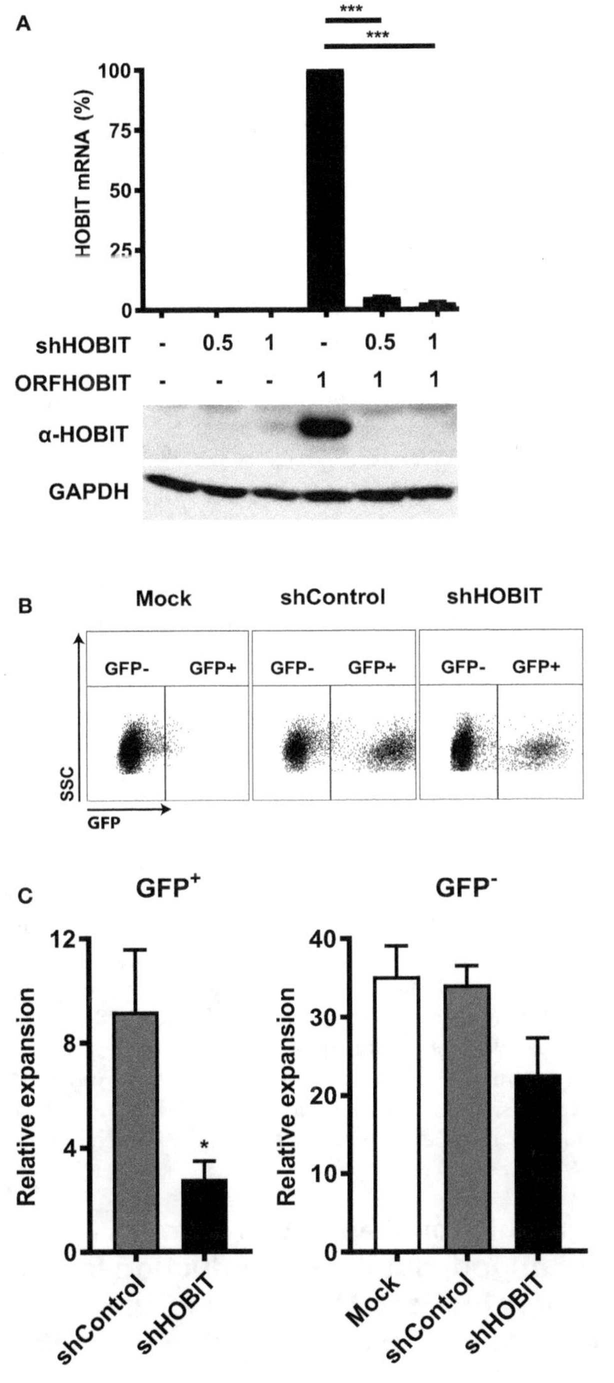

FIGURE 3 | Continued

FIGURE 3 | Continued
Effects of transduction with lentiviruses expressing HOBIT shRNA on total cell expansion. (A) HOBIT shRNA strongly reduces HOBIT expression. Lentiviral vectors expressing HOBIT shRNA were co-transfected with a HOBIT expression construct into HEK293T cells. After 96 h, cells were either used for RNA isolation or lysed in Laemmli sample buffer. The RNA was used for realtime RT-PCR analysis (upper part). Results are derived from two experiments performed in quadruplicates and shown as mean ± SD. The proteins in the lysed samples were separated by SDS-PAGE, Western blotted, and probed with anti-HOBIT antibodies (lower part). As internal control hypoxanthine-guanine phosphoribosyltransferase was detected by respective antibodies. Two experiments with comparable results were performed. **(B,C)** HOBIT shRNA reduces expansion of cells at day 21. Cord blood CD34$^+$ cells were cultured for 5 days, then cells were transduced with lentiviruses expressing either shHOBIT or a scrambled control shRNA (shControl) or were mock-treated. Transduction efficiency was measured 3 days later by flow cytometry scoring GFP-positive cells. Cells were further cultured and differentiated until day 21. Then flow cytometry was performed to evaluate expansion of transduced GFP-expressing cells. Exemplary dot plots for cells transduced with lentiviruses expressing shHOBIT or shControl or mock-transduced controls are shown in **(B)**. The numbers of the transduced GFP$^+$ cells and the non-tranduced GFP$^-$ cells in individual cultures was calculated from the measured cell number and the respective percentages determined by flow cytometry and are shown in **(C)**. The values were normalized to the number of transduced or non-tranduced cells measured at day 8 to establish the expansion rates. Results are calculated from four experiments performed in triplicates and are displayed as mean ± SEM (*$p < 0.05$, ***$p < 0.001$).

indicate that a strong inhibitory effect on a fraction of the culture indirectly affects the whole culture, including the GFP$^-$ part.

The number of CD56$^+$ NK cells was too low to be analyzed in detail at this time point of the culture (**Figures 4A,B**). Together, the data clearly showed that knockdown of ZNF683/HOBIT reduces the expansion of presumptive NKPs already at day 21 of the culture. This suggests a role of ZNF683/HOBIT in the early stages of human NK-cell development before the initiation of CD56 expression.

ZNF683/HOBIT Downmodulation Largely Abrogates the Generation of CD56$^+$ NK Cells

The second part of our analysis focused on the formation of CD56$^+$ cells and their maturation from day 21 onward. At first sight, when we analyzed the percentages of CD56$^+$ NK cells formed during later stages at day 35, both GFP$^+$ populations (shHOBIT and shControl) seemed comparable to regular cultures (>95% CD56$^+$ NK cells, Figure S1 in Supplementary Material). However, when the number of cells was taken into account, it became apparent that the generation of CD56$^+$ NK cells from the GFP$^+$ progenitors in the shHOBIT pool is substantially reduced compared to the GFP$^+$ control shRNA fraction. Whereas the GFP$^+$CD56$^+$ NK cells in the control shRNA cultures showed a strong expansion between days 21 and 35, the GFP$^+$CD56$^+$ NK cells in the shHOBIT cultures displayed almost no increase beyond day 21 and remained low in number (around 20% of the shControl cultures). These cells did not enter the typical expansion phase between days 24 and 35 (**Figures 5A,B**). In contrast, the GFP$^-$ cells in the shHOBIT cultures displayed an expansion phase between days 21 and 30, although this expansion seemed reduced in comparison to the

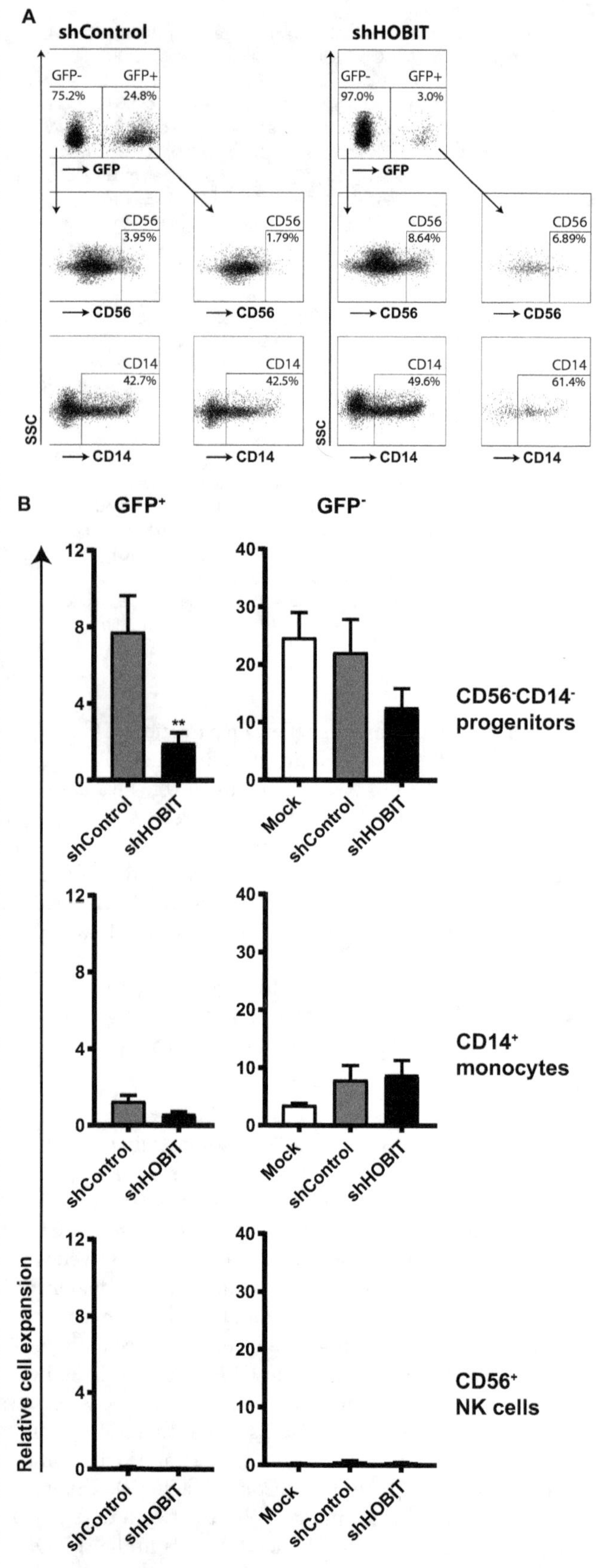

FIGURE 4 | Continued

FIGURE 4 | Continued
HOBIT shRNA strongly reduces the number of CD56⁻CD14⁻ progenitor cells at day 21. Stem cells at day 5 of culture were transduced and further cultured until day 21. Then CD56 and CD14 expression was assessed by flow cytometry within the GFP⁺ and GFP⁻ parts of the culture. **(A)** Exemplary flow cytometry dot plots displaying staining for CD56 and CD14 at day 21. **(B)** The relative expansion of CD56⁻CD14⁻, CD56⁺, and CD14⁺ cells within the GFP⁺ and GFP⁻ fractions is shown. Expansion rates were calculated from the measured cell numbers at day 21 and the percentages of the different subsets obtained by flow cytometry. Differences in transduction efficiency were corrected for by normalizing the obtained values to the number of GFP⁺ and GFP⁻ cells at day 8, respectively. Results are calculated from four independent experiments performed in triplicates using different donor cells and are displayed as mean ± SEM (**$p < 0.01$).

shControl cultures. Again this may suggest an overall indirect effect of the strongly compromised GFP⁺ cells on the GFP⁻ part of the cultures.

Taken together, our data showed that knockdown of ZNF683/HOBIT strongly abrogates the generation of CD56⁺ NK cells. This suggests that ZNF683/HOBIT downmodulation strongly affects NK-cell progenitors shortly before the initiation of CD56 expression and prevents the progression toward CD56⁺ NK cells and/or the further proliferation of the few detected CD56⁺ NK cells.

The Few Developed CD56⁺ NK Cells Display Normal NKG2A and KIR Levels

We were further interested in the phenotype of the few detected CD56⁺ NK cells. To this end, we evaluated if these cells would further differentiate/mature *via* upregulation of NKG2A and KIR. The expression of these important NK-cell receptors was measured after 35 days of cultivation and compared between the GFP⁺ cells of the shHOBIT and the shControl cultures. We observed high levels of NKG2A in all of the GFP⁺ cells of the cultures independent of shHOBIT or control shRNA expression (**Figure 6A**, exemplary dot plots in Figure S2 in Supplementary Material). The GFP⁻ fractions also displayed high NKG2A levels similar to regular control cultures. Similarly, KIR expression was not significantly influenced by the presence or absence of shHOBIT or control shRNA (**Figure 6A**, exemplary dot plots in Figure S2 in Supplementary Material) and was comparable to the typical levels of regular cultures. In this regard, our group previously showed that *ex vivo* generated NK cells express lower levels of KIR than peripheral blood-NK cells (28). Together, these observations demonstrate that cells starting to express CD56 continue to develop normally in regard of NKG2A and KIR expression. Furthermore, it implies ZNF683/HOBIT does not influence the expression of NKG2A or KIR receptors on CD56⁺ NK cells.

Degranulation Capacities Remain Unchanged, but IFN-γ Production Is Strongly Induced Upon Virus Transduction and Further Enhanced by ZNF683/HOBIT Knockdown

Two main functions of NK cells are cytotoxicity, the direct killing of target cells, and IFN-γ production (1, 36). To evaluate whether the CD56⁺ NK cells differentiating under knockdown conditions for ZNF683/HOBIT would functionally mature, we tested degranulation capacity in the CD107a assay and IFN-γ production. Following coculture with the target cell-line K562, we detected similar levels of CD107a on the cell surface of all cells irrespective of the culture conditions (**Figure 6B**, exemplary dot plots in Figure S2 in Supplementary Material). In the same cultures, we also determined the proportion of cells with intracellular IFN-γ. Intriguingly, the transduction with the scrambled control shRNA already caused a substantial increase in IFN-γ producing cells (up to 50%). This was further enhanced upon expression of the ZNF683/HOBIT shRNA leading to 80% IFN-γ-producing cells (**Figure 6C**). In contrast, only a very small fraction of the GFP^- cells was competent to produce IFN-γ (**Figure 6C**). In summary, transduction with shRNA lentiviruses by itself leads to a higher proportion of NK cells with the capacity to produce IFN-γ and this is further increased by knockdown of ZNF683/HOBIT. The ZNF683/HOBIT shRNA effect is restricted to an increase in the number of IFN-γ-producing cells and not reflected in the level of IFN-γ per cell as measured by mean flourescence intensity (MFI). However, virus transduced cells displayed higher levels of

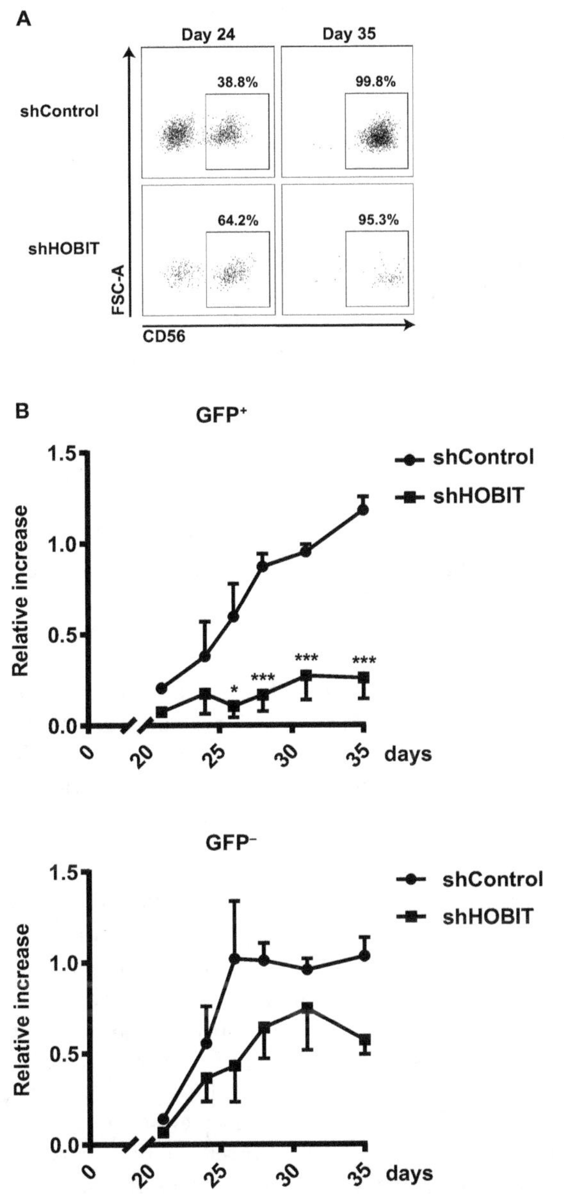

FIGURE 5 | HOBIT shRNA nearly abrogates formation of CD56⁺ natural killer (NK) cells. Stem cell cultures were transduced and cultured until day 35. The generation of CD56⁺ NK cells was monitored by flow cytometry in samples taken at days 24, 26, 28, 31 and 35. **(A)** Exemplary flow cytometry dot plots displaying staining for CD56 within the GFP⁺ fractions at days 24 and 35. **(B)** The formation of CD56⁺ cells within the transduced GFP⁺ and non-transduced GFP⁻ fractions is shown for shHOBIT and shControl cultures. Expansion rates are calculated from measured cell numbers at the respective days and the flow cytometry-derived percentages. Differences in transduction efficiency were corrected for by normalizing the obtained values to the number of GFP⁺ and GFP⁻ cells at day 8, respectively. Results are calculated form four independent experiments, two of them performed in triplicates, using different donor cells and are displayed as mean ± SEM ($^*p < 0.05$, $^{***}p < 0.001$).

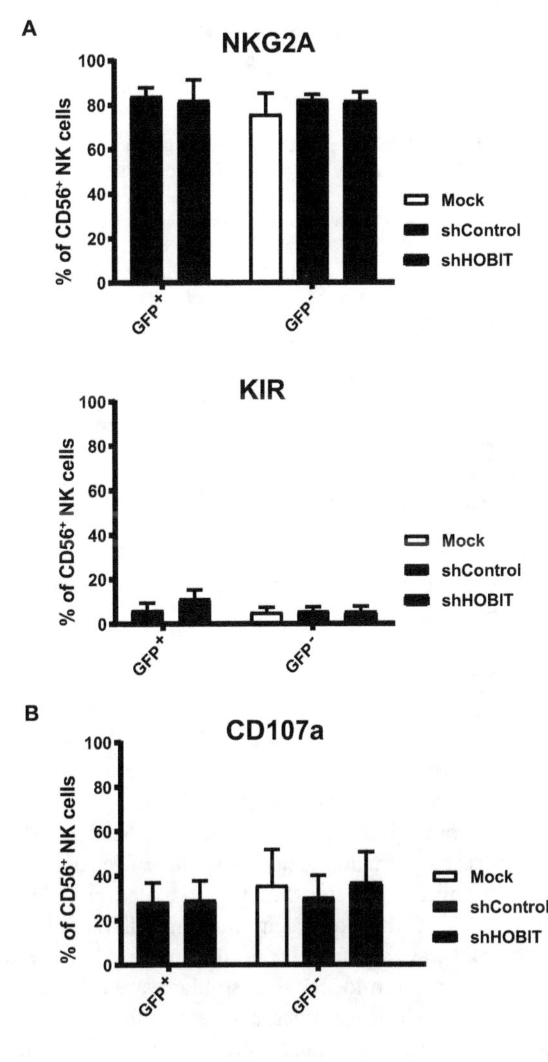

FIGURE 6 | Continued

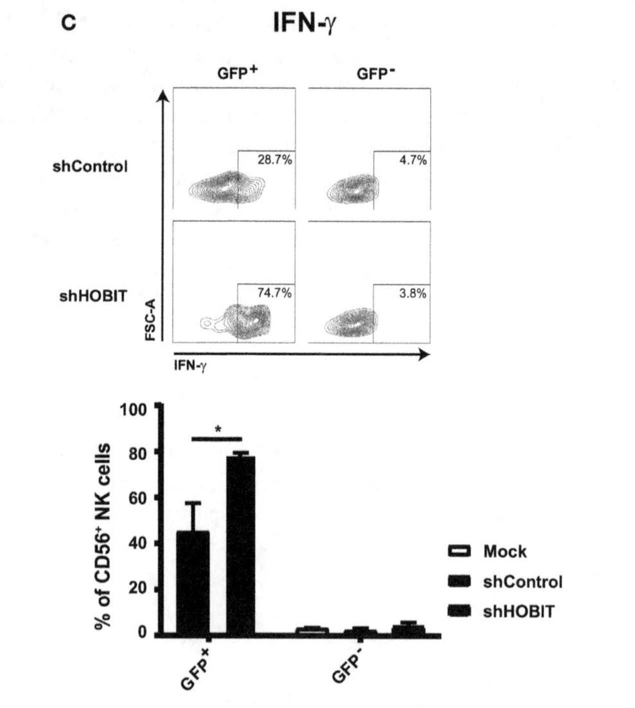

FIGURE 6 | Effects of HOBIT shRNA on phenotype and functions of natural killer (NK) cells at day 35. (A) Expression of NKG2A/CD94 and KIR receptors. Stem cells were transduced and cultured until day 35. Then expression of NKG2A and KIR was measured by flow cytometry. The percentages of cells within the GFP^+ and GFP^- fractions as well as in the mock-treated cultures are displayed. **(B)** Degranulation capacity: separate aliquots of the day 35 cultures were cocultured with K562 target cells and surface CD107a measured after 6 h by flow cytometry. Percentages of CD107a-positive cells within the CD56 population are displayed. **(C)** IFN-γ production: the K562 cocultures were further stained for intracellular IFN-γ. Representative flow cytometry dot plots obtained are shown in the upper part. Percentages of IFN-γ positive cells within the CD56 population are displayed in the lower part. All results are derived from four independent experiments, two of them performed in triplicates, using different donor cells and are shown as mean ± SEM ($^*p < 0.05$).

IFN-γ per cell in comparison to non-transduced cells (Figure S3 in Supplementary Material).

DISCUSSION

The recent decade has witnessed an enormous increase in the understanding of the development of different immune-cell lineages including the various forms of innate and adaptive lymphocytes from HSCs (37). Most of this knowledge is derived from the murine system due to the possibility of genetic manipulation of the animal models. Therefore, detailed knowledge on the specificities of the equivalent human differentiation and maturation pathways and factors are still elusive. Although many factors may function in identical or similar ways in murine and human immune-cell development, certain differences are to be expected based on a much more rapid evolution of the immune system when compared to other tissues. For example, this has led to the convergent evolution of distinct classes of important NK-cell receptors with similar function but encoded by structurally different genes such as the human KIR and murine Ly49 receptors (38, 39). It is conceivable that this is also reflected in the transcription factor circuitry defining human and murine NK cells.

Due to the recent increasing interest in NK cells as therapeutic agents, especially for the treatment of leukemia and potentially also solid cancers (40, 41), a more complete understanding of the control of human NK-cell differentiation and maturation is desirable. Here, we used a feeder cell-free *ex vivo* system for the generation of human NK cells from cord blood HSCs (29). This system can generate therapeutic NK cells that have been proven to be safe in a phase I clinical trial (42). We have, furthermore, demonstrated that the NK cells formed using this system display the typical NK cell receptors, potent ADCC and produce IFN-γ similar to peripheral blood NK cells (28). So this provided an ideal system to study transcription factors during human NK-cell differentiation and maturation.

Identity of cell types and their differentiation and maturation is controlled in large part by the action of transcription factors. Normally, important core transcription factors can be identified by the characteristics of upregulation during differentiation and their high expression in a relatively high cell type-specific fashion (43). In this regard, our transcriptomic profiling has identified a large number of transcription factors strongly upregulated during human NK-cell differentiation. Many of these have previously been described in the context of murine NK-cell development. For example, in earlier stages of NK-cell differentiation, ID2 and ID3 contribute via suppression of E proteins, such as E2A, the B-cell promoting factor. GATA3, TOX, and IKZF3 (AIOLOS) have been reported to promote later maturation stages of NK-cell development (20, 44). From these TOX has also been described specifically to contribute to human NK-cell differentiation, where it seems to control T-BET expression (21, 22).

Upon re-evaluation of the 10 most highly upregulated transcription factor genes from the profiling experiment, it was apparent that ZNF683/HOBIT mRNA was the most strongly upregulated. It displayed a 10-fold higher upregulation compared to GATA-3, the second best factor. Further analysis of different immune-cell types in human peripheral blood showed preferential expression of ZNF683/HOBIT mRNA in $CD56^+$ NK cells when compared to B and T lymphocytes and monocytes. We detected low expression levels in B cells, but it was very low to undetectable in T lymphocytes and monocytes. Furthermore, *ZNF683* is a homolog of the *PRDM1* gene encoding BLIMP-1 (ZNF683 has, therefore, also been termed HOBIT, homolog of Blimp-1 in T cells) (45). In the mouse, Blimp-1 has been shown to be a master regulator of terminal differentiation of $CD8^+$ effector T cells and plasma cells (46) and to play a role in the maturation of peripheral NK cells (18). Some upregulation of *PRDM1* was detected in our profiling analysis (Table S1 in Supplementary Material), but BLIMP-1 mRNA showed a much less dramatic increase during human NK-cell differentiation than ZNF683/HOBIT mRNA (Table S2 in Supplementary Material). Considering its selective upregulation and expression and its homology with BLIMP-1, an established important differentiation factor, ZNF683/HOBIT appeared to fulfill the precondition for a new key transcription factor controlling the development of human NK cells.

We further analysed ZNF683/HOBIT expression in the classical defined NK subsets, CD56^{bright} and CD56^{dim} cells. It is thought that the CD56^{bright} cells are more immature, preferentially produce cytokines, and will mature further into CD56^{dim} cells that display potent cytotoxic activity (1). In our analysis, evaluation of ZNF683/HOBIT mRNA revealed high expression levels in both subsets with somewhat higher levels in CD56^{bright} cells. Together with the kinetic analysis of ZNF683/HOBIT mRNA accumulation, that displayed a parallel increase with the generation of CD56$^+$ NK cells, this initially suggested a preferential activity between acquisition of CD56 and the CD56^{bright} cell stage. However, a closer inspection revealed significantly increased ZNF683/HOBIT mRNA levels already in day 14-progenitors, with an exponential upregulation from day 18 onward. This suggests that ZNF683/HOBIT mRNA starts to be expressed soon after addition of IL-15 in NKPs and continuously increases with highest accumulation rates in parallel to the increase in CD56 expression. In comparison, only minor ZNF683/HOBIT mRNA levels appeared to be present in the CD14$^+$ monocytic subset, which develops between days 10 and 20 and disappears thereafter.

In line with the early expression of ZNF683/HOBIT at the NKP stage, shRNA-mediated downmodulation of ZNF683/HOBIT resulted in a significantly reduced expansion of transduced cells already at day 21 of the culture. The majority of this effect could be traced to the CD56$^-$CD14$^-$ cells containing at least in part the NKPs. CD56$^+$-expressing NK cells were not yet detectable at significant amounts at this time point in the culture. This clearly supports that a first major effect of ZNF683/HOBIT downmodulation is a reduction of proliferation and/or survival of NK-cell progenitors prior to the initiation of CD56 expression. Some reduction seemed also to occur for the CD14$^+$ cells in shHOBIT transduced cultures, although this did not reach significance. As very low levels of ZNF683/HOBIT mRNA appear to be expressed in CD14$^+$ cells, we are unable to rule out a potential specific negative effect also upon this subset following HOBIT knockdown.

The second major effect became apparent when we monitored the acquisition of CD56 expression. CD56 or N-CAM is generally accepted as a major marker for human NK cells, despite that its function remains elusive (47). It is present on about 95% of human NK cells (47), but lacks a clear homolog in murine NK cells (48). The kinetics of accumulation of CD56$^-$CD14$^-$ progenitors and CD56$^+$ cells in the *ex vivo* system are compatible with the start of expression of CD56 by CD56$^-$CD14$^-$ progenitors from day 20 onward. This is indicated by the rapid decline of CD56$^-$CD14$^-$ progenitors and the concomitant appearance of the CD56$^+$ cells. This generation of CD56$^+$ cells was nearly abrogated upon downmodulation of ZNF683/HOBIT. This supports that ZNF683/HOBIT is essential for efficient generation of CD56 expressing cells and/or their further proliferation or survival.

Important receptors in human NK cells are the NKG2A and KIR receptors (11, 49). In the *ex vivo* cultures, the expression of NKG2A starts shortly after CD56 expression (28). NKG2A is further expressed on the majority of the more immature CD56^{bright} cells. According to the current hypothesis, when these cells mature into CD56^{dim} cells, they will reduce NKG2A and increase KIR expression (11). Despite the substantially reduced number of CD56$^+$ cells formed in the shHOBIT cultures, it was astonishing that these cells appeared quite normal in respect of NKG2A expression compared to control cultures. The observed levels of NKG2A expression on about 80% of CD56$^+$ cells were comparable to the levels characteristic for regular cultures (28). Also KIR levels established with a pan-KIR antibody were similar to regular cultures, in the order of 5% of CD56$^+$ cells. As previously discussed (28), in the *ex vivo* cultures NKG2A levels are similarly high as in peripheral CD56^{bright} cells, probably due to the high cytokine levels in the culture. KIR levels are intermediary between CD56^{bright} cells and CD56^{dim} cells. This suggests that *ex vivo*-generated NK cells, in terms of receptor expression, do not fully mature, whereas functionally they display full cytotoxic capabilities. In this regard, we observed no differences in degranulation capabilities against K562 targets between the shHOBIT and control cultures.

In contrast, it was remarkable that virus transduction by itself caused an increase in the proportion of IFN-γ-producing CD56$^+$ NK cells. About 50% of day 35 cells transduced with control shRNA viruses produced IFN-γ upon coculture with K562 target cells, whereas only a few percent for non-transduced cells. Currently, it is unclear whether this is due to a priming process at the time of transduction at day 5 that continues to act into later stages of differentiation or whether it is high expression of shRNAs that can activate the differentiated cells. We can only speculate about the underlying mechanisms for this unexpected finding. For example, it has been described that a uridine-rich part of the HIV RNA (ssRNA40) can activate NK cells *via* TLR7/8 signaling (50). So, maybe the presence of high levels of shRNA or RNA transcribed from the viral sequences could somehow activate the IFN-γ machinery. Irrespective of the cause of this effect, the increase in IFN-γ-producing cells initiated by the virus itself was further enhanced by knockdown of ZNF683/HOBIT. This suggests that ZNF683/HOBIT negatively controls the development of IFN-γ-producing NK cells.

We have to introduce a caveat on the interpretation of the data for the late stages of differentiation in the *ex vivo* sytem. Due to the low numbers of GFP$^+$ cells obtained, that further appeared to be too fragile for preparative sorting, we were unable to exclude that the few late stage GFP$^+$ NK cells generated are derivatives of few cells that failed to express the shRNA and, therefore, continued to differentiate. It is, however, equally conceivable that ZNF683/HOBIT knockdown is effective on proliferation and/or survival only within a certain time window, when the cells normally still proliferate, and is less effective once the cells have passed this stage and do not proliferate but rather further differentiate and mature.

While this work was in progress, a report on ZNF683/HOBIT was published describing that the factor is highly expressed in human effector-type CD8$^+$ T cells, but not in naive or most memory CD8$^+$ T cells or CD4$^+$ helper T cells (51, 52). High levels of ZNF683/HOBIT were expressed in CMV-specific, but not in influenza-specific $CD8^+$ T cells. This may explain why in our experiments we did not detect significant expression levels in peripheral T cells as we only tested the overall population of $CD3^+$ T cells comprising all $CD4^+$ and $CD8^+$ T cells. This report also showed highest expression in human NK cells and low expression in dendritic cells in accordance with our data on peripheral NK cells and monocytes.

Currently available evidence suggests that ZNF683/HOBIT may be special in terms of displaying substantial differences in expression pattern and possibly function in different cell types in mice and humans. In mice, it was reported to be predominantly expressed in NKT cells, a $CD4^+$ T cell subset with immediate effector functions, and to some extent in $CD8^+$ T cells (45, 53). In humans, ZNF683/HOBIT is mainly expressed in NK cells and, as described by others, also in effector-type $CD8^+$ T cells (51). Differences also seem to occur regarding the control of IFN-γ production. In murine NKT cells, ZNF683/HOBIT represses IFN-γ and activates granzyme B production (45). In contrast, in human long-lived effector T cells, ZNF683/HOBIT was reported to induce IFN-γ and to have no effect on granzyme B production (51). Our findings in human NK cells rather support a suppressing effect of ZNF683/HOBIT on IFN-γ production. This correlates with a report on BLIMP-1 describing a similar suppression of IFN-γ production in human NK cells (51). No effect of BLIMP-1 on cytotoxicity was observed, which is also in agreement with our findings on ZNF683/HOBIT. Generally, the homology of ZNF683/HOBIT and BLIMP-1 suggests overlapping and synergistic activities. Both factors display highly conserved zinc finger domains mediating binding to DNA target sequences in enhancer regions of a number of genes. The DNA binding sites of the factors largely overlap and both can bind to target sequences in several identical genes including *TCF7*. Both factors seem to mainly act as transcriptional repressors (44, 54), although for BLIMP-1 activating properties have also been described (55).

Taken together, our data for the first time show a role of ZNF683/HOBIT during differentiation of human NK cells. They strongly support that ZNF683/HOBIT is a key regulatory factor controlling generation, proliferation, and/or survival of NK-cell progenitors ($CD56^-CD14^-$) and is essential for efficient generation of $CD56^+$ cells. Once CD56 expression has been acquired the further maturation including NK receptor expression and development of degranulation capacity seems to be unaffected, whereas IFN-γ production appears to be constrained by ZNF683/HOBIT. Although a more precise identification of the developmental stage(s) affected will need additional investigations, the data are compatible with the possibility that ZNF683/HOBIT may mainly act between early factors, such as NFIL3 (56) with proposed activities at the early committed NKP stage and factors described to promote rather maturation of NK cells at later stages such as GATA-3 (33). It further remains to be established how this is achieved in potential interaction with BLIMP-1 described so far as important mainly for late stage maturation.

ETHICS STATEMENT

Human umbilical cord blood samples were obtained at birth after full-term delivery from the Department of Obstetrics and Gynecology of the University Hospital of Vienna, Austria. This was carried out in accordance with the recommendations of the "Ethical committee of the Medical University of Vienna" with written informed consent from all subjects in accordance with the Declaration of Helsinki. The protocol was approved by the "Ethical committee of the Medical University of Vienna" (protocol number 122/2010).

AUTHOR CONTRIBUTIONS

MP performed experiments and wrote the basic manuscript. AC aided in experimental design, performed experiments, and improved the manuscript. DL provided the basis of the study with initial experiments, including the transcriptional profiling, and MO contributed with part of real-time RT-PCR analyses. UR helped with Western blots and DD with generating viral constructs. BS, WE, AN, JS, and DS contributed in the design of the study, provided cell samples, viral vectors and technology, and corrected the manuscript. EC contributed to the design, supported the progress of the project, and helped with formulating the final manuscript. EH conceived the study, supported MP and AC in all aspects of the experiments, and helped to finish the final version of the manuscript.

ACKNOWLEDGMENTS

The authors acknowledge the collection of cord blood samples by the midwives of the Department of Obstetrics and Gynecology of the University Hospital Vienna and the expert help in preparative cell sorting by the FACS Core Unit Teams from the St. Anna Children's Cancer Research Center. Furthermore, the authors thank the members of the vascular biology group for discussions and support throughout the study.

REFERENCES

1. Caligiuri MA. Human natural killer cells. *Blood* (2008) 112(3):461–9. doi:10.1182/blood-2007-09-077438
2. Zimmer J, editor. *Natural Killer Cells*. Berlin Heidelberg: Springer-Verlag (2010).
3. Wehner R, Dietze K, Bachmann M, Schmitz M. The bidirectional crosstalk between human dendritic cells and natural killer cells. *J Innate Immun* (2011) 3(3):258–63. doi:10.1159/000323923
4. Moretta A, Marcenaro E, Parolini S, Ferlazzo G, Moretta L. NK cells at the interface between innate and adaptive immunity. *Cell Death Differ* (2008) 15(2):226–33. doi:10.1038/sj.cdd.4402170
5. Cerwenka A, Lanier LL. Natural killer cell memory in infection, inflammation and cancer. *Nat Rev Immunol* (2016) 16(2):112–23. doi:10.1038/nri.2015.9
6. Vivier E, Raulet DH, Moretta A, Caligiuri MA, Zitvogel L, Lanier LL, et al. Innate or adaptive immunity? The example of natural killer cells. *Science* (2011) 331(6013):44–9. doi:10.1126/science.1198687

7. Diefenbach A, Colonna M, Koyasu S. Development, differentiation, and diversity of innate lymphoid cells. *Immunity* (2014) 41(3):354–65. doi:10.1016/j.immuni.2014.09.005
8. De Obaldia ME, Bhandoola A. Transcriptional regulation of innate and adaptive lymphocyte lineages. *Annu Rev Immunol* (2015) 33:607–42. doi:10.1146/annurev-immunol-032414-112032
9. Nutt SL, Kee BL. The transcriptional regulation of B cell lineage commitment. *Immunity* (2007) 26(6):715–25. doi:10.1016/j.immuni.2007.05.010
10. Boos MD, Yokota Y, Eberl G, Kee BL. Mature natural killer cell and lymphoid tissue-inducing cell development requires Id2-mediated suppression of E protein activity. *J Exp Med* (2007) 204(5):1119–30. doi:10.1084/jem.20061959
11. Montaldo E, Del Zotto G, Della Chiesa M, Mingari MC, Moretta A, De Maria A, et al. Human NK cell receptors/markers: a tool to analyze NK cell development, subsets and function. *Cytometry A* (2013) 83(8):702–13. doi:10.1002/cyto.a.22302
12. Freud AG, Caligiuri MA. Human natural killer cell development. *Immunol Rev* (2006) 214:56–72. doi:10.1111/j.1600-065X.2006.00451.x
13. Sun JC. Transcriptional control of NK cells. *Curr Top Microbiol Immunol* (2016) 395:1–36. doi:10.1007/82_2015_452
14. Eckelhart E, Warsch W, Zebedin E, Simma O, Stoiber D, Kolbe T, et al. A novel Ncr1-Cre mouse reveals the essential role of STAT5 for NK-cell survival and development. *Blood* (2011) 117(5):1565–73. doi:10.1182/blood-2010-06-291633
15. Colucci F, Samson SI, DeKoter RP, Lantz O, Singh H, Di Santo JP. Differential requirement for the transcription factor PU.1 in the generation of natural killer cells versus B and T cells. *Blood* (2001) 97(9):2625–32. doi:10.1182/blood.V97.9.2625
16. Barton K, Muthusamy N, Fischer C, Ting CN, Walunas TL, Lanier LL, et al. The Ets-1 transcription factor is required for the development of natural killer cells in mice. *Immunity* (1998) 9(4):555–63. doi:10.1016/S1074-7613(00)80638-X
17. Gascoyne DM, Long E, Veiga-Fernandes H, de Boer J, Williams O, Seddon B, et al. The basic leucine zipper transcription factor E4BP4 is essential for natural killer cell development. *Nat Immunol* (2009) 10(10):1118–24. doi:10.1038/ni.1787
18. Kallies A, Carotta S, Huntington ND, Bernard NJ, Tarlinton DM, Smyth MJ, et al. A role for Blimp1 in the transcriptional network controlling natural killer cell maturation. *Blood* (2011) 117(6):1869–79. doi:10.1182/blood-2010-08-303123
19. Lacorazza HD, Miyazaki Y, Di Cristofano A, Deblasio A, Hedvat C, Zhang J, et al. The ETS protein MEF plays a critical role in perforin gene expression and the development of natural killer and NK-T cells. *Immunity* (2002) 17(4):437–49. doi:10.1016/S1074-7613(02)00422-3
20. Leong JW, Wagner JA, Ireland AR, Fehniger TA. Transcriptional and post-transcriptional regulation of NK cell development and function. *Clin Immunol* (2016). doi:10.1016/j.clim.2016.03.003
21. Yun S, Lee SH, Yoon SR, Kim MS, Piao ZH, Myung PK, et al. TOX regulates the differentiation of human natural killer cells from hematopoietic stem cells in vitro. *Immunol Lett* (2011) 136(1):29–36. doi:10.1016/j.imlet.2010.11.008
22. Vong QP, Leung WH, Houston J, Li Y, Rooney B, Holladay M, et al. TOX2 regulates human natural killer cell development by controlling T-BET expression. *Blood* (2014) 124(26):3905–13. doi:10.1182/blood-2014-06-582965
23. Smith MA, Maurin M, Cho HI, Becknell B, Freud AG, Yu J, et al. PRDM1/Blimp-1 controls effector cytokine production in human NK cells. *J Immunol* (2010) 185(10):6058–67. doi:10.4049/jimmunol.1001682
24. Pinho MJ, Marques CJ, Carvalho F, Punzel M, Sousa M, Barros A. Genetic regulation on ex vivo differentiated natural killer cells from human umbilical cord blood CD34+ cells. *J Recept Signal Transduct Res* (2012) 32(5):238–49. doi:10.3109/10799893.2012.700716
25. Mandal A, Viswanathan C. Natural killer cells. In health and disease. *Hematol Oncol Stem Cell Ther* (2015) 8(2):47–55. doi:10.1016/j.hemonc.2014.11.006
26. Tarazona R, Sanchez-Correa B, Casas-Aviles I, Campos C, Pera A, Morgado S, et al. Immunosenescence: limitations of natural killer cell-based cancer immunotherapy. *Cancer Immunol Immunother* (2017) 66(2):233–45. doi:10.1007/s00262-016-1882-x
27. Yoon SR, Kim TD, Choi I. Understanding of molecular mechanisms in natural killer cell therapy. *Exp Mol Med* (2015) 47:e141. doi:10.1038/emm.2014.114
28. Lehmann D, Spanholtz J, Osl M, Tordoir M, Lipnik K, Bilban M, et al. Ex vivo generated natural killer cells acquire typical natural killer receptors and display a cytotoxic gene expression profile similar to peripheral blood natural killer cells. *Stem Cells Dev* (2012) 21(16):2926–38. doi:10.1089/scd.2011.0659
29. Spanholtz J, Tordoir M, Eissens D, Preijers F, van der Meer A, Joosten I, et al. High log-scale expansion of functional human natural killer cells from umbilical cord blood CD34-positive cells for adoptive cancer immunotherapy. *PLoS One* (2010) 5(2):e9221. doi:10.1371/journal.pone.0009221
30. Spanholtz J, Preijers F, Tordoir M, Trilsbeek C, Paardekooper J, de Witte T, et al. Clinical-grade generation of active NK cells from cord blood hematopoietic progenitor cells for immunotherapy using a closed-system culture process. *PLoS One* (2011) 6(6):e20740. doi:10.1371/journal.pone.0020740
31. Bilban M, Ghaffari-Tabrizi N, Hintermann E, Bauer S, Molzer S, Zoratti C, et al. Kisspeptin-10, a KiSS-1/metastin-derived decapeptide, is a physiological invasion inhibitor of primary human trophoblasts. *J Cell Sci* (2004) 117(Pt 8):1319–28. doi:10.1242/jcs.00971
32. Bilban M, Heintel D, Scharl T, Woelfel T, Auer MM, Porpaczy E, et al. Deregulated expression of fat and muscle genes in B-cell chronic lymphocytic leukemia with high lipoprotein lipase expression. *Leukemia* (2006) 20(6):1080–8. doi:10.1038/sj.leu.2404220
33. Samson SI, Richard O, Tavian M, Ranson T, Vosshenrich CA, Colucci F, et al. GATA-3 promotes maturation, IFN-gamma production, and liver-specific homing of NK cells. *Immunity* (2003) 19(5):701–11. doi:10.1016/S1074-7613(03)00294-2
34. Aliahmad P, Seksenyan A, Kaye J. The many roles of TOX in the immune system. *Curr Opin Immunol* (2012) 24(2):173–7. doi:10.1016/j.coi.2011.12.001
35. Weber K, Mock U, Petrowitz B, Bartsch U, Fehse B. Lentiviral gene ontology (LeGO) vectors equipped with novel drug-selectable fluorescent proteins: new building blocks for cell marking and multi-gene analysis. *Gene Ther* (2010) 17(4):511–20. doi:10.1038/gt.2009.149
36. Fauriat C, Long EO, Ljunggren HG, Bryceson YT. Regulation of human NK-cell cytokine and chemokine production by target cell recognition. *Blood* (2010) 115(11):2167–76. doi:10.1182/blood-2009-08-238469
37. Biasco L, Pellin D, Scala S, Dionisio F, Basso-Ricci L, Leonardelli L, et al. In vivo tracking of human hematopoiesis reveals patterns of clonal dynamics during early and steady-state reconstitution phases. *Cell Stem Cell* (2016) 19(1):107–19. doi:10.1016/j.stem.2016.04.016
38. Campbell KS, Purdy AK. Structure/function of human killer cell immunoglobulin-like receptors: lessons from polymorphisms, evolution, crystal structures and mutations. *Immunology* (2011) 132(3):315–25. doi:10.1111/j.1365-2567.2010.03398.x
39. Carrillo-Bustamante P, Kesmir C, de Boer RJ. The evolution of natural killer cell receptors. *Immunogenetics* (2016) 68(1):3–18. doi:10.1007/s00251-015-0869-7
40. Shevtsov M, Multhoff G. Immunological and translational aspects of NK cell-based antitumor immunotherapies. *Front Immunol* (2016) 7:492. doi:10.3389/fimmu.2016.00492
41. Rezvani K, Rouce RH. The application of natural killer cell immunotherapy for the treatment of cancer. *Front Immunol* (2015) 6:578. doi:10.3389/fimmu.2015.00578
42. Dolstra H, Roeven MW, Spanholtz J, Hangalapura BN, Tordoir M, Maas F, et al. Successful transfer of umbilical cord blood CD34^{+} hematopoietic stem and progenitor-derived NK cells in older acute myeloid leukemia patients. *Clin Canc Res* (2017). doi:10.1158/1078-0432.CCR-16-2981
43. D'Alessio AC, Fan ZP, Wert KJ, Baranov P, Cohen MA, Saini JS, et al. A systematic approach to identify candidate transcription factors that control cell identity. *Stem Cell Reports* (2015) 5(5):763–75. doi:10.1016/j.stemcr.2015.09.016
44. Holmes ML, Huntington ND, Thong RP, Brady J, Hayakawa Y, Andoniou CE, et al. Peripheral natural killer cell maturation depends on the transcription factor Aiolos. *EMBO J* (2014) 33(22):2721–34. doi:10.15252/embj.201487900
45. van Gisbergen KP, Kragten NA, Hertoghs KM, Wensveen FM, Jonjic S, Hamann J, et al. Mouse Hobit is a homolog of the transcriptional repressor Blimp-1 that regulates NKT cell effector differentiation. *Nat Immunol* (2012) 13(9):864–71. doi:10.1038/ni.2393
46. Kallies A, Nutt SL. Terminal differentiation of lymphocytes depends on Blimp-1. *Curr Opin Immunol* (2007) 19(2):156–62. doi:10.1016/j.coi.2007.01.003
47. Lanier LL, Phillips JH, Hackett J Jr, Tutt M, Kumar V. Natural killer cells: definition of a cell type rather than a function. *J Immunol* (1986) 137(9):2735–9.

48. Marquardt N, Wilk E, Pokoyski C, Schmidt RE, Jacobs R. Murine CXCR3+CD27bright NK cells resemble the human CD56bright NK-cell population. *Eur J Immunol* (2010) 40(5):1428–39. doi:10.1002/eji.200940056
49. Bjorkstrom NK, Riese P, Heuts F, Andersson S, Fauriat C, Ivarsson MA, et al. Expression patterns of NKG2A, KIR, and CD57 define a process of CD56dim NK-cell differentiation uncoupled from NK-cell education. *Blood* (2010) 116(19):3853–64. doi:10.1182/blood-2010-04-281675
50. Alter G, Suscovich TJ, Teigen N, Meier A, Streeck H, Brander C, et al. Single-stranded RNA derived from HIV-1 serves as a potent activator of NK cells. *J Immunol* (2007) 178(12):7658–66. doi:10.4049/jimmunol.178.12.7658
51. Vieira Braga FA, Hertoghs KM, Kragten NA, Doody GM, Barnes NA, Remmerswaal EB, et al. Blimp-1 homolog Hobit identifies effector-type lymphocytes in humans. *Eur J Immunol* (2015) 45(10):2945–58. doi:10.1002/eji.201545650
52. Braun J, Frentsch M, Thiel A. Hobit and human effector T-cell differentiation: the beginning of a long journey. *Eur J Immunol* (2015) 45(10):2762–5. doi:10.1002/eji.201545959
53. Mackay LK, Minnich M, Kragten NA, Liao Y, Nota B, Seillet C, et al. Hobit and Blimp1 instruct a universal transcriptional program of tissue residency in lymphocytes. *Science* (2016) 352(6284):459–63. doi:10.1126/science.aad2035
54. Rutishauser RL, Martins GA, Kalachikov S, Chandele A, Parish IA, Meffre E, et al. Transcriptional repressor Blimp-1 promotes CD8(+) T cell terminal differentiation and represses the acquisition of central memory T cell properties. *Immunity* (2009) 31(2):296–308. doi:10.1016/j.immuni.2009.05.014
55. Minnich M, Tagoh H, Bonelt P, Axelsson E, Fischer M, Cebolla B, et al. Multifunctional role of the transcription factor Blimp-1 in coordinating plasma cell differentiation. *Nat Immunol* (2016) 17(3):331–43. doi:10.1038/ni.3349
56. Male V, Nisoli I, Kostrzewski T, Allan DS, Carlyle JR, Lord GM, et al. The transcription factor E4bp4/Nfil3 controls commitment to the NK lineage and directly regulates Eomes and Id2 expression. *J Exp Med* (2014) 211(4):635–42. doi:10.1084/jem.20132398

7

KIR3DS1-Mediated Recognition of HLA-*B51: Modulation of KIR3DS1 Responsiveness by Self HLA-B Allotypes and Effect on NK Cell Licensing

Simona Carlomagno[1], Michela Falco[2], Maria Bono[2], Claudia Alicata[1,3], Lucia Garbarino[4], Michela Mazzocco[4], Lorenzo Moretta[5], Alessandro Moretta[1,3*†] and Simona Sivori[1,3†]

[1] Dipartimento di Medicina Sperimentale, Università degli Studi di Genova, Genova, Italy, [2] Istituto Giannina Gaslini, Genova, Italy, [3] Centro di Eccellenza per le Ricerche Biomediche, Università degli Studi di Genova, Genova, Italy, [4] S.C. Laboratorio di Istocompatibilità e IBMDR, E.O. Ospedali Galliera, Genova, Italy, [5] Dipartimento di Immunologia, IRCCS Ospedale Bambin Gesù, Roma, Italy

Edited by:
Gianfranco Pittari,
Hamad Medical Corporation, Qatar

Reviewed by:
Sarah Rowland-Jones,
Oxford University, UK
John Trowsdale,
University of Cambridge, UK

***Correspondence:**
Alessandro Moretta
alemoret@unige.it

†These authors have contributed equally to this work.

Several studies described an association between killer-cell immunoglobulin-like receptor (KIR)/HLA gene combinations and clinical outcomes in various diseases. In particular, an important combined role for KIR3DS1 and HLA-B Bw4-I80 in controlling viral infections and a higher protection against leukemic relapses in donor equipped with activating KIRs in haplo-HSCT has been described. Here, we show that KIR3DS1 mediates positive signals upon recognition of HLA-B*51 (Bw4-I80) surface molecules on target cells and that this activation occurs only in Bw4-I80^{neg} individuals, including those carrying particular KIR/HLA combination settings. In addition, killing of HLA-B*51 transfected target cells mediated by KIR3DS1^{+}/NKG2A^{+} natural killer (NK) cell clones from Bw4-I80^{neg} donors could be partially inhibited by antibody-mediated masking of KIR3DS1. Interestingly, KIR3DS1-mediated recognition of HLA-B*51 could be better appreciated under experimental conditions in which the function of NKG2D was reduced by mAb-mediated blocking. This experimental approach may mimic the compromised function of NKG2D occurring in certain viral infections. We also show that, in KIR3DS1^{+}/NKG2A^{+} NK cell clones derived from an HLA-B Bw4-T80 donor carrying 2 *KIR3DS1* gene copy numbers, the positive signal generated by the engagement of KIR3DS1 by HLA-B*51 resulted in a more efficient killing of HLA-B*51-transfected target cells. Moreover, in these clones, a direct correlation between KIR3DS1 and NKG2D surface density was detected, while the expression of NKp46 was inversely correlated with that of KIR3DS1. Finally, we analyzed KIR3DS1^{+}/NKG2A^{+} NK cell clones from a HLA-B Bw4^{neg} donor carrying cytoplasmic KIR3DL1. Although these clones expressed lower levels of surface KIR3DS1, they displayed responses comparable to those of NK cell clones derived from HLA-B Bw4^{neg} donors that expressed surface KIR3DL1. Altogether these data suggest that, in particular KIR/HLA combinations, KIR3DS1 may play a role in the process of human NK cell education.

Keywords: human natural killer cells, activating killer-immunoglobulin-like receptors, KIR3DS1, natural killer cell education, HLA-B alleles, KIR/KIR-L interaction

INTRODUCTION

Natural killer (NK) cells are lymphocytes of innate immunity that are involved in the host immune defenses against viruses and tumor cells. NK cells can exert cytotoxicity against transformed cells and release soluble factors important for regulating innate and adaptive immune responses (1, 2).

The NK cell function is controlled by an array of activating and inhibitory receptors, including the family of killer-cell immunoglobulin-like receptors (KIRs) (3). Inhibitory KIRs (iKIRs) possess a long cytoplasmic tail, containing ITIM motifs, responsible for the transduction of an inhibitory signal (3, 4), while activating KIRs (aKIRs) are characterized by a short tail and by a positively charged amino acid residue in their transmembrane region, which allows recruitment of the DAP12 signaling adaptor molecule (5, 6).

KIR3DL1/S1 is the *KIR* gene characterized by the highest degree of polymorphism, and it is the only one including alleles coding for either inhibitory (KIR3DL1) or activating (KIR3DS1) receptors. In particular, KIR3DL1 is highly polymorphic, whereas KIR3DS1*013 is the most represented in all examined populations (7). Notably, KIR3DS1 is the only activating receptor with three extracellular domains (8). Its inhibitory counterpart, KIR3DL1, recognizes HLA-A and HLA-B alleles, sharing the Bw4 public epitope (3, 4, 9, 10). Despite the high degree of homology between these two KIR3D receptors, knowledge about KIR3DS1 function and ligand specificity is not completely defined so far.

In this regard, several studies have suggested that certain HLA-B Bw4 alleles characterized by isoleucine in position 80 (Bw4-I80) may be putative ligands for KIR3DS1. In particular, the carriage of a *KIR3DS1* allele in conjunction with *HLA-Bw4-I80* alleles in patients with chronic HIV-1 infection has been associated with a slower progression to AIDS (11, 12). In addition, in individuals affected by acute HIV-1 infections and carrying *HLA-Bw4-I80* alleles, expansion of KIR3DS1$^+$ NK cells (13), killing of HIV-1 infected cells, and inhibition of viral replication have been reported (12).

Carr and colleagues (14) could not detect any KIR3DS1-Fc binding to LCL721.221 cells transfected with HLA-B Bw4-I80 alleles (HLA-B*57:01, HLA-B*58:01) or with HLA-B Bw4-T80 (B*27:05) and HLA-Bw6 (B*15:02). Nevertheless they did not exclude the possibility that binding occurred below their detection limits or required the presence of additional factors, such as the presence of specific peptides in the HLA-B peptide-binding groove. In this regard, it has been shown that KIR3DS1 can interact productively with HLA-Bw4 in the context of HIV infection. Indeed, two HIV-derived peptides have been described to enable HLA-B*57:01/KIR3DS1 interaction (15).

Recent studies have also reported that HLA-F open conformers (OCs) are high-affinity ligands of KIR3DS1 and ligands of lower affinity for the inhibitory receptors KIR3DL1 and KIR3DL2 (16, 17). However, KIR3DS1/HLA-F interaction cannot fully explain the control of HIV infection in KIR3DS1$^+$ HLA-B Bw4-I80$^+$ patients only (16).

A protective role of KIR3DS1 in controlling certain tumors promoted by chronic viral infections has also been observed. For example, a protective effect of *KIR3DS1* in combination with *HLA-B Bw4-I80* alleles has been observed against hepatocellular carcinomas developed in chronically HCV-infected patients (18).

Remarkably, despite several attempts to define the specificity of KIR3DS1, the role of this receptor in the process of NK cell education has not been considered yet. In the present study, by the analysis of distinct NK clones, we show that certain self HLA-B allotypes can modify the functional responsiveness of KIR3DS1, thus providing evidence for an effect on the education of NK cells expressing this aKIR.

MATERIALS AND METHODS

KIR Gene Profile and *KIR*-Ligand (*KIR*-L) Analyses

DNA of the tested donors was extracted using QIAamp DNA Blood Mini Kit (Qiagen, GmbH, Germany). The *KIR* gene profile and *KIR*-L analyses were performed using sequence-specific primer PCR (SSP-PCR) KIR genotyping kit and KIR ligand kit, respectively (GenoVision, Saltsjöbaden, Sweden) following the manufacture's instruction. SSP-PCR analysis of *KIR* gene repertoire has been integrated with sequence of *KIR3DL1* codon 86 in order to distinguish *KIR3DL1* alleles coding for surface receptors from those coding for polypeptide retained into the cytoplasm (19).

KIR3DL1 and *KIR3DS1* Gene Copy Number (GCN)

KIR3DL1 and *KIR3DS1* GCN was measured using a quantitative PCR method and a comparative C_t method ($\Delta\Delta C_t$). The used amplification protocol, as well as primer and probe sequences, has been published by Jiang and coworkers (20). *RNaseP* was used as two copies reference gene (TaqMan Copy Number Reference Assay, human RNase P; Applied Biosystems).

HLA-B High-Resolution Typing

Genomic DNA was used to perform *HLA-B* high-resolution typing. Some *HLA* typings were performed by sequence-based typing (SBT) using ATRIA kits according to the manufacturer's instructions (Abbott–Celera Corporation, Alameda, CA, USA). Exons 2, 3, and 4 were bidirectionally sequenced using an ABI 3130xl Genetic Analyzer (Applied Biosystems, Foster City, CA, USA), and the sequences were analyzed by Assign 3.5+ HARPS software (Conexio Genomics, Applecross, Australia). Some *HLA* typings were performed using sequence-specific oligonucleotide probes (PCR-SSOP; One Lambda, Canoga Park, CA, USA), and the results were analyzed by the software Fusion 3.0 (One Lambda, Canoga Park, CA, USA).

Antibodies and Flow Cytometry

The following mAbs, all produced in our lab, were used in this study: c218 (IgG1, anti-CD56), c127 (IgG1, anti-CD16), AZ20 and F252 (IgG1 and IgM, respectively, anti-NKp30), BAB281 and KL247 (IgG1 and IgM, respectively, anti-NKp46), Z231 (IgG1, anti-NKp44), ON72 and BAT221 (IgG1, anti-NKG2D), KRA236 and F5 (IgG1 and IgM, respectively, anti-DNAM-1), 11PB6 (IgG1, anti-KIR2DL1/S1), GL183 (IgG1, anti-KIR2DL2/

L3/S2), ECM41 (IgM, anti-KIR2DL3), DF200 (IgG1, anti-KIR2DL1/L2/L3/S1/S2/S5), FES172 (IgG2a, anti-KIR2DS4), z27 (IgG1, anti-KIR3DL1/S1), AZ158 (IgG2a, anti-KIR3DL1/L2/S1), Q66 (IgM, anti-KIR3DL2), z199 and Y9 (IgG2b and IgM, respectively, anti-NKG2A), 6A4 and A6/136 (IgG1 and IgM, respectively, anti-HLA-class I), D1/12 (IgG2a, anti-HLA-DR), 5A10 (IgG1, anti-PVR), L14 (IgG2a, anti-Nectin-2), and BAM195 (IgG1, anti-MICA). F278 (IgG1, anti-LIR-1/ILT2) mAb was kindly provided by Dr. Daniela Pende. Anti-NKG2C (IgG2b, 134522 clone), anti-ULBP-1 (IgG2a, 170818 clone), anti-ULBP-2 (IgG2a, 165903 clone), anti-ULBP-3 (IgG2a, 166510 clone), and anti-KIR2DL1-PE, -FITC, -APC, or non-conjugated (IgG1, 143211 clone) mAbs were purchased from R&D System Inc. (Abingdon, UK). Anti-KIR2DL5-PE or non-conjugated (UP-R1 clone), anti-KIR3DL1-FITC and -APC (DX9 clone) mAbs were purchased from Miltenyi Biotec (Bergisch Gladbach, Germany). Anti-CD3-FITC (UCHT-1 clone), anti-CD56-PC7 (N901 clone), anti-NKG2A-APC (z199 clone), IgG1-PE (679.1Mc7 clone, isotype control), anti-KIR3DL1/S1-PE (z27 clone), and anti-KIR2DL2/L3/S2-PC7 (GL183 clone) mAbs were purchased from Beckman Coulter Immunotech (Marseille, France). Anti-KIR2DL2/L3-S2-FITC (CHL-clone) mAb was obtained from BD Bioscience Pharmingen (San Diego, CA, USA). Anti-HLA-Bw6-FITC and anti-HLA-Bw4-FITC mAbs were purchased from One Lambda (Canoga Park, CA, USA). Anti-human HLA-E (IgG1, 3D12 clone) and anti-human HLA-G (IgG1, MEM-G/9 clone) mAbs were purchased from BioLegend (San Diego, CA, USA) and Abnova (Taipei, Taiwan), respectively.

For cytofluorimetric analyses, cells were incubated with appropriate mAbs, followed by PE-, FITC-, or APC-conjugated isotype-specific goat anti-mouse secondary reagents (Southern Biotechnology Associated, Birmingham, AL, USA; Jackson ImmunoResearch Laboratories, Suffolk, UK) and/or fluorochrome-conjugated mAbs. Cytofluorimetric analyses were performed on FACSCalibur (Becton Dickinson & Co., Mountain View, CA, USA), and data were analyzed by the CellQuest Pro software.

Mean fluorescence intensity ratios (MFIRs) were calculated by dividing the mean fluorescence intensity of stained molecules by the mean fluorescence of the respective isotype control.

Generation of Resting NK Cells and NK Cell Clones

Buffy coats from healthy donors were obtained from the Immunohematology and Transfusion Center at the S. Martino Hospital (Genova, Italy). Approval was obtained by the ethical committee of IRCCS S. Martino-IST (39/2012) of Genova (Italy). Informed consent was provided according to the Declaration of Helsinki. Human peripheral blood mononuclear cells (PBMCs) were isolated by Ficoll/Hypaque gradients. PBMCs from 120 healthy donors were screened for their KIR3DS1 expression by cytofluorimetric analysis. Donors characterized by the expression of a clearly detectable KIR3DS1^{+} NK cell subset (i.e., Z27^{+} DX9^{neg}) were further typed for their KIR repertoires and KIR-Ls (n = 40 donors). Based on KIR-L analysis, three groups of donors were selected: Bw4-I80 donors (n = 12), Bw4-T80 donors (n = 7), and Bw4^{neg} donors (n = 8). Individuals carrying both HLA-B Bw4-I80- and Bw4-T80-coding alleles were not considered in this study.

Highly purified NK cells (97–99% purity) were isolated by depletion of non-NK cells, using Miltenyi NK Cell Isolation Kit (Miltenyi Biotec, Bergisch Gladbach, Germany) from some of the selected KIR3DS1^{+} donors (two Bw4-I80, two Bw4-T80, and three Bw4^{neg} donors). NK cells were cultured on irradiated feeder cells in the presence of 100 U/ml rhIL-2 (Proleukin; Chiron Corp., Emeryville, CA, USA) and 2 μg/ml phytoemagglutinin (PHA; Life Technologies, Paisley, UK) in round-bottomed 96-well microtiter plates to obtain activated polyclonal NK cell populations or, after limiting dilution, NK cell clones as previously described (21). After 2–4 weeks of culture, the expanded NK cells were used for the phenotypic analysis and NK cell cytotoxicity experiments. The relatively limited number of donors used to generate NK cell clones reflects the difficult collection of individuals with similar characteristics in terms of KIR and KIR-Ls. Only KIR3DS1^{+} NK cell clones expressing the inhibitory receptor NKG2A as HLA-specific receptor and characterized by a sufficient growth to perform phenotypic and functional tests were selected and analyzed in the study. The number of KIR3DS1^{+}/NKG2A^{+} NK cell clones used is indicated in the legends to the figures.

Analysis of *HLA-F* Transcript

Total RNA was extracted from LCL721.221, C1R, C1R-B51, and JA3 cell lines using RNeasy mini kit (Qiagen) according to the manufacturer's instruction, and cDNA synthesis was performed using oligo-dT primers. *HLA-F* transcript analysis was performed using Hs04193807_g1: HLA-F human kit (Applied Biosystems, Foster City, CA, USA). *GAPDH* transcript was used to normalize the *HLA-F* quantity (Human GAPDH Endogenous Control Kit, Applied Biosystems, Foster City, CA, USA). The normalized *HLA-F* mRNA transcript of the tested samples was calculated as time-fold mRNA detected in the LCL721.221 cell line (chosen as reference in this study). Each cell line was analyzed in four independent experiments, and each reaction was performed at least in triplicate.

^{51}Cr Cytolytic Assays

The NK-mediated cytotoxicity was assessed in a 4-h ^{51}Cr-release assay as previously described (22). Cells used as targets in the various cytolytic assays were the following: P815 (murine mastocytoma cell line), C1R (human EBV-transformed lymphoblastoid cell line), and C1R transfected with HLA-B*51 (Bw4-I80) allele (Figure S1 in Supplementary Material). For redirected killing assays, P815 were used as target cells in the presence of mAbs of IgG isotype at a concentration of 0.5 μg/mL. For masking experiments, NK cells were pre-incubated with mAbs specific to the various NK receptors 10 min before addition of target cells; mAb concentration was 10 μg/mL. The E:T ratios are indicated in the figure legends. Δ and Δ-B51 indicate the variations of C1R or C1R-B51 lysis in the absence or presence of anti-KIR3DS1 mAb calculated for each NK cell clone.

Three different types of NK cell clones were used as effector cells for cytolytic assays: (1) NKG2A^{+}, KIR3DS1^{+}, KIR2DL2/S2/L3^{neg}, KIR2DL1/S1^{neg}, KIR3DL1^{neg}, and NKG2C^{neg}; (2) KIR3DL1^{+}, NKG2A^{+}, KIR3DS1^{neg}, KIR2DL2/S2/L3^{neg}, KIR2DL1/S1^{neg}, and

NKG2C^{neg}; and (3) KIR3DL1^{neg}, NKG2A^{+}, KIR3DS1^{neg}, KIR2DL2/S2/L3^{neg}, KIR2DL1/S1^{neg}, and NKG2C^{neg}.

Statistical Analysis

Wilcoxon–Mann–Whitney non-parametric tests were employed. The statistical significances (p value: $^{*}p < 0.1$, $^{**}p < 0.01$, $^{***}p < 0.001$) are indicated. Graphic representations and statistical analysis were performed using GraphPad Prism 6 (GraphPad Software, La Jolla, CA, USA).

RESULTS

KIR3DS1 Surface Expression on Resting NK Cells Is Not Affected by Differences in HLA-B Allotypes

Healthy donors were screened for their KIR3DS1 expression by cytofluorimetric analysis. To distinguish between KIR3DS1^{+} and KIR3DL1^{+} cells, peripheral blood NK cells were stained in double fluorescence analysis with two different mAbs: an anti-KIR3DL1-specific mAb (clone DX9) and an mAb specific for both KIR3DL1 and KIR3DS1 (clone Z27) (**Figure 1A**). Donors characterized by the expression of a KIR3DS1^{+} NK cell subset (i.e., Z27^{+} DX9^{neg}) were further typed for their *KIR* repertoires and *KIR*-Ls (Figure S2 in Supplementary Material).

KIR gene analysis allowed us to define the telomeric regions of the tested donors. In order to compare donors characterized by similar *KIR3DL1/3DS1* locus, we restricted our analysis to TelA/TelB donors, namely, individuals characterized by one *KIR3DL1* and one *KIR3DS1* gene copy. *KIR3DL1*^{neg} donors (i.e., individuals characterized by two B telomeric regions) were excluded from this study according to recent evidences showing hyporesponsive and less frequent KIR3DS1^{+} NK cells in this type of donors (23). Subsequently, based on KIR-L analysis, three groups of donors were selected: (a) Bw4-I80 donors, typed as Bw4-I80^{pos} and Bw4-T80^{neg} (including both Bw4-I80/Bw4-I80 and Bw4-I80/Bw6 donors); (b) Bw4-T80 donors, typed as Bw4-T80^{pos} and Bw4-I80^{neg} (including both Bw4-T80/Bw4-T80 and Bw4-T80/Bw6 donors); and (c) Bw4^{neg} donors, lacking both Bw4-I80 and Bw4-T80. Individuals carrying both HLA-B Bw4-I80- and Bw4-T80-coding alleles were not considered in this study. Also some HLA-A molecules are characterized by a Bw4-80I motif (e.g., HLA-A*23, -A*24, -A*25 or -A*32) (24), and their frequency in the European population is ~16% (http://www.ncbi.nlm.nih.gov/gv/mhc/ihwg.cgi?ID=9&cmd=PRJOV). Nevertheless, since not all HLA-A Bw4-I80 molecules are KIR3DL1 ligands (10, 25), we restricted our analyses to HLA-B alleles with Bw4-I80.

Comparison of KIR3DS1 surface expression on resting NK cells derived from these three donor groups did not show any significant difference in terms of percentage, MFIR, and median values (**Figure 1B**). These data are in line with previous results showing no substantial differences in *KIR3DS1* mRNA levels between HLA-B Bw4-I80 and HLA-Bw6 healthy individuals (13).

Thus, similar to what had already been demonstrated for HLA-C alleles in relation to KIR2DS1 (26), expression of different HLA-B allotypes does not influence the overall frequency of KIR3DS1^{+} NK cells and the KIR3DS1 surface expression density in peripheral blood NK cells.

Phenotypic and Functional Differences in KIR3DS1^{+}/NKG2A^{+} NK Cell Clones Derived from Donors Carrying Various HLA-B Allotypes

Several NK cell clones were generated from donors belonging to each of the three above-mentioned groups (Bw4-I80: SiCa and J76 donors, Bw4-T80: J13 donor, and Bw4^{neg}: GL115 and U58 donors). These donors were all characterized by a TelA/TelB *KIR* genotype, nevertheless, since *KIR* haplotype characterized by *KIR3DL1/3DS1* duplication has been described (20, 27),

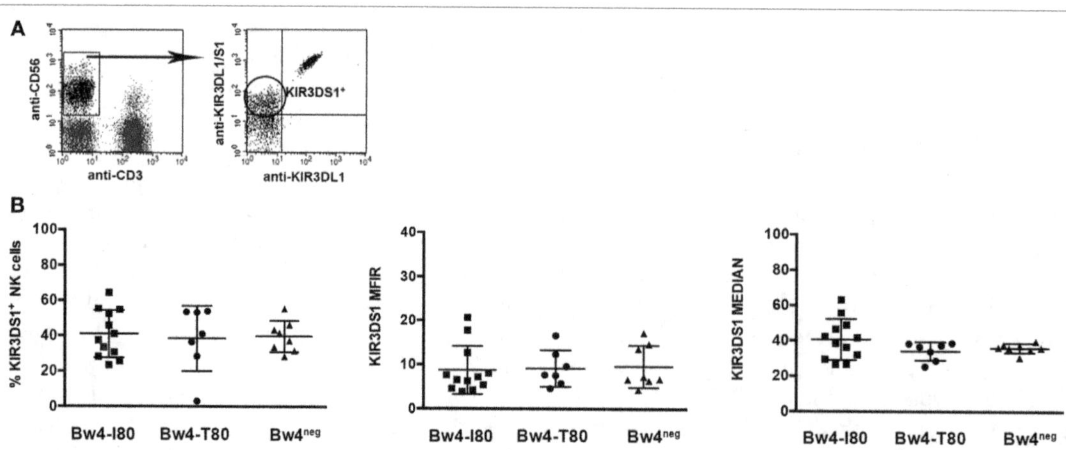

FIGURE 1 | KIR3DS1 surface expression on resting peripheral blood natural killer (NK) cells. (A) Gate strategy for selecting KIR3DS1^{+} NK cell subset (CD3^{neg} CD56^{+} KIR3DS1^{+} KIR3DL1^{neg}) in peripheral blood mononuclear cell. For staining, the following mAbs were used in combination: anti-CD56 (N901), anti-CD3 (UCHT-1), anti-KIR3DL1/S1 (z27), and anti-KIR3DL1 (DX9). **(B)** KIR3DS1 surface expression was evaluated on peripheral blood resting NK cells from HLA-B Bw4-I80 ($n = 12$, full squares), Bw4-T80 ($n = 7$, full circles), or Bw4^{neg} ($n = 8$, full triangles) healthy donors in terms of frequency, mean fluorescence intensity ratio, and median values. Summarized results are compared in three scatter dot plots, respectively. In each plot, average and standard deviation are represented.

GCN analysis was performed to verify, in these five donors, the presence of one *KIR3DL1* and *KIR3DS1* copy (**Figure 2**). To exclude the potential educational effects of different KIR/KIR-L interactions on KIR3DS1 expression and function, KIR3DS1$^+$ NK cell clones expressing the inhibitory receptor NKG2A as the only known HLA-specific receptor were selected.

As shown in **Figure 2A** (left panel), cytofluorimetric analysis revealed that KIR3DS1 expression was significantly heterogeneous among NK cell clones derived from individuals with different HLA-B allotypes. Indeed, NK cell clones derived from Bw4-I80 donors, and to a lower extent those from Bw4-T80 donors, showed a reduced expression of KIR3DS1 as compared to NK cell clones derived from Bw4^{neg} donors (***$p < 0.0001$ and **$p = 0.0018$, respectively). On the contrary, no differences in NKG2A surface expression could be detected among the same series of NK cell clones analyzed (not shown).

KIR3DS1$^+$/NKG2A$^+$ NK cell clones were then analyzed in redirected killing assays using mAbs specific for several activating or inhibitory receptors. Notably, the magnitude of cytolytic responses to KIR3DS1 mAb-mediated triggering correlated with the levels of KIR3DS1 surface expression. In particular, as shown in **Figure 2B** (left panel), NK cell clones from Bw4-I80 donors expressing a significantly lower KIR3DS1 MFIR displayed a reduced increment of cytotoxicity upon KIR3DS1 mAb-mediated cross-linking as compared to those derived from Bw4^{neg} donors (*$p = 0.0287$). On the contrary, mAb-mediated triggering of other activating receptors, including NKp46, CD16 (**Figure 2B**, left panel), NKp30, NKp44, NKG2D, and DNAM-1 (not shown), did not display any significant variation between KIR3DS1$^+$ NK cell clones derived from Bw4-I80 and those derived from Bw4^{neg} donors. Notably, the expression of these (non-HLA-specific) activating receptors did not display any significant difference in KIR3DS1$^+$ NK cell clones derived from Bw4-I80 or Bw4^{neg} donors (**Figure 3**).

Phenotypic and Functional Analysis of KIR3DS1$^+$/NKG2A$^+$ NK Cell Clones from Donors with Peculiar *KIR3DL1/S1* and *HLA-B* Combinations

Considering that it has been shown that the number of *KIR3DL1* and *KIR3DS1* gene copies plays an important role in modulating the HIV-1 control and that this effect seems to be detectable only after epistatic interactions between HLA molecules and KIRs (23, 28), we extended the phenotypic and functional analyses to additional NK cell clones derived from donors with peculiar *KIR3DL1/S1* and *HLA-B* combinations [**Figures 2A,B** (right panels) and **Figure 2C**]. In particular, NK cell clones were derived from: (a) an HLA-B Bw4-T80/Bw6 donor equipped with 2 *KIR3DS1* and 1 *KIR3DL1* GCN (P61, referred to as T80-3DS1-2GCN); (b) an HLA-B Bw4-T80/Bw6 donor carrying HLA-B*37:01, a particular Bw4-T80 allotype characterized by D77-T80 sequence (K9, referred to as HLA-B*37:01); and (c) an HLA-Bw4^{neg} donor equipped with a *KIR3DL1* allele coding for a polypeptide retained into the cytoplasm (K7, referred to as Bw4^{neg} KIR3DL1^{intra}) (29). As shown in **Figure 2A** (right panel), comparison of KIR3DS1 MFIR among NK cell clones of this second set (T80-3DS1-2GCN, HLA-B*37:01, Bw4^{neg} KIR3DL1^{intra} donors) did not reveal significant differences. On the contrary, some differences could be detected by comparing the first and the second set of donors analyzed. As shown in **Figure 2A**, KIR3DS1$^+$ NK cell clones derived from T80-3DS1-2GCN donor displayed a more heterogeneous KIR3DS1 surface expression as compared to NK cell clones derived from T80 donor carrying one *KIR3DS1* GCN. Nevertheless, NK cell clones derived from T80-3DS1-2GCN donor (but also those from B*37:01 donor) did not display significant differences in terms of KIR3DS1 MFIR as compared to NK clones of the first set.

On the contrary, NK cell clones from the Bw4^{neg} KIR3DL1^{intra} donor displayed a significantly lower KIR3DS1 surface expression (***$p < 0.0001$) (**Figure 2A**) but a similar cytotoxicity upon KIR3DS1 mAb-mediated triggering (**Figure 2B**) as compared to NK cell clones derived from Bw4^{neg} donors expressing surface KIR3DL1.

Moreover, the comparison of the cytotoxicity in redirected killing assays between the first and second sets of donors revealed that NK cell clones derived from the second set were characterized by higher increments of cytotoxicity upon KIR3DS1 mAb-mediated triggering than NK cell clones from Bw4-I80 donors (*$p = 0.0243$ with T80-3DS1-2GCN donor, *$p = 0.0014$ with HLA-B*37:01 donor, and **$p = 0.0059$ with Bw4^{neg} KIR3DL1^{intra} donor) (**Figure 2B**).

Correlation Analysis between Expression of KIR3DS1 and Non-HLA-Specific Activating Receptors in Donors Characterized by Different HLA-B Allotypes

The existence of a possible correlation between the expression of KIR3DS1 and that of relevant non-HLA-specific activating receptors (i.e., NCRs, NKG2D and DNAM-1) was analyzed in all KIR3DS1$^+$/NKG2A$^+$ NK cell clones considered in the present study. Of interest, only in T80-3DS1-2GCN donor, KIR3DS1 surface density correlated inversely with NKp46 and directly with NKG2D MFIR (*$p = 0.0154$ and *$p = 0.0107$, respectively) (**Figure 4**). Notably, this result was detected only in NK cell clones from these donors characterized by higher and heterogeneous values of KIR3DS1 MFIR (**Figure 2A**, right panel). Based on this observation, it cannot be ruled out that, in given HLA/KIR haplotype settings, KIR3DS1 may influence the surface density of NKp46 and NKG2D.

KIR3DS1-Mediated Recognition of HLA-B*51 Allele on Transfected Cells

According to previous data suggesting a possible interaction between KIR3DS1 and HLA-B Bw4-I80 alleles (30) and considering the high level of homology between the extracellular domains of KIR3DS1 and KIR3DL1 as well as the ability of KIR3DL1 to recognize in most instances HLA-B Bw4-I80 alleles with higher efficiency than HLA-B Bw4-T80 alleles (9, 25), all KIR3DS1$^+$/NKG2A$^+$ NK cell clones were assessed for their capability of killing HLA-B Bw4-I80 target cells. In particular, the killing activity of KIR3DS1$^+$/NKG2A$^+$ NK cell clones was assessed against the C1R cell line transfected or not with HLA-B*51 allele (31).

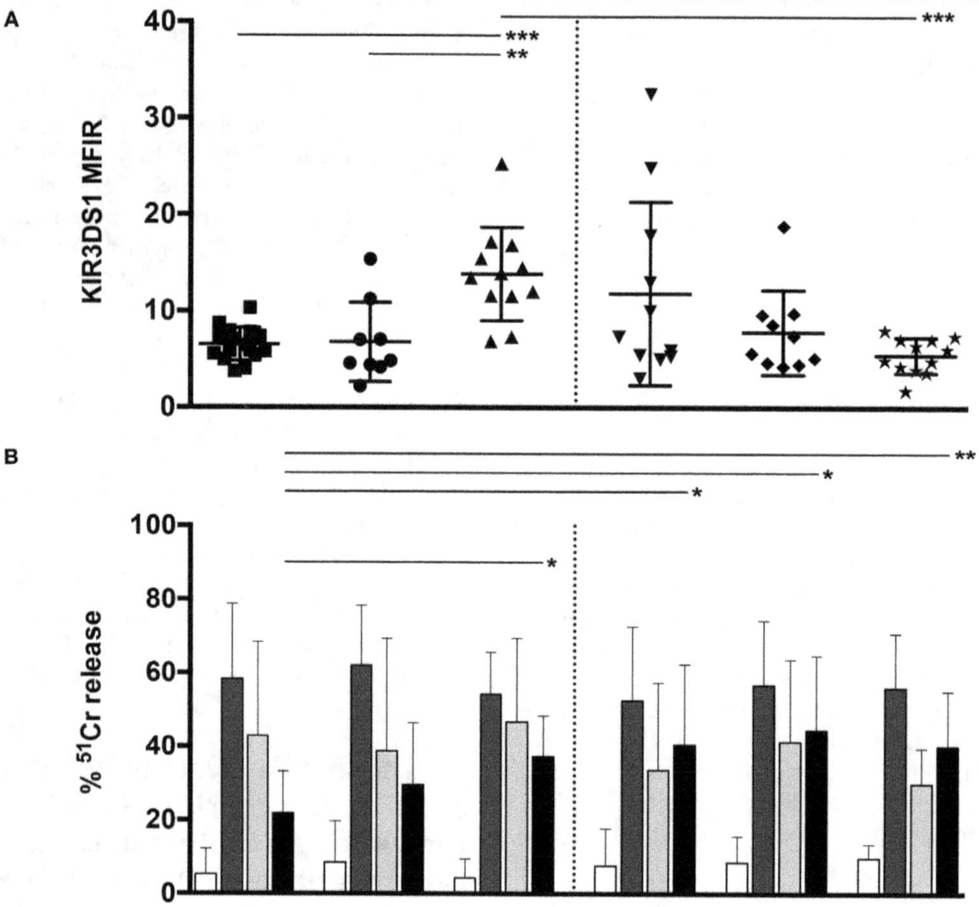

HLA-B	Bw4-I80	Bw4-T80	Bw4^{neg}	Bw4-T80	B*37:01	Bw4^{neg}
KIR3DS1 GCN	1	1	1	2	1	1
KIR3DL1	surface	surface	surface	surface	surface	intra

C

Donor	HLA-B Bw4	HLA-B Bw6		*KIR3DL1* GCN	*KIR3DS1* GCN
SICA	B*51:01P[&]	pos		1	1
J76	B*51:01P,*57:01P	neg		1	1
J13	B*44:03P	pos		1	1
GL115	neg	pos		1	1
U58	neg	pos		1	1
K7	neg	pos		1	1
P61	B*44:02	pos		1	2
K9	B*37:01	pos		1	1

FIGURE 2 | KIR3DS1 surface expression and mAb-mediated cross-linking of KIR3DS1 in KIR3DS1$^+$/NKG2A$^+$ natural killer (NK) cell clones. KIR3DS1$^+$/NKG2A$^+$ NK cell clones derived from HLA-B-typed donors were analyzed for KIR3SD1 surface expression **(A)** and for ^{51}Cr release in redirected killing assay against P815 cell line **(B)**. Data related to HLA typing and surface/intra *KIR3DL1* presence are indicated in the box below panel **(B)**. **(A)** The NK cell clones of the first set (left panel) were derived from Bw4-I80 (n = 10, full square), Bw4-T80 (n = 9, full circle), and Bw4^{neg} (n = 12, full triangles) donors, whereas NK cell clones of the second set (right panel) were derived from T80-3DS1-2GCN (n = 11, overturned full triangle), HLA-B*37:01 (n = 10, full rhombus), and Bw4^{neg} KIR3DL1^{intra} (n = 12, full star) donors. **(B)** The cytolytic activity of the first (left panel) and second (right panel) set of KIR3DS1$^+$/NKG2A$^+$ NK cell clones was evaluated in redirected killing assays upon mAb-mediated cross-linking of different NK receptors. Cytolytic assays were performed in the absence of mAbs (white bar) or in the presence of anti-CD16 (c127, dark gray bar), anti-NKp46 (BAB281, light gray bar), or anti-KIR3DS1 (z27, black bar) mAbs. Average, standard deviation, and p values are indicated ($^*p < 0.1$, $^{**}p < 0.01$, and $^{***}p < 0.001$). **(C)** HLA-B typing, *KIR3DL1*, and *KIR3DS1* gene copy number (GCN) analyses of the indicated donors are shown.
[&]P: indicated a group of alleles with identical sequence at exons 2 and 3 and therefore sharing the same antigen-binding domains.

Considering recent studies showing that HLA-F OCs are high-affinity ligands of KIR3DS1 (16, 17), the *HLA-F* transcript was analyzed in C1R and C1R-B51 target cells to assess whether these cells would express this KIR3DS1 ligand. JA3 (a Jurkat clone) (32) and LCL 721.221 cell lines were used as negative control and positive control, respectively (16, 33). As shown in **Figure 5** and Figure S3 in Supplementary Material, C1R and C1R-B51 expressed very low levels of *HLA-F* transcript similar to JA3 cells, whereas LCL 721.221 expressed *HLA-F* mRNA five times more than C1R-B51. Thus, possible differences of lysis between C1R and C1R-B51 may not be attributed to recognition of HLA-F by KIR3DS1.

A representative cytolytic experiment against C1R/C1R-B51 target cells performed using as effector cell a KIR3DS1+/NKG2A+ NK cell clone derived from the T80-3DS1-2GCN donor is shown in **Figure 6**. As controls, KIR3DS1neg/NKG2A+ and KIR3DL1+/NKG2A+ NK cell clones derived from the same donor were analyzed. In the absence of mAbs, C1R-B51 cells were killed slightly more efficiently than un-transfected C1R by the KIR3DS1+/NKG2A+ NK cell clone (*$p = 0.0159$). Notably, this difference of lysis was abolished upon mAb-mediated masking of KIR3DS1, suggesting a possible positive recognition of HLA-B*51 by KIR3DS1 (**Figure 6A**). Interestingly, this result was more evident when the experiment was performed in the presence of anti-NKG2D mAb. Thus, as shown in **Figure 6B**, the difference between C1R and C1R-B51 killing was amplified by NKG2D mAb-mediated blocking (**$p = 0.0022$). Importantly, further masking of KIR3DS1 abrogated this difference (**$p = 0.0065$) (**Figure 6B**). Similar data were obtained upon additional mAb-mediated masking of NKG2A.

A similar experimental approach was applied to the NK cell clones generated from the other donors analyzed (**Figure 7**). These clones were characterized by a substantially homogeneous NKG2D expression. The cytolytic assays were performed not only in the presence of blocking NKG2D but also upon mAb-mediated masking of NKG2A in order to further reduce possible functional variations caused by different intensities of inhibitory signals consequent to NKG2A/HLA-E interaction. Then, cytotoxicity was evaluated upon additional blocking of KIR3DS1 (**Figure 7**). In this set of experiments, in order to better appreciate the KIR3DS1 contribution to target killing, each NK cell clone was also analyzed for possible variations in the lysis of C1R or C1R-B51 targets in the absence or in the presence of KIR3DS1 mAb-mediated blocking (Δ_{C1R} and $\Delta_{C1R\text{-}B51}$).

In Bw4-I80 donors, lysis of C1R and C1R-B51 was not affected by the addition of anti-KIR3DS1 mAb (as shown by the raw data as well as Δ_{C1R} and $\Delta_{C1R\text{-}B51}$ values) (**Figure 7A**). On the other hand, in Bw4neg donors, the addition of anti-KIR3DS1 mAb slightly decreased C1R-B51 killing and a significant gap between Δ_{C1R} and $\Delta_{C1R\text{-}B51}$ could be detected (*$p = 0.0188$).

KIR3DS1+/NKG2A+ NK cell clones from the T80-3DS1-2GCN donor killed more efficiently C1R-B51 than C1R. Moreover, the addition of anti-KIR3DS1 mAb abolished this difference. Remarkably, in this donor, the comparison between Δ_{C1R} and $\Delta_{C1R\text{-}B51}$ showed a highly significant difference (***$p < 0.0001$, **Figure 7B**). Significant differences between Δ_{C1R} and $\Delta_{C1R\text{-}B51}$ could also be observed in KIR3DS1+/NKG2A+ NK cell clones derived from both HLA-B B*37:01/Bw6 and Bw4neg KIR3DL1intra donors

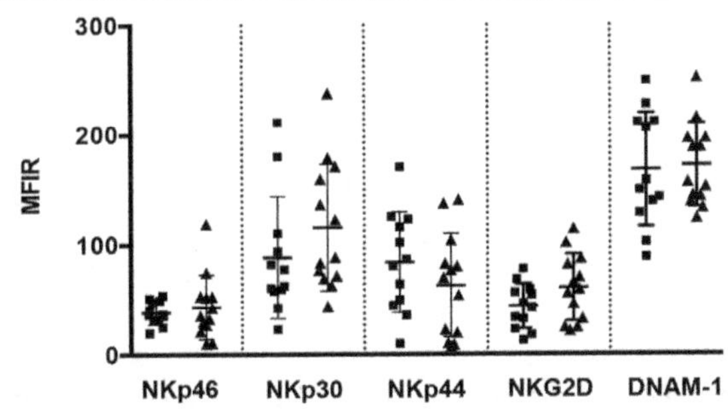

FIGURE 3 | Surface expression of several non-HLA-specific activating receptors. Staining of NKp46 (BAB281), NKp30 (AZ20), NKp44 (z231), NKG2D (ON72), and DNAM-1 (KRA236) molecules on KIR3DS1+/NKG2A+ natural killer (NK) cell clones derived from Bw4-I80 (*n* = 12, full squares) or Bw4neg carrying surface KIR3DL1 (*n* = 13, full triangles) donors were compared in terms of mean fluorescence intensity ratio (MFIR) in scatter dot plot representation. Average and standard deviation are shown.

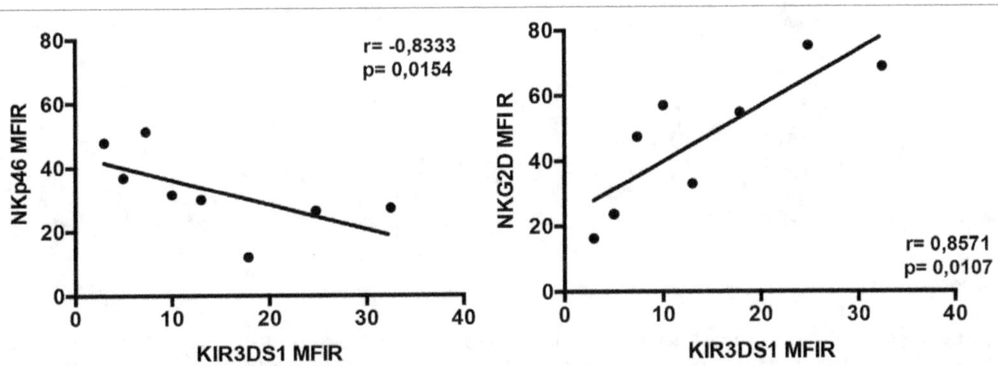

FIGURE 4 | Phenotypic analyses of KIR3DS1+/NKG2A+ natural killer (NK) cell clones derived from T80-3DS1-2GCN donor. Correlation analyses among NKp46 or NKG2D activating receptors and KIR3DS1 surface expression in several KIR3DS1+/NKG2A+ NK cell clones (*n* = 8) derived from a T80-3DS1-2GCN donor are represented. Linear regression values (*r*) and *p* values are shown.

(**$p = 0.0073$ and *$p = 0.0110$, respectively). All together, these data suggest a direct involvement of KIR3DS1 in the recognition of HLA-B*51 (Bw4-I80) target cells. Remarkably, this result can be detected only when NK cells were generated in particular KIR/HLA combination settings. Indeed, KIR3DS1-mediated recognition of HLA-B*51 occurs only when NK cell clones were derived from Bw4-I80^{neg} donors (Bw4-T80-2GCN, B*37:01/Bw6, or Bw4^{neg}), suggesting a role for HLA-B/KIR3DS1 interaction in the process of NK cell education.

Notably, when comparing KIR3DS1^{+}/NKG2A^{+} and KIR3DS1^{neg}/NKG2A^{+} NK cell clones, a significant difference in NKp46 surface expression was observed only in Bw4-I80 donors. Thus, KIR3DS1^{+}/NKG2A^{+} NK cell clones expressed lower levels of NKp46 than KIR3DS1^{neg}/NKG2A^{+} NK cell clones (***$p = 0.0005$), further corroborating the possibility that HLA-B/KIR3DS1 interaction may be involved in the process of NK cell education (**Figure 8**).

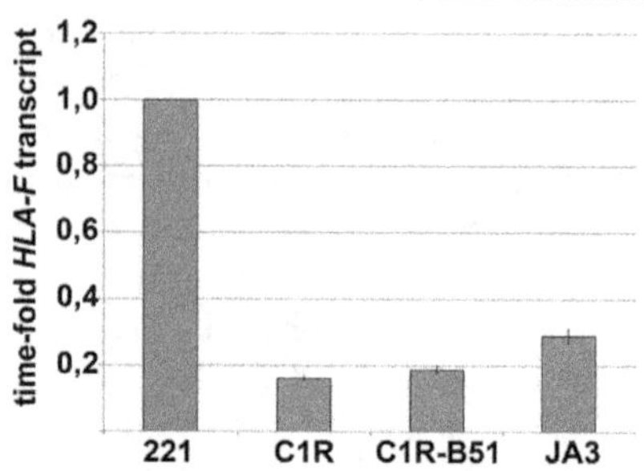

FIGURE 5 | C1R and C1R-B51 express low level of *HLA-F* transcript. *HLA-F* mRNA amount detected in the analyzed cell lines has been normalized to *GAPDH* transcript. The normalized *HLA-F* mRNA transcript of the tested cell lines was calculated as time-fold the mRNA detected in LCL721.221 cell line (chosen as reference). Mean values obtained in four different experiments and their standard deviations are reported.

DISCUSSION

A number of studies described an association between given KIR/HLA gene combinations and clinical outcome in various immune challenges and reported a possible perturbation of KIR/HLA interactions by the presented peptide. In particular, an important combined role played by KIR3DS1 and HLA-B Bw4-I80 in controlling HIV infection (11, 12) and the recognition of specific HIV-derived peptides associated with HLA-B*57 alleles by KIR3DS1 have been described (15).

In the present study, we provided the first evidence of a direct involvement of KIR3DS1 in the NK-mediated recognition of HLA-B*51 surface molecules expressed on target cells. This capability is manifested only when KIR3DS1 is expressed by NK cells derived from individuals carrying particular KIR/HLA combinations. Although our study was performed on a limited number of donors (due to the difficulty in the collection of individuals with similar characteristics in terms of KIR and KIR-Ls as well as in the generation and expansion of appropriate NK cell clones), our results suggest a role for this activating receptor in the process of NK cell education. In particular, the KIR3DS1-mediated positive recognition of HLA-B*51 (Bw4-I80) could be detected in NK cell clones derived from Bw4-I80^{neg} donors (one Bw4-T80-2GCN, one B*37:01/Bw6, and three Bw4^{neg}) but not in those from Bw4-I80 donors (two donors). Thus, in a manner reminiscent of that previously described for KIR2DS1 (26, 34), the interaction of KIR3DS1 with its self-HLA-B class I ligand (Bw4-I80 alleles) would affect the subsequent response mediated by this aKIR. Indeed, in HLA-B Bw4-I80 donors, the process of NK cell education that is taking place via NKG2A/self-HLA-E

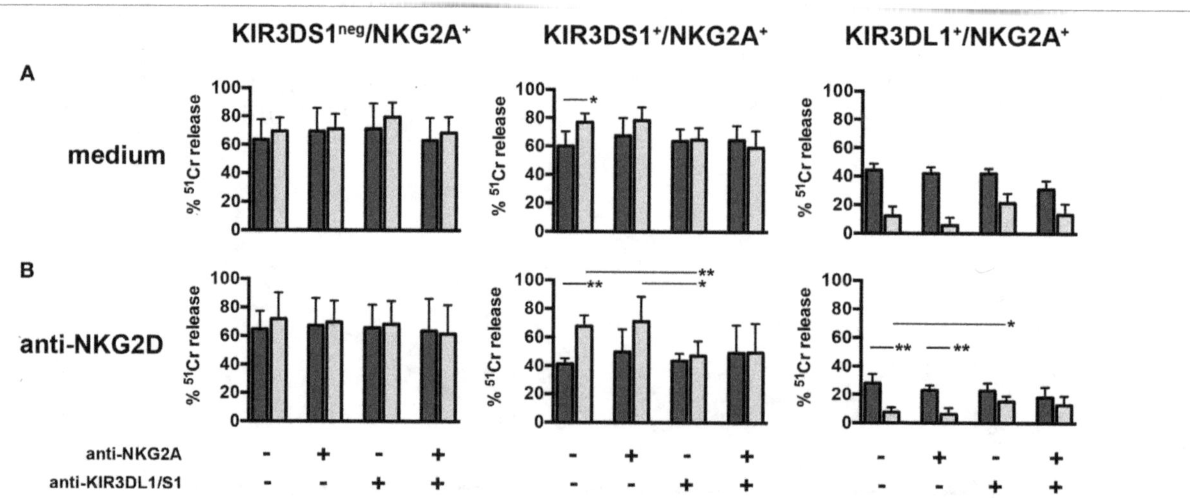

FIGURE 6 | Killing of C1R-B51 target cells by natural killer (NK) cell clones derived from a T80-3DS1-2GCN donor. Three representative NK cell clones (one KIR3DS1^{neg}/NKG2A^{+}, one KIR3DS1^{+}/NKG2A^{+} and one KIR3DL1^{+}/NKG2A^{+}) derived from the T80-3DS1-2GCN donor were tested for cytotoxic activity against C1R (dark gray histograms) and C1R-B51 (light gray histograms) target cell lines. Experiments were performed in the absence **(A)** or in the presence of anti-NKG2D masking mAb **(B)**. Additional mAb-mediated maskings were carried out as indicated below panel B. Histograms summarized results of three independent experiments in duplicate. Average, standard deviation, and *p* values are shown (*$p < 0.1$ and **$p < 0.01$).

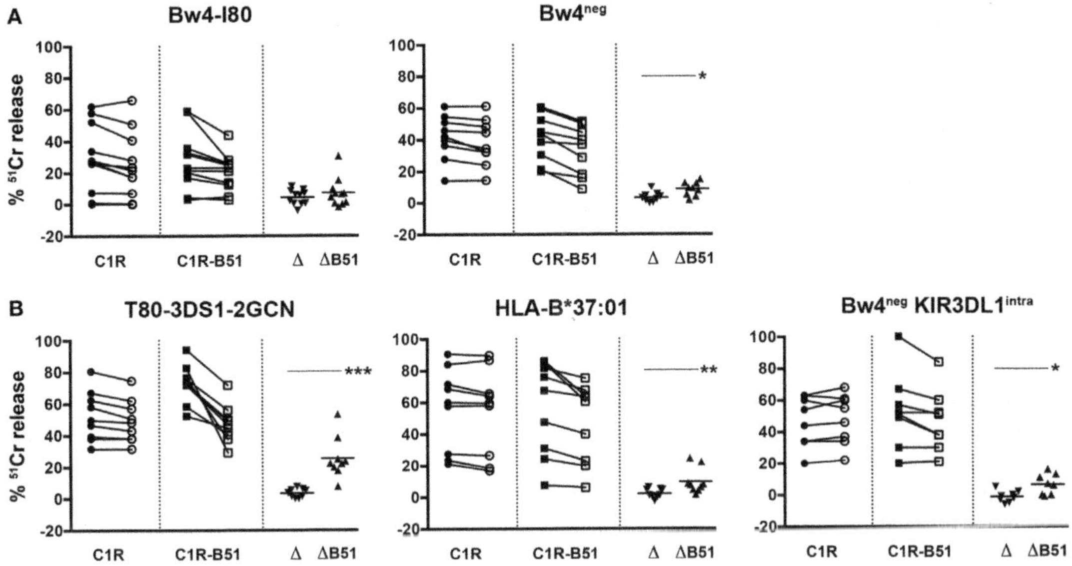

FIGURE 7 | C1R-B51 killing by KIR3DS1⁺/NKG2A⁺ natural killer (NK) cell clones. NK cell clones were tested in ^{51}Cr-release cytotoxicity assay against C1R (circle) and C1R-B51 (square) target cell lines in the presence of anti-NKG2D + anti-NKG2A masking mAbs (full symbols) or anti-NKG2D + anti-NKG2A + anti-KIR3DS1 masking mAbs (empty symbols). Δ and Δ-B51 indicate the variations of C1R or C1R-B51 lysis in the absence or presence of anti-KIR3DS1 mAb for each NK cell clone. NK cell clones derived from Bw4-I80 ($n = 11$) and Bw4^{neg} ($n = 9$) donors are represented in panel **(A)**, whereas those derived from T80-3DS1-2GCN ($n = 9$), HLA-B*37:01 ($n = 9$), and Bw4^{neg} KIR3DL1^{intra} ($n = 8$) donors are represented in panel **(B)**. Plots summarized results of three independent experiments in duplicate. p values indicate a statistically significant difference between the groups (*$p < 0.1$, **$p < 0.01$, and ***$p < 0.001$).

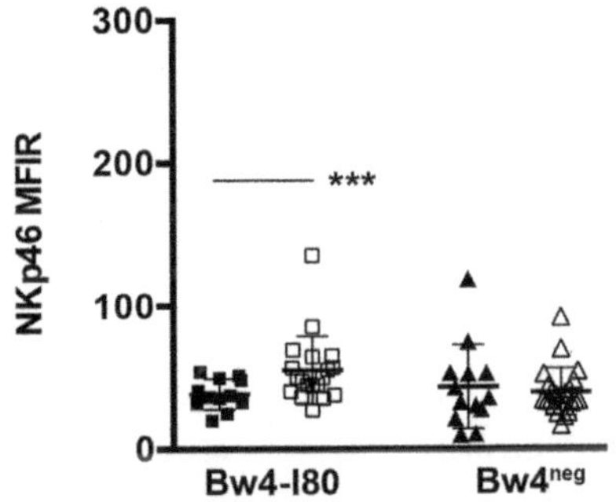

FIGURE 8 | NKp46 surface expression on KIR3DS1$^+$/NKG2A$^+$ and KIR3DS1^{neg}/NKG2A$^+$ natural killer (NK) cell clones derived from Bw4-I80 or Bw4^{neg} donors. KIR3DS1$^+$/NKG2A$^+$ (full symbols) and KIR3DS1^{neg}/NKG2A$^+$ (empty symbols) NK cell clones derived from Bw4-I80 (squares, $n = 12$ and $n = 19$, respectively) and Bw4^{neg} (triangles, $n = 13$ and $n = 22$ respectively) donors were stained with anti-NKp46 mAb (BAB281). Results are summarized in scatter plot analysis. Average, standard deviation, and p values are indicated (***$p < 0.001$).

interaction is characterized by a decrement in NK cell responsiveness upon KIR3DS1 engagement, thus ensuring self-tolerance. In this context, it has to be considered that KIR3DS1 responsiveness in redirected killing assays was lower in Bw4-I80 than in Bw4-I80^{neg} donors. In particular, among those analyzed, KIR3DS1 was the only receptor showing a significant differential expression and responsiveness between Bw4-I80 and Bw4^{neg} donors.

The analysis performed in the present study was limited to KIR3DS1$^+$/NKG2A$^+$ NK cell clones in order to evaluate whether KIR3DS1 expression could confer alloreactivity to non-alloreactive NKG2A$^+$ NK cells and to minimize the effect that other self KIR/HLA interactions would exert on NK cell education. Several studies have suggested that, in given donor/recipient pairs, the expression of aKIRs, such as KIR2DS1, can amplify the size of the alloreactive NK cell subset. This effect is particularly relevant in the successful therapy of high-risk acute leukemias in the haplo-HSCT setting (35–39). Indeed, it has been shown that the "non-alloreactive" NKG2A$^+$ iKIRs^{neg} NK cells can display alloreactivity against HLA-C2$^+$ recipient cells when they co-express KIR2DS1 (40). A similar effect could be true, at least in part, also for KIR3DS1. In this context, the definition of the specificity/function of KIR3DS1 would have important implications not only to identify donors capable of generating alloreactive NK cells (on the basis of the existence of a KIR/HLA-class I mismatch) but also to select the best donor, according to the size of the alloreactive NK cell subset.

On the basis of a recent study (23) showing that KIR3DL1-dependent licensing of NK cells could be involved in shaping a strong antiviral response mediated by KIR3DS1$^+$ NK cells, we also analyzed KIR3DS1 expression and responsiveness in donors characterized by peculiar KIR/HLA combinations. In this context, we showed that KIR3DS1 mAb-mediated triggering of cytotoxicity in NK cell clones from a Bw4^{neg} donor carrying KIR3DL1^{intra} is similar to that detectable in NK cell clones from Bw4^{neg} donors expressing surface KIR3DL1. It is noteworthy that this increment of cytotoxicity occurred despite the low surface expression of KIR3DS1 (**Figure 2**). Remarkably, NK cell clones from the Bw4^{neg} donor carrying KIR3DL1^{intra} were able to kill HLA-B*51$^+$ target cells with efficacy comparable to that of NK cell clones derived from Bw4^{neg} donors with surface KIR3DL1.

Moreover, different from KIR3DS1+ NK cell clones of classical HLA-B Bw4-T80 donors, clones derived from a donor carrying the HLA-B*37 allele (a particular HLA-Bw4-T80 allotype characterized by the D77-T80 sequence) were characterized by higher KIR3DS1 responsiveness in redirected killing assays and higher efficiency of killing HLA-B51+ target cells as compared to NK clones from HLA-B Bw4-I80 donors.

However, the highest increment of cytotoxicity against C1R-B51 cells was detected in NK cell clones from a T80-Bw4 donor carrying two copies of *KIR3DS1* and one copy of *KIR3DL1*. Interestingly, in this donor, an inverse correlation between KIR3DS1 and NKp46 MFIR could also be detected. This finding is relevant in the context of NK cell education. Indeed, a low surface expression of NKp46 combined with high KIR3DS1 surface density on NK cell clones may prevent possible KIR3DS1-mediated autoreactive responses in non-pathological conditions. In the same donor, a direct correlation between KIR3DS1 and NKG2D surface expression was also observed. Notably, these correlations could be observed only in NK clones from this donor in which KIR3DS1 surface expression was more heterogeneous and higher as compared to Bw4-T80 donor carrying one copy of *KIR3DS1* and one copy of *KIR3DL1*.

The direct correlation between KIR3DS1 and NKG2D surface expression is particularly interesting if we consider that NKG2D pathway is less functional in certain viral infections. Indeed, previous studies have described that several viral immune evasion strategies possibly evolved to elude NKG2D-mediated immune-surveillance (30, 41–44). For example, the HIV Nef protein prevents the expression of some NKG2D ligands at the surface of infected cells (45, 46). According to these data, in an attempt to simulate the compromised NKG2D function occurring during certain viral infections, we analyzed the effect of mAb-mediated NKG2D blocking in KIR3DS1+ NK clones. Thanks to this experimental approach, we could show that KIR3DS1 function may be crucial, primarily during viral infections in which other triggering signals, such as those through NKG2D, are compromised. In this context, it is also important to take into consideration that the expression of HLA-B can be modified only marginally during certain viral infections. Thus, in HIV-1-infected viremic patients, while HLA-A and -Bw6 surface molecules were significantly downmodulated in T cell blasts, HLA-B Bw4 alleles were not (47).

Considering the "discontinuity theory for immunity" (48), it is also possible that KIR3DS1+ NK cells in donors expressing specific HLA class I (Bw4-I80) may be less responsive under normal conditions. However, when a non-physiologic triggering signal is given to the cells (e.g., viral infections), the equilibrium that is maintaining tolerance could be disrupted and may become an important component of an efficient immune response (34). This mechanism could explain the association between the expression of *KIR3DS1* in conjunction with *HLA-B Bw4-I80* in patients with chronic HIV-1 infection and a slower progression to AIDS.

KIR3DS1 associates DAP12 ITAM-bearing molecules in its cytoplasmic tail to enable signal transduction. Surprisingly, in recent years, different studies showed that ITAMs can also generate an inhibitory signal in addition to the activating ones. In particular, the same ITAM-coupled receptors can generate both positive and negative signals (49–51). The molecular basis of this dual function is not well understood at the present; however, it has been suggested that the avidity of receptor ligation may define the nature of response. In this regard, one may speculate that the peptide in the HLA-Bw4-I80 groove could determine change of affinity of KIR3DS1 ligation. Thus, healthy self peptides could mediate a low-affinity, tolerogenic signal, whereas viral peptides (i.e., HIV derived) would allow high-affinity activating signals. This hypothesis would be in line with the protective effect exerted by the combined presence of *KIR3DS1* and *HLA-B Bw4-I80* in patients with chronic HIV-1 infection.

All these considerations are consistent with an important role of KIR3DS1 in the control of viral infections (8, 30), primarily in post-transplantation settings (52). Since transplantation from donors displaying NK-cell alloreactivity and expressing *KIR2DS1* and/or *KIR3DS1* has been associated with a reduced risk of no relapse mortality, that is largely infection related, and with significantly better event-free survival (38, 53), our results could further improve the selection of the most suitable donor, taking into account the expression of not only *KIR3DS1* but also the self HLA-B allotype expressed.

ETHICS STATEMENT

Buffy coats from healthy donors were obtained from the Immunohematology and Transfusion Center of the S. Martino Hospital (Genova, Italy). Approval was obtained by the ethical committee of IRCCS S. Martino-IST (39/2012) of Genova (Italy). Informed consent was provided according to the Declaration of Helsinki.

AUTHOR CONTRIBUTIONS

SC and SS designed and performed research, interpreted data, and wrote the paper; MF performed research, interpreted data, and wrote the paper; MB, CA, LG, and MM performed research and analyzed data; LM interpreted and critically revised the paper; and AM interpreted, critically revised data, and wrote the paper.

REFERENCES

1. Caligiuri MA. Human natural killer cells. *Blood* (2008) 112(3):461–9. doi:10.1182/blood-2007-09-077438
2. Vivier E, Raulet DH, Moretta A, Caligiuri MA, Zitvogel L, Lanier LL, et al. Innate or adaptive immunity? The example of natural killer cells. *Science* (2011) 331(6013):44–9. doi:10.1126/science.1198687
3. Parham P. MHC class I molecules and KIRs in human history, health and survival. *Nat Rev Immunol* (2005) 5(3):201–14. doi:10.1038/nri1570
4. Moretta A, Bottino C, Vitale M, Pende D, Biassoni R, Mingari MC, et al. Receptors for HLA class-I molecules in human natural killer cells. *Annu Rev Immunol* (1996) 14:619–48. doi:10.1146/annurev.immunol.14.1.619

5. Moretta A, Sivori S, Vitale M, Pende D, Morelli L, Augugliaro R, et al. Existence of both inhibitory (p58) and activatory (p50) receptors for HLA-C molecules in human natural killer cells. *J Exp Med* (1995) 182(3):875–84. doi:10.1084/jem.182.3.875
6. Tomasello E, Olcese L, Vely F, Geourgeon C, Blery M, Moqrich A, et al. Gene structure, expression pattern, and biological activity of mouse killer cell activating receptor-associated protein (KARAP)/DAP-12. *J Biol Chem* (1998) 273(51):34115–9. doi:10.1074/jbc.273.51.34115
7. Parham P, Norman PJ, Abi-Rached L, Guethlein LA. Variable NK cell receptors exemplified by human KIR3DL1/S1. *J Immunol* (2011) 187(1):11–9. doi:10.4049/jimmunol.0902332
8. Korner C, Altfeld M. Role of KIR3DS1 in human diseases. *Front Immunol* (2012) 3:326. doi:10.3389/fimmu.2012.00326
9. Cella M, Longo A, Ferrara GB, Strominger JL, Colonna M. NK3-specific natural killer cells are selectively inhibited by Bw4-positive HLA alleles with isoleucine 80. *J Exp Med* (1994) 180(4):1235–42. doi:10.1084/jem.180.4.1235
10. Stern M, Ruggeri L, Capanni M, Mancusi A, Velardi A. Human leukocyte antigens A23, A24, and A32 but not A25 are ligands for KIR3DL1. *Blood* (2008) 112(3):708–10. doi:10.1182/blood-2008-02-137521
11. Martin MP, Gao X, Lee JH, Nelson GW, Detels R, Goedert JJ, et al. Epistatic interaction between KIR3DS1 and HLA-B delays the progression to AIDS. *Nat Genet* (2002) 31(4):429–34. doi:10.1038/ng934
12. Alter G, Martin MP, Teigen N, Carr WH, Suscovich TJ, Schneidewind A, et al. Differential natural killer cell-mediated inhibition of HIV-1 replication based on distinct KIR/HLA subtypes. *J Exp Med* (2007) 204(12):3027–36. doi:10.1084/jem.20070695
13. Alter G, Rihn S, Walter K, Nolting A, Martin M, Rosenberg ES, et al. HLA class I subtype-dependent expansion of KIR3DS1+ and KIR3DL1+ NK cells during acute human immunodeficiency virus type 1 infection. *J Virol* (2009) 83(13):6798–805. doi:10.1128/JVI.00256-09
14. Carr WH, Rosen DB, Arase H, Nixon DF, Michaelsson J, Lanier LL. Cutting edge: KIR3DS1, a gene implicated in resistance to progression to AIDS, encodes a DAP12-associated receptor expressed on NK cells that triggers NK cell activation. *J Immunol* (2007) 178(2):647–51. doi:10.4049/jimmunol.178.2.647
15. O'Connor GM, Vivian JP, Gostick E, Pymm P, Lafont BA, Price DA, et al. Peptide-dependent recognition of HLA-B*57:01 by KIR3DS1. *J Virol* (2015) 89(10):5213–21. doi:10.1128/JVI.03586-14
16. Garcia-Beltran WF, Holzemer A, Martrus G, Chung AW, Pacheco Y, Simoneau CR, et al. Open conformers of HLA-F are high-affinity ligands of the activating NK-cell receptor KIR3DS1. *Nat Immunol* (2016) 17(9):1067–74. doi:10.1038/ni.3513
17. Burian A, Wang KL, Finton KA, Lee N, Ishitani A, Strong RK, et al. HLA-F and MHC-I open conformers bind natural killer cell Ig-like receptor KIR3DS1. *PLoS One* (2016) 11(9):e0163297. doi:10.1371/journal.pone.0163297
18. Lopez-Vazquez A, Rodrigo L, Martinez-Borra J, Perez R, Rodriguez M, Fdez-Morera JL, et al. Protective effect of the HLA-Bw4I80 epitope and the killer cell immunoglobulin-like receptor 3DS1 gene against the development of hepatocellular carcinoma in patients with hepatitis C virus infection. *J Infect Dis* (2005) 192(1):162–5. doi:10.1086/430351
19. Alicata C, Pende D, Meazza R, Canevali P, Loiacono F, Bertaina A, et al. Hematopoietic stem cell transplantation: improving alloreactive Bw4 donor selection by genotyping codon 86 of KIR3DL1/S1. *Eur J Immunol* (2016) 46(6):1511–7. doi:10.1002/eji.201546236
20. Jiang W, Johnson C, Jayaraman J, Simecek N, Noble J, Moffatt MF, et al. Copy number variation leads to considerable diversity for B but not A haplotypes of the human KIR genes encoding NK cell receptors. *Genome Res* (2012) 22(10):1845–54. doi:10.1101/gr.137976.112
21. Moretta A, Tambussi G, Bottino C, Tripodi G, Merli A, Ciccone E, et al. A novel surface antigen expressed by a subset of human CD3- CD16+ natural killer cells. Role in cell activation and regulation of cytolytic function. *J Exp Med* (1990) 171(3):695–714. doi:10.1084/jem.171.3.695
22. Moretta A, Vitale M, Sivori S, Bottino C, Morelli L, Augugliaro R, et al. Human natural killer cell receptors for HLA-class I molecules. Evidence that the Kp43 (CD94) molecule functions as receptor for HLA-B alleles. *J Exp Med* (1994) 180(2):545–55. doi:10.1084/jem.180.2.545
23. Pelak K, Need AC, Fellay J, Shianna KV, Feng S, Urban TJ, et al. Copy number variation of KIR genes influences HIV-1 control. *PLoS Biol* (2011) 9(11):e1001208. doi:10.1371/journal.pbio.1001208
24. Foley BA, De Santis D, Van Beelen E, Lathbury LJ, Christiansen FT, Witt CS. The reactivity of Bw4+ HLA-B and HLA-A alleles with KIR3DL1: implications for patient and donor suitability for haploidentical stem cell transplantations. *Blood* (2008) 112(2):435–43. doi:10.1182/blood-2008-01-132902
25. Saunders PM, Pymm P, Pietra G, Hughes VA, Hitchen C, O'Connor GM, et al. Killer cell immunoglobulin-like receptor 3DL1 polymorphism defines distinct hierarchies of HLA class I recognition. *J Exp Med* (2016) 213(5):791–807. doi:10.1084/jem.20152023
26. Fauriat C, Ivarsson MA, Ljunggren HG, Malmberg KJ, Michaelsson J. Education of human natural killer cells by activating killer cell immunoglobulin-like receptors. *Blood* (2010) 115(6):1166–74. doi:10.1182/blood-2009-09-245746
27. Pyo CW, Wang R, Vu Q, Cereb N, Yang SY, Duh FM, et al. Recombinant structures expand and contract inter and intragenic diversification at the KIR locus. *BMC Genomics* (2013) 14:89. doi:10.1186/1471-2164-14-89
28. Beziat V, Traherne JA, Liu LL, Jayaraman J, Enqvist M, Larsson S, et al. Influence of KIR gene copy number on natural killer cell education. *Blood* (2013) 121(23):4703–7. doi:10.1182/blood-2012-10-461442
29. Pando MJ, Gardiner CM, Gleimer M, McQueen KL, Parham P. The protein made from a common allele of KIR3DL1 (3DL1*004) is poorly expressed at cell surfaces due to substitution at positions 86 in Ig domain 0 and 182 in Ig domain 1. *J Immunol* (2003) 171(12):6640–9. doi:10.4049/jimmunol.171.12.6640
30. Carrington M, Alter G. Innate immune control of HIV. *Cold Spring Harb Perspect Med* (2012) 2(7):a007070. doi:10.1101/cshperspect.a007070
31. Vitale M, Sivori S, Pende D, Augugliaro R, Di Donato C, Amoroso A, et al. Physical and functional independency of p70 and p58 natural killer (NK) cell receptors for HLA class I: their role in the definition of different groups of alloreactive NK cell clones. *Proc Natl Acad Sci U S A* (1996) 93(4):1453–7. doi:10.1073/pnas.93.4.1453
32. Moretta A, Poggi A, Olive D, Bottino C, Fortis C, Pantaleo G, et al. Selection and characterization of T-cell variants lacking molecules involved in T-cell activation (T3 T-cell receptor, T44, and T11): analysis of the functional relationship among different pathways of activation. *Proc Natl Acad Sci U S A* (1987) 84(6):1654–8. doi:10.1073/pnas.84.6.1654
33. Lee N, Geraghty DE. HLA-F surface expression on B cell and monocyte cell lines is partially independent from tapasin and completely independent from TAP. *J Immunol* (2003) 171(10):5264–71. doi:10.4049/jimmunol.171.10.5264
34. Ivarsson MA, Michaelsson J, Fauriat C. Activating killer cell Ig-like receptors in health and disease. *Front Immunol* (2014) 5:184. doi:10.3389/fimmu.2014.00184
35. Pende D, Marcenaro S, Falco M, Martini S, Bernardo ME, Montagna D, et al. Anti-leukemia activity of alloreactive NK cells in KIR ligand-mismatched haploidentical HSCT for pediatric patients: evaluation of the functional role of activating KIR and redefinition of inhibitory KIR specificity. *Blood* (2009) 113(13):3119–29. doi:10.1182/blood-2008-06-164103
36. Venstrom JM, Gooley TA, Spellman S, Pring J, Malkki M, Dupont B, et al. Donor activating KIR3DS1 is associated with decreased acute GVHD in unrelated allogeneic hematopoietic stem cell transplantation. *Blood* (2010) 115(15):3162–5. doi:10.1182/blood-2009-08-236943
37. Verheyden S, Schots R, Duquet W, Demanet C. A defined donor activating natural killer cell receptor genotype protects against leukemic relapse after related HLA-identical hematopoietic stem cell transplantation. *Leukemia* (2005) 19(8):1446–51. doi:10.1038/sj.leu.2403839
38. Mancusi A, Ruggeri L, Urbani E, Pierini A, Massei MS, Carotti A, et al. Haploidentical hematopoietic transplantation from KIR ligand-mismatched donors with activating KIRs reduces nonrelapse mortality. *Blood* (2015) 125(20):3173–82. doi:10.1182/blood-2014-09-599993
39. Gabriel IH, Sergeant R, Szydlo R, Apperley JF, DeLavallade H, Alsuliman A, et al. Interaction between KIR3DS1 and HLA-Bw4 predicts for progression-free survival after autologous stem cell transplantation in patients with multiple myeloma. *Blood* (2010) 116(12):2033–9. doi:10.1182/blood-2010-03-273706
40. Sivori S, Carlomagno S, Falco M, Romeo E, Moretta L, Moretta A. Natural killer cells expressing the KIR2DS1-activating receptor efficiently kill T-cell blasts and dendritic cells: implications in haploidentical HSCT. *Blood* (2011) 117(16):4284–92. doi:10.1182/blood-2010-10-316125
41. Matusali G, Tchidjou HK, Pontrelli G, Bernardi S, D'Ettorre G, Vullo V, et al. Soluble ligands for the NKG2D receptor are released during HIV-1 infection and impair NKG2D expression and cytotoxicity of NK cells. *FASEB J* (2013) 27(6):2440–50. doi:10.1096/fj.12-223057

42. Guo H, Kumar P, Moran TM, Garcia-Sastre A, Zhou Y, Malarkannan S. The functional impairment of natural killer cells during influenza virus infection. *Immunol Cell Biol* (2009) 87(8):579–89. doi:10.1038/icb.2009.60
43. Sene D, Levasseur F, Abel M, Lambert M, Camous X, Hernandez C, et al. Hepatitis C virus (HCV) evades NKG2D-dependent NK cell responses through NS5A-mediated imbalance of inflammatory cytokines. *PLoS Pathog* (2010) 6(11):e1001184. doi:10.1371/journal.ppat.1001184
44. Li Y, Wang JJ, Gao S, Liu Q, Bai J, Zhao XQ, et al. Decreased peripheral natural killer cells activity in the immune activated stage of chronic hepatitis B. *PLoS One* (2014) 9(2):e86927. doi:10.1371/journal.pone.0086927
45. Cerboni C, Neri F, Casartelli N, Zingoni A, Cosman D, Rossi P, et al. Human immunodeficiency virus 1 Nef protein downmodulates the ligands of the activating receptor NKG2D and inhibits natural killer cell-mediated cytotoxicity. *J Gen Virol* (2007) 88(Pt 1):242–50. doi:10.1099/vir.0.82125-0
46. Raulet DH. Roles of the NKG2D immunoreceptor and its ligands. *Nat Rev Immunol* (2003) 3(10):781–90. doi:10.1038/nri1199
47. Fogli M, Mavilio D, Brunetta E, Varchetta S, Ata K, Roby G, et al. Lysis of endogenously infected CD4+ T cell blasts by rIL-2 activated autologous natural killer cells from HIV-infected viremic individuals. *PLoS Pathog* (2008) 4(7):e1000101. doi:10.1371/journal.ppat.1000101
48. Pradeu T, Jaeger S, Vivier E. The speed of change: towards a discontinuity theory of immunity? *Nat Rev Immunol* (2013) 13(10):764–9. doi:10.1038/nri3521
49. Pinheiro da Silva F, Aloulou M, Benhamou M, Monteiro RC. Inhibitory ITAMs: a matter of life and death. *Trends Immunol* (2008) 29(8):366–73. doi:10.1016/j.it.2008.05.001
50. Ivashkiv LB. Cross-regulation of signaling by ITAM-associated receptors. *Nat Immunol* (2009) 10(4):340–7. doi:10.1038/ni.1706
51. Turnbull IR, Colonna M. Activating and inhibitory functions of DAP12. *Nat Rev Immunol* (2007) 7(2):155–61. doi:10.1038/nri2014
52. Stern M, Hadaya K, Honger G, Martin PY, Steiger J, Hess C, et al. Telomeric rather than centromeric activating KIR genes protect from cytomegalovirus infection after kidney transplantation. *Am J Transplant* (2011) 11(6):1302–7. doi:10.1111/j.1600-6143.2011.03516.x
53. Venstrom JM, Pittari G, Gooley TA, Chewning JH, Spellman S, Haagenson M, et al. HLA-C-dependent prevention of leukemia relapse by donor activating KIR2DS1. *N Engl J Med* (2012) 367(9):805–16. doi:10.1056/NEJMoa1200503

8

Killer Immunoglobulin-Like Receptor Allele Determination Using Next-Generation Sequencing Technology

*Bercelin Maniangou[1,2], Nolwenn Legrand[1,2], Mehdi Alizadeh[3], Ulysse Guyet[4], Catherine Willem[1,2], Gaëlle David[1,2], Eric Charpentier[4], Alexandre Walencik[5], Christelle Retière[1,2] and Katia Gagne[1,2,5,6]**

[1] Etablissement Français du Sang Pays de la Loire, Nantes, France, [2] CRCINA, INSERM U1232 CNRS, Université d'Angers, Université de Nantes, Nantes, France, [3] Laboratoire de Recherche et Développement, EFS Rennes, Rennes, France, [4] L'institut du thorax, INSERM, CNRS, UNIV Nantes, Nantes, France, [5] Laboratoire d'Histocompatibilité, EFS Nantes, Nantes, France, [6] LabeX Transplantex, Université de Strasbourg, Strasbourg, France

Edited by:
Gianfranco Pittari,
Hamad Medical Corporation, Qatar

Reviewed by:
Daniel Olive,
Institut national de la santé et de la recherche médicale (INSERM), France
Dean Anthony Lee,
The Research Institute at Nationwide Children's Hospital, USA
Mary Carrington,
Leidos Biomedical Research Inc., USA

***Correspondence:**
Katia Gagne
katia.gagne@efs.sante.fr

The impact of natural killer (NK) cell alloreactivity on hematopoietic stem cell transplantation (HSCT) outcome is still debated due to the complexity of graft parameters, HLA class I environment, the nature of killer cell immunoglobulin-like receptor (KIR)/KIR ligand genetic combinations studied, and KIR+ NK cell repertoire size. KIR genes are known to be polymorphic in terms of gene content, copy number variation, and number of alleles. These allelic polymorphisms may impact both the phenotype and function of KIR+ NK cells. We, therefore, speculate that polymorphisms may alter donor KIR+ NK cell phenotype/function thus modulating post-HSCT KIR+ NK cell alloreactivity. To investigate KIR allele polymorphisms of all KIR genes, we developed a next-generation sequencing (NGS) technology on a MiSeq platform. To ensure the reliability and specificity of our method, genomic DNA from well-characterized cell lines were used; high-resolution KIR typing results obtained were then compared to those previously reported. Two different bioinformatic pipelines were used allowing the attribution of sequencing reads to specific KIR genes and the assignment of KIR alleles for each KIR gene. Our results demonstrated successful long-range KIR gene amplifications of all reference samples using intergenic KIR primers. The alignment of reads to the human genome reference (hg19) using BiRD pipeline or visualization of data using Profiler software demonstrated that all KIR genes were completely sequenced with a sufficient read depth (mean 317× for all loci) and a high percentage of mapping (mean 93% for all loci). Comparison of high-resolution KIR typing obtained to those published data using exome capture resulted in a reported concordance rate of 95% for centromeric and telomeric KIR genes. Overall, our results suggest that NGS can be used to investigate the broad KIR allelic polymorphism. Hence, these data improve our knowledge, not only on KIR+ NK cell alloreactivity in HSCT but also on the role of KIR+ NK cell populations in control of viral infections and diseases.

Keywords: high-resolution killer cell immunoglobulin-like receptor typing, allele polymorphism, next-generation sequencing, International Histocompatibility Workshop DNA samples, Natural killer cells

INTRODUCTION

Hematopoietic stem cell transplantation (HSCT) provides a curative therapy for many patients with hematological malignancies (1). Donors for HSCT are currently selected based on the level of matching for HLA-A, -B, -C, -DRB1, and -DQB1 loci. Siblings, 10/10 HLA matched, remain the gold standard. However, substantial risks of morbidity and mortality caused by disease relapse (2), graft-vs-host-disease (GvHD) (3), and infection (4) are still prevalent after related, or unrelated HSCT. Natural killer (NK) cells are the first post-HSCT cells, reconstituting antiviral and antitumoral activity (5). NK cells are able to recognize the missing-self *via* killer cell immunoglobulin-like receptors (KIRs) (6). Ruggeri et al. (7) were first to report the beneficial effect of KIR ligand mismatched donor NK cell alloreactivity after T cell-depleted HLA haplo-identical HSCT resulting in less relapse, less GvHD, and better overall survival in patients with acute myeloid leukemia. The impact of KIR$^+$ NK cell alloreactivity on HSCT outcome is still controversial due to the heterogeneity of graft parameters, HLA class I environment, nature of KIR/KIR ligand genetic combinations studied, and KIR$^+$ NK cell repertoire size (8–12).

As HLA class I genes, KIR genes are highly polymorphic (13). In humans, 16 KIR genes have been described including eight inhibitory genes (2DL1/L2/L3/L4/L5, 3DL1/L2/L3), 6 activating genes (2DS1/S2/S3/S4/S5, 3DS1), 2 two pseudogenes (2DP1, 3DP1). These genes are located within the leukocyte receptor cluster found on chromosome 19q13.4, spanning a region of 150 kb. Within a population, the genotypic diversity of KIR genes occurs at different levels. First, the number and nature of KIR genes vary between individuals defining different KIR haplotypes. KIR haplotypes are classified into group A and group B (14). The group A haplotype is defined by a fixed set of nine KIR genes: four framework KIR genes (3DL3, 3DP1, 3DL2, and 2DL4) that form the centromeric and telomeric part of KIR locus, three inhibitory KIR (2DL1, 2DL3, and 3DL1), a pseudogene (2DP1), and a single activating KIR gene (2DS4). The group B haplotype is defined as having a variable number of KIR genes (7–14) including the four framework KIR genes and specific KIR genes (2DS2, 2DL2, 2DL5, 2DS3, and 2DS1). Second, a variable number of copies [copy number variation (CNV)] of the gene generated by recombination and replication have also been described for some KIR genes particularly those of the B haplotype (15–17). The CNV seems to influence the licensing of KIR$^+$ NK cells (18). Overall, various KIR genotypes can be observed in a population. All KIR genes, and especially for inhibitory KIR, a high degree of allelic polymorphism has been described. The latest KIR Immuno Polymorphism Database (IPD–KIR) describes 753 KIR alleles. KIR allele polymorphisms need to be investigated throughout the exon and the intron regions, and regulatory regions as shown for KIR3DL1 (19). In contrast to HLA class I genes, structure and length of KIR genes vary. KIR allele polymorphisms impact both KIR$^+$ NK cell phenotype and function, as we and other groups having described for KIR3DL1 (20–25) and for KIR2DL2/L3 (26). Differences in the intensity of expression (strong, weak, or null) have been described for the KIR3DL1 receptor, defining different allotypes according to the KIR3DL1/3DS1 allele combinations present in healthy individuals (21, 27). Furthermore, the nature of KIR3DL1 alleles does not only impact the KIR3DL1 cell density but also the strength of the KIR3DL1–HLA interactions which in turn can affect NK cell functions (28, 29). The recognition of KIR allotypes using anti-KIR monoclonal antibodies also varies depending on the KIR allele polymorphism (30).

Taking these points into account, it is therefore necessary to thoroughly investigate the phenotypic and functional impact of KIR allele polymorphisms. Until now, potential KIR$^+$ NK cell alloreactivity in HSCT was mainly evaluated depending on the KIR/KIR ligand genetic combinations present and analyzed only at a generic level (i.e., presence or absence of KIR genes and KIR ligand). We speculate that KIR allele polymorphisms may alter donor KIR$^+$ NK cell phenotype/function, and thus modulate their alloreactivity affecting HSCT outcome. However, the impact of KIR allele polymorphisms on HSCT outcome remains difficult to assess due to the lack of suitable allele typing methods for all KIR genes. Until recently, several standard methods are used to type KIR genes at allelic level. Those methods include sequence-specific oligoprobe hybridization (31–37), sequence-specific primer (SSP) typing (22), SNP assay (38), Sanger sequence-based typing (SBT) (20, 39–42), high-resolution melting (43), and also combined SSP/SBT (21, 44). KIR allelic polymorphisms have been investigated for a few functional KIR genes (KIR2DL1/2DL2/2DL3/2DS1 /3DL1/3DS1). Standard methods to type KIR genes at allelic level are usually single KIR locus specific and/or target a limited polymorphism. In addition, the constant increase in the number of KIR alleles described generates more and more ambiguous KIR typing in heterozygous samples since KIR polymorphism can extend over the entire gene. Recent advances in high-throughput sequencing technology [next-generation sequencing (NGS)], especially in immunology and hematology (45), enable determination of KIR alleles and KIR gene CNV. The extent of KIR allele polymorphisms, as demonstrated by exome capture, reported 37 new KIR alleles from 15 healthy South African individuals (46). Recently, whole KIR genome sequencing by NGS was used as a control method to validate CNV genotyping in the KIR locus (17). An exome capture that focused on KIR and HLA class I loci was also recently described (47). In this study, we developed a reliable NGS method for high quality DNA samples and easily implemented for the study of KIR allele polymorphisms.

MATERIALS AND METHODS

Samples

Thirty B-EBV cell lines from the 10th International Histocompatibility Workshop (IHW) were selected from a well-characterized panel known for their KIR gene content. KIR genotype information, including KIR allele typing of some KIR genes for all these B-EBV cell lines, was obtained either from the IPD/KIR database or from literature for specific KIR loci. Known KIR genotypes and allele typing of these 30 B-EBV cell lines are provided in the Table S1 in Supplementary Material.

KIR Long-Range (LR) PCR and Primers

DNA genomic extractions were performed from B-EBV cell lines using a Nucleospin blood kit (Macherey-Nagel, Duren, Germany). The concentration and the purity of all DNA samples were checked on a NanoDrop 2000C spectrophotometer (ThermoFisher, Wilmington, DE, USA) by measuring the ratio of absorbance at 260 and 280 nm. In parallel, 1.5 μg of each DNA sample was loaded on an agarose gel to check the DNA integrity. For KIR LR PCR, five intergenic KIR primers already described (17) and one additional *in-house* designed primer including four forward primers (#1, 5′-gccaaataacatcctgtgcgctgctcagct-3′; #2, 5′-ctcacaacatcctgtgtgctgctaactga-3′; #4, 5′-acggctgcctgtctgcacagacagcacc-3′, #6, 5′-cacatcgtctgcaccggtcagtcgagccga-3′) and two reverse primers (#3, 5′-ttggagaggtgggcaggggtcaagtg-3′; #5, 5′-ctccatctgagggtcccctgaatgtg-3′) were used to amplify the whole KIR genome.

The KIR LR-PCR protocol was optimized using the method described by Vendelbosch et al. (17). Briefly, KIR LR-PCR was performed with 2.5 U of PrimeSTAR GXL DNA Polymerase (Ozyme, Saint-Quentin en Yvelines, France), 1× PrimeSTAR GXL buffer, 200 μM of dNTP mixture (Ozyme) and 0.2 μM final concentration of each KIR primer. The LR-PCR reaction was performed in a C1000 Touch™ Thermal Cycler (Biorad, Marnes la Coquette, France) consisted of an initial denaturation of 2 min at 94°C followed by 30 cycles of 20 s at 94°C, 12 min at 68°C and 1 cycle of final elongation of 10 min at 72°C in the final 50 μL volume. This protocol enables amplification of each KIR gene from 5′ to 3′ untranslated regions (UTR). The final KIR LR-PCR product was run on 0.7% Seakem agarose gel in TBE1X (Lonza, Verviers, Belgium) and visualized by staining with the SYBR® safe (Invitrogen, Villebon sur Yvette, France) using the SimplyLoad™ Tandem DNA ladder size marker (Ozyme) to confirm the amplification and correct fragment size as well as to check for non-specific amplification.

Library Preparation and Sequencing

Qubit dsDNA High Sensitivity Assay Kit (Life Technologies, Villebon sur Yvette, France) was used to quantify the starting DNA library in the Qubit® fluorometer (Life Technologies). The library preparation was performed using the NGSgo GENDX kit (Bedia Genomics, Chavenay, France). To achieve the optimal insert size and library concentration, 250 ng of each genomic DNA was randomly fragmented according to the manufacturer's instructions. Briefly, 8.25 μL of NGSgo master mix (prepared from 2 μL of NGSgo-LibrX Fragmentase buffer plus 3.25 μL of NGSgo-LibrX End Prep buffer plus 1.5 μL of NGSgo-LibrX Fragmentase Enzyme plus 1.5 μL NGSgo-LibrX End Prep Enzyme) (Bedia Genomics, Chavenay, France) was added to each genomic DNA in a final volume of 32.5 μL. The fragmentation, end-repair, and dA-tailing reactions were performed in a T100™ Thermal Cycler (Biorad, France) consisted of 20 min of fragmentation and end-repair at 25°C followed by 10 min of dA-tailing at 70°C. The dA-tailed DNA fragments of each sample were then subjected to adapter ligation in 9.25 μL of an NGSgo master mix containing 7.5 μL of NGSgo-LibrX Ligase mix, 0.5 μL of NGSgo-LibrX Ligation Enhancer, 0.25 μL of NGSgo-Indx adapter for Illumina, and 1 μL of nuclease free water. The adapter ligation reaction took place in a T100™ Thermal Cycler (Biorad, France) for 15 min at 20°C followed by a cooling step at 15°C. The first cleaning and size selecting of the samples after adapter ligation were performed in a 0.45× beads:DNA ratio by using the Agencourt® Ampure XP (Beckman Coulter, Villepinte, France) according to the manufacturer's instructions and eluted in 12.5 μL of 0.1× elution buffer (Lonza Rockland, USA). The size-selected, adapter-ligated DNA fragments of each DNA sample were then dual indexed with 15 μL of NGSgo reaction mix made from 12.5 μL of NGSgo-LibrX HiFi PCR mix plus 1.25 μL of NGsgo-Indx IN-5 and 1.25 μL of NGSgo-Indx IN-7 in a final volume of 25 μL followed by a PCR reaction in a T100™ Thermal Cycler (Biorad). PCR cycling was performed as follows: an initial denaturation of 30 s at 98°C followed by 10 cycles of 10 s at 98°C, 30 s at 65°C, 30 s at 72°C and 1 cycle of final elongation step of 5 min at 72°C in the final volume of 25 μL. A second DNA cleaning and size selecting was performed in a 0.6× beads: DNA ratio by using the Agencourt® Ampure XP beads according to the manufacturer instructions and eluted in 16.5 μL of 0.1× elution buffer (Lonza Rockland, USA).

Quality control procedure for the library preparation included verification of fragment size before and after purification by using the QiAxcel Advanced System (QiAgen, Courtaboeuf, France). The pooled and barcoded libraries were denaturated with 0.2 M of NaOH and diluted in the pre-chilled HT1-buffer to obtain a final library concentration of 12 pM. The final denatured library was subsequently sequenced by using the MiSeq sequencer (Illumina, Biogenouest Genomics Platform Core Facility, Nantes, France; HLA Laboratory, EFS Nantes, France) with 500 cycles v2 kits, which generated 250-bp end sequence reads.

Sequencing Data Analysis and KIR Allele Assignment

The quality of the Illumina raw data sequences obtained was monitored by using the Sequencing Analysis Viewer Illumina software. The quality of the base calling from images and sequences was determined by the quality score (Q30) which must be ≥75% for 2 × 250 bp reads. KIR reads were mapped to the human genome reference sequence hg19 (GRCh37) by using the Burrows–Wheeler Aligner Memory Efficient Mapping (BWA-MEM) tool. The binary alignment map (BAM) files containing mapped reads were then visualized on Integrative Genomics Viewer (IGV) algorithm (48).

In parallel, raw KIR sequences were aligned and visualized using the Profiler software version 1.70, initially developed by Dr. M. Alizadeh (Research Laboratory, Blood Bank, Rennes, France) for NGS-based HLA typing (49). A flowchart for data analysis using the Profiler software is provided in Figure S1 in Supplementary Material. The first step of analysis consists by merging R1 and R2 sequences to each other when at least 10 complementary bases were found between R1 and R2 of the same cluster. During this phase, for each inconsistency of base calling, the quality value was used to select the best assignment. All sequences issued from a cluster for which we could not determine complementary between R1 and R2 remained unchanged. All sequences were transformed to FASTA format at the end of this step. The second step of analysis consists of phasing each of the sequences obtained in step one by using Blast algorithm. The third step of analysis consists by merging all sequences together using Blast information. In this step, the depth for each position and the number of sequences for each allele were calculated. The first three steps are managed in a Linux environment. The last step is presentation and assignment of each construction based on database information in a friendly interface for user, all mismatches and differences to the database are extracted and presented to the user.

For KIR allele assignment, a manual bioinformatic pipeline was first used in the absence of available softwares. This consisted of exporting from IGV, all exon sequences of each KIR gene and comparing polymorphic bases with those referenced from the IPD–KIR database. Then, two different bioinformatics algorithms were used: the first one, hereafter called "BiRD," was developed by the BiRD platform (E. Charpentier, U. Guyet, Genomics and Bioinformatics Core Facility GenoBiRD, Nantes, France) and consists of an analysis pipeline built with Snakemake on the same logic as the manual method. A flowchart for data analysis using the BiRD software is provided in Figure S2 in Supplementary Material.

Harvesting KIR-Specific Reads

First, raw sequences from fastq files are processed through cutadapt (v1.8.1) in order to remove Illumina adapter sequences. The cleaned reads are then mapped to hg19 (GRCh37) reference genome using BWA-MEM (v0.7.12) with the default parameters.

Determining Presence/Absence or KIR Genes

Absence or presence of KIR genes is evaluated using GATK DepthOfCoverage on the BAM and using a browser extensible data (BED) file describing the chromosome position of each gene (except KIR2DP1 and KIR3DP1). Coverage mean is calculated on each gene position, and a threshold of 10 is applied in order to ascertain its absence or presence. Presence/absence of KIR genes defined by NGS is concordant to the KIR genotype of the 30 IHW samples, stratified by AA vs Bx genotypes, previously validated in our laboratory by PCR-SSP multiplex method (data not shown).

Determining KIR Alleles

Allele-specific nucleotide positions are extracted manually using IPD–KIR alignment tool.[1] For every gene, the Nucleotide—CDS of all alleles are aligned against the default reference allele. A python script is then used to reformat the multipage alignments in order to have one allele alignment per line. A second python script is utilized to extract all variations from the default reference allele and map the exon position number of these variations to the chromosome position. A file is created for each gene listing all the variations found for every allele. Bases at these positions are then called using SAMTools (v1.2-2) mpileup for all samples. Finally, KIR alleles are determined by calculating the percentage of nucleotide matches between the base calls and the allele variations for each KIR allele, the highest percentage giving the most confident allele.

The second algorithm used for KIR allele assignment was the Profiler software, previously described in Figure S1 in Supplementary Material, version 1.70 (49), which permits to directly assign KIR alleles at the highest level resolution (seven digits) since full intron and exon sequences were considered and also provides quality data such as mean coverage for each KIR locus. The fragment size percentage of sequences for each allele/locus was also considered as well as percentage of mapping for each KIR gene.

Overall, KIR allele assignment for each locus and for all samples corresponds to the combined KIR results obtained using manual pipelines, BiRD, and Profiler softwares. KIR alleles were assigned on the basis of the known DNA sequences identity within the IPD/KIR database.[2] KIR alleles are named in an analogous fashion as the nomenclature used for HLA class I alleles. After the gene name, an asterisk is used as a separator before a numerical allele designation. The first three digits of the numerical designation are used to indicate alleles that differ in the sequences of their encoded proteins. The next two digits are used to distinguish alleles that only differ by synonymous (non-coding) differences within the coding sequence. The final two digits are used to distinguish alleles that only differ by substitutions in an intron, promoter, or other non-coding region of the sequence.

RESULTS

LR KIR Gene Amplifications

Thirty reference IHW samples with known KIR genotyping (Table S1 in Supplementary Material) were used to validate our NGS method for typing of each KIR gene at allelic resolution. DNA integrity, checked by loading each sample on an agarose gel, confirmed high quality for all samples (data not shown). In order to amplify all KIR genes from the 5′ UTR to the 3′ UTR, six intergenic KIR primers were chosen to allow the amplification of framework KIR genes. These intergenic primers also amplify KIR genes located either in the centromeric or telomeric region, which belong to the A and/or B specific KIR haplotype genes (**Figure 1A**). A robust LR amplification of KIR genes was obtained for all samples as illustrated for three representative IHW samples (**Figure 1B**). One specific band between 4 and 5 kb for the KIR3DP1 pseudogene and another specific band between 9 and 17 kb corresponding to a cluster of all other KIR genes were observed, irrespective of KIR AA or AB genotype (**Figure 1B**) as KIR genomic length varies depending on KIR genes (Table S2 in Supplementary Material). For some IHW samples such as BOB, two specific bands at 4 and 5 kb were observed for the KIR3DP1 gene corresponding to KIR3DP1*003 and KIR3DP1*001 variants, respectively, whereas only one band at 4 kb specific of KIR3DP1*003 variant was observed for OLGA and SPO010 samples (**Figure 1B**).

Complete Sequencing of All KIR Genes

In order to check the specificity of KIR LR-PCR obtained, amplicons were further fragmented and sequenced on paired end 2 × 250 bp from Illumina MiSeq platform. The sequencing of all amplicons yielded a total of 6.3 Gb, which was generated from a 755 ± 31 K/mm^2 cluster density (data not shown). Approximately 88.2% of the clusters passed QC filters and on

[1]https://www.ebi.ac.uk/ipd/kir/align.html.

[2]http://www.ebi.ac.uk/ipd/kir/alleles.html

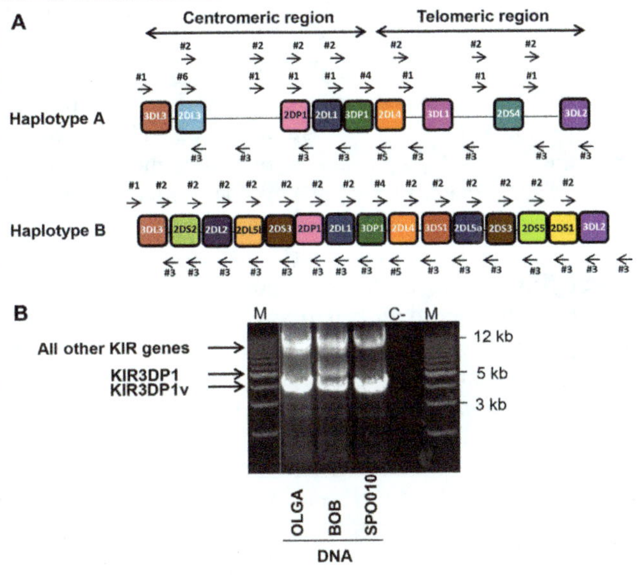

FIGURE 1 | Long-range (LR) killer cell immunoglobulin-like receptor (KIR) gene amplification. (A) Six intergenic KIR primers (four forward primers: #1, #2, #4, and #6 and two reverse primers: #3, and #5) were used to perform LR PCR amplifications. These primers were able to amplify full length KIR genes in both the centromeric and telomeric regions belonging either to the A or B KIR haplotype. **(B)** Illustrative 0.7% agarose gel electrophoresis of LR PCR amplifications spanning the KIR genome of three representative International Histocompatibility Workshop (IHW) DNA samples. One IHW sample with an AA KIR genotype (i.e., SPO010) and two DNA samples with an AB KIR genotype (i.e., OLGA and BOB) were used. Amplicon lengths vary from 4 to 5 kb for the KIR3DP1 gene to 9–17 kb for all other KIR genes. Two specific bands of 4 and 5 kb corresponding to two KIR3DP1 variants were observed for BOB sample. M: Tandem ladder Lonza Seakem size marker; C⁻: H_2O, negative control.

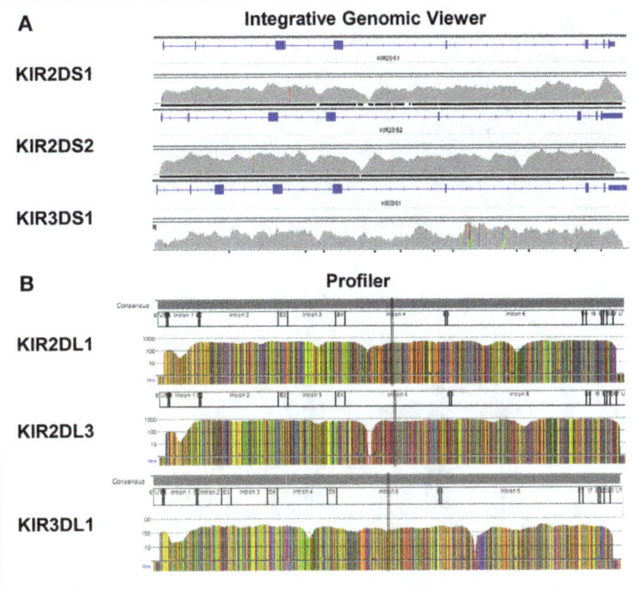

FIGURE 2 | Full sequencing of all killer cell immunoglobulin-like receptor (KIR) genes. Reads were mapped to the human genome reference sequence hg19 using the Burrows–Wheeler Aligner Memory Efficient Mapping tool. The binary alignment map files containing mapped reads were then visualized on the Integrative Genomics Viewer as illustrated for KIR2DS1, KIR2DS2, and KIR3DS1 genes **(A)** or using Profiler software **(B)** as illustrated for KIR2DL1, KIR2DL3, and KIR3DL1 genes from one representative International Histocompatibility Workshop DNA sample.

average, 82.4% of both reads passed with a Q30 > 82% (data not shown). Thus, analysis of FastQ data obtained from all IHW samples reported an excellent quality control. The entire length of KIR genes was sequenced with good coverage as illustrated for KIR2DS1, KIR2DS2, and KIR3DS1 (Figure 2A) activating genes, and for KIR2DL1, KIR2DL3, and KIR3DL1 (**Figure 2B**) inhibitory genes using either IGV or Profiler software, respectively. For all genes, the depth of coverage varies most at the beginning and at the end of the amplicons, but all key regions were sufficiently covered. In particular, we observed that mean coverage ranged from 62.5× (KIR2DS4) to 2,373.3× (KIR3DP1) leading to a mean coverage of 316.55× for all KIR genes except for KIR2DL5A genes since not analyzed using Profiler (Table S3 in Supplementary Material). A significant correlation was observed between mean coverage and genomic KIR length ($r = 0.85$, $p < 0.0001$) as illustrated **Figure 3A**. Indeed, the lower the genomic length, the higher the mean coverage is as illustrated for the KIR3DP1 gene. The mean percentage of mapping, established by the coverage of amplicon, ranged from 86.2% (KIR3DL2) to 98.2% (KIR2DP1 and KIR3DP1) (**Figure 3B**; Table S3 in Supplementary Material) suggesting that sufficient read depth was obtained for determination of all KIR genes. However, KIR2DL5A reads could have been mapped only using BWA-MEM software and BiRD algorithm. Overall, these results demonstrate the efficiency of our NGS-KIR allele typing approach to capture the full KIR genomic locus and the uniformity of coverage for each KIR locus confers assurance for KIR allele assignment.

Specificity of NGS-Based KIR Allele Typing

Due to the high degree of KIR polymorphisms and the fact that NGS technology generates a lot of sequencing reads, three different algorithms were evaluated to increase the reliability of KIR allele assignment as reported for NGS-based HLA typing (50). KIR allele assignment was first done manually and then confirmed using both BiRD pipeline and Profiler software. Overall, resulting KIR allele assignments of the 30 reference IHW samples were feasible for all loci and for the majority of samples without remaining ambiguities (**Table 1**).

We further evaluated the strength of our NGS-based method for KIR allele assignment. For all IHW reference samples tested ($N = 30$), the number of KIR alleles previously known in the IPD/KIR database and those obtained by our NGS-KIR based typing approach was compared for each KIR locus. As an example, from the 30 IHW samples tested, only 5 KIR3DL3 alleles out of 60 expected alleles for this framework gene were previously known in the IPD/KIR database (Table S1 in Supplementary Material), 54 KIR3DL3 alleles from 24 heterozygous and 6 homozygous samples (**Table 1**) were assigned

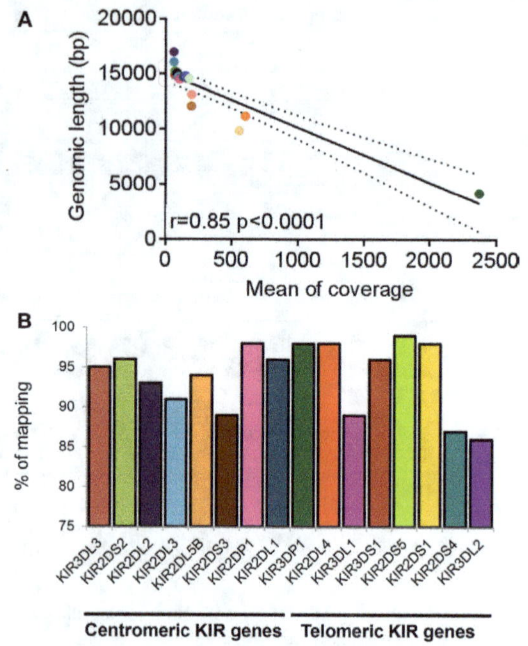

FIGURE 3 | Coverage and percentage of mapping obtained for killer cell immunoglobulin-like receptor (KIR) genes. (A) Correlation graph representing mean coverage for each KIR gene and KIR genomic length. Mean coverage was estimated for each KIR gene present for all International Histocompatibility Workshop (IHW) samples using Profiler software. Statistical significance was determined using the Pearson's rank coefficient using GraphPad Prism version 6 software (GraphPad Software, La Jolla, CA, USA). A significant *p*-value between mean coverage and genomic KIR length was observed ($p < 0.0001$). **(B)** Bars representing the percentage of mapping of each centromeric and telomeric KIR gene present for all IHW samples determined using Profiler software. KIR2DL5A locus was not included since not analyzed using Profiler.

by our NGS-based KIR allele typing approach (**Figure 4**). Our NGS-based KIR allele typing approach permits identification of additional framework KIR alleles, e.g., KIR3DP1 ($n = 43$), KIR2DL4 ($n = 48$), and KIR3DL2 ($n = 44$) (**Table 1**; **Figure 4**). NGS-based KIR typing method also allows the identification of polymorphisms of well-functionally characterized KIR by increasing the number of assigned KIR alleles of the 30 IHW samples available in the IPD/KIR database (Table S1 in Supplementary Material), e.g., KIR2DL1 ($n = 30$), KIR2DL2 ($n = 10$), KIR2DS1 ($n = 11$), KIR2DS2 ($n = 12$), KIR3DL1 ($n = 35$), and KIR3DS1 ($n = 11$) (**Table 1**; **Figure 4**). The number of activating KIR2DS1, KIR2DS2, and KIR3DS1 assigned alleles by NGS remained low because only IHW samples with the corresponding activating KIR gene were included in this analysis. Overall, a higher number of KIR alleles were identified from these 30 IHW samples by our shotgun NGS methodology compared to those previously characterized by other less sensitive methods, as referred to in the IPD/KIR database ($N = 422$ vs $N = 233$, respectively).

The knowledge of KIR allele typing of IHW samples, recently updated thanks to an exome capture (47), permits to evaluate the concordance of our NGS-based KIR allele results (**Table 1**) with those of Norman et al. since 22 IHW samples were commonly used in both methods (Table S1 in Supplementary Material). In this case, a large number of allelic KIR typing for all loci was compared ensuring the reliability of our NGS-based KIR allele typing method. For each KIR locus and for the 22 IHW concerned samples, KIR allele typing results were divided into: concordant (one KIR allele matched for homozygous samples or two KIR alleles matched for heterozygous samples), semi-concordant (one KIR allele matched and one KIR allele mismatched), and discordant (one KIR allele mismatched for homozygous sample or two KIR alleles mismatched for heterozygous sample). For each KIR allele, only the first three digits were taken into account for the assessment of concordance. Complete concordance (100%) of KIR allele typing was demonstrated in 11 KIR genes. The concordant genes were KIR2DS2 (8 samples out of 8), KIR2DL5B (2 out of 2), KIR2DS3 (1 out of 1), KIR2DL1 (18 out of 18), KIR2DL4 (22 out of 22), KIR3DL1 (17 out of 17), KIR3DS1 (7 out of 7), KIR2DL5A (7 out of 7), KIR2DS5 (6 out of 6), KIR2DS1 (6 out of 6), and KIR2DS4 (16 out of 16) (**Figure 5**). Concordant results were observed, but at a lesser frequency for KIR3DL3 (20 out of 22, i.e., 91%), KIR2DL2 (4 out of 5, i.e., 80%), KIR2DL3 (7 out of 8, i.e., 88%), KIR2DP1 (16 out of 20, i.e., 80%), KIR3DP1 (16 out of 18, i.e., 89%), and KIR3DL2 (16 out of 18, i.e., 89%) (**Figure 5**).

Ten semi-discordant KIR allele results and two discordant KIR allele results between our NGS-based method and exome data were identified (**Table 2**). Except for the pseudogene KIR2DP1, with four IHW samples, these discrepancies were limited to 1 or 2 out of 22 IHW samples per locus (**Table 2**). KIR allele determinations using manual, BiRD algorithm, and different versions (the latest one Rev 2.0.188) of Profiler software were carefully reviewed. Only IHW samples sequenced on different runs and with the same KIR allelic results were reported (data not shown). These potential discrepancies (5%), possibly linked to the design and implementation of each algorithm, need to be further validated by another typing method such as SSP or sequencing.

Overall, our NGS-based method and exome data showed a rate of concordance of 95% for all loci, established for all KIR genes on 22 IHW samples, suggesting a reliable method.

DISCUSSION

In this study, we developed an NGS-based KIR allele typing approach to characterize the sequence of all polymorphic KIR genes. Our method of typing all KIR genes at high resolution provides an alternative, easily implemented method practice, to study the KIR allele polymorphisms. It may be a cheaper method than exome capture (47). This tool is currently adapted to the KIR gene large-scale analysis. Using our approach, the majority of KIR alleles previously uncharacterized by standard methods were clearly identified from genomic DNA of 30 B-EBV cell lines from the 10th IHW. High quality DNA samples, high fidelity of enzyme polymerase, and a reliable library preparation were needed since evaluation of different Taq polymerase enzymes

TABLE 1 | Next-generation sequencing-based killer cell immunoglobulin-like receptor (KIR) allele typings of 30 reference B-EBV cell lines from the 10th International Histocompatibility Workshop.

ID	Centromeric KIR genes									Telomeric KIR genes								
	3DL3	2DS2	2DL2	2DL3	2DL5B	2DS3	2DP1	2DL1	3DP1	2DL4	3DL1	3DS1	2DL5A	2DS3	2DS5	2DS1	2DS4	3DL2
AMAI	*013 *041			+			*00301 *004	+	*003 *006	*0080101 *0080102	*001						*00301	*00101
AMALA	*00402 *00802	*00101	*00301	*001			*00201	*00302	*007 *00901	*00102 *00501	*01502	*01301	*001		*00201	*00201	*001	*0020105 *0070102
BOB	*00101 *01303	*00101	*00301	*00201			*00301	*00302	*002 *00302	*001 *005	*002	*01301	*00101		*00201	*00201	*001	*0020101 *0070102
BRIP	*00801	*00104 *004	*00301	+		*00103 *00201	*0010201 *0020101	*00302	+	*0010305 *00501	*008	*01301	*00103 *00501	*00103 *00201		*002	*003	*0070102 *0070103
CALOG ERO	*00207 *01001			+			*00201	*00302	*00302 *010	*008	*001 *004						*00301 *00601	*00101 *00301
COX	*00102 *00103			*00201 *007			*00301	*00201	*005 *006	*00501* *011	*005010	*055	*00101		*00201	*00201	*010	*00103 *007
DEU	*00101 *01402	*00101	*001	*00201			*00301	*00201	*001 *006	*00801 *011	*00101 *00501						*003 *010	*01001 *01101
DKB	*00101 *006			+			*00301	*00201	*00302 *006	*0010201 *00103	*002 *02001						*00101	*0020101 *00902
HO301	*014	*00101 *002	*00101 *00301		*010	*00103 *00201	*00102	*004 *010	*003010 *004	*00102	*002			*00103 *00201			*001	*00201
HID	*01402 *018			*00101			*00201	*00302	*00302 *010	*00102	*01502						*00101	*00201
HOM-2	*00101 *0090101			+			*00201 *005	*00302	*00302 *006	*00801 *00802	*001 *004						*00301 *00601	*0010102 *00501
HOR	*001 *048			+			*00301	*00201	+	*00501		*01301	*00101		*002	*00201		*007 *021
JHAF	*00901 *026			*00101			*002	*00302	*00302	*011	*00501						*010	*001 *01001
JVM	*007 *00801	*00101	*00301	+			*005	*00302	*001 *00302	*00103 *00801	*00101 *008						*003010	*00101 *009
KAS011	*00901 *01302			+			*002 *00301	*00201 *00302	*00302 *006	*00103 *005	*008	*01301	*00101		*00201	*00201	*00301	*01001 *019
KAS116	*013 *01501			+			*002	*00302	+	*011	*00501						*010	*0103 *010
LBUF	*00301 *0090101	+	+	*001			*002	+	*00302 *0090102	*00102 *011	+						+	+
LUY	*001 *02701			*00101 *00501			*00201 *00301	*00302	*00302	*00801 *011	*00401 *00501						*00601 *010	*001 *00501
MOU	*00207 *00801			*001			*00201 *005	*00302	*00302	*00801	*00101 *00401						*00301 *00601	*010 *01101

(Continued)

TABLE 1 | Continued

ID	Centromeric KIR genes									Telomeric KIR genes								
	3DL3	2DS2	2DL2	2DL3	2DL5B	2DS3	2DP1	2DL1	3DP1	2DL4	3DL1	3DS1	2DL5A	2DS3	2DS5	2DS1	2DS4	3DL2
OLGA	*00201 *00902			*00101			*00201 *006	*00302	*00302	*005 *011	*001 *00501	*01301	*00103		*002	*002	*010	*00701
PE117	*00101 *01002			*00101 *00201			*00201 *00301	*00201 *00302	*00901	*00501 *00802	*00401	*01301	*001		*00201	*00201	*00601	*00701 *018
PF04015	*01402	*00101	*00101 *003						*001	*011	*00501						*010	*00103
RSH	*0040202 *00901	*00101	+	+	*004		*00201 *009	*00302 *01201	*00304 *008	*0010307 *011	*00501 *017				*006		+	+
SAVC	*00801 *00202			*00101			*008 *00201	*00302	*00302	*00102 *00802	*00401 *01502						*006010	2*00202 *00301
SPO010	*00206			+			*00201	*00302	+	*011	*0050101						*010	*001
T7526	*0090101			*00101			*00201	*00302	*00302	*00501 *00102	*01502	*013	*00101		*00201	*002	*001	*0020105 *0070102
VAVY	*002 *017			*00101 *00201			*002 *003	*002 *00302	*00302 *006	*011	*00501						*010	*0010302
WT51	*00103 *036	*00101	+	+	*00201	+	*001 *004	+	+	*00501		*01301	*00101 *00501	+	*002	*002		+
WDV	*00301 *0090101	*00101	*003	+		*002	*002	*00302 *00901	*00302	*00501		*01301	*00501	*002		*00201		*0070103
YAR	*00102			+			*002 *003	+	*00302 *006	*0010201 *011	+						+	+

Results are presented according to the centromeric or telomeric localization of KIR genes or human genome. KIR alleles were named according to the last nomenclature available on the IPD/KIR database (http://www.ebi.ac.uk/ipd/kir/). + indicates the presence of a specific KIR gene. ID, sample identification.

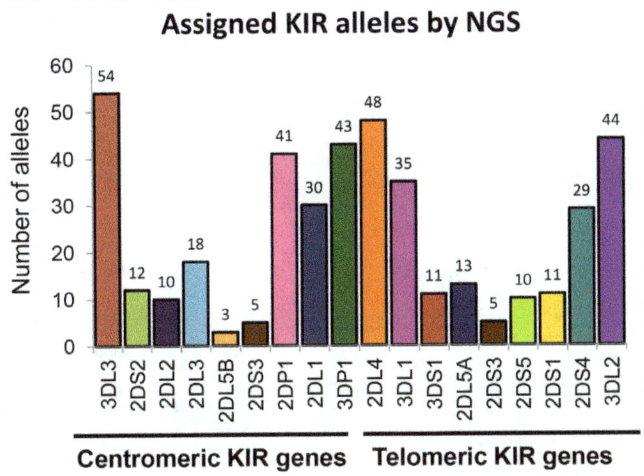

FIGURE 4 | Number of killer cell immunoglobulin-like receptor (KIR) allele assigned by next-generation sequencing (NGS). Number of KIR alleles assigned for each centromeric and telomeric KIR gene of 30 International Histocompatibility Workshop samples obtained by our NGS method. KIR allele assignment for each locus and for all samples corresponds to the combined KIR allelic results obtained using manual pipelines, BiRD, and Profiler software.

and library preparation kits gave conflicting results (data not shown).

Our study evaluated the performance of different algorithms for KIR allele assignment. Reliability of the manual, BiRD pipeline, and Profiler software was tested since neither algorithm alone was able to provide 100% accuracy for all KIR loci. Our results showed that the Profiler software was reliable to assign KIR alleles through the full length of each KIR gene, excluding KIRDL5A variants. In this case, KIR2DL5A and KIR2DL5B sequences were too closed, and Profiler software failed to accurately analyze both sequences. Since all coding, non-coding, and regulatory regions were explored, one could expect that a lot of new KIR alleles will soon be described. Analysis with Profiler consists of two distinct parts. The first part is performed in three steps in a Linux environment: the first step corresponds to the merging of each R1 and R2 issued from the same cluster to each other, each time that a complementarity of at least 10 bases is found, with correction or base calling inconsistencies using a quality value for each nucleotide. There are two interests in this step: longer sequences and lower sequences number were analyzed. The second step corresponds to the phasing of each sequence based on KIR databases using Blast algorithm. Third, the data file from Blast was used to merge all sequences together to construct each allele. In this step, calculation for depth of each position and the number of sequences used for each allele are determined. The second part is done on a Windows environment. A friendly interface presents graphics of all sequences for all studied loci. Assignment of all sequences is done using a database, highlighting all mismatches compared to reference and also differences between KIR alleles selected. Each allele is scored for quality control as per the European Federation for Immunogenetics guideline.

Killer cell immunoglobulin-like receptor alleles of all genes including KIR2DL5A, but excluding the pseudogenes KIR2DP1 and KIR3DP1, were assigned using BiRD algorithm. However, many allelic ambiguities remained when this pipeline was used alone (data not shown). It is likely that this is due to the fact that only coding regions (CDS) were taken into account for allele variation comparison. Analysis of all exon/intron polymorphisms, CNV detection, summary statistics of call accuracy for KIR gene content (presence/absence) and for KIR allele identification needs to be completed. Furthermore, the two pseudogenes KIR2DP1 and KIR3DP1 could be manually added to the BED file describing the gene positions on the genome in order to include them in the analysis pipeline.

Due to the time-consuming nature of manual KIR allele assignment, two different algorithms are needed to ensure the reliability of NGS-based typing methods for the identification of KIR allele polymorphisms.

Until now, KIR genetic population studies have often been restricted to the identification of KIR gene content, or of A and/or B KIR haplotypes (51, 52). Determination of KIR alleles in healthy individuals of a given population may provide a better definition of KIR haplotypes (52) and KIR gene linkage disequilibrium (53) and will considerably increase the IPD/KIR database.

The implementation of our suitable NGS.KIR method will enable analysis of all allelic polymorphism within KIR genes extending to all coding, non-coding, and regulatory regions. A link between KIR allelic polymorphism and the expression level and/or function of the corresponding KIR$^+$ NK cells is necessary for all KIR genes as previously established for the expression level of HLA-A and HLA-Cw molecules (54–56). We speculate that KIR allelic polymorphisms may affect not only the distribution and function of these gene products but also the licensing of NK subpopulations as described for HLA class I molecules (57, 58). Deep analysis of KIR$^+$ NK cell phenotype and function depending on KIR and HLA class I alleles present is needed to assess the diversity of KIR$^+$ NK cell repertoire (21, 59), as well as the specificity of anti-KIR antibodies (30, 60). Overall, the analysis of KIR allelic polymorphisms combined with the autologous HLA class I environment will enable better evaluation of KIR$^+$ NK cell functional subpopulations (61). This functional KIR$^+$ NK cell repertoire will be better defined by taking into account the nature of KIR alleles present in addition to the autologous HLA class I environment.

Investigation of KIR allelic polymorphism may be of an immunological interest in the context of viral infections such as those related to CMV (62), HIV (63), HCV (64), and of human reproduction (65). In the context of HSCT, inclusion of KIR allele typing in addition to HLA typing may provide a better evaluation of HSC donor's KIR$^+$ NK cell repertoire (21, 59, 60, 66, 67). An identification of those with the best antileukemic potential will provide a potential tool to determine an early posttransplant hematopoietic chimerism when donor and recipient have identical KIR genotypes (68) as well as the impact of KIR$^+$ NK cell alloreactivity on HSCT outcome (69–73). The functional relevance of typing both KIR and HLA genes at

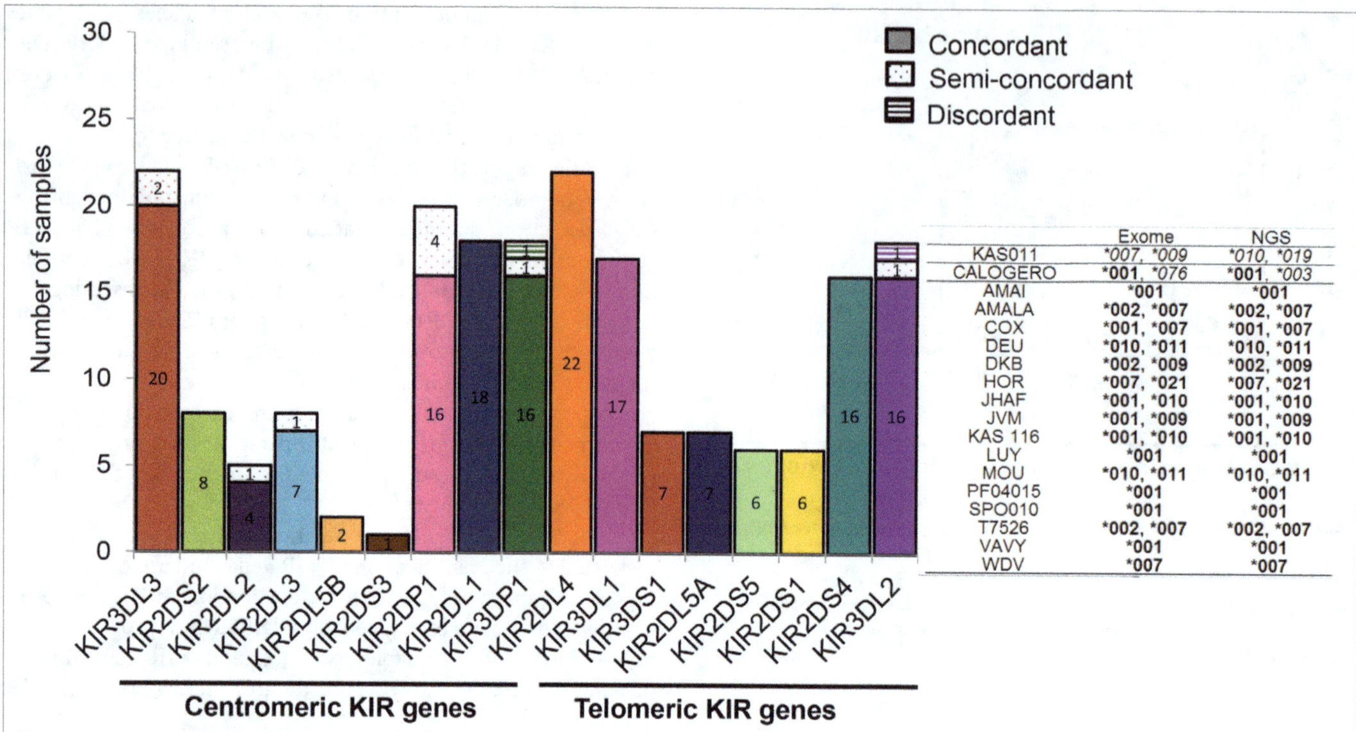

FIGURE 5 | Specificity of next-generation sequencing (NGS)-based killer cell immunoglobulin-like receptor (KIR) allele typing. Bars representing the numbers of KIR allele typing obtained by our NGS-based method from 22 International Histocompatibility Workshop (IHW) samples compared with those assigned by exome analysis (47). In this case, KIR allele typing for each locus and for all samples corresponds to the combined KIR allelic results obtained using manual pipelines, BiRD, and Profiler softwares. Each bar represents one specific centromeric or telomeric KIR gene. Concordant (one KIR allele matched for homozygous typing or two KIR alleles matched for heterozygous typing), semi-discordant (one KIR allele mismatched for heterozygous typing), and discordant KIR allele typing (one KIR allele mismatched for homozygous typing or two KIR alleles mismatched for heterozygous typing) were highlighted by a specific color code. Representative KIR3DL2 typing of IHW samples obtained by exome analysis compared to those assigned by NGS method is provided in the right of the graphs. Concordances are highlighted in bold and discordances in italics.

TABLE 2 | Discordant killer cell immunoglobulin-like receptor (KIR) typing of International Histocompatibility Workshop samples observed between typing obtained by exome capture[a] (47) and those obtained by next generation sequencing (NGS) in this study.[b]

ID	Centromeric KIR genes						Telomeric KIR genes					
	3DL3		2DL2		2DL3		2DP1		3DP1		3DL2	
	Exome KIR typing[a]	NGS typing[b]	Exome KIR typing[a]	NGS typing[b]	Exome KIR typing[a]	NGS typing[b]	Exome KIR typing[a]	NGS typing[b]	Exome KIR typing[a]	NGS typing[b]	Exome KIR typing[a]	NGS typing[b]
AMAI							__*003__, _*013_	__*00301__, _*004_				
CALOGERO	__*00207__, _*017_	__*00207__, _*01001_							_*002_, __*00302__	__*00302__, _*010_	__*00101__, _*076_	__*00101__, _*00301_
COX					__*002__	__*00201__, _*007_						
DKB									_*015_	_*00302_, _*006_		
KAS011							__*002__	__*002__, _*00301_			_*00701_, _*00902_	_*01001_, _*019_
LUY							__*002__, _*016_	__*00201__, _*00301_				
PF04015			__*003__	_*00101_, __*003__								
WT51							__*004__, _*018_	__*004__, _*001_				
YAR	__*00102__, _*044_	__*00102__										

Discrepancies are shown in italics for each KIR locus concerned. Allelic typing in bold represent concordant alleles. KIR alleles were named according to the last KIR nomenclature. ID, sample identification.

high resolution may help determine their combined effects on outcome of HSCT.

AUTHOR CONTRIBUTIONS

BM performed KIR allele typing by next-generation sequencing, KIR allele assignment, interpretation of data, and wrote the manuscript. NL performed DNA extractions from EBV-B cell lines, KIR genotyping and KIR allele typing by next-generation sequencing and commented on the manuscript. MA upgraded Profiler software integrating a KIR module, analyzed data, and commented on the manuscript. UG developed a pipeline for KIR allele assignment, analyzed data, and commented on the manuscript. CW and GD performed DNA extractions from EBV-B cell lines and commented on the manuscript. EC supervised the development of a pipeline for KIR allele assignment, provided a bioinformatic help on KIR read mapping, and commented on the manuscript. AW provided advices on the library construction setting and commented on the manuscript. CR designed the study, analyzed and interpreted data, commented on the manuscript, and contributed to writing the manuscript. KG designed the study, analyzed data, and wrote the paper. All the authors have approved the manuscript for publication.

ACKNOWLEDGMENTS

The authors are most grateful to Audrey Donnart, Dr. Audrey Bihouée, and Dr. Richard Redon of the Genomics and Bioinformatics Core Facility of Nantes (GenoBiRD, Biogenouest, Nantes, France) for their technical support. The authors would also like to thank Dr. Anne Cesbron (HLA laboratory, EFS Nantes, France) for the access of MiSeq platform, Dr. Kenza Belhaj (EFS La Plaine Saint Denis, France) for her contribution in KIR pipeline development by GenoBIRD genomics platform, Britt House Europe (Hegenheim, France), and Dr. David Senitzer (Division of Hematology and Bone Marrow Transplantation, Duarte, CA, USA) for help in the editing of the manuscript. This work was supported by EFS Pays de la Loire and by grants from International Research Group on unrelated Hematopoietic stem cell Transplantation (IRGHET 2015), Etablissement Français du Sang (EFS 2016-09), and Association Leucémie Espoir Atlantique Famille (LEAF). BM is a PhD student supported by a CIFRE/EFS Pays de la Loire/Nantes Université no. 481/2015 grant.

REFERENCES

1. Juric MK, Ghimire S, Ogonek J, Weissinger EM, Holler E, van Rood JJ, et al. Milestones of hematopoietic stem cell transplantation from first human studies to current developments. *Front Immunol* (2016) 7:470. doi:10.3389/fimmu.2016.00470
2. Barrett AJ, Battiwalla M. Relapse after allogeneic stem cell transplantation. *Expert Rev Hematol* (2010) 3(4):429–41. doi:10.1586/ehm.10.32
3. Petersdorf EW. The major histocompatibility complex: a model for understanding graft-versus-host disease. *Blood* (2013) 122(11):1863–72. doi:10.1182/blood-2013-05-355982
4. Ogonek J, Kralj Juric M, Ghimire S, Varanasi PR, Holler E, Greinix H, et al. Immune reconstitution after allogeneic hematopoietic stem cell transplantation. *Front Immunol* (2016) 7:507. doi:10.3389/fimmu.2016.00507
5. Ullah MA, Hill GR, Tey SK. Functional reconstitution of natural killer cells in allogeneic hematopoietic stem cell transplantation. *Front Immunol* (2016) 7:144. doi:10.3389/fimmu.2016.00144
6. Thielens A, Vivier E, Romagné F. NK cell MHC class I specific receptors (KIR): from biology to clinical intervention. *Curr Opin Immunol* (2012) 24(2):239–45. doi:10.1016/j.coi.2012.01.001
7. Ruggeri L, Capanni M, Urbani E, Perruccio K, Shlomchik WD, Tosti A, et al. Effectiveness of donor natural killer cell alloreactivity in mismatched hematopoietic transplants. *Science* (2002) 295(5562):2097–100. doi:10.1126/science.1068440
8. Shilling HG, Young N, Guethlein LA, Cheng NW, Gardiner CM, Tyan D, et al. Genetic control of human NK cell repertoire. *J Immunol* (2002) 169(1):239–47. doi:10.4049/jimmunol.169.1.239
9. Yawata M, Yawata N, Draghi M, Partheniou F, Little AM, Parham P. MHC class I-specific inhibitory receptors and their ligands structure diverse human NK-cell repertoires toward a balance of missing self-response. *Blood* (2008) 112(6):2369–80. doi:10.1182/blood-2008-03-143727
10. Kim S, Sunwoo JB, Yang L, Choi T, Song YJ, French AR, et al. HLA alleles determine differences in human natural killer cell responsiveness and potency. *Proc Natl Acad Sci U S A* (2008) 105(8):3053–8. doi:10.1073/pnas.0712229105
11. Babor F, Fischer JC, Uhrberg M. The role of KIR genes and ligands in leukemia surveillance. *Front Immunol* (2013) 4:27. doi:10.3389/fimmu.2013.00027
12. Shaffer BC, Hsu KC. How important is NK alloreactivity and KIR in allogeneic transplantation? *Best Pract Res Clin Haematol* (2016) 29(4):351–8. doi:10.1016/j.beha.2016.10.010
13. Trowsdale J. Genetic and functional relationships between MHC and NK receptor genes. *Immunity* (2001) 15(3):363–74. doi:10.1016/S1074-7613(01)00197-2
14. Uhrberg M, Valiante NM, Shum BP, Shilling HG, Lienert-Weidenbach K, Corliss B, et al. Human diversity in killer cell inhibitory receptor genes. *Immunity* (1997) 7(6):753–63. doi:10.1016/S1074-7613(00)80394-5
15. Pontikos N, Smyth DJ, Schuilenburg H, Howson JM, Walker NM, Burren OS, et al. A hybrid qPCR/SNP array approach allows cost efficient assessment of KIR gene copy numbers in large samples. *BMC Genomics* (2014) 15:274. doi:10.1186/1471-2164-15-274
16. Traherne J, Martin M, Ward R, Ohashi M, Pellett F, Gladman D, et al. Mechanisms of copy number variation and hybrid gene formation in the KIR immune gene complex. *Hum Mol Genet* (2010) 19(5):737–51. doi:10.1093/hmg/ddp538
17. Vendelbosch S, de Boer M, Gouw RA, Ho CK, Geissler J, Swelsen WT, et al. Extensive variation in gene copy number at the killer immunoglobulin-like receptor locus in humans. *PLoS One* (2013) 8(6):e67619. doi:10.1371/journal.pone.0067619
18. Béziat V, Traherne JA, Liu LL, Jayaraman J, Enqvist M, Larsson S, et al. Influence of KIR gene copy number on natural killer cell education. *Blood* (2013) 121(3):4703–7. doi:10.1182/blood-2012-10-461442
19. Sun JY, Oki A, Senitzer D. Alleles and intron polymorphism of KIR3DL1 shown by combination of allele group-specific primers and sequencing. *Tissue Antigens* (2008) 72(6):578–80. doi:10.1111/j.1399-0039.2008.01141.x
20. Norman PJ, Abi-Rached L, Gendzekhadze K, Korbel D, Gleimer M, Rowley D, et al. Unusual selection on the KIR3DL1/S1 natural killer cell receptor in Africans. *Nat Genet* (2007) 39(9):1092–9. doi:10.1038/ng2111
21. Gagne K, Willem C, Legrand N, Djaoud Z, David G, Rettman P, et al. Both the nature of KIR3DL1 alleles and the KIR3DL1/S1 allele combination affect the KIR3DL1 NK-cell repertoire in the French population. *Eur J Immunol* (2013) 43(4):1085–98. doi:10.1002/eji.201243007
22. Gardiner CM, Guethlein LA, Shilling HG, Pando M, Carr WH, Rajalingam R, et al. Different NK cell surface phenotypes defined by the DX9 antibody are

due to KIR3DL1 gene polymorphism. *J Immunol* (2001) 166(5):2992–3001. doi:10.4049/jimmunol.166.5.2992
23. O'Connor GM, Guinan KJ, Cunningham RT, Middleton D, Parham P, Gardiner CM. Functional polymorphism of the KIR3DL1/S1 receptor on human NK cells. *J Immunol* (2007) 178(1):235–41. doi:10.4049/jimmunol.178.1.235
24. Thomas R, Yamada E, Alter G, Martin MP, Bashirova AA, Norman PJ, et al. Novel KIR3DL1 alleles and their expression levels on NK cells: convergent evolution of KIR3DL1 phenotype variation? *J Immunol* (2008) 180(10):6743–50. doi:10.4049/jimmunol.180.10.6743
25. Saunders PM, Pymm P, Pietra G, Hughes VA, Hitchen C, O'Connor GM, et al. Killer cell immunoglobulin-like receptor 3DL1 polymorphism defines distinct hierarchies of HLA class I recognition. *J Exp Med* (2016) 213(5):791–807. doi:10.1084/jem.20152023
26. Bari R, Thapa R, Bao J, Li Y, Zheng J, Leung W. KIR2DL2/2DL3-E(35) alleles are functionally stronger than Q(35) alleles. *Sci Rep* (2016) 6:23689. doi:10.1038/srep23689
27. Parham P, Norman PJ, Abi-Rached L, Guethlein LA. Variable NK cell receptors exemplified by human KIR3DL1/S1. *J Immunol* (2011) 187(1):11–9. doi:10.4049/jimmunol.0902332
28. Carr WH, Pando MJ, Parham P. KIR3DL1 polymorphisms that affect NK cell inhibition by HLA-Bw4 ligand. *J Immunol* (2005) 175(8):5222–9. doi:10.4049/jimmunol.175.8.5222
29. Mulrooney TJ, Zhang AC, Goldgur Y, Boudreau JE, Hsu KC. KIR3DS1-specific D0 domain polymorphisms disrupt KIR3DL1 surface expression and HLA binding. *J Immunol* (2015) 195(3):1242–50. doi:10.4049/jimmunol.1500243
30. Falco M, Romeo E, Marcenaro S, Martini S, Vitale M, Bottino C, et al. Combined genotypic and phenotypic killer cell Ig-like receptor analyses reveal KIR2DL3 alleles displaying unexpected monoclonal antibody reactivity: identification of the amino acid residues critical for staining. *J Immunol* (2010) 185(1):433–41. doi:10.4049/jimmunol.0903632
31. Halfpenny IA, Middleton D, Barnett YA, Williams F. Investigation of killer cell immunoglobulin-like receptor gene diversity: IV. KIR3DL1/S1. *Hum Immunol* (2004) 65(6):602–12. doi:10.1016/j.humimm.2004.03.003
32. Keaney L, Williams F, Meenagh A, Sleator C, Middleton D. Investigation of killer cell immunoglobulin-like receptor gene diversity III. KIR2DL3. *Tissue Antigens* (2004) 64(2):188–94. doi:10.1111/j.1399-0039.2004.00263.x
33. Maxwell LD, Williams F, Gilmore P, Meenagh A, Middleton D. Investigation of killer cell immunoglobulin-like receptor gene diversity: II. KIR2DS4. *Hum Immunol* (2004) 65(6):613–21. doi:10.1016/j.humimm.2004.02.028
34. Williams F, Meenagh A, Sleator C, Middleton D. Investigation of killer cell immunoglobulinlike receptor gene diversity: I. KIR2DL4. *Hum Immunol* (2004) 65(1):31–8. doi:10.1016/j.humimm.2003.10.009
35. Meenagh A, Williams F, Sleator C, Halfpenny IA, Middleton D. Investigation of killer cell immunoglobulin-like receptor gene diversity V. KIR3DL2. *Tissue Antigens* (2004) 64(3):226–34. doi:10.1111/j.1399-0039.2004.00272.x
36. Meenagh A, Gonzalez A, Sleator C, McQuaid S, Middleton D. Investigation of killer cell immunoglobulin-like receptor gene diversity, KIR2DL1 and KIR2DS1. *Tissue Antigens* (2008) 72(4):383–91. doi:10.1111/j.1399-0039.2008.01093.x
37. Gonzalez A, Meenagh A, Sleator C, Middleton D. Investigation of killer cell immunoglobulin-like receptor (KIR) gene diversity: KIR2DL2, KIR2DL5 and KIR2DS5. *Tissue Antigens* (2008) 72(1):11–20. doi:10.1111/j.1399-0039.2008.01050.x
38. Bari R, Leung M, Turner VE, Embrey C, Rooney B, Holladay M, et al. Molecular determinant-based typing of KIR alleles and KIR ligands. *Clin Immunol* (2011) 138(3):274–81. doi:10.1016/j.clim.2010.12.002
39. Zhu FM, Jiang K, Lv QF, He J, Yan LX. Investigation of killer cell immunoglobulin-like receptor KIR2DL4 diversity by sequence-based typing in Chinese population. *Tissue Antigens* (2006) 67(3):214–21. doi:10.1111/j.1399-0039.2006.00562.x
40. Thananchai H, Gillespie G, Martin MP, Bashirova A, Yawata N, Yawata M, et al. Cutting edge: allele-specific and peptide-dependent interactions between KIR3DL1 and HLA-A and HLA-B. *J Immunol* (2007) 178(1):33–7. doi:10.4049/jimmunol.178.1.33
41. Schellekens J, Tilanus MG, Rozemuller EH. The elucidation of KIR2DL4 gene polymorphism. *Mol Immunol* (2008) 45(7):1900–6. doi:10.1016/j.molimm.2007.10.038
42. Hou L, Chen M, Steiner N, Kariyawasam K, Ng J, Hurley CK. Killer cell immunoglobulin-like receptors (KIR) typing by DNA sequencing. *Methods Mol Biol* (2012) 882:431–68. doi:10.1007/978-1-61779-842-9_25
43. Gonzalez A, McErlean C, Meenagh A, Shovlin T, Middleton D. Killer cell immunoglobulin-like receptor allele discrimination by high-resolution melting. *Hum Immunol* (2009) 70(10):858–63. doi:10.1016/j.humimm.2009.07.006
44. Witt CS, Martin A, Christiansen FT. Detection of KIR2DL4 alleles by sequencing and SSCP reveals a common allele with a shortened cytoplasmic tail. *Tissue Antigens* (2000) 56(3):248–57. doi:10.1034/j.1399-0039.2000.560307.x
45. Mori A, Deola S, Xumerle L, Mijatovic V, Malerba G, Monsurrò V. Next generation sequencing: new tools in immunology and hematology. *Blood Res* (2013) 48(4):242–9. doi:10.5045/br.2013.48.4.242
46. Kidd JM, Sharpton TJ, Bobo D, Norman PJ, Martin AR, Carpenter ML, et al. Exome capture from saliva produces high quality genomic and metagenomic data. *BMC Genomics* (2014) 15:262. doi:10.1186/1471-2164-15-262
47. Norman PJ, Hollenbach JA, Nemat-Gorgani N, Marin WM, Norberg SJ, Ashouri E, et al. Defining KIR and HLA class I genotypes at highest resolution via high-throughput sequencing. *Am J Hum Genet* (2016) 99(2):375–91. doi:10.1016/j.ajhg.2016.06.023
48. Robinson JT, Thorvaldsdóttir H, Winckler W, Guttman M, Lander ES, Getz G, et al. Integrative genomics viewer. *Nat Biotechnol* (2011) 29(1):24–6. doi:10.1038/nbt.1754
49. Alizadeh M, Frassati C, Santiago S, Regnault T, Abi-Rached L, Picard C, et al. Validation of profiler software: new analyser of HLA sequences from NGS. *HLA* (2016) 87:270.
50. Duke JL, Lind C, Mackiewicz K, Ferriola D, Papazoglou A, Gasiewski A, et al. Determining performance characteristics of an NGS-based HLA typing method for clinical applications. *HLA* (2016) 87(3):141–52. doi:10.1111/tan.12736
51. Denis L, Sivula J, Gourraud PA, Kerdudou N, Chout R, Ricard C, et al. Genetic diversity of KIR natural killer cell markers in populations from France, Guadeloupe, Finland, Senegal and Reunion. *Tissue Antigens* (2005) 66(4):267–76. doi:10.1111/j.1399-0039.2005.00473.x
52. Vierra-Green C, Roe D, Hou L, Hurley CK, Rajalingam R, Reed E, et al. Allele-level haplotype frequencies and pairwise linkage disequilibrium for 14 KIR loci in 506 European-American individuals. *PLoS One* (2012) 7(11):e47491. doi:10.1371/journal.pone.0047491
53. Gourraud P, Meenagh A, Cambon-Thomsen A, Middleton D. Linkage disequilibrium organization of the human KIR superlocus: implications for KIR data analyses. *Immunogenetics* (2010) 62(11–12):729–40. doi:10.1007/s00251-010-0478-4
54. Petersdorf EW, Gooley TA, Malkki M, Bacigalupo AP, Cesbron A, Du Toit E, et al. HLA-C expression levels define permissible mismatches in hematopoietic cell transplantation. *Blood* (2014) 124(26):3996–4003. doi:10.1182/blood-2014-09-599969
55. Tiercy JM. HLA-C incompatibilities in allogeneic unrelated hematopoietic stem cell transplantation. *Front Immunol* (2014) 5:216. doi:10.3389/fimmu.2014.00216
56. René C, Lozano C, Villalba M, Eliaou JF. 5' and 3' untranslated regions contribute to the differential expression of specific HLA-A alleles. *Eur J Immunol* (2015) 45(12):3454–63. doi:10.1002/eji.201545927
57. Sips M, Liu Q, Draghi M, Ghebremichael M, Berger CT, Suscovich TJ, et al. HLA-C levels impact natural killer cell subset distribution and function. *Hum Immunol* (2016) 77(12):1147–53. doi:10.1016/j.humimm.2016.08.004
58. Horowitz A, Djaoud Z, Nemat-Gorgani N, Blokhuis J, Hilton HG, Béziat V, et al. Class I HLA haplotypes form two schools that educate NK cells in different ways. *Sci Immunol* (2016) 1(3):eaag1672. doi:10.1126/sciimmunol.aag1672
59. David G, Djaoud Z, Willem C, Legrand N, Rettman P, Gagne K, et al. Large spectrum of HLA-C recognition by KIR2DL2 and KIR2DL3 and restricted C1 specificity of KIR2DS2: dominant impact of KIR2DL2/2DS2 on KIR2D NK cell repertoire formation. *J Immunol* (2013) 191(9):4778–88. doi:10.4049/jimmunol.1301580
60. David G, Morvan M, Gagne K, Kerdudou N, Willem C, Devys A, et al. Discrimination between the main activating and inhibitory killer cell immunoglobulin-like receptor positive natural killer cell subsets using newly characterized monoclonal antibodies. *Immunology* (2009) 128(2):172–84. doi:10.1111/j.1365-2567.2009.03085.x

61. Béziat V, Hilton H, Norman PJ, Traherne JA. Deciphering the KIR system at super-resolution for NK and T cell biology. *Immunology* (2016) 150(3):248–64. doi:10.1111/imm.12684
62. Djaoud Z, David G, Bressollette C, Willem C, Rettman P, Gagne K, et al. Amplified NKG2C+ NK cells in cytomegalovirus (CMV) infection preferentially express killer cell Ig-like receptor 2DL: functional impact in controlling CMV-infected dendritic cells. *J Immunol* (2013) 191(5):2708–16. doi:10.4049/jimmunol.1301138
63. Boudreau JE, Mulrooney TJ, Le Luduec JB, Barker E, Hsu KC. KIR3DL1 and HLA-B density and binding calibrate NK education and response to HIV. *J Immunol* (2016) 196(8):3398–410. doi:10.4049/jimmunol.1502469
64. Khakoo SI, Thio CL, Martin MP, Brooks CR, Gao X, Astemborski J, et al. HLA and NK cell inhibitory receptor genes in resolving hepatitis C virus infection. *Science* (2004) 305(5685):872–4. doi:10.1126/science.1097670
65. Penman BS, Moffett A, Chazara O, Gupta S, Parham P. Reproduction, infection and killer-cell immunoglobulin-like receptor haplotype evolution. *Immunogenetics* (2016) 68(10):755–64. doi:10.1007/s00251-016-0935-9
66. Morvan M, Willem C, Gagne K, Kerdudou N, David G, Sébille V, et al. Phenotypic and functional analyses of KIR3DL1+ and KIR3DS1+ NK cell subsets demonstrate diffrential regulation by Bw4 molecules and induced KIR3DS1 expression on stimulated NK cells. *J Immunol* (2009) 182(11):6727–35. doi:10.4049/jimmunol.0900212
67. Rettman P, Willem C, David G, Riou R, Legrand N, Esbelin J, et al. New insights on the natural killer cell repertoire from a thorough analysis of cord blood cells. *J Leukoc Biol* (2016) 100(3):374–85. doi:10.1189/jlb.1HI0116-036R
68. Rettman P, Legrand N, Willem C, Lodé L, Chevallier P, Cesbron A, et al. Use of killer cell immunoglobulin-like receptor (KIR) genes as early markers of haematopoietic chimerism after double-umbilical cord blood transplantation. *Haematologica* (2015) 100(11):475–9. doi:10.3324/haematol.2015.127993
69. Alicata C, Pende D, Meazza R, Canevali P, Loiacono F, Bertaina A, et al. Hematopoietic stem cell transplantation: improving alloreactive Bw4 donor selection by genotyping codon 86 of KIR3DL1/S1. *Eur J Immunol* (2016) 46(6):1511–7. doi:10.1002/eji.201546236
70. Gagne K, Busson M, Balere-Appert ML, Absi L, Jollet I, Bignon JD, et al. Relevance of KIR gene matching in unrelated hematopoietic stem cell transplantations. *Tissue Antigens* (2007) 69(Suppl 1):118–22. doi:10.1111/j.1399-0039.2006.76210.x
71. Gagne K, Busson M, Bignon JD, Balere-Appert ML, Loiseau P, Dormoy A, et al. Donor KIR3DL1/3DS1 gene and recipient Bw4 KIR ligand as prognostic markers for outcome in unrelated hematopoietic stem cell transplantation. *Biol Blood Marrow Transplant* (2009) 15(11):1366–75. doi:10.1016/j.bbmt.2009.06.015
72. Rettman P, Willem C, Volteau C, Legrand N, Chevallier P, Lodé L, et al. Impact of graft-versus-graft NK cell alloreactivity on single unit dominance after double umbilical cord blood transplantation. *Transplantation* (2016). doi:10.1097/TP.0000000000001545
73. Rettman P, Malard F, Legrand N, Avinens O, Eliaou J, Picard C, et al. Impact of KIR/HLA genetic combinations on double umbilical cord blood transplantation outcomes. Results of a French multicentric retrospective study on behalf of the Société Francophone de Greffe de Moelle et de Thérapie Cellulaire (SFGM-TC) and the Société Francophone d'Histocompatibilité et d'Immunogénétique (SFHI). *Bone Marrow Transplant* (2016) 51(11):1499–503. doi:10.1038/bmt.2016.151

9

Development of Three Different NK Cell Subpopulations during Immune Reconstitution after Pediatric Allogeneic Hematopoietic Stem Cell Transplantation: Prognostic Markers in GvHD and Viral Infections

Sabine Huenecke[1], Claudia Cappel[1], Ruth Esser[2], Verena Pfirrmann[1], Emilia Salzmann-Manrique[1], Sibille Betz[1], Eileen Keitl[1], Julia Banisharif-Dehkordi[1], Shahrzad Bakhtiar[1], Christoph Königs[1], Andrea Jarisch[1], Jan Soerensen[1], Evelyn Ullrich[1,3], Thomas Klingebiel[1], Peter Bader[1] and Melanie Bremm[1]*

[1] Clinic for Pediatric and Adolescent Medicine, University Hospital, Frankfurt, Germany, [2] GMP Development Unit, Hannover Medical School, Institute of Cellular Therapeutics, Hannover, Germany, [3] LOEWE Center for Cell and Gene Therapy, Goethe University, Frankfurt, Germany

Edited by:
Gianfranco Pittari, Hamad Medical Corporation, Qatar

Reviewed by:
Nicholas D. Huntington, Walter and Eliza Hall Institute of Medical Research, Australia
Anne Caignard, Institut National de la Santé et de la Recherche Médicale, France

***Correspondence:**
Sabine Huenecke
sabine.huenecke@kgu.de

Natural killer (NK) cells play an important role following allogeneic hematopoietic stem cell transplantation (HSCT) exerting graft-versus-leukemia/tumor effect and mediating pathogen-specific immunity. Although NK cells are the first donor-derived lymphocytes reconstituting post-HSCT, their distribution of $CD56^{++}CD16^{-}$ ($CD56^{bright}$), $CD56^{++}CD16^{+}$ ($CD56^{intermediate=int}$), and $CD56^{+}CD16^{++}$ ($CD56^{dim}$) NK cells is explicitly divergent from healthy adults, but to some extent comparable to the NK cell development in early childhood. The proportion of $CD56^{bright}$/$CD56^{int}$/$CD56^{dim}$ changed from 15/8/78% in early childhood to 6/4/90% in adults, respectively. Within this study, we first compared the NK cell reconstitution post-HSCT to reference values of NK cell subpopulations of healthy children. Afterward, we investigated the reconstitution of NK cell subpopulations post-HSCT in correlation to acute graft versus host disease (aGvHD) and chronic graft versus host disease (cGvHD) as well as to viral infections. Interestingly, after a HSCT follow-up phase of 12 months, the distribution of NK cell subpopulations largely matched the 50th percentile of the reference range for healthy individuals. Patients suffering from aGvHD and cGvHD showed a delayed reconstitution of NK cells. Remarkably, within the first 2 months post-HSCT, patients suffering from aGvHD had significantly lower levels of $CD56^{bright}$ NK cells compared to patients without viral infection or without graft versus host disease (GvHD). Therefore, the amount of $CD56^{bright}$ NK cells might serve as an early prognostic factor for GvHD development. Furthermore, a prolonged and elevated peak in $CD56^{int}$ NK cells seemed to be characteristic for the chronification of

GvHD. In context of viral infection, a slightly lower CD56 and CD16 receptor expression followed by a considerable reduction in the absolute $CD56^{dim}$ NK cell numbers combined with reoccurrence of $CD56^{int}$ NK cells was observed. Our results suggest that a precise analysis of the reconstitution of NK cell subpopulations post-HSCT might indicate the occurrence of undesired events post-HSCT such as severe aGvHD.

Keywords: NK cells, immune reconstitution, CD56, CD16, allogeneic transplantation, children, reference values

INTRODUCTION

The reconstitution of natural killer (NK) cells following allogeneic hematopoietic stem cell transplantation (HSCT) plays an important role in the response against residual malignant cells and the control of viral infections (1, 2). Independent of the graft source, NK cells typically regenerate within the first month following HSCT (3). However, there is an over-representation of $CD56^{bright}CD16^{neg}$ NK cells in the early phase post-HSCT compared to healthy individuals (4, 5), where NK cells are composed of about 90% $CD56^{dim}CD16^{++}$ and 10% $CD56^{bright}CD16^{-}$ cells. Whereas $CD56^{dim}CD16^{++}$ NK cells mediate cytotoxicity and antibody-dependent cellular cytotoxicity, the $CD56^{bright}CD16^{-}$ subpopulation, which is mainly present in the early period post-HSCT, primarily secretes immunoregulatory cytokines. Presumably, the development from $CD56^{bright}$ to $CD56^{dim}$ NK cells corresponds to sequential steps of NK cell differentiation (6). In most patients, the ratio between $CD56^{bright}$ and $CD56^{dim}$ NK cells normalizes within the first 6 months post-HSCT influenced by the patient's age and events following HSCT. A correlation between the reconstitution of NK cells and overall survival was described by few studies emphasizing their essential role in the defense of infections when T cell immunity is mainly absent (7, 8). Furthermore, NK cells were successfully applied as immunotherapy for patients with high-risk malignancies suffering from impending relapse following HSCT. In clinical studies, adoptively infused NK cells induced graft-versus-leukemia/tumor effect without concomitant severe graft versus host disease (GvHD) (9–11). In addition, recent studies ascribe GvHD reduction to NK cell function post-HSCT (12). In this work, we focused on the regeneration of $CD56^{bright}$ and $CD56^{dim}$ NK cells with a special regard on the population shifting between these subpopulations. We divided NK cells in three NK cell subpopulations. In the first step, we established reference values of $CD56^{++}CD16^{-}$ ($CD56^{bright}$), $CD56^{++}CD16^{+}$ ($CD56^{int}$), and $CD56^{+}CD16^{++}$ ($CD56^{dim}$) NK cells of healthy children and adolescents (n = 174). These reference values were matched and compared to the NK cell reconstitution of patients, who did not suffer from any viral infection or GvHD and are still alive after allogenic stem cell transplantation. In the next step, we investigated the associations between the reconstitution of the three different NK cell subpopulations in regard to the occurrence of events such as acute graft versus host disease (aGvHD) or chronic graft versus host disease (cGvHD) and severe viral infections in the first year post-HSCT in contrast to a cohort of patients without severe events assumed as control group.

MATERIALS AND METHODS

Reference Cohort of Children and Adolescents

In this cross-sectional monocentric study (approval ethic committee Ref. No. 139/09), 174 donors (64 females and 110 males) were included. Residual peripheral blood samples of hematologically healthy children aged 1 month to 18 years were analyzed (patients were hospitalized, e.g., for cleft lip and palate correction). Inclusion criteria involved no incidence of immunodeficiency or infection (defined as >2 severe infections/year, >8 infections/year, persistent fungal infections, post-vaccinal complications, no evidence of acute bleeding, negative for CRP and normal leukocytes, lymphocytes, and neutrophil granulocytes).

Patients and Grafts

The reconstitution of NK cells and their subpopulation $CD56^{bright}$, $CD56^{int}$, and $CD56^{dim}$ was analyzed in n = 74 patients (n = 25 females and n = 49 males) transplanted from 2010 to 2016 (see **Table 1**). Indications for HSCT were high-risk acute lymphoblastic leukemia (n = 51), acute myeloid leukemia (n = 12), and myelodysplastic syndrome (n = 11). Median age at HSCT was 10.4 years (range: 1.3–24.4 years). Grafts were received from matched family donors (MSD; n = 20), matched unrelated donors (MUD) (n = 39), and haploidentical mismatched family donors with <8/10 HLA matches (MMFD; n = 15). No significant differences in the occurrence of GvHD were available in the different donor groups (MSD: 6/20, MUD: 10/39, and MMFD: 3/15). A second or third transplant was administered to n = 9 and n = 2 patients, respectively, because of relapse after first or second HSCT. Stem cell sources consisted of (bone marrow; n = 48), unmanipulated PBSC (n = 11), and T cell-depleted PBSC (n = 15). Post-transplant aGvHD occurred in n = 31 patients with grades I (n = 9), II (n = 3), III (n = 16), and IV (n = 3), cGvHD in 22 patients. Severe viral infections occurred in 18 patients including primary infection or reactivation with cytomegalovirus (CMV) (n = 8), adenovirus (ADV) (n = 5), and Epstein–Barr virus (EBV) (n = 5).

Study Design

This study was carried out in accordance with the Declaration of Helsinki and was approved by the Medical Ethics Committee of the Frankfurt University Hospital (Ref. No. #198/16). Peripheral blood was collected within the framework of a post-HSCT routine sampling for clinical follow-up from 2010 to 2016. In engrafted patients, samples were collected starting at day 15, within the first

TABLE 1 | Patient's characteristics.

Groups	All	aGvHD	cGvHD	Viral infection	Control group
Patients	74	19	22 (8)	18	23
Diagnosis					
ALL	51	13	12 (4)	13	17
AML	12	3	3 (2)	5	3
MDS	11	3	7 (2)	0	3
Age at HSCT					
Median (range)	10.4 (1.3–24.4)	8.9 (1.9–17.6)	9.4 (1.3–18.2)	10.3 (3.4–17.7)	12.1 (2.3–24.4)
Sex					
Male	49	11	15 (6)	13	16
Female	25	8	7 (2)	5	7
Donor type					
MSD	20	6	11 (4)	1	6
MUD	39	10	9 (2)	9	13
MMFD	15	3	2 (2)	8	4
Stem cell source					
BM	48	12	17	7	16
Unmanipulated PBSC	11	4	3	3	3
T cell depleted	15	3	2	8	4
GvHD					
aGVHD	31	19	20	0	0
Grade I/II/III/IV	9/3/16/3	0/0/16/3	9/3/7/1	0/0/0/0	0/0/0/0
cGvHD	22	8	22	0	0
Viral infections					
ADV/CMV/EBV	5/8/5	0/0/0	0/0/0	5/8/5	0/0/0
Survival (%)	87	74	96	78	100
Follow-up (months)					
Median (range)	35 (3–105)	33 (3–77)	43 (6–92)	28 (2–103)	48 (24–105)

ALL, acute lymphoblastic leukemia; AML, acute myeloid leukemia; MDS, myelodysplastic syndrome; HSCT, hematopoietic stem cell transplantation; MSD, matched sibling donor; MUD, matched unrelated donor; MMFD, mismatched family donor; BM, bone marrow; PBSC, peripheral blood stem cells; GvHD, graft versus host disease; aGVHD, acute graft versus host disease; cGVHD, chronic graft versus host disease; ADV, adenovirus; CMV, cytomegalovirus; EBV, Epstein–Barr virus.

year monthly, within the second year three monthly, and afterward every 6 months until 36 months post-HSCT, respectively. In total, $n = 925$ measurements of $n = 74$ patients were included in this analysis.

Depending on the occurrence of unexpected events such as aGvHD (only grades III and IV, grades I and II excluded), cGvHD, and severe viral infections, the study group was partitioned. The patient group without aGvHD/cGvHD or any viral infections was chosen for comparison as control group.

Patients showing a distinct CD56^{int} population in a sufficient absolute amount post-HSCT were elected for more detailed analysis applying a 10-color flow cytometry.

Assessment of NK Cells and Their Subpopulations

Natural killer cell subpopulations were analyzed on a FC500 flow-cytometer (Beckman Coulter, Krefeld, Germany) applying a five-color panel to estimate CD3$^+$ T cells, CD19$^+$ B cell, and CD3$^-$CD56$^+$ NK cells, including the differentiation into CD56^{++}CD16$^-$, CD56^{++}CD16$^+$, CD56$^+$CD16^{++} NK cells, abbreviated as CD56^{bright}, CD56^{int}, and CD56^{dim} NK cells, respectively. Absolute cell numbers were estimated from peripheral blood samples in a dual platform lyse-no-wash procedure as described previously (13). In brief, a tube of 100 µl EDTA-peripheral blood was labeled with CD45/CD56/CD19/CD3 tetraCHROME (clones B3821F4A/N901/J3-119/UCHT1) multi-color mAb conjugated with FITC, phycoerythrin (PE), phycoerythrin texas red (ECD), and phycoerythrin-cyanine 5 (PC5). CD16 phycoerythrin-cyanine 7 (PC7, clone: 6607118) was additionally added. For the measurement of T cells including T helper and cytotoxic T cells, we applied the tetraCHROME multi-color reagent CD45/CD4/CD8/CD3 (clones B3821F4A/SFCI12T4D11/SFCI21Thy2D3/UCHT1) conjugated with FITC, PE, ECD, and PC5. All reagents were acquired from Beckman Coulter Immutech (Marseille, France).

Differences in the expression profiles of CD56^{bright}, CD56^{int}, and CD56^{dim} cells were analyzed applying two panels on a Navios™ 10-color flow cytometer (Beckman Coulter, Krefeld, Germany). Panel 1: CD226 = DNAM-1 (FITC; clone: KRA236), NKG2A (PE; clone: Z199), DUMP = CD3&CD14&CD19 (ECD; clones: UCHT1, RMO52, J3-119), CD117 (PC5.5; clone: 104D2D1), CD27 (PC7; clone: 1A4CD27), CD56 [allophycocyanin (APC); clone: N901], CD127 (APC-A700; clone: R34.34), CD16 (APC-A750; clone: 3G8), CD57 [Pacific Blue (PB); clone: NC1], CD45 [Krome Orange (KrO); clone: J.33]. Panel 2: CCR5 (FITC; clone: 2D7), killer cell immunoglobulin-like receptor (KIR) mix (CD158 + CD158b + CD158e1; clones: EB6B, GL183, Z27.3.7), DUMP = CD3 + CD14 + CD19 (ECD; clones: UCHT1, RMO52, J3-119), CD117 (PC5.5; clone: 104D2D1), CX3CR1 (PC7; clone: 2A9-1), CD56 (APC; clone: N901), CD62L (APC-A700; clone: DREG56), CD16 (APC-A750; clone: 3G8), CCR7 (PB; clone: C013I17), and CD45 (KrO, clone: J.33). All antibodies were purchased from Beckman Coulter Immutech except CCR5 (BD Biosciences, Heidelberg, Germany) and CX3CR1 (Biolegend, San Diego, CA, USA). Staining was performed, using 100 µl of peripheral blood for each tube followed by 15 min of incubation at room temperature and erythrocyte lysis applying NH_4Cl reagent (Beckman Coulter, Marseille, France).

Stained Cyto-Comp™ cells were applied to compensate the fluorescence overlap. The flow cytometer fluidic stability and the optical alignment were daily tested using Flow-Check™ Fluorospheres (Beckman Coulter, Krefeld, Germany). For verification, Immunotrol cells (Beckman Coulter) were applied three times a day. Furthermore, we participate in an external quality assessment for the detection of T-, B-, and NK cells (INSTANT e. V.—provider for German round robin test, No 213).

Data evaluation was performed using Kaluza and CXP-software (Beckman Coulter).

Statistical Analysis

Statistical analysis was performed using GraphPad Prism 6 (GraphPad Software, San Diego, CA, USA). The NK subpopulation reference values of healthy children and adolescents were calculated with non-linear exponential regression analysis (equation: one-phase decay, least square fit; function: $Y = (Y0 - \text{Plateau}) \times \exp(-K \times X) + \text{Plateau}$). Significant differences between groups were assessed by a non-paired two-tailed Mann–Whitney *U* test. *p*-Values <0.05 were regarded as significant and are indicated as $^{*}p < 0.05$, $^{**}p < 0.01$, $^{***}p < 0.001$, and $^{****}p < 0.0001$.

RESULTS

Normalization of CD56^{bright} to CD56^{dim} Ratio Post-HSCT within the First Year Post-HSCT

The distribution of NK cell subpopulations post-HSCT is divergent from healthy individuals showing a high proportion of CD56^{bright} cells shifting toward CD56^{dim} NK cells. To estimate the reconstitution time to a normal CD56^{bright}, CD56^{int}, and CD56^{dim} NK cell distribution and absolute cell counts post-HSCT, we first generated age-matched reference values of NK cell subpopulation frequencies and absolute NK cell counts for healthy children and adolescents (Figure S1 in Supplementary Material). Interestingly, the proportion of CD56^{bright}, CD56^{int}, and CD56^{dim} in early childhood changed from 15, 8, and 78 to 6, 4, and 90% in adults, respectively. Subsequently, we matched the NK cell reconstitution of all patients of the control group without severe events (e.g., GvHD, infections) with the newly generated reference values (**Figure 1**). The frequency of CD56^{bright} NK cells is elevated directly after HSCT but reaches the upper reference range following 2 months. However, the levels of CD56^{bright} appear slightly increased remaining between the 50th and 90th percentile until 12 months post-HSCT (**Figure 1A**). Surprisingly, the percentage of CD56^{int} cells matches the reference range within the first 2 months post-HSCT but increases starting from 3 months exceeding the 90th percentile of the reference range following 5 months post-HSCT. After this period, the CD56^{int} fraction declines and converges to the 50th percentile of the reference values 12 months after HSCT (**Figure 1B**). By contrast, the CD56^{dim} fraction remains below the 10th percentile until 8 months and reaches the 50th percentile of normal reference values after 12 months post-HSCT (**Figure 1C**). Comparable reconstitution profiles are also apparent for absolute values of CD56^{bright}, CD56^{int}, and CD56^{dim} NK cells. However, absolute

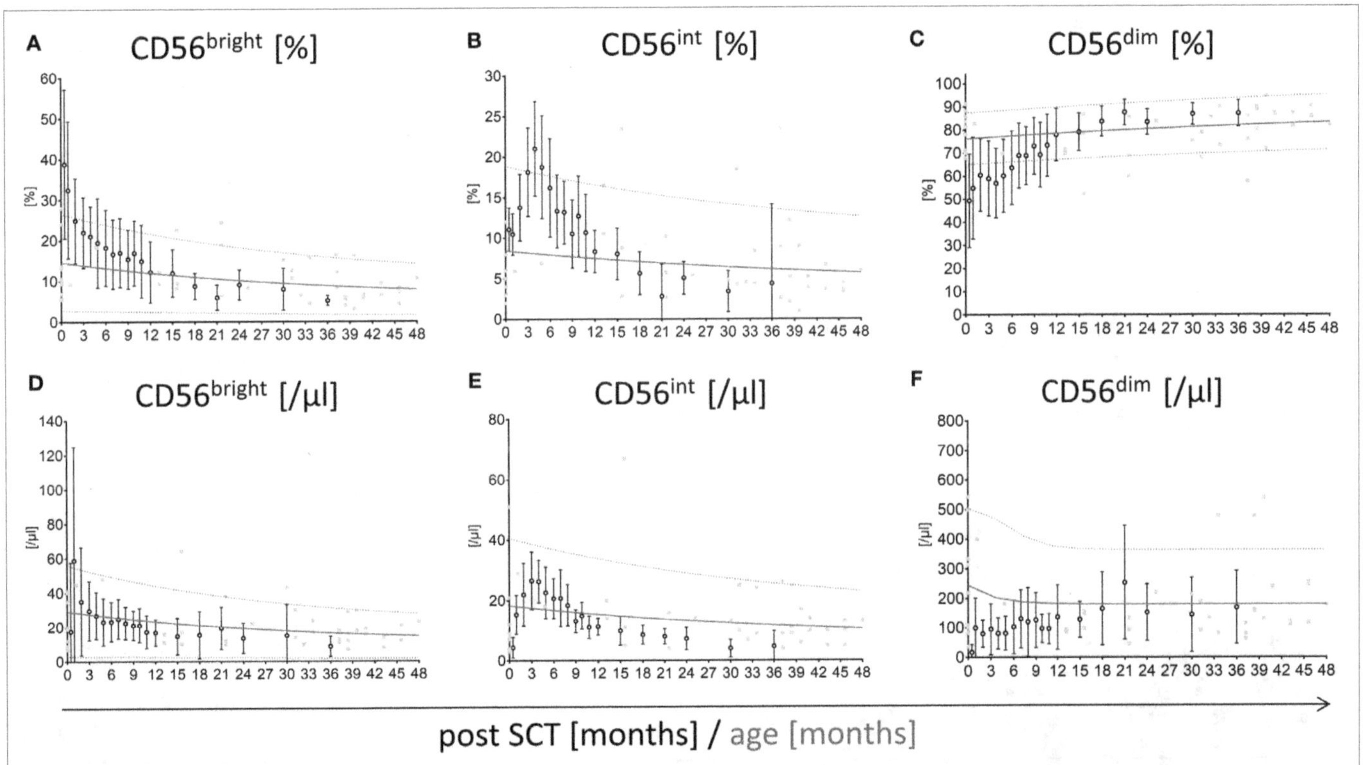

FIGURE 1 | Reconstitution of CD56^{bright}, CD56^{int}, and CD56^{dim} natural killer (NK) cells with respect to reference percentiles of healthy children. Reconstitution of CD56^{bright}, CD56^{int}, and CD56^{dim} NK cells plotted into a graph showing the 10th, 90th (···, dotted line), and 50th (▬, solid line) percentile of NK cell reference values of healthy children (*n* = 174) as well as their underlying measurements (■, gray squares). Measurements of stem cell transplanted patients without graft-versus-host disease or viral infections (*n* = 23; mean with SD) were plotted into the reference model (○, black rimmed circles) at 19 time points until 36 months after hematopoietic stem cell transplantation (HSCT). Time points were 15 days, 1, 2, 3, 4, 5, 6, 7, 8, 9, 10, 11, 12, 15, 18, 21, 14, 30, and 36 months after SCT. The CD56^{bright} and CD56^{dim} NK cells post-HSCT needed around 12 months to reach the distribution of healthy individuals NK cells. At that time point, the CD56^{bright}, CD56^{int}, and CD56^{dim} NK cells met the 50th percentile of the reference range **(A–C)**. CD56^{bright} NK cells were elevated post-HSCT but already reached the reference range after 2 months **(A)**. Intermediate NK cells matched the reference range directly after HSCT but showed a fast proliferation until 5 months followed by a decline until 12 months post-HSCT **(B)**. CD56^{dim} NK cells were underrepresented directly after HSCT, reaching the lower reference range after 8 months post-HSCT **(C)**. The development of absolute cell counts of CD56^{bright}, CD56^{int}, and CD56^{dim} NK cells is shown in subfigures **(D–F)**.

values after HSCT seem to be lower than NK cells of healthy children (**Figures 1D–F**).

Reconstitution of NK Cells with Regard to GvHD Development

Patients suffering from aGvHD with grade higher than III show significant differences in the reconstitution of NK cell subpopulations compared to patients without any severe events post-HSCT. The median time of first symptoms of aGvHD was 22 days post-HSCT (range: 13–84). Within the first, second, and third month 68, 89, and 100% of the affected patients showed first signs of aGvHD, respectively. Patients without events show conspicuously higher frequency of $CD56^{bright}$ NK cells within the first 3 months following HSCT compared to patients suffering from aGvHD with the most significant difference already 15 days following engraftment ($p < 0.0001$; **Figure 2A**). This could also be shown for the absolute $CD56^{bright}$ NK cell amount 1 month after HSCT ($p < 0.01$; **Figure 2D**). This tendency was less pronounced when analyzing NK cells in patients with lower grades aGvHD (data not shown). Furthermore, the reconstitution of NK cell subpopulations seems to be delayed in patients suffering from aGvHD. For the $CD56^{int}$ NK cell population, a displacement in time could be shown, which leads to a longer increase in $CD56^{int}$ frequency and absolute amount (**Figures 2B,E**). To reach the 50th percentile of normal reference values, patients with aGvHD need a prolonged reconstitution time taking at least two times longer compared to patients without events. Patients with lower GvHD grades were lying in between (data not shown). Even after 3 years of monitoring, a trend toward higher frequency of $CD56^{bright}$ and $CD56^{int}$ NK cells concomitant with lower $CD56^{dim}$ was seen in patients suffering from severe aGvHD (not significant, **Figures 2A–C**). These differences in the NK cell development post-HSCT were also found evaluating the absolute amounts of NK cell subpopulations (**Figures 2D–F**). Noteworthy, patients affected with aGvHD following HSCT also show a reduced absolute amount of $CD56^{dim}$ NK cells after 3 years post-HSCT (**Figure 2F**). Analyzing absolute NK cell count (including all three subgroups), we did not see a correlation between patients with and without aGvHD (Figure S2A in Supplementary Material). Analyzing cytotoxic T cells, we detected that patients reaching levels of cytotoxic T cells above 1,500/μl within the first year post-HSCT developed in almost all cases an aGvHD (Figure S2B in Supplementary Material). But this fact was only true for less than 25% of the total aGvHD patient cohort. Furthermore, we evaluated the NK cell regeneration of

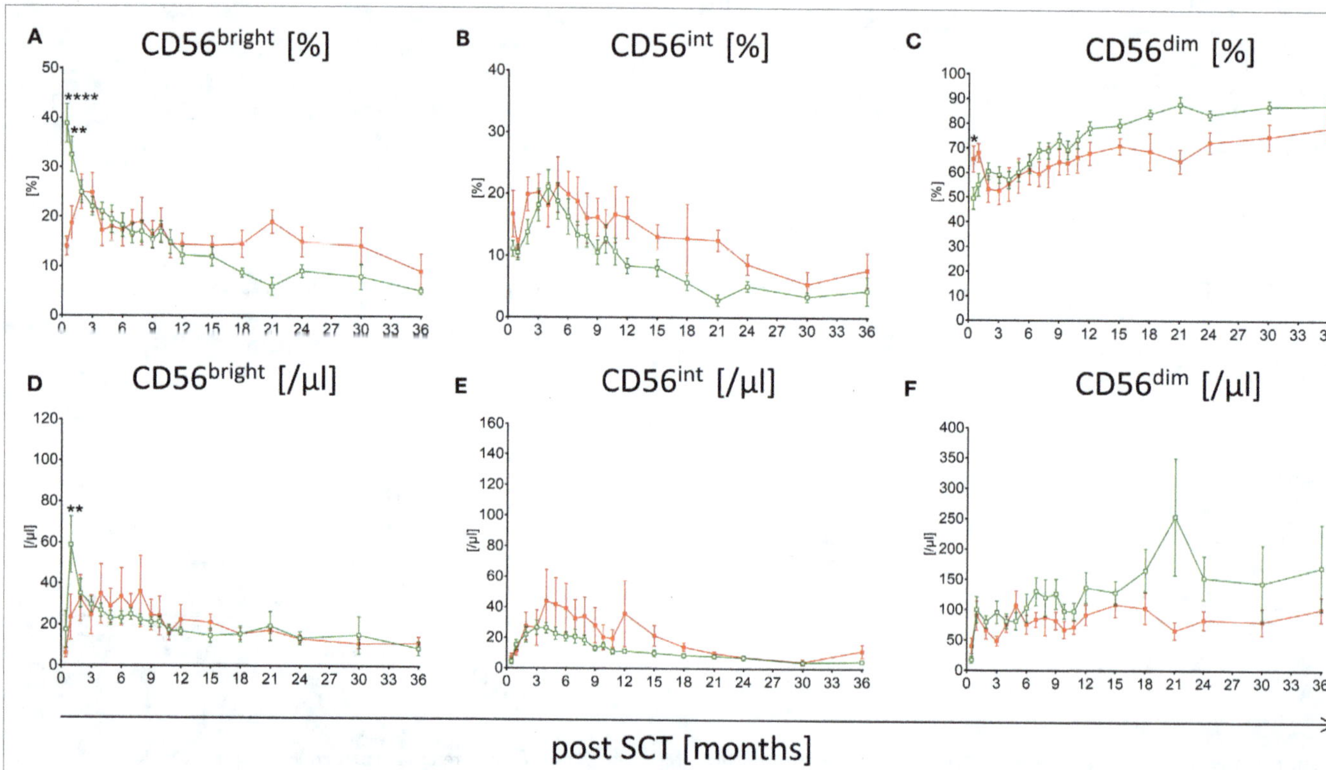

FIGURE 2 | Comparison of natural killer (NK) cell reconstitution in patients suffering from acute graft versus host disease (aGvHD) versus patients without events. Regeneration of $CD56^{bright}$, $CD56^{int}$, and $CD56^{dim}$ NK cells in patients affected with severe aGvHD with grade III and IV (■, red squares, $n = 19$) and patients without events post-hematopoietic stem cell transplantation (HSCT) (□, green rimmed squares, $n = 23$) at time point of 15 days, 1, 2, 3, 4, 5, 6, 7, 8, 9, 10, 11, 12, 15, 18, 21, 14, 30, and 36 months after SCT. Especially, in the first 3 months, post-HSCT aGvHD patients showed significantly lower $CD56^{bright}$ NK cells in both percentage **(A)** ($p < 0.0001$) and absolute amounts **(D)** ($p < 0.01$). $CD56^{int}$ NK cells of aGvHD patients seem delayed in their development as patients without events show an increased intermediate population over a period of at least 18 months **(B,E)**. Except in the beginning, $CD56^{dim}$ NK cell levels of GvHD patients remain below those of patients without events during the whole monitoring interval **(C,F)**. Measurements were available at almost all time points, except for five patients of the aGvHD group who died 2, 3, 9, 9, and 18 months post-HSCT.

patients with primary aGvHD grade III or IV that became chronic. Thereby, we detected important differences in the development of NK cell subpopulations from CD56^{bright} above CD56^{int} to CD56^{dim} NK cells. The development from CD56^{int} to CD56^{dim} NK cells is delayed for at least 2 years. Furthermore, patients with chronification of aGvHD with grade >III have a markedly elevated CD56^{int} frequency (**Figure 3A**), which is also clearly visible in absolute cell count of CD56^{int} NK cells (**Figure 3B**).

Receptor Expression of CD56^{bright}, CD56^{int}, and CD56^{dim} NK Cells Post-HSCT

As part of the patients had low absolute cell counts post-HSCT, a detailed phenotyping of the CD56^{bright}, CD56^{int}, and CD56^{dim} NK cell populations was only possible for an elected cohort of patients. To get an understanding of the function of CD56^{int} NK cells, we analyzed surface molecules linked to NK cell cytotoxicity, adhesion, and immune regulatory functions (e.g., chemokine and cytokine receptors) and compared the expression of KIRs, CD62L, NKG2A, CD127, CD117, CX3CR1, CD226, and CD57 on all three NK subpopulations. The CD56^{dim} population showed a higher expression of KIRs, whereas the CD56^{int} and CD56^{bright} population did not. However, not all KIRs applied within the mix were expressed with equal density (**Figure 4**). Regarding the homing receptor CD62L, the CD56^{int}, and the CD56^{bright} fraction showed an increased expression compared to CD56^{dim} population. The expression of NKG2A was highest on CD56^{int} and CD56^{bright} cells but bipartite in CD56^{dim} NK cells. CX3CR1 is involved in adhesion and migration of NK cells and to a small extent higher presented on CD56^{dim} NK cells. As already described, we could also show that CD57 was only detectable on the CD56^{dim} subpopulation. Low expression of CD127 (IL7α chain), CD117 (c-Kit), and CD226 (DNAM-1) could be seen on all NK cell subpopulations; however, CD56^{bright} intend to have higher expression than CD56^{dim}, whereas CD56^{int} was always lying in between. In summary, the expression profiles of CD56^{int} and CD56^{bright} NK cells were nearly congruent, but differed to CD56^{dim} cells in KIR, CD62L, NKG2A, CX3CR1, and CD57 expression (**Figure 4**).

Influence of Viral Infection Post-HSCT on NK Cell Reconstitution

The immune reconstitution of NK cell subpopulations post-HSCT was analyzed in patients without events and patients suffering from ADV (n = 5), EBV (n = 5), and CMV infection (n = 8). Infection was detected by the routine analysis of DNA copies in peripheral blood. Patients with elevated viral load at the day of transplantation were excluded from the study, resulting in a cohort of patients with occurrence of a positive viral load between 30 and 90 days post-HSCT. Interestingly, we observed a slight reduction in CD56 and CD16 expression in patients suffering from viral infection in between day 30 and day 60 post-HSCT measured by mean fluorescence intensity (**Figure 5**). After viral clearance in most patients, a considerable loss in absolute CD56^{dim} NK cell count occurred followed by continued regeneration of CD56^{int} NK cells, which was lower in patients without events post-HSCT on day 150 post-HSCT (**Figure 5**).

DISCUSSION

Especially in the early phase following allogeneic HSCT, together with neutrophils, NK cells are the first line of immune defense. Their immune reconstitution is of crucial importance for transplantation outcome with special regard to the occurrence of GvHD and viral infections. In this project, we analyzed NK cell subpopulations in detail not only focusing on CD56^{bright} and CD56^{dim} cells but also the fraction in between those subsets. With regard to NK cell subpopulation development in young healthy

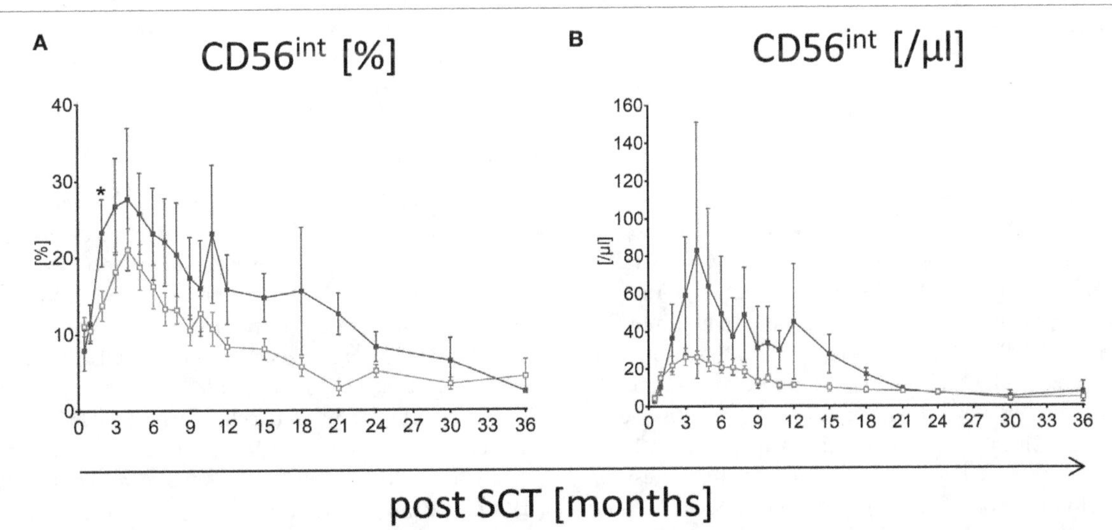

FIGURE 3 | Natural killer (NK) cell reconstitution in patients suffering from chronic graft versus host disease (cGvHD). Development of CD56^{int} NK cells of patients suffering from cGvHD after severe acute graft versus host disease (grade III or IV) is shown for frequency **(A)** and absolute count **(B)**. Over a period of at least 18 months, a higher amount of CD56^{int} NK cells for patients suffering from cGvHD (■, dark red squares, n = 8) was detectable compared to patients without severe events post-hematopoietic stem cell transplantation (□, green rimmed squares, n = 23).

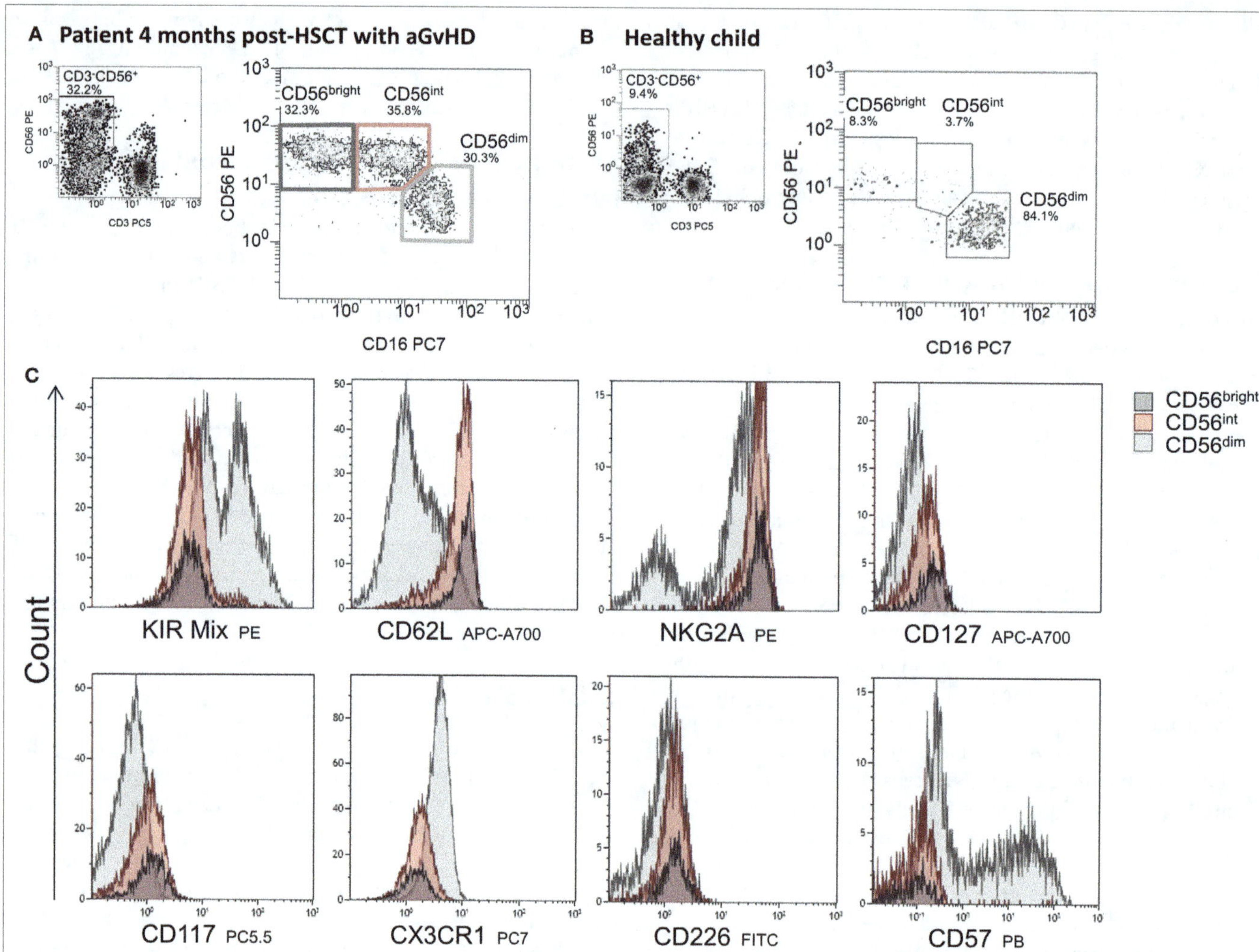

FIGURE 4 | Immunophenotyping of CD56^{bright}, CD56^{int}, and CD56^{dim} cells. Exemplary flow cytometric plots of natural killer (NK) cell subpopulations of a patient 4 months post-hematopoietic stem cell transplantation (HSCT) with GvHD showing an almost equal distribution of CD56^{bright}, CD56^{int}, and CD56^{dim} cells **(A)** and of a healthy child aged 3 years with mainly CD56^{dim} cytotoxic NK cells **(B)**. NK cell subpopulations were characterized in detail by flow cytometric measurement of surface antigen expression of several ligands involved in adhesion, chemotaxis, and cytotoxicity. Expression of the corresponding receptors on NK cells is shown by overlay plots with MFI on the *x*-axis of respective antigen in allocation to CD56^{bright} (dark gray), CD56^{int} (red) compared to CD56^{dim} NK cells (light gray). CD56^{bright} and CD56^{int} NK cells show comparable receptor expression, but differ to CD56^{dim} cells in KIR-, CD62L, NKG2A, CX3CR1, and CD57 expression **(C)**.

children, we found that it takes around 12 months until CD56^{bright}, CD56^{int}, and CD56^{dim} NK cells of patients post-HSCT reach the 50th percentile of age-matched reference range. Comparable results were published by Pical-Izard et al. describing that rapidly re-emerging NK cells remain immature for more than 6 months (8). Directly after HSCT we detected a highly increased frequency of CD56^{bright}, whereas CD56^{int} NK cells correspond to the reference range, but considerably expand within the first 3 months post-HSCT. In contrast, CD56^{dim} NK cells deserve around 8 months to enter the reference range. These results confirm the hypothesis of sequential development of NK cells with CD56^{int} NK cells representing an intermediate state from CD56^{bright} to CD56^{dim} NK cells (14, 15). These CD56^{bright} NK cells in peripheral blood are closely related to those NK cells populating secondary lymphoid tissues (16). Further evidence supporting this hypothesis was published by Freud et al. describing that the CD56^{bright} subset is the major NK cell population that is derived early *in vitro* when CD34^{+} HPC are cultured in NK development supportive conditions, whereas CD56^{dim} NK cells develop later (17). Furthermore, CD56^{bright} NK cells display longer telomeres than the CD56^{dim} NK cells, indicating lower proliferation capacity (6).

We further characterized all three NK cell subpopulations with the finding, that CD56^{int} presented antigen expressions among CD56^{bright} and CD56^{dim} NK cells, even so CD56^{bright} and CD56^{int} NK cells showed rather equal expression profiles and seemed related more to CD56^{bright}. However, differential expression of KIRs, CD62L, NKG2A, and CD57 was observed on CD56^{dim} NK cells. This is in parallel to other findings describing an increased expression of NKG2A, the IL-7 receptor (CD127) and the lymph node homing receptor CCR7 on CD56^{bright} cells (2, 5, 8, 18, 19) whereas CD56^{dim} NK cells acquire KIR, NKG2C, and CD57 expression (20).

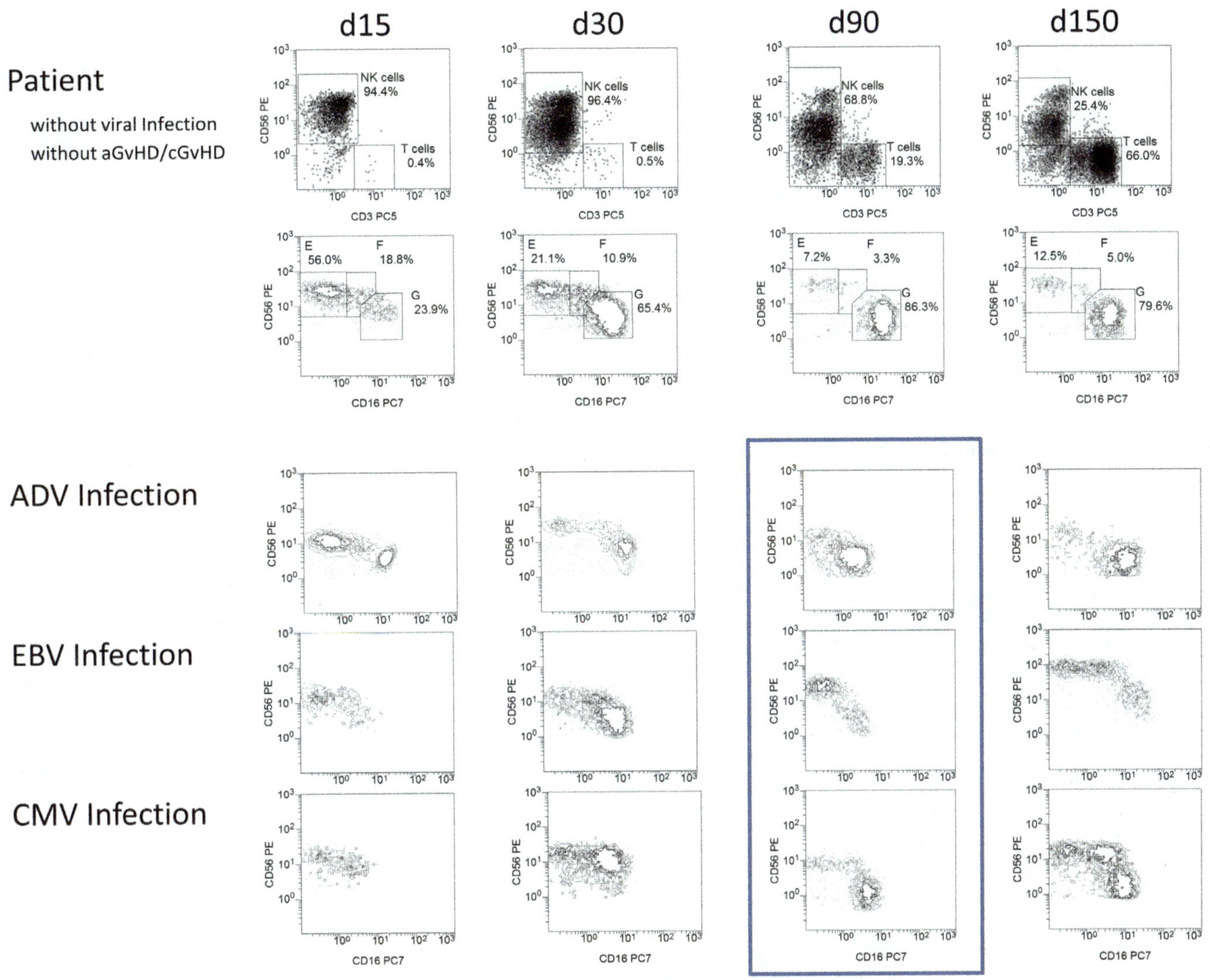

FIGURE 5 | Viral infections influencing natural killer (NK) cell reconstitution and NK marker expression density. Exemplary flow cytometric plots of a patient without events and patients suffering from adenovirus (ADV), Epstein–Barr virus (EBV), and cytomegalovirus (CMV) infection between 15 and 150 days post-HSCT. Patients with high viral load (e.g., ADV above 40,000 genome equivalent/ml and CMV above 90,000 IU/ml) or qualitatively confirmed EBV infection with disease onset between day 30 and day 60 were exemplarily presented. NK cells of patients suffering from viral infection showed a slight reduction in CD56 and CD16 receptor expression 90 days post-HSCT (frame). After 150 days post-HSCT, the NK cells of the patient without events consisted of a major population of $CD56^{dim}$, a small amount of $CD56^{bright}$ and only few $CD56^{int}$ cells whereas patients after EBV and CMV infection showed a distinct population of $CD56^{int}$ NK cells again. Patients suffering from infections: ADV (n = 5), EBV (n = 5), CMV (n = 8).

Promoted by the IL-15 rich cytokine milieu post-transplant, NK cells are known to be one of the first lymphocyte subpopulation recovering post-HSCT (21). Therefore, NK cell reconstitution might be the basis for generating early prognostic markers regarding the occurrence of severe events and transplantation outcome. Kim et al. published that NK cell counts after allo-HSCT, especially on day 30, were predictive markers for GvHD, non-relapse mortality, and survival (22). Furthermore, there is evidence that the speed of NK cell reconstitution correlates with transplant outcome, suggesting their important role in the early period when specific T cell immunity is absent (7, 8). Our and other findings suggest that the monitoring of NK cell subsets in the early phase post-HSCT might provide first signs of aGvHD development (23). Interestingly, within the first 2 months post-HSCT patients without aGvHD or viral infections had significantly elevated levels of $CD56^{bright}$ NK cells compared to patients suffering from aGvHD. This might be an early prognostic factor regarding GvHD development; however, it needs to be confirmed in a prospective study. Likewise results were also published by Kheav et al. showing an impaired reconstitution of $CD56^{dim}$ NK cells 3 months post-HSCT (24). We also found a comparable trend for NK cell regeneration in patients suffering from cGvHD, although not significant (data not shown). This might be explained by the fact, that for aGvHD analysis, only patients suffering from GvHD grades III and IV were considered, whereas no differentiation was available regarding cGvHD (e.g., chronification of primary aGvHD grades I and II).

Literature is discordant whether steroids/immunosuppression have a negative impact on NK cell reconstitution. Giebel et al. proposed that the use of steroids for GvHD prophylaxis negatively affects quantitative reconstitution of NK cells after allo-HSCT (25). Although, patients suffering from GvHD grades III and IV normally receive steroids in our transplantation unit, we did not see any significant differences in the quantitative reconstitution of absolute NK cell numbers. Interestingly, Wang et al. described that CSA suppresses the *in vitro* proliferation of NK cells, especially the $CD56^{dim}CD16^{+}KIR^{+}$ NK cells, resulting in a relative increase in the number of immature $CD56^{bright}CD16^{-}KIR^{-}$ NK cells (26). This might also contribute to the delayed NK cell development in patients suffering from higher grade aGvHD that we observed within this study. However, this remains controversial as other studies analyzed the effect of CSA on NK cell function in short-term cultures and their cytokine production without finding significant differences between NK cells with and without CSA treatment (27, 28).

In patients suffering from viral infection post-HSCT, we observed a slight reduction in CD56 and CD16 expression. Notably, other publications already described the existence of $CD56^{-}CD16^{+}$ NK cells ($CD56^{negative}$) NK cells in viral infections (e.g., HIV, hepatitis C), where NK cells undergo numerous phenotypic and functional changes (29). This $CD56^{negative}$ subset has been associated with high HIV viral load and has been reported to have an impaired cytolytic function and cytokine production (30). This increase occurred primarily at the expense of $CD56^{dim}$ NK cells, whereas numbers of $CD56^{bright}$ NK cells remained stable (31). Furthermore, we observed a considerable loss in absolute $CD56^{dim}$ NK cells followed by continued regeneration of $CD56^{int}$ NK cells. Alteration of NK cells upon viral infection has already been shown by other research groups, for example Pical-Izard et al. showed that in patients being affected by CMV reactivation, NK cells showed lower degranulation and TNF-α production compared to patients without CMV reactivation post-HSCT (8). In addition, it was shown that CMV reactivation is followed by an increase in the proportion of $NKG2C^{+}$ NK cells within 2–4 weeks, which persist for at least a year (32, 33).

In conclusion, only after around 12 months, NK cells post-HSCT reconstitute to a distribution of the subpopulations $CD56^{bright}$, $CD56^{int}$, and $CD56^{dim}$ comparable to age-matched healthy controls. The expression profiles of $CD56^{int}$ and $CD56^{bright}$ NK cells resemble each other but differed in KIR, CD62L, NKG2A, CX3CR1, and CD57 expression to $CD56^{dim}$ NK cells. We observed elevated levels of $CD56^{bright}$ directly after and $CD56^{int}$ NK cells 3 months post-HSCT accompanied by reduced $CD56^{dim}$ NK cells supporting the hypothesis of sequential NK cell development. Furthermore, we analyzed alterations in NK cell development in patients with severe viral infections and GvHD. Following viral infection, there was a slight reduction in CD56 and CD16 receptor expression followed by a considerable loss in absolute $CD56^{dim}$ NK cells and continued regeneration of $CD56^{int}$ NK cells. Most important, within the first 2 months, post-HSCT patients without severe events had significantly elevated levels of $CD56^{bright}$ NK cells compared to patients suffering from aGvHD. While first measurements performed as early as 15 days following HSCT revealed the most significant differences, clinical occurrence of aGvHD was observed in median on day 22 post-HSCT. Therefore, we recommend immunophenotyping of NK cell subpopulations directly following engraftment, which might be an early prognostic factor regarding GvHD development.

AUTHOR CONTRIBUTIONS

Study design: MB, SH, RE, and EK. Performed the experiments: JB-D and SBe. Provided clinical data: JS, AJ, SBa, and PB. Analyzed the data: SH and EK. Coordinated the research: SH and CC. Contributed reagents/materials/analysis tools: MB, SH, VP, JB-D, and SBe. Performed statistical analyses: SH and ES-M. Wrote the manuscript: MB and SH. Discussed data and revised the manuscript: CC, PB, VP, EU, and CK. Supervised the research: PB and TK. All the authors read and approved the final manuscript.

ACKNOWLEDGMENTS

The authors would like to thank Stephanie Erben, Olga Zimmermann, Regine Müller, Andrea Quaiser, Tanja Gardlowski, and Michael Marburger for the excellent technical support.

REFERENCES

1. Farag SS, Caligiuri MA. Human natural killer cell development and biology. *Blood Rev* (2006) 20(3):123–37. doi:10.1016/j.blre.2005.10.001
2. Vivier E, Tomasello E, Baratin M, Walzer T, Ugolini S. Functions of natural killer cells. *Nat Immunol* (2008) 9(5):503–10. doi:10.1038/ni1582
3. Ullah MA, Hill GR, Tey SK. Functional reconstitution of natural killer cells in allogeneic hematopoietic stem cell transplantation. *Front Immunol* (2016) 7:144. doi:10.3389/fimmu.2016.00144
4. Jacobs R, Stoll M, Stratmann G, Leo R, Link H, Schmidt RE. CD16- CD56+ natural killer cells after bone marrow transplantation. *Blood* (1992) 79(12):3239–44.
5. Giebel S, Dziaczkowska J, Czerw T, Wojnar J, Krawczyk-Kulis M, Nowak I, et al. Sequential recovery of NK cell receptor repertoire after allogeneic hematopoietic SCT. *Bone Marrow Transplant* (2010) 45(6):1022–30. doi:10.1038/bmt.2009.384
6. Romagnani C, Juelke K, Falco M, Morandi B, D'Agostino A, Costa R, et al. CD56brightCD16- killer Ig-like receptor- NK cells display longer telomeres and acquire features of CD56dim NK cells upon activation. *J Immunol* (2007) 178(8):4947–55. doi:10.4049/jimmunol.178.8.4947
7. Savani BN, Mielke S, Adams S, Uribe M, Rezvani K, Yong AS, et al. Rapid natural killer cell recovery determines outcome after T-cell-depleted HLA-identical stem cell transplantation in patients with myeloid leukemias but not with acute lymphoblastic leukemia. *Leukemia* (2007) 21(10):2145–52. doi:10.1038/sj.leu.2404892
8. Pical-Izard C, Crocchiolo R, Granjeaud S, Kochbati E, Just-Landi S, Chabannon C, et al. Reconstitution of natural killer cells in HLA-matched HSCT after

reduced-intensity conditioning: impact on clinical outcome. *Biol Blood Marrow Transplant* (2015) 21(3):429–39. doi:10.1016/j.bbmt.2014.11.681

9. Ruggeri L, Capanni M, Urbani E, Perruccio K, Shlomchik WD, Tosti A, et al. Effectiveness of donor natural killer cell alloreactivity in mismatched hematopoietic transplants. *Science* (2002) 295(5562):2097–100. doi:10.1126/science.1068440
10. Koehl U, Brehm C, Huenecke S, Zimmermann SY, Kloess S, Bremm M, et al. Clinical grade purification and expansion of NK cell products for an optimized manufacturing protocol. *Front Oncol* (2013) 3:118. doi:10.3389/fonc.2013.00118
11. Stern M, Passweg JR, Meyer-Monard S, Esser R, Tonn T, Soerensen J, et al. Pre-emptive immunotherapy with purified natural killer cells after haploidentical SCT: a prospective phase II study in two centers. *Bone Marrow Transplant* (2013) 48(3):433–8. doi:10.1038/bmt.2012.162
12. Meinhardt K, Kroeger I, Bauer R, Ganss F, Ovsiy I, Rothamer J, et al. Identification and characterization of the specific murine NK cell subset supporting graft-versus-leukemia- and reducing graft-versus-host-effects. *Oncoimmunology* (2015) 4(1):e981483. doi:10.4161/2162402X.2014.981483
13. Koenig M, Huenecke S, Salzmann-Manrique E, Esser R, Quaritsch R, Steinhilber D, et al. Multivariate analyses of immune reconstitution in children after allo-SCT: risk-estimation based on age-matched leukocyte sub-populations. *Bone Marrow Transplant* (2010) 45(4):613–21. doi:10.1038/bmt.2009.204
14. Beziat V, Duffy D, Quoc SN, Le Garff-Tavernier M, Decocq J, Combadiere B, et al. CD56brightCD16+ NK cells: a functional intermediate stage of NK cell differentiation. *J Immunol* (2011) 186(12):6753–61. doi:10.4049/jimmunol.1100330
15. Gismondi A, Stabile H, Nisti P, Santoni A. Effector functions of natural killer cell subsets in the control of hematological malignancies. *Front Immunol* (2015) 6:567. doi:10.3389/fimmu.2015.00567
16. Vukicevic M, Chalandon Y, Helg C, Matthes T, Dantin C, Huard B, et al. CD56bright NK cells after hematopoietic stem cell transplantation are activated mature NK cells that expand in patients with low numbers of T cells. *Eur J Immunol* (2010) 40(11):3246–54. doi:10.1002/eji.200940016
17. Freud AG, Yu J, Caligiuri MA. Human natural killer cell development in secondary lymphoid tissues. *Semin Immunol* (2014) 26(2):132–7. doi:10.1016/j.smim.2014.02.008
18. Cooper MA, Fehniger TA, Caligiuri MA. The biology of human natural killer-cell subsets. *Trends Immunol* (2001) 22(11):633–40. doi:10.1016/S1471-4906(01)02060-9
19. Nguyen S, Dhedin N, Vernant JP, Kuentz M, Al Jijakli A, Rouas-Freiss N, et al. NK-cell reconstitution after haploidentical hematopoietic stem-cell transplantations: immaturity of NK cells and inhibitory effect of NKG2A override GvL effect. *Blood* (2005) 105(10):4135–42. doi:10.1182/blood-2004-10-4113
20. Bjorkstrom NK, Riese P, Heuts F, Andersson S, Fauriat C, Ivarsson MA, et al. Expression patterns of NKG2A, KIR, and CD57 define a process of CD56dim NK-cell differentiation uncoupled from NK-cell education. *Blood* (2010) 116(19):3853–64. doi:10.1182/blood-2010-04-281675
21. Boyiadzis M, Memon S, Carson J, Allen K, Szczepanski MJ, Vance BA, et al. Up-regulation of NK cell activating receptors following allogeneic hematopoietic stem cell transplantation under a lymphodepleting reduced intensity regimen is associated with elevated IL-15 levels. *Biol Blood Marrow Transplant* (2008) 14(3):290–300. doi:10.1016/j.bbmt.2007.12.490
22. Kim SY, Lee H, Han MS, Shim H, Eom HS, Park B, et al. Post-transplantation natural killer cell count: a predictor of acute graft-versus-host disease and survival outcomes after allogeneic hematopoietic stem cell transplantation. *Clin Lymphoma Myeloma Leuk* (2016) 16(9):527.e–35.e. doi:10.1016/j.clml.2016.06.013
23. Ullrich E, Salzmann-Manrique E, Bakhtiar S, Bremm M, Gerstner S, Herrmann E, et al. Relation between acute GVHD and NK cell subset reconstitution following allogeneic stem cell transplantation. *Front Immunol* (2016) 7:595. doi:10.3389/fimmu.2016.00595
24. Kheav VD, Busson M, Scieux C, Peffault de Latour R, Maki G, Haas P, et al. Favorable impact of natural killer cell reconstitution on chronic graft-versus-host disease and cytomegalovirus reactivation after allogeneic hematopoietic stem cell transplantation. *Haematologica* (2014) 99(12):1860–7. doi:10.3324/haematol.2014.108407
25. Giebel S, Dziaczkowska J, Wojnar J, Krawczyk-Kulis M, Markiewicz M, Kruzel T, et al. The impact of immunosuppressive therapy on an early quantitative NK cell reconstitution after allogeneic haematopoietic cell transplantation. *Ann Transplant* (2005) 10(2):29–33.
26. Wang H, Grzywacz B, Sukovich D, McCullar V, Cao Q, Lee AB, et al. The unexpected effect of cyclosporin A on CD56+CD16- and CD56+CD16+ natural killer cell subpopulations. *Blood* (2007) 110(5):1530–9. doi:10.1182/blood-2006-10-048173
27. Shao-Hsien C, Lang I, Gunn H, Lydyard P. Effect of in vitro cyclosporin. A treatment on human natural and antibody-dependent cell-mediated cytotoxicity. *Transplantation* (1983) 35(2):127–9. doi:10.1097/00007890-198302000-00004
28. Derniame S, Perazzo J, Lee F, Domogala A, Escobedo-Cousin M, Alnabhan R, et al. Differential effects of mycophenolate mofetil and cyclosporine A on peripheral blood and cord blood natural killer cells activated with interleukin-2. *Cytotherapy* (2014) 16(10):1409–18. doi:10.1016/j.jcyt.2014.05.010
29. Fauci AS, Mavilio D, Kottilil S. NK cells in HIV infection: paradigm for protection or targets for ambush. *Nat Rev Immunol* (2005) 5(11):835–43. doi:10.1038/nri1711
30. Mavilio D, Lombardo G, Benjamin J, Kim D, Follman D, Marcenaro E, et al. Characterization of CD56-/CD16+ natural killer (NK) cells: a highly dysfunctional NK subset expanded in HIV-infected viremic individuals. *Proc Natl Acad Sci U S A* (2005) 102(8):2886–91. doi:10.1073/pnas.0409872102
31. Bjorkstrom NK, Ljunggren HG, Sandberg JK. CD56 negative NK cells: origin, function, and role in chronic viral disease. *Trends Immunol* (2010) 31(11):401–6. doi:10.1016/j.it.2010.08.003
32. Beziat V, Dalgard O, Asselah T, Halfon P, Bedossa P, Boudifa A, et al. CMV drives clonal expansion of NKG2C+ NK cells expressing self-specific KIRs in chronic hepatitis patients. *Eur J Immunol* (2012) 42(2):447–57. doi:10.1002/eji.201141826
33. Foley B, Cooley S, Verneris MR, Pitt M, Curtsinger J, Luo X, et al. Cytomegalovirus reactivation after allogeneic transplantation promotes a lasting increase in educated NKG2C+ natural killer cells with potent function. *Blood* (2012) 119(11):2665–74. doi:10.1182/blood-2011-10-386995

10

Impact of the MICA-129Met/Val Dimorphism on NKG2D-Mediated Biological Functions and Disease Risks

Antje Isernhagen[1†], Dörthe Malzahn[2†], Heike Bickeböller[2] and Ralf Dressel[1,3]*

[1] Institute of Cellular and Molecular Immunology, University Medical Center Göttingen, Göttingen, Germany, [2] Institute of Genetic Epidemiology, University Medical Center Göttingen, Göttingen, Germany, [3] DZHK (German Center for Cardiovascular Research), Partner Site Göttingen, Göttingen, Germany

Edited by:
Ulrike Koehl,
Hannover Medical School, Germany

Reviewed by:
Ofer Mandelboim,
Hebrew University of Jerusalem, Israel
Alexander Steinle,
Goethe University Frankfurt, Germany

***Correspondence:**
Ralf Dressel
rdresse@gwdg.de

†Antje Isernhagen and Dörthe Malzahn share first authorship.

The major histocompatibility complex (MHC) class I chain-related A (*MICA*) is the most polymorphic non-classical MHC class I gene in humans. It encodes a ligand for NKG2D (NK group 2, member D), an activating natural killer (NK) receptor that is expressed mainly on NK cells and CD8+ T cells. The single-nucleotide polymorphism (SNP) rs1051792 causing a valine (Val) to methionine (Met) exchange at position 129 of the MICA protein is of specific interest. It separates MICA into isoforms that bind NKG2D with high (Met) and low affinities (Val). Therefore, this SNP has been investigated for associations with infections, autoimmune diseases, and cancer. Here, we systematically review these studies and analyze them in view of new data on the functional consequences of this polymorphism. It has been shown recently that the MICA-129Met variant elicits a stronger NKG2D signaling, resulting in more degranulation and IFN-γ production in NK cells and in a faster costimulation of CD8+ T cells than the MICA-129Val variant. However, the MICA-129Met isoform also downregulates NKG2D more efficiently than the MICA-129Val isoform. This downregulation impairs NKG2D-mediated functions at high expression intensities of the MICA-Met variant. These features of the MICA-129Met/Val dimorphism need to be considered when interpreting disease association studies. Particularly, in the field of hematopoietic stem cell transplantation, they help to explain the associations of the SNP with outcome including graft-versus-host disease and relapse of malignancy. Implications for future disease association studies of the MICA-129Met/Val dimorphism are discussed.

Keywords: NK cells, T cells, activating NK receptor, costimulation, single-nucleotide polymorphism, autoimmune diseases, cancer, hematopoietic stem cell transplantation

INTRODUCTION

The major histocompatibility complex (MHC) class I chain-related A (*MICA*) is the most polymorphic non-classical MHC class I gene in humans, and 105 alleles are known encoding for 82 protein variants (http://www.ebi.ac.uk/imgt/hla/, release 3.25.0). *MICA* is encoded within the human leukocyte antigen (HLA) complex close to HLA-B (1, 2). The protein structure is similar to classical class I molecules, but MICA is not associated with β2-microglobulin and does not present peptides. MICA is constitutively expressed only on a few cell types, including gastrointestinal epithelium, but is induced due to cellular and genotoxic stress (3, 4), malignant transformation, or virus infection

(5, 6). MICA is a ligand for NKG2D (NK group 2, member D), an activating natural killer (NK) receptor encoded by the *KLRK1* gene (7). NKG2D is expressed on most human NK cells, CD8$^+$ αβ T cells, γδ T cells, iNKT cells, and subsets of effector or memory CD4$^+$ T cells (8, 9). On NK cells, NKG2D signaling elicits killing of target cells (10) and secretion of IFN-γ (11). On CD8$^+$ αβ T cells, NKG2D provides a costimulatory signal to activate naïve cytotoxic T lymphocytes (12). NKG2D contributes to the elimination of tumor cells (13) and plays a role in the defense against pathogens (14, 15). In addition to MICA, MICB and the UL16-binding proteins (ULBP) encoded by the retinoic acid early transcript 1 (*RAET1*) family function as ligands for NKG2D. *MICB* is also very polymorphic with 42 alleles encoding 28 protein variants (http://www.ebi.ac.uk/imgt/hla/, release 3.25.0). The *RAET1* gene family is localized on chromosome 6 outside the HLA complex and six loci encode functional proteins (16). *RAET1* genes are less polymorphic than *MICA* and *MICB*.

Polymorphisms of *MICA* have been investigated for their role in infections, autoimmune diseases, and cancer (17–21). The single-nucleotide polymorphism (SNP) rs1051792 (G/A) causing a valine (Val) to methionine (Met) exchange at position 129 in the α2 domain of the MICA protein has gained specific interest. It separates *MICA* alleles into two groups (22). MICA isoforms containing a methionine at position 129 bind NKG2D with high affinity, whereas those with a valine bind NKG2D with low affinity. High-affinity alleles include *MICA*001*, **002*, **007*, and **017*; among the low-affinity alleles are *MICA*004*, **006*, **008*, **009*, and **010* (23). Due to its functional consequences, the MICA-129Met/Val dimorphism has been investigated in several disease association studies. Here, we review these studies in view of new data on the functional consequences of this amino acid variation elicited after binding to NKG2D.

MICA-129Met/Val DISEASE ASSOCIATION STUDIES

In September 2016, we searched Pubmed for MICA-129Met/Val disease association studies using the key words rs1051792, MICA-129, MICA AND polymorphism AND Met, and MICA AND polymorphism AND Val. Moreover, we exchanged polymorphism by SNP, Met by methionine, and Val by valine. We identified 17 publications, in which an association of the MICA-129Met/Val dimorphism with a disease or disease complication has been investigated. One study in Chinese language (24) appeared to be not independent of a larger study published in English (25). Thus, we analyzed 16 independent studies published between 2005 and 2015 (Table S1 in Supplementary Material). Three studies are small with less than 100 cases. All others are of a medium size with more than 100 but less than 1,000 patients included, and most studies used a case–control design.

Eight studies investigated associations with autoimmune diseases, i.e., ankylosing spondylitis (AS) (26), rheumatoid arthritis (RA) (27–29), inflammatory bowel disease (IBD) (25, 30) [including ulcerative colitis (UC) and Crohn's disease], systemic lupus erythematosus (SLE) (28), type I diabetes (31), latent autoimmune diabetes in adults (LADA) (31), and psoriasis (32). In one study, the *MICA-129* SNP has not been determined directly. Instead, the SNP rs1051794 was typed and reported to be in complete linkage disequilibrium with the rs1051792 (27). Five studies reported on malignancies, i.e., nasopharyngeal cancer (33), hepatitis B virus (HBV)-induced hepatocellular carcinoma (HCC) (34), cutaneous malignant melanoma (35), and relapse of malignancy after hematopoietic stem cell transplantation (HSCT) (36, 37). Three studies investigated infections or their complications, i.e., HBV infection and HBV-induced HCC (34), left ventricular systolic dysfunction (LVSD) in chronic Chargas heart disease (38), and ocular toxoplasmosis (39). One study investigated an association of the MICA-129Met/Val dimorphism with recurrent miscarriage (40). The two studies on HSCT (36, 37) investigated besides relapse also other outcomes including graft-versus-host disease (GVHD).

Three studies, on recurrent miscarriage (40), ocular toxoplasmosis (39), and malignant melanoma (35), failed to demonstrate an association with the SNP. Thus, 81% of the studies showed an association at least for a subgroup, e.g., juvenile AS, whereas in all patients with AS, the association was dependent on HLA-B27 (26), or a sub-phenotype, e.g., severe LVSD (38). However, we must assume that other negative association studies have not been published. In seven studies, a *MICA-129* allele and the corresponding homozygous genotype were both associated with a disease risk (25, 28, 29, 31, 32, 34, 38). The odds ratio (OR) was then always higher for the genotype than the allele. In six studies, the *Met* allele and/or the *Met/Met* genotype were found to be associated with a risk, including autoimmune diseases [juvenile AS (26), UC (30), SLE (28), and psoriasis (32)], a malignancy (HBV-induced HCC) (34), and a complication of an infection (severe LVSD in chronic Chargas disease) (38). In three studies, the *Val* allele and/or the *Val/Val* genotype has been identified to confer a risk for autoimmune diseases [including RA (27), UC (25), and diabetes (31)] and for nasopharyngeal carcinoma (NPC) (33). Moreover, rheumatoid factor (RF) positivity in RA patients has been associated with the *Val* allele and the *Val/Val* genotype (29). In the studies on HSCT, different outcomes showed different associations. In one study (36), the *Met/Met* genotype was associated with an increased risk of relapse and the *Val/Val* genotype with an increased risk of chronic GVHD. In our recent study (37), the *Met/Met* genotype conferred a risk of acute GVHD, whereas having *Met* alleles reduced the risk to die from acute GVHD. Overall, the *Val* allele was associated with a higher mortality after HSCT (37).

The results of these disease association studies do not allow for a simple unifying interpretation, such as the high-affinity MICA-129Met variant being associated with an activation of the immune system resulting in a lower risk of infections and cancer but higher risk of autoimmunity (**Figures 1A,B**). Autoimmune diseases are associated with both variants even within the same disease entity. UC, e.g., has been associated with the *Met/Met* genotype in a small study from Spain (30) but with the *Val* allele and *Val/Val* genotype in a larger study from China (25). RA has been associated with the *Val* allele in a study from France and Germany (27), but no association was found in cohorts from Japan (28) and Tunisia (29). Notably, a role of the NKG2D pathway has been reported for the pathogenesis of RA (41) and SLE (42), although this has not been linked to

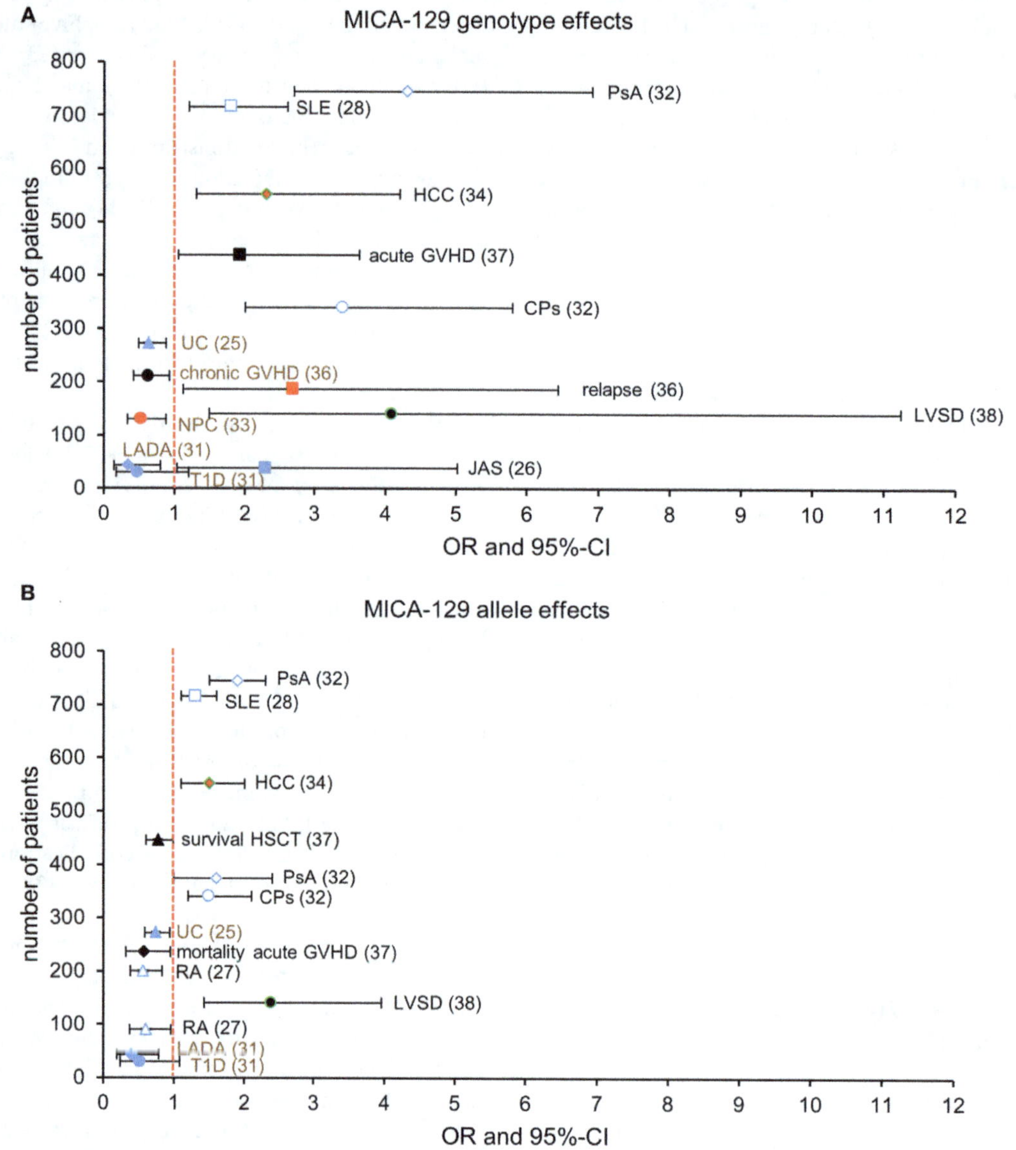

FIGURE 1 | Reported associations of the homozygous *MICA-129* genotypes (A) and the *MICA-129* alleles (B) with health risks [odds ratio (OR) > 1] or advantages (OR < 1). Shown are ORs with 95% confidence intervals (CI) or hazard ratios reported by Boukouaci et al. (36) and Isernhagen et al. (37) (overall survival) in event-time data. The number of patients analyzed in the studies is indicated at the *y*-axis. Studies reporting on autoimmune diseases are shown by open and closed blue symbols and malignancies by red symbols; studies reporting complications of infections (LVSD, Chargas disease; HCC, hepatitis B virus infection) are shown by green frames, and others are displayed by black symbols. The investigated diseases or complications and the references for the studies are indicated. **(A)** *MICA-129Met/Met* genotype effects are directly displayed. For studies that reported *MICA-129Val/Val* genotype effects [chronic GVHD (36), NPC (33), UC (25), RA (27), T1D, and LADA (31), indicated by brown font], the graph displays the corresponding effect of the pooled *MICA-129 Met/Met* and *MICA-129Met/Val* genotypes to allow for a direct comparison. **(B)** *MICA-129Met* allele effects are directly displayed. For studies that reported *MICA-129Val* allele effects [UC (25), T1D, and LADA (31), indicated by brown font], the graph displays the corresponding effect of the *MICA-129Met* allele; ORMet = 1/ORVal and 95%-CIMet = (1/CIVal, upper, 1/CIVal, lower). Abbreviations: CPs, cutaneous psoriasis; GVHD, graft-versus-host disease; HCC, hepatocellular carcinoma; JAS, juvenile ankylosing spondylitis; LADA, latent autoimmune diabetes in adults; LVSD, left ventricular systolic dysfunction; NPC, nasopharyngeal carcinoma; PsA, psoriatic arthritis; RA, rheumatoid arthritis; SLE, systemic lupus erythematosus; T1D, type 1 diabetes; UC, ulcerative colitis.

polymorphisms. Juvenile AS has been associated in a small study with the *Met/Met* genotype (26), and a larger sequencing study identified the *MICA*007:01* allele that encodes a methionine at position 129 as a risk allele for AS in both Caucasian and Han Chinese populations (43). However, *MICA*019*, encoding a valine-129, has been identified as the major risk allele in Han Chinese (43). Malignancies were found to be associated with *Val/Val* genotype in the case of NPC (33) but with the *Met/Met* genotype in the case of relapse after HSCT (36). These different associations could suggest that the observed associations are random or dependent on the population studied. However, since the MICA-129Met/Val dimorphism is functional, it could also indicate that we need to better understand this function to predict its consequences in the pathophysiology of different diseases in various populations, which might be exposed to different interfering environmental factors. This assumption is supported by genome-wide association studies (GWAS), which have assigned disease risks for NPC (44), HCC (45, 46), cervical cancer (47, 48), and asthma (49) or advantages, such as HIV long-term non-progression (50) to the *MICA* gene region in an unbiased manner.

FUNCTIONAL CONSEQUENCES OF THE MICA-129Met/Val DIMORPHISM

It has been shown by Steinle and colleagues that MICA-129Met isoforms bind NKG2D with high affinity in contrast to MICA-129Val isoforms that bind with low affinity (22). Yoshida and colleagues combined the MICA-129Met variant with the A9 variant of a microsatellite polymorphism in the transmembrane (TM) region and the MICA-129Val variant with the A5-TM variant in GST-fusion proteins (28). NK92MI cells showed a reduced NKG2D expression and killed K562 cells less efficiently when exposed to the MICA-129Met-A9-TM variant, but IFN-γ production was increased (28). We recently studied the consequences of binding of the two MICA-129 variants to NKG2D on primary NK cells and $CD8^+$ T cells using cell lines transfected with expression constructs and recombinant Fc-fusion proteins differing only in amino acid 129 (37, 51). The recombinant MICA-129Met variant stimulated a stronger phosphorylation of SRC family kinases in NK cells than the MICA-129Val variant. Subsequently, the MICA-129Met ligand triggered more degranulation and IFN-γ production than the MICA-129Val ligand (**Figure 2A**). We then exposed NK cells to target cells expressing different amounts of the MICA-129 variants. The extent of degranulation and IFN-γ secretion correlated positively with the MICA expression intensity on the target cells but only for the MICA-129Val isoform. The expression intensity of the MICA-129Met isoform, in contrast, had either none or even a negative effect on the extent of degranulation, target cell killing, and IFN-γ release (37). On $CD8^+$ T cells, the MICA-129Met isoform induced an earlier costimulatory activation than the MICA-129Val isoform (**Figure 2B**). Importantly, the MICA-129Met ligand induced also a stronger downregulation of NKG2D on both NK and $CD8^+$ T cells than the MICA-129Val ligand. This downregulation of NKG2D impaired the capability of NK and $CD8^+$ T cells to receive signals *via* NKG2D (37). Thus, MICA-129Met ligands, which elicit strong NKG2D responses, stimulate in parallel a robust negative feedback signal by downregulation of NKG2D that limits the initially stronger effects of MICA-129Met ligands. These data show that the biological effect of the MICA-129Met/Val dimorphism changes with the MICA expression intensity. Variant MICA-129Met triggers more NKG2D signals at low expression intensities, whereas variant MICA-129Val elicits more NKG2D effects at high expression, at which the MICA-129Met variant already downregulates NKG2D leading to impaired function. Thus, the biological effect of the SNP can hardly be predicted without information on the expression intensity of MICA.

It is known that expression intensities vary for certain *MICA* alleles (52, 53). The *G* allele of the SNP at -1878 (rs2596542) in the promoter region of the *MICA* gene region, e.g., was found to have a higher transcriptional activity (54). Biological effects of the MICA-129Met/Val dimorphism can be expected to be modified by polymorphisms affecting *MICA* gene expression. We have investigated whether the Met/Val dimorphism itself affects MICA expression. In transfected cells, more of the MICA-129Met variant was retained in intracellular compartments (51). A similar alteration of the intracellular transport has been described for *MICA-A5.1* variants (55). Thus, the combination of polymorphisms affecting transcription and intracellular transport of MICA could modify the effect of the Met/Val dimorphism.

Another important aspect of MICA is the generation of soluble MICA (sMICA) by proteolytic shedding. sMICA can induce NKG2D downregulation (56, 57) resulting in tumor immune escape (58). Some *MICA* polymorphisms have been reported to affect the amounts of sMICA in sera of patients including the SNP at -1878 (rs2596542) in the promoter region (34, 45, 59) that affects transcription (54), a microsatellite in exon 5 encoding the TM region (60, 61), and the MICA-129Met/Val dimorphism. In patients with UC, the *MICA-129Val/Val* genotype was associated with higher sMICA serum levels (25), and the *MICA-129Val* allele was also associated with higher sMICA serum levels in HBV patients and controls (34). In transfected cells, we found that the MICA-129Met isoform was more susceptible to shedding than the MICA-129Val isoform (51). However, due to the intracellular retention of the MICA-129Met variant (51), less sMICA might appear in sera (25, 34). Notably, intracellular retention and preferred shedding both appear to limit the expression of the high-affinity MICA-129Met isoform at the plasma membrane.

MICA-129Met/Val DISEASE ASSOCIATIONS IN VIEW OF BIOLOGICAL FUNCTIONS

Recent data on the MICA-129Met/Val variation demonstrate the complexity of the functional consequences of this exchange of a single amino acid (37, 51). There are several layers of this complexity, which are as follows: (1) the function of the variant is not constant but dynamic (37); it depends on the MICA expression intensity, and the direction of the biological effect can invert for the MICA-129Met variant at higher expression. (2) Epistatic effects must be expected for this SNP as polymorphisms affecting the expression of MICA will modify the functional effects of the MICA-129Met/Val isoforms. Moreover, the expression intensity of NKG2D can be modified by SNPs in the *KLRK1* gene (62) and those might interact with the *MICA-129* variants. Other genes within the NKG2D pathway including other ligands might also show epistatic effects (63). (3) MICA can target NKG2D on several cell types, and biological effects on different cell types might be synergistic or antagonistic. An activation of NK cells and a costimulation of $CD8^+$ T cells both can promote antitumor

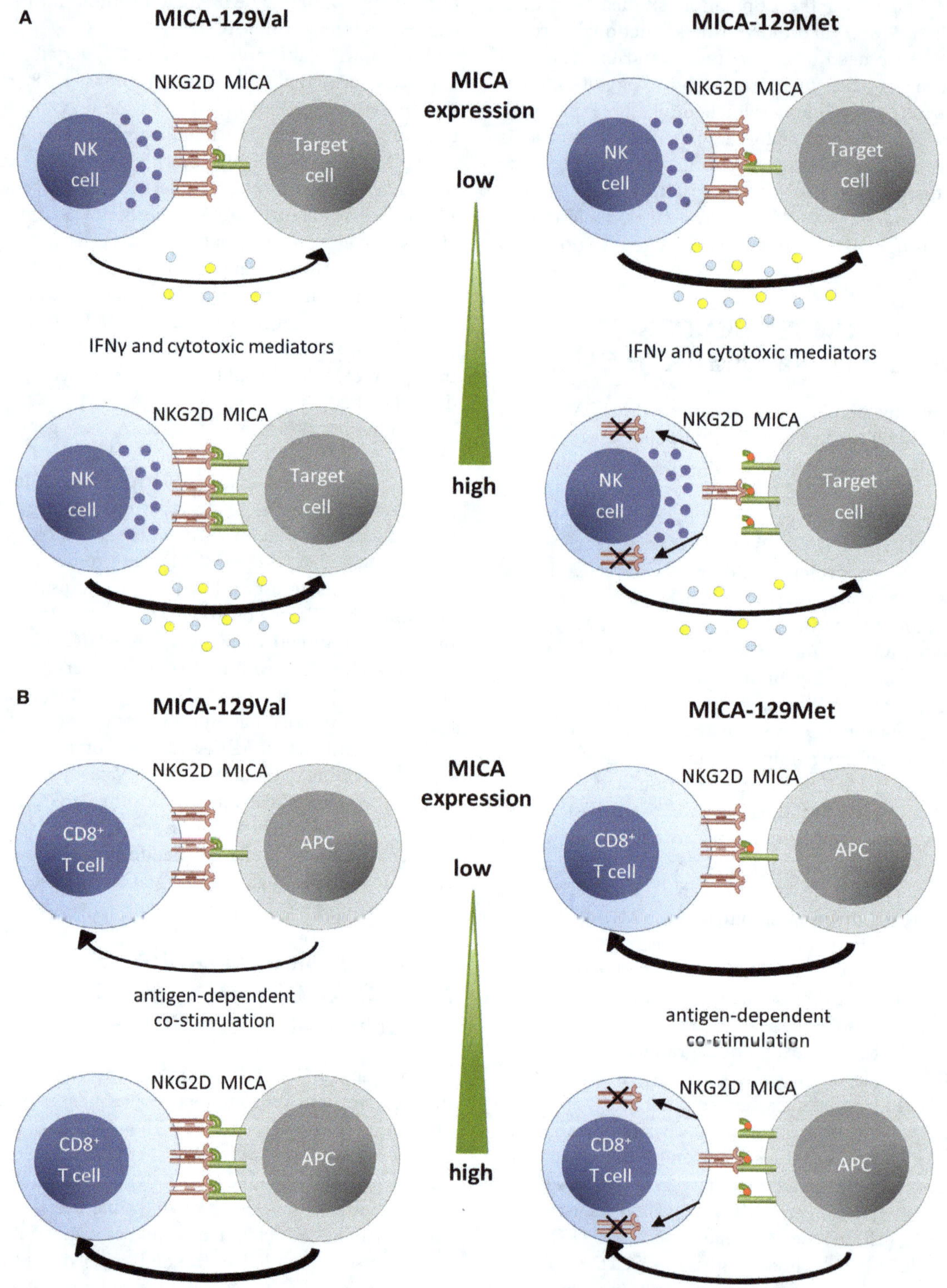

FIGURE 2 | **Summary of functional effects of MICA-129 variants depending on expression intensity. (A)** For target cells expressing the MICA-129Val variant, the degree of natural killer (NK) cell cytotoxicity and IFN-γ production increases steadily with the MICA expression intensity. Augmented expression of the high-affinity MICA-129Met isoform, in contrast, has none or even a negative effect on these NK cell functions due to a rapid downregulation of NKG2D on NK cells. **(B)** Antigen-dependent costimulation of $CD8^+$ T cells with the MICA-129Met variant allows for an earlier antigen-dependent activation than costimulation with the MICA-129Val variant. However, the downregulation of NKG2D in response to MICA-129Met ligands impairs any subsequent NKG2D-dependent costimulation and T cell activation. The downregulation of NKG2D on $CD8^+$ T cells is augmented with MICA-129Met expression intensity. The figure is reproduced from Isernhagen et al. (37).

immunity. By contrast, a strong activation of NK cells might polarize an immune response to a Th_1 reaction and reduce the risk to develop a Th_2-mediated autoimmune disease. (4) Additional factors, such as sMICA or anti-MICA antibodies (36) that might neutralize sMICA, have been shown to be functionally important and have been determined in some of the disease association studies (25, 34, 36).

Currently, we mostly have not sufficient clinical and biological information to interpret the MICA-129Met/Val disease association studies in view of the complex function of this polymorphism. However, the two HSCT studies do provide more information and illustrate the clinical effects of the MICA-129Met/Val dimorphism as explained previously in detail (37). In our study (37), the homozygous carriers of *Met* alleles had an increased risk to experience acute GVHD, possibly due to immediate strong effects of MICA-129Met variants on NKG2D signaling. Having at least one *Met* allele reduced the risk to die from acute GVHD likely due to a rapid downregulation of NKG2D on alloreactive $CD8^+$ T cells mediated by engagement of a high-affinity MICA-129Met variant. Carrying a *MICA-129Met* allele increased in consequence the chance of survival in all patients and in patients receiving a *MICA-129*-matched graft (37). Boukouaci and colleagues reported an increased risk of chronic GVHD for recipients with the *Val/Val* genotype, whereas the *Met/Met* genotype was associated with the risk of relapse (36). Sustained NKG2D-mediated activation of alloreactive $CD8^+$ T cells would be expected if only MICA-129Val variants are present that fail to efficiently downregulate NKG2D, and this could increase the risk of chronic GVHD but reduce the risk of relapse. Thus, the different risk associations reported in the two studies are not arguing against the relevance of the MICA-129 dimorphism for the outcome of HSCT. The principal relevance of the NKG2D pathway for HSCT is further emphasized by studies showing an effect of the genotype of the NKG2D ligand *RAET1L* (64) and NKG2D itself (65) on the survival of patients. Moreover, matching for *MICA* alleles (66–69) and specifically for the *MICA-129* polymorphism (70) is beneficial in HSCT. The huge effect of *MICA-129* matching appears hardly explainable solely by the avoidance of a potential minor histocompatibility antigen. A "tuning" of the threshold of NKG2D signaling toward the affinity of NKG2D ligands present in an individual (52) and disturbance of this balance by mismatching could be considered as an alternative explanation.

Despite the functional relevance of the *MICA-129* SNP, it cannot be excluded that some of the associations reported are random or caused by linkage disequilibrium with classical HLA genes. The association of MICA-129 with psoriasis (32) has been disproven in large GWAS cohorts (71). However, associations with NPC (33) and HCC (34) are supported by GWAS data pointing to the *MICA* gene region (44–46).

CONCLUSION

Information on functional consequences of a polymorphism is indispensable for understanding disease associations. The variation in the disease associated allele or genotype of *MICA-129* in the published studies must not indicate random associations. For *MICA-129*, the biological function can change with expression intensity, epistatic interactions can be expected, the effect on different lymphocytes can vary, and modifying factors, such as sMICA, have to be considered. Notably, as expected for a functional SNP with a minor allele (*MICA-129Met*) frequency ranging from 48% in Africans to 30% in Asians (72), and being even the major allele reported in one of the analyzed studies (26), both alleles appear to confer advantages and disadvantages in specific situations suggesting balancing evolution of the *MICA* alleles. Since the MICA-129 dimorphism is considered as decisive for distinguishing low- and high-affinity variants (22), the frequency of alleles encoding high-affinity MICA variants is expected to match the frequency of the MICA-129Met variant. However, other *MICA* polymorphisms and their interaction need to be studied further (73).

In future studies, the MICA-129Met/Val dimorphism should be analyzed in larger cohorts. Detailed clinical information would help to understand why associations might differ in cohorts. Additional biological information should be obtained in parallel to genetic data. Most important would be data on MICA expression intensities in relevant tissues at relevant time points. Due to the complexity of MICA-129Met/Val effects, this polymorphism is unlikely to become a simple genetic biomarker for prediction of disease risks. However, it still may provide highly important information. We found that *Val/Val* genotype carriers undergoing HSCT specifically profited from a treatment with antithymocyte globulin to deplete T cells (37). This might be explained by a lack of a high-affinity MICA variant that efficiently downregulates NKG2D on alloreactive donor $CD8^+$ T cells. Moreover, the *MICA-129* dimorphism might be relevant when considering therapies aiming at upregulation of MICA on tumor cells to sensitize them for NK cells (74, 75). Increasing the expression of MICA-129Met variants could result in opposite effects than intended.

AUTHOR CONTRIBUTIONS

RD searched the literature; RD and AI interpreted the functional data; RD and DM interpreted the genetic association data; RD drafted the manuscript; AI, DM, and HB commented the draft; and all the authors approved the final version.

ACKNOWLEDGMENTS

The authors thank all the colleagues who contributed to their studies on the MICA-129Met/Val dimorphism.

REFERENCES

1. Bahram S, Bresnahan M, Geraghty DE, Spies T. A second lineage of mammalian major histocompatibility complex class I genes. *Proc Natl Acad Sci U S A* (1994) 91:6259–63. doi:10.1073/pnas.91.14.6259
2. Leelayuwat C, Townend DC, Degli-Esposti MA, Abraham LJ, Dawkins RL. A new polymorphic and multicopy MHC gene family related to nonmammalian class I. *Immunogenetics* (1994) 40:339–51. doi:10.1007/BF01246675

3. Groh V, Bahram S, Bauer S, Herman A, Beauchamp M, Spies T. Cell stress-regulated human major histocompatibility complex class I gene expressed in gastrointestinal epithelium. *Proc Natl Acad Sci U S A* (1996) 93:12445–50. doi:10.1073/pnas.93.22.12445
4. Gasser S, Orsulic S, Brown EJ, Raulet DH. The DNA damage pathway regulates innate immune system ligands of the NKG2D receptor. *Nature* (2005) 436:1186–90. doi:10.1038/nature03884
5. Champsaur M, Lanier LL. Effect of NKG2D ligand expression on host immune responses. *Immunol Rev* (2010) 235:267–85. doi:10.1111/j.0105-2896.2010.00893.x
6. Raulet DH, Gasser S, Gowen BG, Deng W, Jung H. Regulation of ligands for the NKG2D activating receptor. *Annu Rev Immunol* (2013) 31:413–41. doi:10.1146/annurev-immunol-032712-095951
7. Glienke J, Sobanov Y, Brostjan C, Steffens C, Nguyen C, Lehrach H, et al. The genomic organization of NKG2C, E, F, and D receptor genes in the human natural killer gene complex. *Immunogenetics* (1998) 48:163–73. doi:10.1007/s002510050420
8. Zafirova B, Wensveen FM, Gulin M, Polic B. Regulation of immune cell function and differentiation by the NKG2D receptor. *Cell Mol Life Sci* (2011) 68:3519–29. doi:10.1007/s00018-011-0797-0
9. Lanier LL. NKG2D receptor and its ligands in host defense. *Cancer Immunol Res* (2015) 3:575–82. doi:10.1158/2326-6066.CIR-15-0098
10. Billadeau DD, Upshaw JL, Schoon RA, Dick CJ, Leibson PJ. NKG2D-DAP10 triggers human NK cell-mediated killing via a Syk-independent regulatory pathway. *Nat Immunol* (2003) 4:557–64. doi:10.1038/ni929
11. Andre P, Castriconi R, Espeli M, Anfossi N, Juarez T, Hue S, et al. Comparative analysis of human NK cell activation induced by NKG2D and natural cytotoxicity receptors. *Eur J Immunol* (2004) 34:961–71. doi:10.1002/eji.200324705
12. Groh V, Rhinehart R, Randolph-Habecker J, Topp MS, Riddell SR, Spies T. Costimulation of CD8alphabeta T cells by NKG2D via engagement by MIC induced on virus-infected cells. *Nat Immunol* (2001) 2:255–60. doi:10.1038/85321
13. Guerra N, Tan YX, Joncker NT, Choy A, Gallardo F, Xiong N, et al. NKG2D-deficient mice are defective in tumor surveillance in models of spontaneous malignancy. *Immunity* (2008) 28:571–80. doi:10.1016/j.immuni.2008.02.016
14. Fang M, Lanier LL, Sigal LJ. A role for NKG2D in NK cell-mediated resistance to poxvirus disease. *PLoS Pathog* (2008) 4:e30. doi:10.1371/journal.ppat.0040030
15. Wesselkamper SC, Eppert BL, Motz GT, Lau GW, Hassett DJ, Borchers MT. NKG2D is critical for NK cell activation in host defense against *Pseudomonas aeruginosa* respiratory infection. *J Immunol* (2008) 181:5481–9. doi:10.4049/jimmunol.181.8.5481
16. Radosavljevic M, Cuillerier B, Wilson MJ, Clement O, Wicker S, Gilfillan S, et al. A cluster of ten novel MHC class I related genes on human chromosome 6q24.2-q25.3. *Genomics* (2002) 79:114–23. doi:10.1006/geno.2001.6673
17. Choy MK, Phipps ME. MICA polymorphism: biology and importance in immunity and disease. *Trends Mol Med* (2010) 16:97–106. doi:10.1016/j.molmed.2010.01.002
18. Chen D, Gyllensten U. MICA polymorphism: biology and importance in cancer. *Carcinogenesis* (2014) 35:2633–42. doi:10.1093/carcin/bgu215
19. Goto K, Kato N. MICA SNPs and the NKG2D system in virus-induced HCC. *J Gastroenterol* (2015) 50:261–72. doi:10.1007/s00535-014-1000-9
20. Ji M, Wang J, Yuan L, Zhang Y, Zhang J, Dong W, et al. MICA polymorphisms and cancer risk: a meta-analysis. *Int J Clin Exp Med* (2015) 8:818–26.
21. Wang Q, Zhou X. Associations of MICA polymorphisms with inflammatory rheumatic diseases. *Open Rheumatol J* (2015) 9:94–100. doi:10.2174/1874312901409010094
22. Steinle A, Li P, Morris DL, Groh V, Lanier LL, Strong RK, et al. Interactions of human NKG2D with its ligands MICA, MICB, and homologs of the mouse RAE-1 protein family. *Immunogenetics* (2001) 53:279–87. doi:10.1007/s002510100325
23. Stastny P. Introduction: MICA/MICB in innate immunity, adaptive immunity, autoimmunity, cancer, and in the immune response to transplants. *Hum Immunol* (2006) 67:141–4. doi:10.1016/j.humimm.2006.02.019
24. Zhao J, Jiang Y, Lei Y, Chen LP, Yi FM, Wang CG, et al. [The relationship between major histocompatibility complex class I chain-related antigens A (MICA)-129 gene polymorphism, soluble MICA level and ulcerative colitis]. *Zhonghua Nei Ke Za Zhi* (2011) 50:311–5.
25. Zhao J, Jiang Y, Lei Y, Zou K, Wang C, Huang S, et al. Functional MICA-129 polymorphism and serum levels of soluble MICA are correlated with ulcerative colitis in Chinese patients. *J Gastroenterol Hepatol* (2011) 26:593–8. doi:10.1111/j.1440-1746.2010.06524.x
26. Amroun H, Djoudi H, Busson M, Allat R, El Sherbini SM, Sloma I, et al. Early-onset ankylosing spondylitis is associated with a functional MICA polymorphism. *Hum Immunol* (2005) 66:1057–61. doi:10.1016/j.humimm.2005.09.004
27. Kirsten H, Petit-Teixeira E, Scholz M, Hasenclever D, Hantmann H, Heider D, et al. Association of MICA with rheumatoid arthritis independent of known HLA-DRB1 risk alleles in a family-based and a case control study. *Arthritis Res Ther* (2009) 11:R60. doi:10.1186/ar2683
28. Yoshida K, Komai K, Shiozawa K, Mashida A, Horiuchi T, Tanaka Y, et al. Role of the MICA polymorphism in systemic lupus erythematosus. *Arthritis Rheum* (2011) 63:3058–66. doi:10.1002/art.30501
29. Achour Y, Kammoun A, Ben Hamad M, Mahfoudh N, Chaabane S, Marzouk S, et al. Association study of MICA gene polymorphisms with rheumatoid arthritis susceptibility in south Tunisian population. *Int J Immunogenet* (2014) 41:486–92. doi:10.1111/iji.12146
30. Lopez-Hernandez R, Valdes M, Lucas D, Campillo JA, Martinez-Garcia P, Salama H, et al. Association analysis of MICA gene polymorphism and MICA-129 dimorphism with inflammatory bowel disease susceptibility in a Spanish population. *Hum Immunol* (2010) 71:512–4. doi:10.1016/j.humimm.2010.02.003
31. Raache R, Belanteur K, Amroun H, Benyahia A, Heniche A, Azzouz M, et al. Association of major histocompatibility complex class 1 chain-related gene a dimorphism with type 1 diabetes and latent autoimmune diabetes in adults in the Algerian population. *Clin Vaccine Immunol* (2012) 19:557–61. doi:10.1128/CVI.05473-11
32. Pollock RA, Chandran V, Pellett FJ, Thavaneswaran A, Eder L, Barrett J, et al. The functional MICA-129 polymorphism is associated with skin but not joint manifestations of psoriatic disease independently of HLA-B and HLA-C. *Tissue Antigens* (2013) 82:43–7. doi:10.1111/tan.12126
33. Douik H, Ben Chaaben A, Attia Romdhane N, Romdhane HB, Mamoghli T, Fortier C, et al. Association of MICA-129 polymorphism with nasopharyngeal cancer risk in a Tunisian population. *Hum Immunol* (2009) 70:45–8. doi:10.1016/j.humimm.2008.10.008
34. Tong HV, Toan NL, Song LH, Bock CT, Kremsner PG, Velavan TP. Hepatitis B virus-induced hepatocellular carcinoma: functional roles of MICA variants. *J Viral Hepat* (2013) 20:687–98. doi:10.1111/jvh.12089
35. Campillo JA, Lopez-Hernandez R, Martinez-Banaclocha H, Boların JM, Gimeno L, Mrowiec A, et al. MHC class I chain-related gene a diversity in patients with cutaneous malignant melanoma from southeastern Spain. *Dis Markers* (2015) 2015:831864. doi:10.1155/2015/831864
36. Boukouaci W, Busson M, Peffault De Latour R, Rocha V, Suberbielle C, Bengoufa D, et al. MICA-129 genotype, soluble MICA, and anti-MICA antibodies as biomarkers of chronic graft-versus-host disease. *Blood* (2009) 114:5216–24. doi:10.1182/blood-2009-04-217430
37. Isernhagen A, Malzahn D, Viktorova E, Elsner L, Monecke S, von Bonin F, et al. The MICA-129 dimorphism affects NKG2D signaling and outcome of hematopoietic stem cell transplantation. *EMBO Mol Med* (2015) 7:1480–502. doi:10.15252/emmm.201505246
38. Ayo CM, Oliveira AP, Camargo AV, Mattos CC, Bestetti RB, Mattos LC. Association of the functional MICA-129 polymorphism with the severity of chronic Chagas heart disease. *Clin Infect Dis* (2015) 61:1310–3. doi:10.1093/cid/civ540
39. Ayo CM, Camargo AV, Frederico FB, Siqueira RC, Previato M, Murata FH, et al. MHC class I chain-related gene a polymorphisms and linkage disequilibrium with HLA-B and HLA-C alleles in ocular toxoplasmosis. *PLoS One* (2015) 10:e0144534. doi:10.1371/journal.pone.0144534
40. Hizem S, Mtiraoui N, Massaoudi S, Fortier C, Boukouaci W, Kahina A, et al. Polymorphisms in genes coding for the NK-cell receptor NKG2D and its ligand MICA in recurrent miscarriage. *Am J Reprod Immunol* (2014) 72:577–85. doi:10.1111/aji.12314
41. Groh V, Bruhl A, El-Gabalawy H, Nelson JL, Spies T. Stimulation of T cell autoreactivity by anomalous expression of NKG2D and its MIC ligands in

rheumatoid arthritis. *Proc Natl Acad Sci U S A* (2003) 100:9452–7. doi:10.1073/pnas.1632807100
42. Dai Z, Turtle CJ, Booth GC, Riddell SR, Gooley TA, Stevens AM, et al. Normally occurring NKG2D+CD4+ T cells are immunosuppressive and inversely correlated with disease activity in juvenile-onset lupus. *J Exp Med* (2009) 206:793–805. doi:10.1084/jem.20081648
43. Zhou X, Wang J, Zou H, Ward MM, Weisman MH, Espitia MG, et al. MICA, a gene contributing strong susceptibility to ankylosing spondylitis. *Ann Rheum Dis* (2014) 73:1552–7. doi:10.1136/annrheumdis-2013-203352
44. Low JS, Chin YM, Mushiroda T, Kubo M, Govindasamy GK, Pua KC, et al. A genome wide study of copy number variation associated with nasopharyngeal carcinoma in Malaysian Chinese identifies CNVs at 11q14.3 and 6p21.3 as candidate loci. *PLoS One* (2016) 11:e0145774. doi:10.1371/journal.pone.0145774
45. Kumar V, Kato N, Urabe Y, Takahashi A, Muroyama R, Hosono N, et al. Genome-wide association study identifies a susceptibility locus for HCV-induced hepatocellular carcinoma. *Nat Genet* (2011) 43:455–8. doi:10.1038/ng.809
46. Chen K, Shi W, Xin Z, Wang H, Zhu X, Wu X, et al. Replication of genome wide association studies on hepatocellular carcinoma susceptibility loci in a Chinese population. *PLoS One* (2013) 8:e77315. doi:10.1371/journal.pone.0077315
47. Chen D, Juko-Pecirep I, Hammer J, Ivansson E, Enroth S, Gustavsson I, et al. Genome-wide association study of susceptibility loci for cervical cancer. *J Natl Cancer Inst* (2013) 105:624–33. doi:10.1093/jnci/djt051
48. Chen D, Cui T, Ek WE, Liu H, Wang H, Gyllensten U. Analysis of the genetic architecture of susceptibility to cervical cancer indicates that common SNPs explain a large proportion of the heritability. *Carcinogenesis* (2015) 36:992–8. doi:10.1093/carcin/bgv083
49. Nieuwenhuis MA, Siedlinski M, Van Den Berge M, Granell R, Li X, Niens M, et al. Combining genomewide association study and lung eQTL analysis provides evidence for novel genes associated with asthma. *Allergy* (2016) 71(12):1712–20. doi:10.1111/all.12990
50. Le Clerc S, Delaneau O, Coulonges C, Spadoni JL, Labib T, Laville V, et al. Evidence after imputation for a role of MICA variants in nonprogression and elite control of HIV type 1 infection. *J Infect Dis* (2014) 210:1946–50. doi:10.1093/infdis/jiu342
51. Isernhagen A, Schilling D, Monecke S, Shah P, Elsner L, Walter L, et al. The MICA-129Met/Val dimorphism affects plasma membrane expression and shedding of the NKG2D ligand MICA. *Immunogenetics* (2016) 68:109–23. doi:10.1007/s00251-015-0884-8
52. Shafi S, Vantourout P, Wallace G, Antoun A, Vaughan R, Stanford M, et al. An NKG2D-mediated human lymphoid stress surveillance response with high interindividual variation. *Sci Transl Med* (2011) 3:113ra124. doi:10.1126/scitranslmed.3002922
53. Shi C, Li H, Couturier JP, Yang K, Guo X, He D, et al. Allele specific expression of MICA variants in human fibroblasts suggests a pathogenic mechanism. *Open Rheumatol J* (2015) 9:60–4. doi:10.2174/1874312901409010060
54. Lo PH, Urabe Y, Kumar V, Tanikawa C, Koike K, Kato N, et al. Identification of a functional variant in the MICA promoter which regulates MICA expression and increases HCV-related hepatocellular carcinoma risk. *PLoS One* (2013) 8:e61279. doi:10.1371/journal.pone.0061279
55. Ashiru O, Lopez-Cobo S, Fernandez-Messina L, Pontes-Quero S, Pandolfi R, Reyburn HT, et al. A GPI anchor explains the unique biological features of the common NKG2D-ligand allele MICA*008. *Biochem J* (2013) 454:295–302. doi:10.1042/BJ20130194
56. Groh V, Wu J, Yee C, Spies T. Tumour-derived soluble MIC ligands impair expression of NKG2D and T-cell activation. *Nature* (2002) 419:734–8. doi:10.1038/nature01112
57. Salih HR, Rammensee HG, Steinle A. Cutting edge: down-regulation of MICA on human tumors by proteolytic shedding. *J Immunol* (2002) 169:4098–102. doi:10.4049/jimmunol.169.8.4098
58. El-Gazzar A, Groh V, Spies T. Immunobiology and conflicting roles of the human NKG2D lymphocyte receptor and its ligands in cancer. *J Immunol* (2013) 191:1509–15. doi:10.4049/jimmunol.1301071
59. Kumar V, Yi Lo PH, Sawai H, Kato N, Takahashi A, Deng Z, et al. Soluble MICA and a MICA variation as possible prognostic biomarkers for HBV-induced hepatocellular carcinoma. *PLoS One* (2012) 7:e44743. doi:10.1371/journal.pone.0044743
60. Tamaki S, Kawakami M, Yamanaka Y, Shimomura H, Imai Y, Ishida J, et al. Relationship between soluble MICA and the MICA A5.1 homozygous genotype in patients with oral squamous cell carcinoma. *Clin Immunol* (2009) 130:331–7. doi:10.1016/j.clim.2008.09.004
61. Jiang X, Zou Y, Huo Z, Yu P. Association of major histocompatibility complex class I chain-related gene A microsatellite polymorphism and hepatocellular carcinoma in South China Han population. *Tissue Antigens* (2011) 78:143–7. doi:10.1111/j.1399-0039.2011.01693.x
62. Hayashi T, Imai K, Morishita Y, Hayashi I, Kusunoki Y, Nakachi K. Identification of the NKG2D haplotypes associated with natural cytotoxic activity of peripheral blood lymphocytes and cancer immunosurveillance. *Cancer Res* (2006) 66:563–70. doi:10.1158/0008-5472.CAN-05-2776
63. Isernhagen A, Malzahn D, Monecke S, Schilling D, Shah P, Multhoff G, et al. Functional consequences of genetic polymorphisms in the NKG2D receptor signaling pathway and putative gene interactions. *Receptors Clin Investig* (2016) 3:e1269. doi:10.14800/rci.1269
64. Antoun A, Vekaria D, Salama RA, Pratt G, Jobson S, Cook M, et al. The genotype of RAET1L (ULBP6), a ligand for human NKG2D (KLRK1), markedly influences the clinical outcome of allogeneic stem cell transplantation. *Br J Haematol* (2012) 159:589–98. doi:10.1111/bjh.12072
65. Espinoza JL, Takami A, Onizuka M, Sao H, Akiyama H, Miyamura K, et al. NKG2D gene polymorphism has a significant impact on transplant outcomes after HLA-fully-matched unrelated bone marrow transplantation for standard risk hematologic malignancies. *Haematologica* (2009) 94:1427–34. doi:10.3324/haematol.2009.008318
66. Kitcharoen K, Witt CS, Romphruk AV, Christiansen FT, Leelayuwat C. MICA, MICB, and MHC beta block matching in bone marrow transplantation: relevance to transplantation outcome. *Hum Immunol* (2006) 67:238–46. doi:10.1016/j.humimm.2006.02.012
67. Parmar S, Del Lima M, Zou Y, Patah PA, Liu P, Cano P, et al. Donor-recipient mismatches in MHC class I chain-related gene A in unrelated donor transplantation lead to increased incidence of acute graft-versus-host disease. *Blood* (2009) 114:2884–7. doi:10.1182/blood-2009-05-223172
68. Askar M, Sun Y, Rybicki L, Zhang A, Thomas D, Kalaycio M, et al. Synergistic effect of major histocompatibility complex class I-related chain a and human leukocyte antigen-DPB1 mismatches in association with acute graft-versus-host disease after unrelated donor hematopoietic stem cell transplantation. *Biol Blood Marrow Transplant* (2014) 20:1835–40. doi:10.1016/j.bbmt.2014.07.019
69. Carapito R, Jung N, Kwemou M, Untrau M, Michel S, Pichot A, et al. Matching for the nonconventional MHC-I MICA gene significantly reduces the incidence of acute and chronic GVHD. *Blood* (2016) 128:1979–86. doi:10.1182/blood-2016-05-719070
70. Fuerst D, Neuchel C, Niederwieser D, Bunjes D, Gramatzki M, Wagner E, et al. Matching for the MICA-129 polymorphism is beneficial in unrelated hematopoietic stem cell transplantation. *Blood* (2016). doi:10.1182/blood-2016-05-716357
71. Okada Y, Han B, Tsoi LC, Stuart PE, Ellinghaus E, Tejasvi T, et al. Fine mapping major histocompatibility complex associations in psoriasis and its clinical subtypes. *Am J Hum Genet* (2014) 95:162–72. doi:10.1016/j.ajhg.2014.07.002
72. Cunningham F, Amode MR, Barrell D, Beal K, Billis K, Brent S, et al. Ensembl 2015. *Nucleic Acids Res* (2015) 43:D662–9. doi:10.1093/nar/gku1010
73. Carapito R, Bahram S. Genetics, genomics, and evolutionary biology of NKG2D ligands. *Immunol Rev* (2015) 267:88–116. doi:10.1111/imr.12328
74. Kato N, Tanaka J, Sugita J, Toubai T, Miura Y, Ibata M, et al. Regulation of the expression of MHC class I-related chain A, B (MICA, MICB) via chromatin remodeling and its impact on the susceptibility of leukemic cells to the cytotoxicity of NKG2D-expressing cells. *Leukemia* (2007) 21:2103–8. doi:10.1038/sj.leu.2404862

11

Regulation of the Functions of Natural Cytotoxicity Receptors by Interactions with Diverse Ligands and Alterations in Splice Variant Expression

*Tatiana Pazina[1,2], Avishai Shemesh[3], Michael Brusilovsky[4], Angel Porgador[3] and Kerry S. Campbell[1]**

[1] Blood Cell Development and Function Program, Institute for Cancer Research, Fox Chase Cancer Center, Philadelphia, PA, USA, [2] Federal State Budgetary Scientific Institution "Institute of Experimental Medicine", St. Petersburg, Russia, [3] Faculty of Health Sciences, The Shraga Segal Department of Microbiology, Immunology and Genetics, Ben-Gurion University of the Negev, Beer Sheva, Israel, [4] Division of Allergy and Immunology, Cincinnati Children's Hospital Medical Center, Cincinnati, OH, USA

Edited by:
Gianfranco Pittari, Hamad Medical Corporation, Qatar

Reviewed by:
Eric Vivier, Centre d'immunologie de Marseille-Luminy, France
Alessandro Moretta, University of Genova, Italy

***Correspondence:**
Kerry S. Campbell
kerry.campbell@fccc.edu

The natural cytotoxicity receptor (NCR) family is constituted by NKp46, NKp44, and NKp30 in humans, which are expressed mainly on natural killer (NK) cells and are encoded by the *ncr1*, *ncr2*, and *ncr3* genes, respectively. NCRs have classically been defined as activating receptors that trigger cytotoxicity and cytokine responses by NK cells upon engaging with ligands on tumor cells. Several new findings, however, have challenged this model and identified alternative mechanisms regulating the function of NCRs. Recent reports indicate that ligand matters, since the interaction of NKp44 with distinct ligands on target cells can either activate or inhibit NK cells. Also, the NCRs have been found to interact with distinct specificities to various heparan sulfate glycosaminoglycans, which are complex polysaccharides found in extracellular matrix or on cell surface heparan sulfate proteoglycans (HSPGs). The NCRs can engage with HSPGs in trans as a co-ligand on the target cells or in cis on the NK cell surface to regulate receptor–ligand interactions and NK cell activation. A number of splice variants of *ncr2* and *ncr3* have also been identified, and a predominant expression of certain variants results in inhibitory signaling through NKp44 and NKp30. Several recent studies have found that the selective expression of some of these inhibitory splice variants can significantly influence outcome in the contexts of cancer, infection, and pregnancy. These findings establish that NCR functions are more diverse than originally thought, and better understanding of their splice variant expression profiles and ligand interactions are needed to establish their functional regulation in the context of human health.

Keywords: natural cytotoxicity receptors, natural killer cells, RNA splice variants, cytotoxicity, cancer immunology, virus immunity, human immunology, pregnancy

INTRODUCTION

It has become increasingly clear that human natural killer (NK) cells use an array of germline-encoded cell surface receptors to spontaneously recognize and respond to the abnormal status of tumor cells, virus-infected cells, and stressed cells (1). Different NK cell surface receptors transduce either activating or inhibitory signals, directly or through associated adaptor proteins, to dynamically regulate the activation state of NK cells (2–4). Inhibitory receptors, such as killer cell Ig-like receptors (KIRs) and CD94/NKG2A, provide NK cells with a dominant, tolerizing immune checkpoint through their recognition of the MHC class I (MHC-I) molecules ubiquitously expressed on the surfaces of most normal cells. The loss of MHC-I by many transformed cells, however, overcomes this inhibitory checkpoint to favor activation. The loss of inhibitory restraint allows activating receptor signals to predominate and triggers exocytosis of perforin and granzymes to induce targeted apoptosis of the MHC-I-deficient cell and localized secretion of pro-inflammatory cytokines [especially interferon (IFN)-γ, TNF-α, and several chemokines].

The activation signals transduced in NK cells are derived from adhesion molecules (especially LFA-1), co-stimulatory receptors (such as NKG2D, DNAM-1, and SLAM family receptors), and several activating receptors physically linked to the immunoreceptor tyrosine-based activation motif (ITAM)-containing transmembrane proteins, DAP12, TCR-ζ, and/or FcεRI-γ (2–4). The key ITAM-coupled activating receptors on human NK cells, include CD16 (FcγRIIIa), an activating subfamily of KIR (KIR2DS or KIR3DS receptors), CD94/NKG2C, and the natural cytotoxicity receptors (NCRs).

The human NCRs consist of three receptors, named NKp46 (NCR1, CD335), NKp44 (NCR2, CD336), and NKp30 (NCR3, CD337). These NCR were classically defined as germline-encoded receptors that play important roles in the activation of human NK cells toward transformed target cells (5, 6). Recent work, however, has established that the NCRs can also generate inhibitory responses under certain circumstances. Here, we will review our current understanding of the expression and function of NCRs on NK cells, particularly in humans, although it is important to note that the NCRs are also expressed on other innate lymphoid cells (ILCs) and a subset of T cells, which has been previously reviewed elsewhere (7, 8).

NCRs AND THEIR STRUCTURES

The NCRs were initially discovered and characterized by the laboratories of Alessandro and Lorenzo Moretta in the late 1990s (9–13). They are type I transmembrane glycoproteins that were originally recognized as activating receptors and named in accordance with their molecular weight on SDS-PAGE (NKp30, NKp44, and NKp46). NKp46 is the only NCR also expressed in mice, although a receptor analogous to NKp30 has been shown to be expressed in 1 of 13 mouse strains examined (14) and in rats (15, 16). While the *ncr2* and *ncr3* genes encoding NKp44 and NKp30, respectively, are localized to human MHC class III locus on chromosome 6, the NKp46 encoding gene, *ncr1*, is found near the leukocyte regulatory complex on human chromosome 19 (10, 12, 13).

NKp46 has been shown to be a highly selective marker of all NK cells in mouse and man, although surface expression can be low on some NK cells, particularly in humans, and the receptor is also expressed on some ILCs and a small subset of T cells (9, 17, 18). Importantly, NKp46 is not expressed by CD1d-restricted invariant NKT cells in mice and humans (17). NKp46 has been shown to provide NK cells with the capacity to recognize and kill a variety of tumor target cells (19–21). NKp46 ligands have been reported to be enriched in areas of high malignant potential and high proliferation within melanoma lesions, whereas surrounding normal melanocytes were found to lack NKp46 ligands (22). Evidence in mice also suggests that NKp46 also contributes to the development of type 1 diabetes by interacting with an uncharacterized ligand on pancreatic islet beta cells (21). As a tumor immunosuppressive mechanism, the surface expression of NKp46 on NK cells can be down-modulated by exposure to L-kynurenine, which is a tryptophan catabolism product generated by the indoleamine 2,3-dioxygenase (IDO) enzyme in tumor microenvironments (23).

NKp30, similar to NKp46, is expressed on nearly all human NK cells (13). This NCR has been shown to play important roles in crosstalk between NK cells and dendritic cells (DCs) through promoting both the maturation of and the cytotoxicity of immature DC (24, 25). Surface expression levels of NKp30 and NKp46 can be upregulated by IFN-α, IL-2, and prolactin and downregulated by cortisol and methylprednisolone (26–28). In addition, both receptors are also commonly downregulated in "adaptive" or "memory-like" NK cells that are found in some cytomegalovirus-infected individuals (29, 30). TGF-β has been shown to selectively down-modulate the expression of NKp30, but not NKp46 on NK cells (31).

NKp44 is distinct among NCRs, since it is unique to humans and only expressed constitutively on some CD56^{bright} NK cells in a subset of individuals, but expression can be upregulated on essentially all NK cells after culture with IL-2, IL-15, or IL-1β (11, 32). Similarly, NKp44 can be upregulated on plasmacytoid DCs upon culture with IL-3 (33). Therefore, NKp44 may also be considered a marker of cytokine-activated NK cells in humans. IL-2-induced upregulation of NKp44 on NK cells can be inhibited by prostaglandin E2, which is readily produced by tumor-associated fibroblasts, especially when exposed to NK cells in culture (34). Similarly, prednisolone can suppress IL-2-mediated upregulation of NKp44 (26).

The extracellular domains of NCRs consist of one (NKp30 and NKp44) or two (NKp46) Ig-like domains that are responsible for ligand binding (10, 12, 13). Ligand binding and signaling function by NKp30 is highly dependent upon integrity of the membrane proximal stalk region (35). Crystal structures have revealed that NKp30 and NKp44 can form homodimeric structures with NKp30 dimerizing in a head-to-tail fashion to form an I-type Ig-like fold and two NKp44 V-type Ig-like domains form a saddle-shaped dimer with unique disulfide bridging (36, 37). On the other hand, the crystal structure of NKp46 demonstrates two C2-type Ig-like domains that are folded and oriented similar to the Ig-like domains of KIRs (38). Evidence for a homodimerization interface within the membrane proximal Ig-like domain of NKp46 has also been reported, and disruption of this dimerization interaction prevented ligand binding and activating function of the receptor (39). In conclusion, the three NCRs have unrelated structures, and grouping these receptors together is based more on their shared functional properties than related structure or genetic evolution (13).

The transmembrane domains of all three NCRs contain a positively charged lysine (NKp44) or arginine (NKp30 and NKp46) residue that interacts with acidic aspartic acid residues found in the transmembrane regions of the adaptor proteins DAP12 (NKp44) or TCR-ζ and/or FcεRI-γ (NKp30 and NKp46) (10). Physical association with DAP12 *via* these transmembrane charged residues is essential for surface expression of NKp44 (40). The reductions in surface expression levels of NKp30 and NKp46 on "adaptive" or "memory-like" NK cells is associated with the lack of FcεRI-γ expression in these cells (29, 30), exemplifying the importance of associating with this specific adaptor to transport a functional receptor to the cell surface. In addition to promoting surface expression, physical association with these associated transmembrane adaptors provides potent activation signaling function to the NCRs, since the tyrosine phosphorylation of their cytoplasmic ITAM domains results in the recruitment and activation of the Syk and ZAP-70 protein tyrosine kinases (2, 41). A unique activation signaling crosstalk has been reported between the NCRs, in which engagement of one NCR appears to initiate signaling through the others (41). Curiously, while several mRNA splice variants encoding NKp44 have been described, the major protein product or isoform was found to also contain a cytoplasmic ITIM-like domain. Although early work suggested that this domain was incapable of providing inhibitory signaling function in an NK-like cell line (40), more recent work has demonstrated ITIM-mediated inhibitory function by NKp44 upon recognition of a specific ligand, proliferating cell nuclear antigen (PCNA), as detailed below (42).

LIGANDS OF THE NCRs

Despite a great deal of work by numerous research groups, our understanding of the ligands for NCRs is still not clearly established. A diverse array of molecules have been report to interact with the extracellular domains of NCRs, including carbohydrate-based contacts, cell surface proteins, and surprisingly, several intracellular-localized proteins that appear to reach the surface of infected or transformed cells. While engagement with most of these reported ligands stimulates activation of NK cells, some have been found to inhibit their functions. Our current understanding of putative ligands for NCRs and their functions are described below and summarized in **Figure 1** and **Table 1**.

Viral Ligands

Several viral-derived NCR ligands have been reported. Early work showed that the viral hemagglutinin (HA) of influenza virus on the surface of infected cells can readily engage with branched α-2,3- and α-2,6-sialylated *O*-glycan sequences conjugated on NKp46, and influenza-infected target cells can be killed by human NK cells in a NKp46-dependent manner (65–68). In fact, exposure of NK cells to influenza virions or free HA has been found to decrease NCR-mediated cytotoxicity, which was associated with loss of TCR-ζ protein expression (71). In addition, influenza virus-infected DCs stimulate IFN-γ production by NK cells in an NKp46- and HA-dependent manner (69). Furthermore, it has been demonstrated that NKp46-deficient mice are more susceptible to death after infection with influenza virus (72). These interactions are consistent with the known sialic acid-binding properties of viral HAs. HAs from influenza and Sendai viruses have further been shown to also interact with NKp44, but not NKp30, and NKp44$^+$ NK cells can kill cells infected with these viruses (54, 55). Similarly, both NKp46 and NKp44 were found to interact with the HA from avian Newcastle disease virus and this interaction potentiates cytotoxicity of target cells infected with this paramyxovirus (56).

Hemagglutinin from the orthopox family viruses, human vaccinia virus, and murine ectromelia virus has been shown to interact with NKp46 and NKp30 (43). Late-stage vaccinia virus-infected target cells were further shown to be less susceptible to NK cell cytotoxicity compared to uninfected targets, and this reduced killing was dependent upon viral HA in the target cells and NKp30 in the NK cells (43). The results from this study suggest that the HA on the surface of vaccinia virus-infected cells interacts with NKp30 to either block its activating function or to mediate inhibitory signaling in NK cells, whereas NKp46 engagement with vaccinia virus-derived HA on target cell surfaces stimulates cytotoxicity responses.

NKp44 has been shown to recognize the envelope glycoproteins from West Nile and Dengue flaviviruses (57). In particular, NKp44 was found to directly bind domain III of WNV envelope protein, but does not appear to involve viral HA, since independent of sialylation of oligosaccharides on NKp44. Consistent with this finding, West Nile virus-infected cells more readily bind a soluble recombinant form of NKp44 and stimulate NK cells to degranulate and produce IFN-γ in an NKp44-dependent manner (57). It should be noted that expression of Dengue viral non-structural proteins in target cells reduces susceptibility to NK cell cytotoxicity through upregulating MHC-I expression (73).

In addition, NKp30 has been shown to directly interact with pp65, which is the main tegument protein of human cytomegalovirus (HCMV) (44). HCMV infected target cells were found to be less susceptible to NK cell-mediated killing, and this inhibition was lost if the target cells were infected with pp65-deficient HCMV or if anti-NKp30 blocking antibodies were added (44). The authors of this report further provided evidence that treating NK cells with a recombinant soluble form of pp65 resulted in the dissociation of the TCR-ζ signaling adaptor protein from NKp30 (44). In this way, pp65 appears to provide HCMV with a mechanism to avoid NK cell-mediated immunity by disrupting activation signaling through NKp30.

Other Ligands Expressed by Pathogens or Pathological Conditions

Natural cytotoxicity receptors have also been shown to directly recognize bacterial and parasite pathogens. It has been demonstrated that NKp30 (and to a lesser extent, NKp46) can interact with the Duffy binding-like (DBL)-1α domain of *Plasmodium*

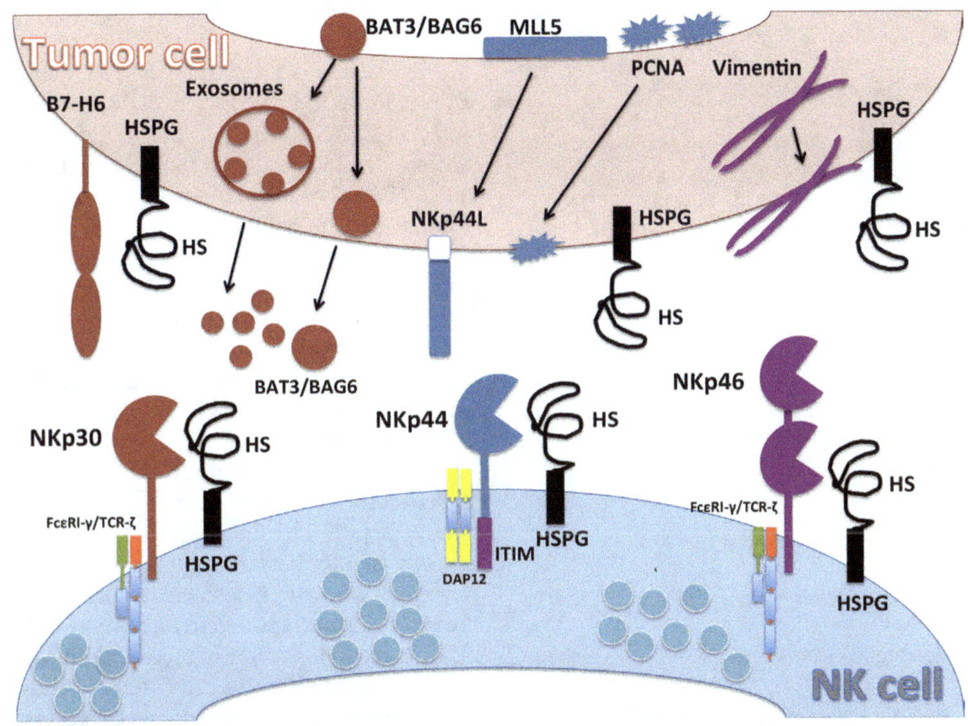

FIGURE 1 | Ligands for natural cytotoxicity receptors (NCRs). Schematic representation of NCR ligands on tumor cell and their interaction with NKp30, NKp44, and NKp46 on natural killer (NK) cells. B7-H6 is an activating ligand for NKp30 upregulated on tumor cells and absent on normal cells. HLA-B-associated transcript 3 (BAT3)/Bcl2-associated anthogene 6 (BAG6) expressed in the nucleus moves to the plasma cell membrane or is released in exosomes. NKp44L is a splice variant isoform of the nuclear protein Mixed-lineage leukemia-5 protein that localizes to the tumor cell plasma membrane to serve as an activating ligand for NKp44. Proliferating cell nuclear antigen (PCNA) is a nuclear protein involved in DNA replication and repair mechanisms that relocalizes to the plasma membrane to serve as an NKp44 inhibitory ligand. Cytoskeleton type III filamentous vimentin is an intracellular protein but can be upregulated on the cell surface of infected cells, where it serves as a ligand for NKp46. Heparan sulfate proteoglycans (HSPGs) can interact with all NCRs. Heparan sulfate (HS) expressed on NK cell surface (cis interaction) can mask interactions with HSPG or other ligands on target cells (trans interactions).

falciparum erythrocyte membrane protein-1 to mediate cytolysis of malaria-infected erythrocytes (45). The interaction appears to be direct, since the effect can be inhibited by the addition of recombinant soluble forms of these NCRs or peptides matching the sequence of DBL-1α (45).

It has been reported that *Mycobacterium bovis bacillus* Calmette-Guérin (BCG) can directly interact with NKp44, and exposure to BCG can increase NKp44 expression on CD56^{bright} NK cells (58). This study also found that additional *Mycobacterium* family members can bind NKp44, such as *Nocardia farcinica* and *Pseudomonas aeruginosa*. Interactions of these bacteria with NKp44 did not activate NK cell functions, however, so the relevance of these interactions is currently unclear.

NKp44 has been reported to recognize an uncharacterized ligand on cartilage-derived chondrocytes, and cytotoxicity of primary chondrocytes by long-term IL-2 activated-NK cells was inhibited by an NKp44-blocking antibody (59). These results suggest that NKp44 activation signaling may promote NK cell-mediated auto-immunity in chronic inflammatory cartilaginous disease.

Heparan Sulfates

Heparan sulfate (HS) glycosaminoglycans (GAGs) have also been shown to interact with all of the NCRs, with different affinities for the three receptors (46–48, 60). While these carbohydrate-directed interactions likely do not represent primary ligands for the NCRs, they appear to have the capacity to regulate NCR function or may play a supporting role as co-ligands (61). HS GAGs consist of long, unbranched, anionic polysaccharides that are found on cell surfaces and the extracellular matrix (74). The HS GAG polysaccharides are composed of repeating disaccharide units of uronic acid (iduronic or glucuronic acid) and glucosamine that are differentially sulfated at *N*, 2-*O*, 3-*O*, and 6-*O* positions to generate highly diverse structures with unique protein binding properties (75). HS GAG can be conjugated to a small subset of proteins to form HS proteoglycans, including syndecans and glypicans, and their negative-charged configurations can provide docking sites for basic domains on chemokines, FGF, and wnt ligand family members, thereby "presenting" them to cell surface receptors (76–79).

Interestingly, HS GAG are highly diverse structures, and the three NCRs preferentially recognize highly sulfated HS structures *via* basic amino acid patches on the receptor surfaces, and each NCR demonstrates distinct HS binding specificity (47, 48, 60). Therefore, it is conceivable that each NCR has the capacity to distinguish particular configurations of HS GAG primary and tertiary structures that might be uniquely expressed in the

TABLE 1 | Ligands for natural cytotoxicity receptors.

Receptor	Ligands	Function	Reference
NKp30	Hemagglutinin (HA) of human vaccinia virus	Inhibition	(43)
	pp65, Main tegument protein of human cytomegalovirus	Inhibition	(44)
	(DBL)-1α domain of *Plasmodium falciparum* erythrocyte membrane protein-1	Activation	(45)
	Heparan sulfate (HS) glycosaminoglycans (GAGs)	Activation/ regulation	(46–48)
	BAT3/BAG6	Activation	(49–51)
	B7-H6	Activation	(52, 53)
NKp44	Redirected cytotoxicity and blockade of natural cytotoxicity with NKp44 antibody	Activation	(11)
	HA of influenza and Sendai viruses	Activation	(54, 55)
	HA from avian Newcastle disease	Activation	(56)
	Domain III of WNV envelope protein of West Nile and Dengue virus	Activation	(57)
	Mycobacterium bovis bacillus Calmette-Guérin (BCG)	Unclear	(58)
	Unknown ligand on cartilage-derived chondrocytes	Activation	(59)
	HS GAGs	Activation/ regulation	(47, 60, 61)
	NKp44L	Activation	(62, 63)
	Proliferating cell nuclear antigen	Inhibition	(42, 64)
NKp46	HA of influenza virus	Activation	(65–69)
	HA of avian Newcastle disease	Activation	(56)
	HA of human vaccinia virus	Activation	(43)
	(DBL)-1α domain of *Plasmodium falciparum* erythrocyte membrane protein-1	Activation	(45)
	HS GAGs	Activation/ regulation	(47)
	Vimentin	Activation	(70)

contexts of tumor microenvironments or sites of infection or inflammation. We have also shown the HS GAGs can interact with another NK cell receptor, KIR2DL4, and the binding can modulate receptor function (80). In addition, an interaction of NKp44 with the heparan sulfate proteoglycans (HSPGs), syndecan-4, in cis on the NK cell surface can modulate the surface distribution and function of the receptor (61). Based on this and other reports (47), we have proposed that interactions with HSPGs in cis (on the surface of NK cells) may be impacting KIR2DL4 and NCR functions through masking interactions with HS GAG or other ligands on adjacent target cells (trans interactions) and/or may be affecting the trafficking of NCR to intracellular degradation and recycling pathways upon endocytosis (61, 81). In this way, cis interactions between NCR and HSPGs may provide an allosteric regulation mechanism. It is also intriguing to speculate that treatment of patients with structurally related heparin as a therapeutic agent could impact NK cell functions through binding to NCRs and other NK cell surface receptors, including KIR2DL4.

Intracellular Proteins As Cell Surface Ligands

The expression of a ligand for NKp44, named NKp44L, was first shown to be induced by the HIV-1 envelope protein gp41 on infected CD4$^+$ T cells, and the expression increased in patients with increasing viral load (62). NKp44L is an activating ligand, since NK cell-mediated lysis of HIV-infected CD4$^+$ T cells was inhibited by antibodies to NKp44 or NKp44L (62). The NKp44L was subsequently identified as a unique splice variant isoform of mixed-lineage leukemia-5 (MLL5) protein (63). While full-length MLL5 is a nuclear protein, the NKp44L splice variant is localized near the plasma membrane in the cytoplasm and expressed in several tumor tissues and transformed cell lines, but not in normal tissues (63).

It was reported that NKp46 is involved in NK cell-mediated cytolytic attack of monocytes infected with *Mycobacterium tuberculosis* (20). Subsequent work established that this is due to an interaction with vimentin, which is expressed at high levels in infected monocytes and appears on the cell surface (70). Since vimentin is a type III intermediate filament of the cytoskeleton, however, it is unexpected to find on the cell surface, but this follows an emerging theme of several traditionally intracellular proteins serving as putative cell surface ligands for NCRs. NK cells were more efficient at lysing target cells transfected to overexpress vimentin, and this cytotoxicity was inhibited by antibodies targeting NKp46 or vimentin (70).

In addition, NKp30 has been shown to interact with the HLA-B-associated transcript 3 (BAT3)/Bcl2-associated anthogene 6 (BAG6) protein to stimulate NK cytolytic responses (49). BAT3/BAG6 is predominantly expressed in the nucleus, but can move to the plasma membrane in cells exposed to heat shock and can be secreted in exosomes by tumors and stressed cells (49, 82). BAT3/BAG6-expressing exosomes can stimulate cytokine release from NK cells upon interaction with NKp30, and BAT3/BAG6 expression by DC is responsible for activation NK cells to mediate the crosstalk with DC (49, 50). Similarly, RIG-I stimulation of melanoma cell lines was shown to trigger the extracellular release of BAT3/BAG6-containing vesicles that can stimulate NK cell cytolytic responses (51). In contrast, a soluble form of BAT3/BAG6 has been found at high levels in the plasma of CLL patients and can suppress NK cytolytic responses, apparently by blocking recognition of this and other ligands on tumor cells (82, 83).

It was also reported that NKp44 can interact with PCNA, at target cell surfaces (42). PCNA is highly expressed in proliferating cancer cells, where it is usually tightly associated with DNA and involved in DNA replication and repair mechanisms (84). Surprisingly, PCNA was found to migrate to the plasma membrane of target cells within the immunological synapse with NKp44-expressing NK cells, and this interaction inhibited cytolytic function and IFN-γ production by the NK cells (42). A second report has also described the interaction of PCNA and NKp44 and the association of PCNA with MHC-I molecules at the plasma membrane of tumor cells as a potential surface transport mechanism (64). The PCNA-induced inhibition was found to be mediated through the ITIM-like sequence in the cytoplasmic domain of NKp44 (42), despite earlier work in which the ITIM-containing NKp44 cytoplasmic domain was shown to lack inhibitory function in the context of a chimeric receptor construct (40). It appears that PCNA interaction with the full NKp44 receptor establishes a unique conformation that transduces an inhibitory signal. Inhibitory function was also previously reported for NKp44 expressed on ILCs and plasmacytoid DCs (33, 85).

B7-H6

NKp30 has been found to bind to a cell surface protein member of the B7 family, named B7-H6 (52). B7-H6 is not normally expressed on healthy cells, but can be upregulated on human tumor cells through a Myc-mediated mechanism (86) or upon stimulation of monocytes and neutrophils with TLR ligands or pro-inflammatory cytokines (87). Upon recognition by NKp30, B7-H6 triggers cytotoxicity and cytokine production by NK cells (52). The interaction of B7-H6 with NKp30 is the most rigorously characterized of NCR ligands, since it is the only NCR–ligand interaction so far confirmed in an X-ray crystallography structure (53). It has also been shown that some tumors can escape NKp30 recognition by shedding B7-H6 from their surfaces with the metalloproteases, ADAM-10, and ADAM-17 (88). Soluble and tumor-associated expression of B7-H6 in the peritoneum of ovarian cancer patients has also been shown to correlate with reduced surface expression of NKp30 on peritoneal NK cells, presumably due to chronic interaction with ligand (89).

SPLICE VARIANTS OF NCRs RESULTING IN DISTINCT RECEPTOR ISOFORMS

A variety of mRNA splice variants encoding different isoforms of NCRs have been recognized for many years, but only recent work has established that some of these variant NCR isoforms can facilitate inhibitory functions. Distinct splice variant expression patterns have also been shown to correlate with outcomes in cancer and infectious disease, suggesting potential prognostic value in patients. A summary diagram of current reported functions of distinct isoforms of NKp30 and NKp44 is presented in **Figure 2.**

Splice Variants of *ncr3*

The expression of distinct isoforms of NKp30 is of great interest, because this NCR is involved in DC-to-NK cell crosstalk (25), can facilitate tumor cell recognition (13, 49, 90), and can influence the prognosis of different infectious diseases (91). Six splice variant transcripts have been identified from the *ncr3* gene, which was originally called 1C7 (92, 93). The most highly expressed *ncr3* variants are designated a, b, and c that encode NKp30 proteins with an extracellular V-type Ig domain, while d, e, and f isoforms encode NKp30 receptors possessing a C2-type Ig domain that lacks 25 amino acids (93). The three members of each subgroup share three distinct cytoplasmic domains encoded by splice variations within exon 4. Although the d, e, and f protein isoforms of NKp30 have not been studied to date, several groups have examined the functions and interesting clinical outcomes associated with differential expression of the a, b, and c isoforms, as discussed below.

Alternative splicing of the *ncr3* gene impacts functions of NKp30 isoforms that can be immunosuppressive or immunostimulatory. Delahaye and colleagues expressed isoforms NKp30a, NKp30b, and NKp30c in the human NK cell line NKL to characterize their functions. It was shown that antibody- or B7-H6-mediated engagement of NKp30 on NKL transfected with either NKp30a or NKp30b isoforms stimulated production of large amounts of IFN-γ, degranulation, and cytotoxicity responses. In contrast, engagement of NKp30c on NKL cell transfectants did not result in degranulation or elicit cytotoxicity, but instead produced the inhibitory cytokine IL-10 and very little IFN-γ (94). Similar immunostimulatory functions for NKp30a and NKp30b isoforms were observed when NKp30 transfectants were cocultured with DCs, whereas NKL-NKp30c transfectants demonstrated minimal responsiveness in these assays. In all of these experiments, NKp30a stimulated the most potent activating responses, whereas NKp30c was inhibitory or non-responsive, and NKp30b induced intermediate activation (94).

It was also shown that NKp30a associated more tightly with TCR-ζ upon crosslinking, as compared to NKp30c (94). Surprisingly, p38 MAP kinase activation was more pronounced when NKp30 was engaged in NKL-NKp30c cells then for NKL cells transfected with NKp30a or NKp30b. Furthermore, treatment of NKL-NKp30c cells with a p38 inhibitor produced IFN-γ upon exposure to immature DCs (94). Therefore, NKp30b and especially NKp30a are stimulatory isoforms that can induce cytotoxicity and cytokine production, whereas NKp30c induces an immunosuppressive response that appears to involve activation of p38 and the production of IL-10.

NKp30 Isoforms and Cancer

In addition to characterizing different functions for distinct NKp30 isoforms, Delahaye performed a retrospective analysis of NKp30 expression profiles in 80 patients with gastrointestinal stromal tumors (GIST), which is a malignancy that expresses NKp30 ligands. In that analysis, predominant expression of the immunosuppressive NKp30c isoform over the immunostimulatory NKp30a/b isoforms was found to be associated with reduced overall survival in imatinib-treated patients (94). Moreover, a subset of GIST patients with predominant expression of the NKp30c isoform and a distinct haplotype involving two SNPs in the *ncr3* gene were found to be associated with particularly poor survival (94). For GI carcinomas and variety of other cancers, we observed that both cancerous and matched normal tissues manifested balanced NKp30c inhibitory and NKp30a/b activation profiles; yet, we found skewed NKp30 splice variant profiles in about 50% of a variety of tumor tissues compared to their matched normal tissues (95).

Neuroblastoma is another malignancy where NCRs are involved in the tumor cell recognition (96–98). Semeraro et al., found that neuroblastoma tumor samples express the NKp30 ligand, B7-H6, and some patients with metastatic neuroblastoma had high levels of soluble B7-H6 in their serum, which was associated with reduced expression of NKp30 on the surface of NK cells and higher degree of metastases (99). Furthermore, serum from patients with high soluble B7-H6 suppressed NK cell IFN-γ responses (99). Analysis of NKp30 isoform expression

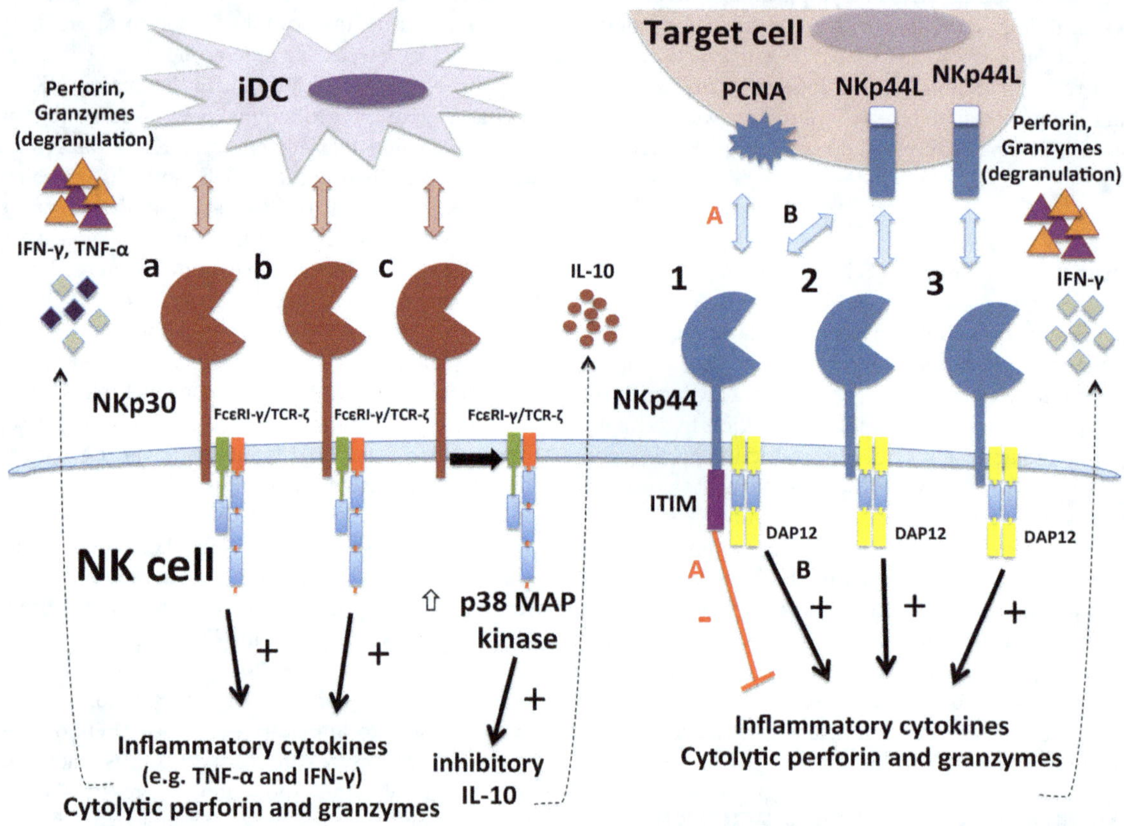

FIGURE 2 | Differential functions of different isoforms of NKp30 and NKp44. Immunostimulatory NKp30a and NKp30b isoforms interaction with ligands on tumor target cells or immature DCs (iDCs) to stimulate degranulation responses and interferon (IFN)-γ production. Inhibitory NKp30c isoform instead produces inhibitory cytokine IL-10 upon engagement. NKp30c is also less tightly associated with TCR-ζ and triggers stronger activation of p38 MAP kinase. The NKp44-1 isoform contains a cytoplasmic ITIM that can transduce inhibitory signals when engaged with proliferating cell nuclear antigen (PCNA) ligand on target cell surface (pathway "A"), leading to decreased cytotoxicity and IFN-γ production. In contrast, all three NKp44 isoforms can engage with NKp44L to transduce activation signaling through DAP12 (pathway "B" for NKp44-1) that triggers cytotoxicity and IFN-γ production.

suggested lower expression of NKp30a/b isoforms in patients with metastatic neuroblastoma compared to patients with localized disease or healthy donors. Comparison of the ratios of expression of mRNA for individual NKp30 isoforms revealed that high expression of the suppressive NKp30c isoform over the activating NKp30b isoform was associated with shorter progression-free survival of patients with metastatic disease (99). Surprisingly, IFN-γ was found to suppress the expression of *B7-H6* and *BAT3* mRNA, whereas IL-10 increased *B7-H6* mRNA expression on neuroblastoma tumors. The results of this study suggest that high expression of IL-10 by NKp30c-expressing NK cells induces B7-H6 expression on tumor, thereby potentiating further expression of this immunosuppressive cytokine.

The potential prognostic value of NKp30 isoforms was also recently investigated in melanoma patients (100). The overall mRNA expression levels of each of the NKp30 isoforms were found to be reduced compared to expression in healthy donors, but expression ratios between the isoforms did not differ. Levels of mRNA expression of the immunosuppressive NKp30c isoform were found to be higher on NK cells from stage IV melanoma patients, but relative expression levels of each NKp30 isoform did not predict overall survival of patients. Interestingly, long-term surviving melanoma patients were found to express higher levels of stimulatory NKp30a transcript. In accordance with this finding, the long-term survivors were more likely to have SNPs associated with reduced expression of suppressive NKp30c and increased expression of NKp30a. Furthermore, NK cells from long-term surviving patients exhibited increased degranulation potential in response to NKp30 stimulation (100). Therefore, higher expression of NKp30a in melanoma patients appears to be beneficial.

NKp30 Isoforms and Infectious Diseases

Surface expression NKp30 on NK cells was found to be downregulated in HIV-1 patients, but expression patterns of NKp30 isoforms do not affect disease progression or survival (101). NK cells from hepatitis C virus (HCV)-infected patients also showed reduced expression of NKp30 on the surface of NK cells, and expression of the immunosuppressive NKp30c transcript was found to be significantly decreased in infected patients compare to healthy controls (102). Also NKp30a/NKp30c ratio was significantly higher compared to healthy individuals suggesting an immunostimulatory profile in infected patients. If stratified according to mRNA expression levels, patients with low expression of all three isoforms had lower surface expression of NKp30 on NK cells, but higher ratio of NKp30a/NKp30c isoform transcripts. Accordingly, NK cells from these patients exhibited more potent degranulation and cytokine production responses upon engagement of the receptor, as compared to healthy controls (102). Interestingly, positive correlations were observed between NKp30a isoform mRNA levels and liver stiffness and between NKp30a/NKp30c ratio and a measure of liver fibrosis, suggesting that reduction of immunosuppressive NKp30c isoform and increase of immunostimulatory NKp30a isoform is associated with advanced liver disease in HCV-infected patients.

NKp30 in the Decidua of Pregnant Women

Natural killer cells from decidual tissue in pregnant uterus (dNK) have unique phenotypical and functional properties compared to NK cells found in peripheral blood (pNK) (103). Mouse studies originally demonstrated that dNK cells function to support embryo implantation by secretion of numerous factors, including IFN-γ, to promote angiogenesis and trophoblast invasion (104). Comparison of freshly isolated pNK and dNK cells from the same pregnant donors showed that pNK cells express high levels of NKp30a and NKp30b isoforms and low levels of NKp30c isoform, while dNK cells had high levels of NKp30c but significantly lower amounts of NKp30a/b (105). While crosslinking NKp30 on IL-15-stimulated pNK cells induced degranulation, freshly isolated dNK cells did not degranulate, and co-crosslinking NKp30 in dNK cells did not inhibit NKp46-mediated degranulation, consistent with the lack of activation by NKp30c in previously described studies by Delahaye et al. (94, 105). Coculturing pNK cells in presence of IL-15 and IL-18 induced increased expression of NKp30a/b isoforms, whereas further addition of TGF-β suppressed the induction of expression of all three isoforms, although expression of NKp30b and NKp30c isoforms was greater than NKp30a upon addition of TGF-β (105). Overall, the combination of IL-15, IL-18, and TGF-β, which are found together in the decidual stromal microenvironment, shifted pNK cells toward higher expression of inhibitory NKp30c isoform and expression of other markers characteristic of dNK cells (105).

Shemesh et al. found significantly increased mRNA encoding the activating NKp30a/b isoforms in the placenta of women who had experienced sporadic or recurrent miscarriage within the first trimester, whereas this shift toward activating isoforms was not evident in the peripheral blood (106). The increase in activating isoforms did not correlate with higher expression of TNF-α, IFN-γ, IL-10, and placental growth factor mRNAs in the placental tissue of those women who experienced sporadic miscarriage, as compared to those that had undergone elective abortions. These results suggest that increased expression of these activating isoforms of NKp30 on dNK cells may be in some way contributing to failed pregnancies through promoting dysregulated cytokine production in the placenta.

Splice Variants of *ncr2*

Three major mRNA splice variants of *ncr2* have been recognized, and one of these (NKp44-1) encodes the classic receptor that possesses a cytoplasmic ITIM, while the others (NKp44-2 and NKp44-3) have alternative sequences in the cytoplasmic region that lack ITIMs. It has long been known that IL-2-cultured NK cells upregulate expression of an activating form of NKp44-1 associated with the transmembrane signaling protein, DAP12, which becomes tyrosine phosphorylated upon receptor engagement with an antibody (11, 40, 41). Recent analysis of the major NKp44 isoforms, however, has demonstrated that isolated human NK cells cultured in IL-2- or IL-15 express predominantly NKp44-1 mRNA and have reduced capacity to kill PCNA-transfected target cells in an NKp44-dependent manner (12, 107). Furthermore, NK-92 cells transduced to overexpress NKp44-1 show suppressed cytotoxicity and diminished immune synapse formation toward PCNA-transfected target cells, as compared to NK-92 transduced to overexpress the other isotypes. These results indicate that the NKp44-1 isoform is an inhibitory receptor when engaged with the PCNA ligand expressed by target cells, whereas NKp44-2 and NKp44-3 do not transduce inhibitory signals (107).

NKp44 Isoforms in Cancer

Shemesh et al. studied the impact of NKp44 isoforms on overall survival of patients with acute myeloid leukemia (AML) using TCGA RNA-Seq data (107). Whereas no survival advantage was found in newly diagnosed patients who expressed mRNA encoding NKp44, as compared to those lacking expression, survival was significantly diminished in patients who exclusively expressed NKp44-1, as compared to patients who also expressed at least some detectable level of NKp44-2 and/or NKp44-3 or lacked NKp44 expression altogether. These results imply that NKp44 plays a key role in NK cell responsiveness toward AML tumors, but exclusive expression of the ITIM-containing NKp44-1 isoform can stifle these beneficial responses, presumably through inhibitory signaling, thereby resulting in poor patient outcome.

The same group found higher incidence of NKp44 mRNA in various solid tumor tissues, as compared to surrounding normal tissues (95). Those tumor samples that expressed NKp44 mRNA were found to consist of predominantly NKp44-1 isoform. Thus, NK cells in the tumor microenvironment predominantly express the inhibitory NKp44 isoform.

NKp44 in the Decidua of Pregnant Women

Siewiera et al. found that while NKp44-2 mRNA is expressed more by freshly isolated pNK cells than NKp44-1 and NKp44-3, dNK cells from women undergoing elective first trimester abortions (healthy pregnancies) express all three isoforms of NKp44 at similar levels (105). While crosslinking NKp44 on IL-15-stimulated pNK cells induces degranulation, crosslinking on freshly isolated dNK cells did not result in degranulation and co-crosslinking suppressed degranulation in response to crosslinking NKp46, demonstrating predominant inhibitory function for NKp44 in decidual/placental NK cells (105).

Shemesh et al. also found that NKp44-2 and NKp44-3 isoforms predominated in decidual tissue obtained from the majority of first trimester spontaneous abortions, while a NKp44-1-dominant (inhibitory) profile was found in dNK cells from most elective abortions or term deliveries (healthy pregnancies), which is consistent with the inhibitory function of dNK cells in the study by Siewiera et al. (95).

Splice Variants of *ncr1*

Recent work by Shemer-Avni et al. has also provided new insights on differential splicing of *ncr1* (108). Five major splice variants of *ncr1* have been described, and three of these encode NKp46 protein isoforms containing both extracellular domains, while two lack the first Ig-like domain (D1, encoded by exon 3). Surprisingly, NK-92 cells transduced with a D1-negative *ncr1* cDNA degranulated significantly more efficiently toward HEK293T target cells, as compared to NK-92 cells transduced to express conventional NKp46 protein containing both Ig-like domains (108). Using NKp46-reactive antibodies, D1-negative NKp46 was not observed in fresh peripheral blood from healthy donors, but a subset expressing the NKp46 D1-negative receptor was found after long-time culturing of NK cells in IL-2 (108). This subset of NK cells containing D1-negative NKp46 degranulated more robustly in response to a combination of plate-bound anti-NKp46 and anti-NKp30 antibodies (108).

This group also studied NKp46 isoform expression in upper airway lavage samples from pediatric patients with respiratory tract viral infections. While most of these samples expressed both NKp30 and NKp46 mRNA, none contained mRNA encoding NKp44. When *ncr1* splice variants were analyzed, most of the lavage samples from virus-infected patients were found to contain D1-negative NKp46 isoform transcripts (108). Taken together, these results suggest that NK cells by IL-2 or viral infection can express isoforms of NKp46 lacking the D1 Ig-like domain, and NK cells expressing these domain-deficient receptors exhibit increased functional capacity.

CONCLUSION

Significant progress has been made in recent years to improve our understanding of the functions and ligand recognition capacities of NCRs. Clearly these receptors play important roles in NK cell recognition of tissue changes in cancer, viral infections, decidual tissues in pregnancy, and immature DC. This new knowledge is crucial for establishing the basis of molecular mechanisms controlling NK cell responses under these diverse conditions. In contrast to the original dogma that NCRs are exclusively activating receptors, several new findings have revealed inhibitory functions for these receptors.

Numerous ligands or co-ligands for NCRs have now been described. Several of these can surprisingly trigger inhibitory signals or induce production of inhibitory cytokines when engaged with NCRs. Furthermore, some of these putative ligands are classically nuclear or cytosolic proteins that appear to relocate to the cell surface in cancer cells, where they can engage with NCRs. Also, heparan sulfates seem to have capacity to interact with all three of the NCRs and may regulate their functions in trans and cis.

The study of differentially spliced isoforms of NCRs has revealed surprising insights, since some of these isoforms elicit inhibitory function. Furthermore, dominant expression of the inhibitory forms has been linked to poor outcome in the context of cancer, but healthy outcome in pregnancy. Nonetheless, our understanding of the complexities of NCR isoforms is still in its infancy and requires a great deal of additional study.

Future work is clearly needed to sort out true NCR ligands and functional mechanisms responsible for the functions of some NCR isoforms, and their complexities are growing. Importantly, a firmer foundation of understanding promises to provide potential opportunities as prognostic indicators of disease status or opportunities to develop therapeutic strategies to manipulate NCRs on NK cells that could be beneficial to treat a wide variety of human pathologies.

AUTHOR CONTRIBUTIONS

TP and KC authored and edited the manuscript. AS, MB, and AP provided critical input and editing.

REFERENCES

1. Campbell KS, Hasegawa J. Natural killer cell biology: an update and future directions. *J Allergy Clin Immunol* (2013) 132:536–44. doi:10.1016/j.jaci.2013.07.006
2. MacFarlane AW 4th, Campbell KS. Signal transduction in natural killer cells. *Curr Top Microbiol Immunol* (2006) 298:23–57.
3. Lanier LL. Up on the tightrope: natural killer cell activation and inhibition. *Nat Immunol* (2008) 9:495–502. doi:10.1038/ni1581
4. Vivier E, Tomasello E, Baratin M, Walzer T, Ugolini S. Functions of natural killer cells. *Nat Immunol* (2008) 9:503–10. doi:10.1038/ni1582
5. Moretta A, Bottino C, Vitale M, Pende D, Cantoni C, Mingari MC, et al. Activating receptors and coreceptors involved in human natural killer cell-mediated cytolysis. *Annu Rev Immunol* (2001) 19:197–223. doi:10.1146/annurev.immunol.19.1.197
6. Kruse PH, Matta J, Ugolini S, Vivier E. Natural cytotoxicity receptors and their ligands. *Immunol Cell Biol* (2014) 92:221–9. doi:10.1038/icb.2013.98
7. Hudspeth K, Silva-Santos B, Mavilio D. Natural cytotoxicity receptors: broader expression patterns and functions in innate and adaptive immune cells. *Front Immunol* (2013) 4:69. doi:10.3389/fimmu.2013.00069
8. Diefenbach A, Colonna M, Koyasu S. Development, differentiation, and diversity of innate lymphoid cells. *Immunity* (2014) 41:354–65. doi:10.1016/j.immuni.2014.09.005
9. Sivori S, Vitale M, Morelli L, Sanseverino L, Augugliaro R, Bottino C, et al. p46, a novel natural killer cell-specific surface molecule that mediates cell activation. *J Exp Med* (1997) 186:1129–36. doi:10.1084/jem.186.7.1129
10. Pessino A, Sivori S, Bottino C, Malaspina A, Morelli L, Moretta L, et al. Molecular cloning of NKp46: a novel member of the immunoglobulin superfamily involved in triggering of natural cytotoxicity. *J Exp Med* (1998) 188:953–60. doi:10.1084/jem.188.5.953
11. Vitale M, Bottino C, Sivori S, Sanseverino L, Castriconi R, Marcenaro E, et al. NKp44, a novel triggering surface molecule specifically expressed by activated natural killer cells, is involved in non-major histocompatibility complex-restricted tumor cell lysis. *J Exp Med* (1998) 187:2065–72. doi:10.1084/jem.187.12.2065
12. Cantoni C, Bottino C, Vitale M, Pessino A, Augugliaro R, Malaspina A, et al. NKp44, a triggering receptor involved in tumor cell lysis by activating human natural killer cells, is a novel member of the immunoglobulin superfamily. *J Exp Med* (1999) 189:787–96. doi:10.1084/jem.189.5.787

13. Pende D, Parolini S, Pessino A, Sivori S, Augugliaro R, Morelli L, et al. Identification and molecular characterization of NKp30, a novel triggering receptor involved in natural cytotoxicity mediated by human natural killer cells. *J Exp Med* (1999) 190:1505–16. doi:10.1084/jem.190.10.1505
14. Hollyoake M, Campbell RD, Aguado B. NKp30 (NCR3) is a pseudogene in 12 inbred and wild mouse strains, but an expressed gene in *Mus caroli*. *Mol Biol Evol* (2005) 22:1661–72. doi:10.1093/molbev/msi162
15. Backman-Petersson E, Miller JR, Hollyoake M, Aguado B, Butcher GW. Molecular characterization of the novel rat NK receptor 1C7. *Eur J Immunol* (2003) 33:342–51. doi:10.1002/immu.200310008
16. Hsieh CL, Ogura Y, Obara H, Ali UA, Rodriguez GM, Nepomuceno RR, et al. Identification, cloning, and characterization of a novel rat natural killer receptor, RNKP30: a molecule expressed in liver allografts. *Transplantation* (2004) 77:121–8. doi:10.1097/01.TP.0000110423.27977.6F
17. Walzer T, Blery M, Chaix J, Fuseri N, Chasson L, Robbins SH, et al. Identification, activation, and selective in vivo ablation of mouse NK cells via NKp46. *Proc Natl Acad Sci U S A* (2007) 104:3384–9. doi:10.1073/pnas.0609692104
18. Tomasello E, Yessaad N, Gregoire E, Hudspeth K, Luci C, Mavilio D, et al. Mapping of NKp46(+) cells in healthy human lymphoid and non-lymphoid tissues. *Front Immunol* (2012) 3:344. doi:10.3389/fimmu.2012.00344
19. Sivori S, Pende D, Bottino C, Marcenaro E, Pessino A, Biassoni R, et al. NKp46 is the major triggering receptor involved in the natural cytotoxicity of fresh or cultured human NK cells. Correlation between surface density of NKp46 and natural cytotoxicity against autologous, allogeneic or xenogeneic target cells. *Eur J Immunol* (1999) 1999:1656–66. doi:10.1002/(SICI)1521-4141(199905)29:05<1656::AID-IMMU1656>3.0.CO;2-1
20. Vankayalapati R, Wizel B, Weis SE, Safi H, Lakey DL, Mandelboim O, et al. The NKp46 receptor contributes to NK cell lysis of mononuclear phagocytes infected with an intracellular bacterium. *J Immunol* (2002) 168:3451–7. doi:10.4049/jimmunol.168.7.3451
21. Gur C, Porgador A, Elboim M, Gazit R, Mizrahi S, Stern-Ginossar N, et al. The activating receptor NKp46 is essential for the development of type 1 diabetes. *Nat Immunol* (2010) 11:121–8. doi:10.1038/ni.1834
22. Cagnano E, Hershkovitz O, Zilka A, Bar-Ilan A, Golder A, Sion-Vardy N, et al. Expression of ligands to NKp46 in benign and malignant melanocytes. *J Invest Dermatol* (2007) 128:972–9. doi:10.1038/sj/jid.5701111
23. Della Chiesa M, Carlomagno S, Frumento G, Balsamo M, Cantoni C, Conte R, et al. The tryptophan catabolite L-kynurenine inhibits the surface expression of NKp46- and NKG2D-activating receptors and regulates NK-cell function. *Blood* (2006) 108:4118–25. doi:10.1182/blood-2006-03-006700
24. Ferlazzo G, Tsang ML, Moretta L, Melioli G, Steinman RM, Munz C. Human dendritic cells activate resting natural killer (NK) cells and are recognized via the NKp30 receptor by activated NK cells. *J Exp Med* (2002) 195:343–51. doi:10.1084/jem.20011149
25. Vitale M, Della Chiesa M, Carlomagno S, Pende D, Arico M, Moretta L, et al. NK-dependent DC maturation is mediated by TNFalpha and IFNgamma released upon engagement of the NKp30 triggering receptor. *Blood* (2005) 106:566–71. doi:10.1182/blood-2004-10-4035
26. Vitale C, Chiossone L, Cantoni C, Morreale G, Cottalasso F, Moretti S, et al. The corticosteroid-induced inhibitory effect on NK cell function reflects down-regulation and/or dysfunction of triggering receptors involved in natural cytotoxicity. *Eur J Immunol* (2004) 34:3028–38. doi:10.1002/eji.200425418
27. Mavoungou E, Bouyou-Akotet MK, Kremsner PG. Effects of prolactin and cortisol on natural killer (NK) cell surface expression and function of human natural cytotoxicity receptors (NKp46, NKp44 and NKp30). *Clin Exp Immunol* (2005) 139:287–96. doi:10.1111/j.1365-2249.2004.02686.x
28. Bozzano F, Picciotto A, Costa P, Marras F, Fazio V, Hirsch I, et al. Activating NK cell receptor expression/function (NKp30, NKp46, DNAM-1) during chronic viraemic HCV infection is associated with the outcome of combined treatment. *Eur J Immunol* (2011) 41:2905–14. doi:10.1002/eji.201041361
29. Lee J, Zhang T, Hwang I, Kim A, Nitschke L, Kim M, et al. Epigenetic modification and antibody-dependent expansion of memory-like NK cells in human cytomegalovirus-infected individuals. *Immunity* (2015) 42:431–42. doi:10.1016/j.immuni.2015.02.013
30. Schlums H, Cichocki F, Tesi B, Theorell J, Beziat V, Holmes TD, et al. Cytomegalovirus infection drives adaptive epigenetic diversification of NK cells with altered signaling and effector function. *Immunity* (2015) 42:443–56. doi:10.1016/j.immuni.2015.02.008
31. Castriconi R, Cantoni C, Della Chiesa M, Vitale M, Marcenaro E, Conte R, et al. Transforming growth factor beta 1 inhibits expression of NKp30 and NKG2D receptors: consequences for the NK-mediated killing of dendritic cells. *Proc Natl Acad Sci U S A* (2003) 100:4120–5. doi:10.1073/pnas.0730640100
32. Mattiola I, Pesant M, Tentorio PF, Molgora M, Marcenaro E, Lugli E, et al. Priming of human resting NK cells by autologous M1 macrophages via the engagement of IL-1beta, IFN-beta, and IL-15 pathways. *J Immunol* (2015) 195:2818–28. doi:10.4049/jimmunol.1500325
33. Bonaccorsi I, Cantoni C, Carrega P, Oliveri D, Lui G, Conte R, et al. The immune inhibitory receptor LAIR-1 is highly expressed by plasmacytoid dendritic cells and acts complementary with NKp44 to control IFNalpha production. *PLoS One* (2010) 5:e15080. doi:10.1371/journal.pone.0015080
34. Balsamo M, Scordamaglia F, Pietra G, Manzini C, Cantoni C, Boitano M, et al. Melanoma-associated fibroblasts modulate NK cell phenotype and antitumor cytotoxicity. *Proc Natl Acad Sci U S A* (2009) 106:20847–52. doi:10.1073/pnas.0906481106
35. Memmer S, Weil S, Beyer S, Zoller T, Peters E, Hartmann J, et al. The stalk domain of NKp30 contributes to ligand binding and signaling of a preassembled NKp30-CD3zeta complex. *J Biol Chem* (2016) 291:25427–38. doi:10.1074/jbc.M116.742981
36. Cantoni C, Ponassi M, Biassoni R, Conte R, Spallarossa A, Moretta A, et al. The three-dimensional structure of the human NK cell receptor NKp44, a triggering partner in natural cytotoxicity. *Structure* (2003) 11:725–34. doi:10.1016/S0969-2126(03)00095-9
37. Joyce MG, Tran P, Zhuravleva MA, Jaw J, Colonna M, Sun PD. Crystal structure of human natural cytotoxicity receptor NKp30 and identification of its ligand binding site. *Proc Natl Acad Sci U S A* (2011) 108:6223–8. doi:10.1073/pnas.1100622108
38. Foster CE, Colonna M, Sun PD. Crystal structure of the human natural killer (NK) cell activating receptor NKp46 reveals structural relationship to other leukocyte receptor complex immunoreceptors. *J Biol Chem* (2003) 278:46081–6. doi:10.1074/jbc.M308491200
39. Jaron-Mendelson M, Yossef R, Appel MY, Zilka A, Hadad U, Afergan F, et al. Dimerization of NKp46 receptor is essential for NKp46-mediated lysis: characterization of the dimerization site by epitope mapping. *J Immunol* (2012) 188:6165–74. doi:10.4049/jimmunol.1102496
40. Campbell KS, Yusa S, Kikuchi-Maki A, Catina TL. NKp44 triggers NK cell activation through DAP12 association that is not influenced by a putative cytoplasmic inhibitory sequence. *J Immunol* (2004) 172:899–906. doi:10.4049/jimmunol.172.2.899
41. Augugliaro R, Parolini S, Castriconi R, Marcenaro E, Cantoni C, Nanni M, et al. Selective cross-talk among natural cytotoxicity receptors in human natural killer cells. *Eur J Immunol* (2003) 33:1235–41. doi:10.1002/eji.200323896
42. Rosental B, Brusilovsky M, Hadad U, Oz D, Appel MY, Afergan F, et al. Proliferating cell nuclear antigen is a novel inhibitory ligand for the natural cytotoxicity receptor NKp44. *J Immunol* (2011) 187:5693–702. doi:10.4049/jimmunol.1102267
43. Jarahian M, Fiedler M, Cohnen A, Djandji D, Hammerling GJ, Gati C, et al. Modulation of NKp30- and NKp46-mediated natural killer cell responses by poxviral hemagglutinin. *PLoS Pathog* (2011) 7:e1002195. doi:10.1371/journal.ppat.1002195
44. Arnon TI, Achdout H, Levi O, Markel G, Saleh N, Katz G, et al. Inhibition of the NKp30 activating receptor by pp65 of human cytomegalovirus. *Nat Immunol* (2005) 6:515–23. doi:10.1038/ni1190
45. Mavoungou E, Held J, Mewono L, Kremsner PG. A Duffy binding-like domain is involved in the NKp30-mediated recognition of *Plasmodium falciparum*-parasitized erythrocytes by natural killer cells. *J Infect Dis* (2007) 195:1521–31. doi:10.1086/515579
46. Bloushtain N, Qimron U, Bar-Ilan A, Hershkovitz O, Gazit R, Fima E, et al. Membrane-associated heparan sulfate proteoglycans are involved in the recognition of cellular targets by NKp30 and NKp46. *J Immunol* (2004) 173:2392–401. doi:10.4049/jimmunol.173.4.2392
47. Hecht ML, Rosental B, Horlacher T, Hershkovitz O, De Paz JL, Noti C, et al. Natural cytotoxicity receptors NKp30, NKp44 and NKp46 bind to

different heparan sulfate/heparin sequences. *J Proteome Res* (2009) 8:712–20. doi:10.1021/pr800747c

48. Hershkovitz O, Jarahian M, Zilka A, Bar-Ilan A, Landau G, Jivov S, et al. Altered glycosylation of recombinant NKp30 hampers binding to heparan sulfate: a lesson for the use of recombinant immuno-receptors as an immunological tool. *Glycobiology* (2008) 18:28–41. doi:10.1093/glycob/cwm125
49. Pogge von Strandmann E, Simhadri VR, Von Tresckow B, Sasse S, Reiners KS, Hansen HP, et al. Human leukocyte antigen-B-associated transcript 3 is released from tumor cells and engages the NKp30 receptor on natural killer cells. *Immunity* (2007) 27:965–74. doi:10.1016/j.immuni.2007.10.010
50. Simhadri VR, Reiners KS, Hansen HP, Topolar D, Simhadri VL, Nohroudi K, et al. Dendritic cells release HLA-B-associated transcript-3 positive exosomes to regulate natural killer function. *PLoS One* (2008) 3:e3377. doi:10.1371/journal.pone.0003377
51. Dassler-Plenker J, Reiners KS, Van Den Boorn JG, Hansen HP, Putschli B, Barnert S, et al. RIG-I activation induces the release of extracellular vesicles with antitumor activity. *Oncoimmunology* (2016) 5:e1219827. doi:10.1080/2162402X.2016.1219827
52. Brandt CS, Baratin M, Yi EC, Kennedy J, Gao Z, Fox B, et al. The B7 family member B7-H6 is a tumor cell ligand for the activating natural killer cell receptor NKp30 in humans. *J Exp Med* (2009) 206:1495–503. doi:10.1084/jem.20090681
53. Xu X, Narni-Mancinelli E, Cantoni C, Li Y, Guia S, Gauthier L, et al. Structural insights into the inhibitory mechanism of an antibody against B7-H6, a stress-induced cellular ligand for the natural killer cell receptor NKp30. *J Mol Biol* (2016) 428:4457–66. doi:10.1016/j.jmb.2016.09.011
54. Arnon TI, Lev M, Katz G, Chernobrov Y, Porgador A, Mandelboim O. Recognition of viral hemagglutinins by NKp44 but not by NKp30. *Eur J Immunol* (2001) 31:2680–9. doi:10.1002/1521-4141(200109)31:9<2680::AID-IMMU2680>3.0.CO;2-A
55. Ho JW, Hershkovitz O, Peiris M, Zilka A, Bar-Ilan A, Nal B, et al. H5-type influenza virus hemagglutinin is functionally recognized by the natural killer-activating receptor NKp44. *J Virol* (2008) 82:2028–32. doi:10.1128/JVI.02065-07
56. Jarahian M, Watzl C, Fournier P, Arnold A, Djandji D, Zahedi S, et al. Activation of natural killer cells by Newcastle disease virus hemagglutinin-neuraminidase. *J Virol* (2009) 83:8108–21. doi:10.1128/JVI.00211-09
57. Hershkovitz O, Rosental B, Rosenberg LA, Navarro-Sanchez ME, Jivov S, Zilka A, et al. NKp44 receptor mediates interaction of the envelope glycoproteins from the West Nile and dengue viruses with NK cells. *J Immunol* (2009) 183:2610–21. doi:10.4049/jimmunol.0802806
58. Esin S, Batoni G, Counoupas C, Stringaro A, Brancatisano FL, Colone M, et al. Direct binding of human NK cell natural cytotoxicity receptor NKp44 to the surfaces of mycobacteria and other bacteria. *Infect Immun* (2008) 76:1719–27. doi:10.1128/IAI.00870-07
59. Bialoszewska A, Baychelier F, Niderla-Bielinska J, Czop A, Debre P, Vieillard V, et al. Constitutive expression of ligand for natural killer cell NKp44 receptor (NKp44L) by normal human articular chondrocytes. *Cell Immunol* (2013) 285:6–9. doi:10.1016/j.cellimm.2013.08.005
60. Hershkovitz O, Jivov S, Bloushtain N, Zilka A, Landau G, Bar-Ilan A, et al. Characterization of the recognition of tumor cells by the natural cytotoxicity receptor, NKp44. *Biochemistry* (2007) 46:7426–36. doi:10.1021/bi7000455
61. Brusilovsky M, Radinsky O, Cohen L, Yossef R, Shemesh A, Braiman A, et al. Regulation of natural cytotoxicity receptors by heparan sulfate proteoglycans in -cis: a lesson from NKp44. *Eur J Immunol* (2015) 45:1180–91. doi:10.1002/eji.201445177
62. Vieillard V, Strominger JL, Debre P. NK cytotoxicity against CD4+ T cells during HIV-1 infection: a gp41 peptide induces the expression of an NKp44 ligand. *Proc Natl Acad Sci U S A* (2005) 102:10981–6. doi:10.1073/pnas.0504315102
63. Baychelier F, Sennepin A, Ermonval M, Dorgham K, Debre P, Vieillard V. Identification of a cellular ligand for the natural cytotoxicity receptor NKp44. *Blood* (2013) 122:2935–42. doi:10.1182/blood-2013-03-489054
64. Horton NC, Mathew SO, Mathew PA. Novel interaction between proliferating cell nuclear antigen and HLA I on the surface of tumor cells inhibits NK cell function through NKp44. *PLoS One* (2013) 8:e59552. doi:10.1371/journal.pone.0059552
65. Mandelboim O, Lieberman N, Lev M, Paul L, Arnon TI, Bushkin Y, et al. Recognition of haemagglutinins on virus-infected cells by NKp46 activates lysis by human NK cells. *Nature* (2001) 409:1055–60. doi:10.1038/35059110
66. Arnon TI, Achdout H, Lieberman N, Gazit R, Gonen-Gross T, Katz G, et al. The mechanisms controlling the recognition of tumor- and virus-infected cells by NKp46. *Blood* (2004) 103:664–72. doi:10.1182/blood-2003-05-1716
67. Achdout H, Meningher T, Hirsh S, Glasner A, Bar-On Y, Gur C, et al. Killing of avian and swine influenza virus by natural killer cells. *J Virol* (2010) 84:3993–4001. doi:10.1128/JVI.02289-09
68. Mendelson M, Tekoah Y, Zilka A, Gershoni-Yahalom O, Gazit R, Achdout H, et al. NKp46 O-glycan sequences that are involved in the interaction with hemagglutinin type 1 of influenza virus. *J Virol* (2010) 84:3789–97. doi:10.1128/JVI.01815-09
69. Draghi M, Pashine A, Sanjanwala B, Gendzekhadze K, Cantoni C, Cosman D, et al. NKp46 and NKG2D recognition of infected dendritic cells is necessary for NK cell activation in the human response to influenza infection. *J Immunol* (2007) 178:2688–98. doi:10.4049/jimmunol.178.5.2688
70. Garg A, Barnes PF, Porgador A, Roy S, Wu S, Nanda JS, et al. Vimentin expressed on *Mycobacterium tuberculosis*-infected human monocytes is involved in binding to the NKp46 receptor. *J Immunol* (2006) 177:6192–8. doi:10.4049/jimmunol.177.9.6192
71. Mao H, Tu W, Liu Y, Qin G, Zheng J, Chan PL, et al. Inhibition of human natural killer cell activity by influenza virions and hemagglutinin. *J Virol* (2010) 84:4148–57. doi:10.1128/JVI.02340-09
72. Gazit R, Gruda R, Elboim M, Arnon TI, Katz G, Achdout H, et al. Lethal influenza infection in the absence of the natural killer cell receptor gene Ncr1. *Nat Immunol* (2006) 7:517–23. doi:10.1038/ni1322
73. Hershkovitz O, Zilka A, Bar-Ilan A, Abutbul S, Davidson A, Mazzon M, et al. Dengue virus replicon expressing the nonstructural proteins suffices to enhance membrane expression of HLA class I and inhibit lysis by human NK cells. *J Virol* (2008) 82:7666–76. doi:10.1128/JVI.02274-07
74. Nakato H, Kimata K. Heparan sulfate fine structure and specificity of proteoglycan functions. *Biochim Biophys Acta* (2002) 1573:312–8. doi:10.1016/S0304-4165(02)00398-7
75. Sasisekharan R, Venkataraman G. Heparin and heparan sulfate: biosynthesis, structure and function. *Curr Opin Chem Biol* (2000) 4:626–31. doi:10.1016/S1367-5931(00)00145-9
76. Baeg GH, Lin X, Khare N, Baumgartner S, Perrimon N. Heparan sulfate proteoglycans are critical for the organization of the extracellular distribution of Wingless. *Development* (2001) 128:87–94.
77. Loo BM, Salmivirta M. Heparin/heparan sulfate domains in binding and signaling of fibroblast growth factor 8b. *J Biol Chem* (2002) 277:32616–23. doi:10.1074/jbc.M204961200
78. Laguri C, Arenzana-Seisdedos F, Lortat-Jacob H. Relationships between glycosaminoglycan and receptor binding sites in chemokines – the CXCL12 example. *Carbohydr Res* (2008) 343:2018–23. doi:10.1016/j.carres.2008.01.047
79. Lortat-Jacob H. The molecular basis and functional implications of chemokine interactions with heparan sulphate. *Curr Opin Struct Biol* (2009) 19:543–8. doi:10.1016/j.sbi.2009.09.003
80. Brusilovsky M, Cordoba M, Rosental B, Hershkovitz O, Andrake MD, Pecherskaya A, et al. Genome-wide siRNA screen reveals a new cellular partner of NK cell receptor KIR2DL4: heparan sulfate directly modulates KIR2DL4-mediated responses. *J Immunol* (2013) 191:5256–67. doi:10.4049/jimmunol.1302079
81. Brusilovsky M, Radinsky O, Yossef R, Campbell KS, Porgador A. Carbohydrate-mediated modulation of NK cell receptor function: structural and functional influences of heparan sulfate moieties expressed on NK cell surface. *Front Oncol* (2014) 4:185. doi:10.3389/fonc.2014.00185
82. Reiners KS, Topolar D, Henke A, Simhadri VR, Kessler J, Sauer M, et al. Soluble ligands for NK cell receptors promote evasion of chronic lymphocytic leukemia cells from NK cell anti-tumor activity. *Blood* (2013) 121:3658–65. doi:10.1182/blood-2013-01-476606
83. Binici J, Hartmann J, Herrmann J, Schreiber C, Beyer S, Guler G, et al. A soluble fragment of the tumor antigen BCL2-associated athanogene 6 (BAG-6) is essential and sufficient for inhibition of NKp30 receptor-dependent cytotoxicity of natural killer cells. *J Biol Chem* (2013) 288:34295–303. doi:10.1074/jbc.M113.483602
84. Stoimenov I, Helleday T. PCNA on the crossroad of cancer. *Biochem Soc Trans* (2009) 37:605–13. doi:10.1042/BST0370605

85. Fuchs A, Cella M, Kondo T, Colonna M. Paradoxic inhibition of human natural interferon-producing cells by the activating receptor NKp44. *Blood* (2005) 106:2076–82. doi:10.1182/blood-2004-12-4802
86. Textor S, Bossler F, Henrich KO, Gartlgruber M, Pollmann J, Fiegler N, et al. The proto-oncogene Myc drives expression of the NK cell-activating NKp30 ligand B7-H6 in tumor cells. *Oncoimmunology* (2016) 5:e1116674. doi:10.1080/2162402X.2015.1116674
87. Matta J, Baratin M, Chiche L, Forel JM, Cognet C, Thomas G, et al. Induction of B7-H6, a ligand for the natural killer cell-activating receptor NKp30, in inflammatory conditions. *Blood* (2013) 122:394–404. doi:10.1182/blood-2013-01-481705
88. Schlecker E, Fiegler N, Arnold A, Altevogt P, Rose-John S, Moldenhauer G, et al. Metalloprotease-mediated tumor cell shedding of B7-H6, the ligand of the natural killer cell-activating receptor NKp30. *Cancer Res* (2014) 74:3429–40. doi:10.1158/0008-5472.CAN-13-3017
89. Pesce S, Tabellini G, Cantoni C, Patrizi O, Coltrini D, Rampinelli F, et al. B7-H6-mediated downregulation of NKp30 in NK cells contributes to ovarian carcinoma immune escape. *Oncoimmunology* (2015) 4:e1001224. doi:10.1080/2162402X.2014.1001224
90. Fiegler N, Textor S, Arnold A, Rolle A, Oehme I, Breuhahn K, et al. Downregulation of the activating NKp30 ligand B7-H6 by HDAC inhibitors impairs tumor cell recognition by NK cells. *Blood* (2013) 122:684–93. doi:10.1182/blood-2013-02-482513
91. Marras F, Bozzano F, Ascierto ML, De Maria A. Baseline and dynamic expression of activating NK cell receptors in the control of chronic viral infections: the paradigm of HIV-1 and HCV. *Front Immunol* (2014) 5:305. doi:10.3389/fimmu.2014.00305
92. Nalabolu SR, Shukla H, Nallur G, Parimoo S, Weissman SM. Genes in a 220-kb region spanning the TNF cluster in human MHC. *Genomics* (1996) 31:215–22. doi:10.1006/geno.1996.0034
93. Neville MJ, Campbell RD. A new member of the Ig superfamily and a V-ATPase G subunit are among the predicted products of novel genes close to the TNF locus in the human MHC. *J Immunol* (1999) 162:4745–54.
94. Delahaye NF, Rusakiewicz S, Martins I, Menard C, Roux S, Lyonnet L, et al. Alternatively spliced NKp30 isoforms affect the prognosis of gastrointestinal stromal tumors. *Nat Med* (2011) 17:700–7. doi:10.1038/nm.2366
95. Shemesh A, Kugel A, Steiner N, Yezersky M, Tirosh D, Edri A, et al. NKp44 and NKp30 splice variant profiles in decidua and tumor tissues: a comparative viewpoint. *Oncotarget* (2016) 7:70912–23. doi:10.18632/oncotarget.12292
96. Bottino C, Falco M, Sivori S, Moretta L, Moretta A, Biassoni R. Identification and molecular characterization of a natural mutant of the p50.2/KIR2DS2 activating NK receptor that fails to mediate NK cell triggering. *Eur J Immunol* (2000) 30:3569–74. doi:10.1002/1521-4141(200012)30:12<3569::AID-IMMU3569>3.0.CO;2-E
97. Sivori S, Parolini S, Marcenaro E, Castriconi R, Pende D, Millo R, et al. Involvement of natural cytotoxicity receptors in human natural killer cell-mediated lysis of neuroblastoma and glioblastoma cell lines. *J Neuroimmunol* (2000) 107:220–5. doi:10.1016/S0165-5728(00)00221-6
98. Bottino C, Dondero A, Bellora F, Moretta L, Locatelli F, Pistoia V, et al. Natural killer cells and neuroblastoma: tumor recognition, escape mechanisms, and possible novel immunotherapeutic approaches. *Front Immunol* (2014) 5:56. doi:10.3389/fimmu.2014.00056
99. Semeraro M, Rusakiewicz S, Minard-Colin V, Delahaye NF, Enot D, Vely F, et al. Clinical impact of the NKp30/B7-H6 axis in high-risk neuroblastoma patients. *Sci Transl Med* (2015) 7:283ra255. doi:10.1126/scitranslmed.aaa2327
100. Messaoudene M, Fregni G, Enot D, Jacquelot N, Neves E, Germaud N, et al. NKp30 isoforms and NKp46 transcripts in metastatic melanoma patients: unique NKp30 pattern in rare melanoma patients with favorable evolution. *Oncoimmunology* (2016) 5:e1154251. doi:10.1080/2162402X.2016.1154251
101. Prada N, Antoni G, Commo F, Rusakiewicz S, Semeraro M, Boufassa F, et al. Analysis of NKp30/NCR3 isoforms in untreated HIV-1-infected patients from the ANRS SEROCO cohort. *Oncoimmunology* (2013) 2:e23472. doi:10.4161/onci.23472
102. Mantovani S, Mele D, Oliviero B, Barbarini G, Varchetta S, Mondelli MU. NKp30 isoforms in patients with chronic hepatitis C virus infection. *Immunology* (2015) 146:234–42. doi:10.1111/imm.12495
103. Jabrane-Ferrat N, Siewiera J. The up side of decidual natural killer cells: new developments in immunology of pregnancy. *Immunology* (2014) 141:490–7. doi:10.1111/imm.12218
104. Ashkar AA, Croy BA. Functions of uterine natural killer cells are mediated by interferon gamma production during murine pregnancy. *Semin Immunol* (2001) 13:235–41. doi:10.1006/smim.2000.0319
105. Siewiera J, Gouilly J, Hocine HR, Cartron G, Levy C, Al-Daccak R, et al. Natural cytotoxicity receptor splice variants orchestrate the distinct functions of human natural killer cell subtypes. *Nat Commun* (2015) 6:10183. doi:10.1038/ncomms10183
106. Shemesh A, Tirosh D, Sheiner E, Tirosh NB, Brusilovsky M, Segev R, et al. First trimester pregnancy loss and the expression of alternatively spliced NKp30 isoforms in maternal blood and placental tissue. *Front Immunol* (2015) 6:189. doi:10.3389/fimmu.2015.00189
107. Shemesh A, Brusilovsky M, Hadad U, Teltsh O, Edri A, Rubin E, et al. Survival in acute myeloid leukemia is associated with NKp44 splice variants. *Oncotarget* (2016) 7:32933–45. doi:10.18632/oncotarget.8782
108. Shemer-Avni Y, Kundu K, Shemesh A, Brusilovsky M, Yossef R, Meshesha M, et al. Expression of NKp46 splice variants in nasal lavage following respiratory viral infection: domain 1-negative isoforms predominate and manifest higher activity. *Front Immunol* (2017) 8:161. doi:10.3389/fimmu.2017.00161

12

Disarming Cellular Alarm Systems—Manipulation of Stress-Induced NKG2D Ligands by Human Herpesviruses

*Dominik Schmiedel and Ofer Mandelboim**

Faculty of Medicine, The Lautenberg Center for General and Tumor Immunology, The BioMedical Research Institute Israel-Canada, The Hebrew University Hadassah Medical School, Jerusalem, Israel

Edited by:
Ulrike Koehl,
Hannover Medical School, Germany

Reviewed by:
Karl-Johan Malmberg,
Oslo University Hospital, Norway
Adelheid Cerwenka,
German Cancer Research Center, Germany

****Correspondence:***
Ofer Mandelboim
oferm@ekmd.huji.ac.il

The coevolution of viruses and their hosts led to the repeated emergence of cellular alert signals and viral strategies to counteract them. The herpesvirus family of viruses displays the most sophisticated repertoire of immune escape mechanisms enabling infected cells to evade immune recognition and thereby maintain infection. The herpesvirus family consists of nine viruses that are capable of infecting humans: herpes simplex virus 1 and 2 (HSV-1, HSV-2), varicella zoster virus (VZV), Epstein–Barr virus (EBV), human cytomegalovirus (HCMV), roseoloviruses (HHV-6A, HHV-6B, and HHV-7), and Kaposi's-sarcoma-associated herpesvirus (KSHV). Most of these viruses are highly prevalent and infect a vast majority of the human population worldwide. Notably, research over the past 15 years has revealed that cellular ligands for the activating receptor natural-killer group 2, member D (NKG2D)—which is primarily expressed on natural killer (NK) cells—are common targets suppressed during viral infection, i.e., their surface expression is reduced in virtually all lytic herpesvirus infections by diverse mechanisms. Here, we review the viral mechanisms by which all herpesviruses known to date to downmodulate the expression of the NKG2D ligands. Also, in light of recent findings, we speculate about the importance of the emergence of eight different NKG2D ligands in humans and further allelic diversification during host and virus coevolution.

Keywords: NKG2D ligands, stress-induced ligands, NKG2D, herpesvirus, host-pathogen interaction, immune evasion, coevolution

HERPESVIRUSES—CONSTANT COMPANIONS DURING HUMAN LIFE AND EVOLUTION

Herpesviruses have accompanied humankind since the dawn of evolution. Herpesvirus infections date back at least 6 million years, even before evolutionary split between hominids and chimpanzees (1). From that time on, viral strategies to ensure survival and dissemination coevolved together with the immune system that continuously developed new measures to clear viral infections.

To date, nine different herpesviruses capable of infecting humans have been identified: HSV-1, HSV-2, varicella zoster virus (VZV), Epstein–Barr virus (EBV), human cytomegalovirus (HCMV), HHV-6A, HHV-6B, HHV-7, and Kaposi's-sarcoma-associated herpesvirus (KSHV) (2).

Despite their different life cycle and growth properties, cellular tropisms and although they cause different diseases, all herpesviruses share common features. They are enveloped and contain a linear, double-stranded DNA genome, ranging from 125 kb (for VZV) to 235 kb (for HCMV) (2).

Furthermore, all have the ability not only to infect lytically but also to establish life-long infection in their host, a status called latency, which is a dormant infection lacking pathology and viral replication (2, 3).

Most herpesviruses are widely spread in human populations. Serological tests reveal that HSV-1, VZV, EBV, HHV-6, and HHV-7 have the highest prevalence of the herpesvirus family and infect about 90% of the adult population (4–10). Notably, the prevalence of herpesviruses varies geographically and is influenced by socioeconomic status (2, 11, 12). HCMV prevalence can therefore vary between 50 and 100% dependent on the population studied (13). Some herpesviruses reactivate symptomatically and frequently in healthy individuals for as yet unknown reasons, while others only cause symptomatic reactivation in immunodeficient patients (3). However, research over the past few years revealed that all herpesviruses use common strategies during primary infection, reactivation, and sometimes even during latency, in order to evade the immune surveillance during the different phases of herpesvirus life cycle. The interactions between adaptive immunity and herpesviruses are described elsewhere (14–17). In this review, we will focus on the interaction of herpesviruses and natural-killer group 2, member D (NKG2D)-expressing immune cells. The human-activating receptor NKG2D is expressed on all natural killer (NK) cells as well as on most T cells including γδ T cells and NK T cells (18). Its importance was shown for tumor surveillance (19) and inflammatory diseases (20). The significance of NK cells in herpesvirus immune surveillance becomes clear by looking at NK cell-deficient individuals who suffer from recurrent, severe, potentially life-threatening herpesvirus infections (21, 22).

GENETICS OF NKG2D LIGANDS

In the course of human evolution, eight different, functional ligands for the NKG2D receptor emerged: MHC class I polypeptide-related sequence A and B (MICA and MICB, respectively) and the unique long 16 binding protein 1–6 (ULBP1–6) (23). Also known as "stress-induced ligands," they are barely found on healthy cells in order to avoid auto-reactivity toward normal tissues. These ligands, however, are upregulated and expressed on the cell surface following various stresses including genotoxic stress, oncogene activation or hypoxia that are commonly seen in tumorigenesis, or following viral infection (24, 25).

All NKG2D ligands belong to the MHC class I-like protein family. ULBP family members have an α1/α2 domain structure, whereas the MIC proteins possess an α1/α2/α3 domain structure (26). Interestingly, classical MHC class I proteins serve mainly as inhibitory ligands for NK cells, whereas the NKG2D ligands activate NK cells (27, 28).

Up until now, 16 different allelic variants were identified for the 6 members of the ULBP family (29). More than 100 different MICA alleles and more than 40 MICB alleles were identified to date; a finding that demonstrates the striking superior evolutionary plasticity of the MIC family [http://hla.alleles.org/alleles/classo.html; (30)] (**Figure 1**). The reason behind this enormous diversity of the MIC family is still unknown. MIC genes lack hypervariable regions; point mutations and genetic shuffles occur over all three domains (31). Comparing amino acid sequence homology, MICA and MICB are very similar (about 85% identity), whereas the similarity to ULBP family proteins is comparatively low (only about 20–25% identity between MIC and ULBP proteins). ULBP family members shares about 60% amino acid sequence identity with each other (32–34). Interestingly, MICA, MICB, ULBP4, and ULBP5 contain a transmembrane domain and a cytoplasmic tail, whereas ULBP1, ULBP2, ULBP3, ULBP6, and one particular allelic variant of MICA (allele *008) are glycosylphosphatidylinositol (GPI) anchored (30).

Only recently, post-transcriptional cellular mechanisms that control stress-induced ligand expression by RNA-binding proteins (35–37) and microRNAs (miRNAs) (38, 39) began to be unraveled; however, the regulatory circuits and expression patterns in normal cells remain incompletely understood. By contrast, much information was gathered about the suppression of NKG2D ligands during herpesvirus infection, emphasizing the importance of the receptor NKG2D for anti-viral immunity.

ALL HERPESVIRUSES SUPPRESS EXPRESSION OF STRESS-INDUCED LIGANDS DURING INFECTION

HSV-1 and -2—HHV-1 and -2

Herpes simplex virus 1 and 2 can cause orofacial and genital infections in elsewise healthy individuals with a competent immune system (40). Reactivation is believed to be triggered by stress, sunlight, fever, or skin traumas, e.g., caused by surgery (40, 41).

The effects of HSV-1 infection on the expression of NKG2D ligands were first studied by Schepis et al. [(42); **Figure 1**; **Table 1**]. Both ULBP2 and MICA surface expression levels were found to be decreased following infection with HSV-1 strain F. Concurrent with a loss of surface expression, MICA messenger RNA (mRNA) levels were decreased. Since MICA downregulation was abrogated by inhibiting the viral DNA polymerase, the authors concluded that a late viral gene is responsible for the reduction of MICA expression. However, in this study, none of the cell lines tested expressed other ligands besides MICA and ULBP2. Another study, performed by Campbell et al., confirmed the decrease in MICA and ULBP2, but could additionally show a downregulation of ULBP1 and ULBP3 (43). Interestingly, MICA, ULBP2, and ULBP3 were shown to be reduced at the overall protein level, whereas ULBP1 was retained intracellularly, proving that different mechanisms act on these ligands (43).

In a subsequent study, Enk et al. added some mechanistic detail about the regulation of ULBP2 and ULBP3 (44). They reported that the viral miRNA miR-H8 interferes with the generation of GPI-anchored proteins by targeting PIGT, a key protein in the GPI-anchoring process (45, 46). Consequently, both ULBP2 and ULBP3 levels were reduced in miR-H8 overexpressing cells. Interestingly, ULBP1 is also GPI anchored but not affected by this pathway, explaining the necessity for another mechanism of down-regulation—intracellular retention. However, since both MICA (except the allele MICA*008) and MICB are transmembrane proteins and were not affected by the miR-H8 overexpression,

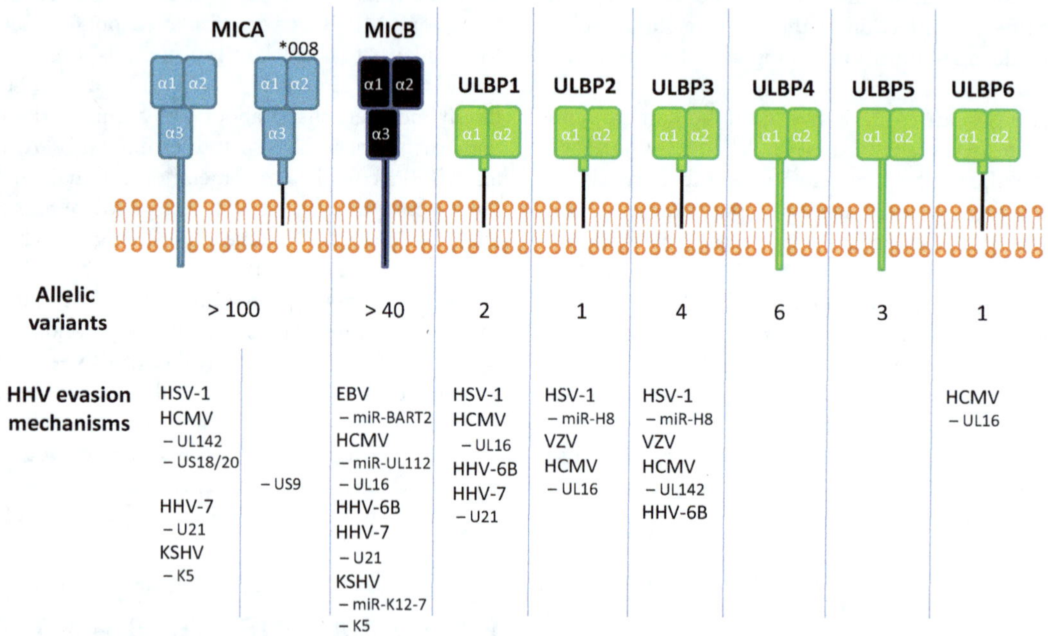

FIGURE 1 | The human genome encodes for eight functional stress-induced ligands, subdivided in the MIC and ULBP family. MICA, MICB, ULBP4, and ULBP5 contain a transmembrane domain, whereas ULBP1, 2, and 3 and one particular allele of MICA, MICA*008, are GPI-anchored. Interestingly, MICA and MICB, genes having a high evolutionary plasticity as reflected by the number of allelic variants, seem to be targeted more frequently by viral immune evasion mechanisms.

the viral mechanism responsible for these downregulations are still unknown.

VZV—HHV-3

Varicella zoster virus is the causative agent of varicella (chickenpox) in primary infection (47). In the elderly or immunosuppressed patients, VZV can reactivate and cause herpes zoster (shingles), which is characterized by painful skin lesions as well as neurological and ocular disorders (47, 48).

By infecting retinal epithelial cells with a clinical VZV strain, Campbell et al. revealed a down-modulation of the surface expression of the NKG2D ligands ULBP2 and ULBP3 [(43); **Figure 1**; **Table 1**]. By contrast, MICA surface expression increased during the course of VZV infection; ULBP1 and MICB were not expressed in the studied cells. Due to the overall reduction of surface expression of NKG2D ligands, reduced activation of NK cells in the presence of VZV-infected cells as compared to mock infected cells was observed. Interestingly, the total protein levels of ULBP2 and ULBP3 were not reduced in infected cells (43), indicating intracellular retention of these ligands by a yet unknown viral factor.

EBV—HHV-4

Epstein–Barr virus is usually acquired asymptomatically in childhood (2, 49). Infection during adolescence can lead to infectious mononucleosis (in about 50% of primary infections), a weakening and sometimes painful but self-limiting disease associated with the occurrence of atypical lymphocytes in the blood stream (2, 49). Reactivation can occur in immunocompromised individuals and is, among others, linked not only to lympho-proliferative diseases such as Burkitt's and Hodgkin's lymphoma but also to nasopharyngeal carcinoma (50, 51).

A sensitization of EBV-infected cells switching from latent to lytic infection to NK cell killing was reported by Pappworth et al. (52). They showed the induction of ULBP1 following this switch in a Burkitt's lymphoma-derived cell line, whereas all other NKG2D ligands were absent from the cell surface. Later on, an overexpression study performed by Nachmani et al. revealed that the latency-associated viral miRNA miR-BART2-5p is capable of binding MICB mRNA and suppressing its translation [(53); **Figure 1**; **Table 1**]. Interestingly, they showed that the binding site in the MICA mRNA sequence was mutated in such a way that prevented the miRNA from suppressing MICA as well.

Remarkably, to the best of our knowledge, there are no immune evasion mechanisms regarding NKG2D ligands during lytic EBV infection described to date. This phenomenon might be explained by a study published by Song et al. (54). They showed that EBV-transformed B cells produce and release the tryptophan-derived metabolite L-kynurenine that downmodulates NKG2D receptor expression on by-stander NK cells. Therefore, the suppression of NKG2D ligands on infected cells might be of little importance if the effector cells themselves are effectively disarmed.

HCMV—HHV-5

While being a harmless pathogen for immunocompetent individuals, HCMV constitutes a major risk for the elderly, patients after organ transplantation and AIDS patients (55). Additionally, primary infection in pregnant women can cause miscarriage, stillbirth, or developmental retardation of the child (55). HCMV possesses the largest genome of all HHVs of about 235 kb (2).

TABLE 1 | Overview of known suppression mechanisms for NKG2D ligands by HHVs.

	Viral effector	Ligand	Mechanism	Reference
HSV-1	?	MICA	?	(42, 43)
	?	ULBP1	Intracellular retention	(43)
	miR-H8	ULBP2	Interferes with protein maturation	(42–44)
	miR-H8	ULBP3	Interferes with protein maturation	(43, 44)
HSV-2	?	?	?	?
Varicella zoster virus	?	ULBP2	Intracellular retention	(43)
	?	ULBP3	Intracellular retention	
Epstein–Barr virus	miR-BART2-5p	MICB	Translational repression	(53)
Human cytomegalovirus	miR-UL112	MICB	Translation repression	(56)
	UL16	MICB	Intracellular retention	(60)
		ULBP1	Intracellular retention	(32, 57)
		ULBP2	Intracellular retention	(32, 57)
		ULBP6	Intracellular retention	(61)
	UL142	MICA	Intracellular retention	(62)
		ULBP3	Intracellular retention	(63)
	US18/US20	MICA	Lysosomal degradation	(64)
	US9	MICA*008	Proteasomal degradation	(68)
HHV-6A	?	?	?	?
HHV-6B	?	MICB	Proteasomal degradation	(71)
	?	ULBP1	Proteasomal degradation	
	?	ULBP3	Proteasomal degradation	
HHV-7	U21	MICA	?	(72)
		MICB	?	
		ULBP1	Lysosomal degradation	
Kaposi's-sarcoma-associated herpesvirus	K5	MICA	Ubiquitinylation/intracellular retention	(75)
		MICB	Ubiquitinylation/intracellular retention	(75)
	miR-K12-7	MICB	Translational repression	(53)

?, no published data available.

Therefore, it might not be surprising that HCMV has the greatest number of viral mechanisms dedicated to the immune evasion by manipulating NKG2D ligands described to date.

The first viral miRNA identified to target immune molecules in general and NKG2D ligands in particular was miR-UL112, discovered by Stern-Ginossar et al. [(56); **Figure 1**; **Table 1**]. By binding to the 3′-UTR of the MICB mRNA, it represses translation, and surface levels are rapidly reduced, leading to decreased NK cell activation. UL16 was the first HCMV viral protein found to bind and retain ULBP1, ULBP2, ULBP6, and MICB intracellularly ("ULBPs" were named for being UL16-binding proteins) (32, 57–61). Later, UL142 was shown to sequester both MICA and ULBP3 intracellularly, they colocalized with markers of the *cis*-Golgi apparatus inside infected cells (62, 63).

Fielding et al. showed that the viral proteins US18 and US20 are capable of both independently and synergistically downregulating MICA expression by targeting it for lysosomal degradation (64).

Notably, the GPI-anchored allele MICA*008 was not found to be targeted by the abovementioned viral mechanisms and was therefore considered as HCMV-resistant escape variant. Since the MICA*008 allele is a highly prevalent in human populations worldwide, the hypothesis was formed that its prevalence is the result of viral selective pressure (65–67). However, Seidel et al. showed that this supposed escape variant is specifically targeted by the HCMV protein US9 during its maturation process, prior to its egress from the ER, instead forcing MICA*008 to proteasomal degradation (68).

Roseoloviruses—HHV-6A, HHV-6B, and HHV-7

HHV-6A, -6B, and -7 have long been neglected in research. Only in the past years have these viruses gained attention since it became obvious that they not only cause a common children's disease (roseola infantum) but might also be involved in severe illnesses, especially in immunoincompetent individuals like neuroinflammatory diseases (HHV-6A), transplant rejection, myocarditis (HHV-6B), or encephalitis (HHV-6A, -6B, and HHV-7) (69, 70). For this reason, immunomodulatory features of these viruses were studied only relatively recently.

We showed that HHV-6B strain Z29 is capable of suppressing the surface expression of the NKG2D ligands ULBP1, ULBP3, and MICB, but not MICA or ULBP2 [(71); **Figure 1**; **Table 1**]. This was true both in primary T cells and in T cell lines. As a cellular response to the viral infection, mRNA levels of all stress-induced ligands rise following infection; however, the virus suppresses the three abovementioned ligands on protein level and degrades them rapidly in a proteasome-dependent pathway shortly after the start of infection. Also, we showed that the degradation of the three ligands is mediated by at least two different viral proteins.

As for HHV-7, Schneider et al. showed that U21, which was previously shown to target HLA class I for lysosomal degradation, also causes lysosomal degradation of ULBP1 resulting in a mild downregulation. Additionally, they observed a major downregulation of MICA and MICB (72). These findings were established using the overexpression of the viral protein U21. However, the exact mechanism for MIC proteins degradation remained unclear. Probably, U21 interferes with proper protein glycosylation rendering the MIC proteins unstable and causing them to be targeted for cellular degradation. Due to the mild loss of ULBP1, this degradation was proposed to be the result of a "side-effect" of U21-mediated HLA class I degradation, since these related proteins were targeted to lysosomal degradation with higher affinity and to a greater extent.

However, since the study was limited to overexpression of a single gene and no studies were conducted using an actual infection model, it is possible that additional stress-induced ligands are affected by HHV-7 or that additional mechanisms targeting the same ligands exist.

KSHV—HHV-8

Kaposi's-sarcoma-associated herpesvirus is the human herpesvirus with the lowest seroprevalence in the Western world with only about 1–3% of individuals infected (73). Still, this virus is a significant cause of cancer, primarily in AIDS patients, whereas immunocompetent individuals do not experience KSHV reactivation (73, 74). In developing countries, seroprevalence is substantially higher (73).

During lytic infection, KSHV evades NK cell recognition by expressing the viral E3 ligase K5. Thomas et al. showed that K5 modifies lysine residues within the cytoplasmic tails of both MICA and MICB with ubiquitin. Consequently, these molecules are internalized from the cell membrane and intracellularly sequestered, but not degraded [(75); **Figure 1**; **Table 1**]. Notably, the fact that the MICA allele *008 as well as ULBP1, ULBP2, and ULBP3 are GPI anchored and therefore lack a cytoplasmic tail, render them resistant to K5-mediated ubiquitinylation. Additionally, Nachmani et al. reported that the viral miRNA miR-K12-7 specifically represses the translation of MICB by binding to the 3′-UTR of its mRNA (53). Interestingly, MICA mRNA was shown not to be targeted by miR-K12-7 since the 3′-UTR is significantly shorter than the MICB equivalent and does not contain the binding site (53).

EIGHT LIGANDS, FURTHER ALLELIC DIVERSIFICATION: HOST–PATHOGEN EVOLUTION AT FULL SPEED

As emphasized above, herpesvirus family members developed numerous mechanisms to interfere with the expression of the stress-induced ligands. However, most of these studies still leave unanswered questions. More mechanisms and viral effectors are still waiting to be discovered. The viral protein repertoire is probably much larger than currently known; by using ribosome profiling of HCMV and KSHV, numerous new open reading frames (ORFs) have been identified (76, 77). The functions of many viral proteins and ORFs are yet unknown and we are just on the verge of understanding the importance of viral non-coding RNAs, including long non-coding RNAs (78, 79).

While the NKG2D receptor itself is conserved among species, its ligands are not. Interestingly, having eight functional ligands of two different families (MIC and ULBP) and various alleles, the human NKG2D ligand repertoire is more complex than that of other species. Mice possess even nine functional ligands (MULT1, Raet1α–ε, H60a–c) (80). However, their domain structure reveals them to be ULBP family homologs with low allelic diversity. Non-human primates were shown to have homologs of the MIC proteins (81, 82). Still, compared to humans with more than 100 allelic variants, even great apes seem to possess lower allelic variation (83).

Herpesviruses might be a major driving force for diversification of stress-induced ligands and further mutagenesis within alleles leading to allelic variations. None of the described viral mechanisms is capable of eliminating the expression of all stress-induced ligands, the evolutionary pressure rendered these ligands so diverse that no single viral protein or RNA is sufficient to regulate all of them.

As described earlier, viral miRNAs of HCMV, EBV, and KSHV target MICB mRNA at different sites of its 3′-UTR and suppress protein translation (53, 56). Despite the high degree of sequence homology in their 3′-UTRs, MICA is not targeted by any of these miRNAs. The binding sites for the viral miRNAs of HCMV (miR-UL112) and EBV (miR-BART2-5p) are modified by a single-nucleotide insertion, thus abolishing miRNA-induced translation repression. The sequence that is targeted by the KSHV encoded miR-K12-7 is completely absent due to a major deletion in the MICA 3′-UTR.

A similar mutagenesis apparently occurred in the MICA protein to escape UL16 binding. UL16 binds to an α-helical structure in the α2 domain of MICB. By substituting single amino acid residues in MICB with their MICA equivalents, Spreu et al. could show that a single substitution (at two different positions) is sufficient to abolish UL16 binding (84); hence, MICA is spared from UL16-mediated intracellular retention by virtue of very few mutations. Additionally, Klumkrathok et al. suggested that even different allelic variants of MICB are bound with different affinities by UL16 due to amino acid substitutions in the α2 domain (85).

A third piece of evidence for a herpesvirus-driven coevolution is the emergence of MICA*008, a highly prevalent, GPI-anchored MICA variant. MICA*008 is not targeted by UL142, by US18, or by US20. Only recently, it was discovered that the evolutionary relatively novel US9 is capable of targeting solely this distinct allele, but none of the full-length alleles containing a transmembrane domain (68).

These few examples illustrate well the human capability to adjust to viral immune evasion strategies. Accordingly, this strong selective, coevolutionary pressure necessitates modification of viral effector molecules targeting the immune surveillance system as well. By comparing different isolates from HCMV-infected individuals, Renzette et al. and Sijmons et al. indeed showed on a global level that genes involved in immune evasion within the HCMV genome are strongly diversified and contain high numbers of single-nucleotide polymorphisms (86, 87). Among others, one particular mutable gene was found to be UL142, which interacts with NKG2D ligands as pointed out before (87).

DIVERSIFICATION: AN EVOLUTIONARY NECESSITY?

It seems obvious to conclude that herpesviruses and ligands for NKG2D continuously shape each other during coevolution, whereas the NKG2D receptor itself remains conserved.

Particularly MICA took the lead in this race on the human side, its 3′-UTR became shortened and modified and numerous allelic variations emerged to withstand herpesvirus infection. The diversity of MICA alleles might thereby even create a "population level resistance" by making it difficult for newly emerging viral mechanisms to successfully target all MICA variants at once.

However, in contradiction to this theory and the supposed importance of stress–ligand evolution, several reports showed a wide distribution of a MICA–MICB null haplotype (also described as MICA-del–MICB-null), a phenotype that occurs mainly, but not exclusively, in East Asia (88–91), apparently with no major evolutionary disadvantage or clinical manifestations. In fact, there are several known MICA-null alleles also independent of this haplotype. If and how MICA and MICB functions are compensated in these individuals, e.g., by the redundancy of the other NKG2D ligands that are still present, has yet to be elucidated; however, this phenomenon teaches us that we are still far from a complete understanding of the complex families of NKG2D ligands.

AUTHOR CONTRIBUTIONS

DS outlined, wrote, referenced the manuscript, and prepared the figure and table. OM supervised and carefully edited the work.

REFERENCES

1. Wertheim JO, Smith MD, Smith DM, Scheffler K, Kosakovsky Pond SL. Evolutionary origins of human herpes simplex viruses 1 and 2. *Mol Biol Evol* (2014) 31:2356–64. doi:10.1093/molbev/msu185
2. Arvin A, Campadelli-Fiume G, Mocarski E, Moore PS, Roizman B, Whitley R, et al. *Human Herpesviruses: Biology, Therapy, and Immunoprophylaxis*. Cambridge: Cambridge University Press (2007).
3. Grinde B. Herpesviruses: latency and reactivation – viral strategies and host response. *J Oral Microbiol* (2013) 5. doi:10.3402/jom.v5i0.22766
4. Cohen JI. Epstein-Barr virus infection. *N Engl J Med* (2000) 343:481–92. doi:10.1056/NEJM200008173430707
5. Kowitdamrong E, Pancharoen C, Thammaborvorn R, Bhattarakosol P. The prevalence of varicella-zoster virus infection in normal healthy individuals aged above 6 months. *J Med Assoc Thai* (2005) 88(Suppl 4):S7–11.
6. Wald A, Corey L. Persistence in the population: epidemiology, transmission. In: Arvin A, Campadelli-Fiume G, Mocarski E, Moore PS, Roizman B, Whitley R, et al., editors. *Human Herpesviruses: Biology, Therapy, and Immunoprophylaxis*. Cambridge: Cambridge University Press (2007).
7. Hjalgrim H, Friborg J, Melbye M. The epidemiology of EBV and its association with malignant disease. In: Arvin A, Campadelli-Fiume G, Mocarski E, Moore PS, Roizman B, Whitley R, et al., editors. *Human Herpesviruses: Biology, Therapy, and Immunoprophylaxis*. Cambridge: Cambridge University Press (2007).
8. Emery VC, Clark DA. HHV-6A, 6B, and 7: persistence in the population, epidemiology and transmission. In: Arvin A, Campadelli-Fiume G, Mocarski E, Moore PS, Roizman B, Whitley R, et al., editors. *Human Herpesviruses: Biology, Therapy, and Immunoprophylaxis*. Cambridge: Cambridge University Press (2007).
9. Lee BW. Review of varicella zoster seroepidemiology in India and Southeast Asia. *Trop Med Int Health* (1998) 3:886–90. doi:10.1046/j.1365-3156.1998.00316.x
10. Wutzler P, Farber I, Wagenpfeil S, Bisanz H, Tischer A. Seroprevalence of varicella-zoster virus in the German population. *Vaccine* (2001) 20:121–4. doi:10.1016/S0264-410X(01)00276-6
11. Hofscheier A, Ponciano A, Bonzheim I, Adam P, Lome-Maldonado C, Vela T, et al. Geographic variation in the prevalence of Epstein-Barr virus-positive diffuse large B-cell lymphoma of the elderly: a comparative analysis of a Mexican and a German population. *Mod Pathol* (2011) 24:1046–54. doi:10.1038/modpathol.2011.62
12. Balfour HH Jr, Sifakis F, Sliman JA, Knight JA, Schmeling DO, Thomas W. Age-specific prevalence of Epstein-Barr virus infection among individuals aged 6-19 years in the United States and factors affecting its acquisition. *J Infect Dis* (2013) 208:1286–93. doi:10.1093/infdis/jit321
13. Manicklal S, Emery VC, Lazzarotto T, Boppana SB, Gupta RK. The "silent" global burden of congenital cytomegalovirus. *Clin Microbiol Rev* (2013) 26:86–102. doi:10.1128/CMR.00062-12
14. Loewendorf A, Benedict CA. Modulation of host innate and adaptive immune defenses by cytomegalovirus: timing is everything. *J Intern Med* (2010) 267:483–501. doi:10.1111/j.1365-2796.2010.02220.x
15. Kinchington PR, Leger AJ, Guedon JM, Hendricks RL. Herpes simplex virus and varicella zoster virus, the house guests who never leave. *Herpesviridae* (2012) 3:5. doi:10.1186/2042-4280-3-5
16. Hu Z, Usherwood EJ. Immune escape of gamma-herpesviruses from adaptive immunity. *Rev Med Virol* (2014) 24:365–78. doi:10.1002/rmv.1791
17. Wang FZ, Pellett PE. HHV-6A, 6B, and 7: immunobiology and host response. In: Arvin A, Campadelli-Fiume G, Mocarski E, Moore PS, Roizman B, Whitley R, et al., editors. *Human Herpesviruses: Biology, Therapy, and Immunoprophylaxis*. Cambridge: Cambridge University Press (2007).
18. Ullrich E, Koch J, Cerwenka A, Steinle A. New prospects on the NKG2D/NKG2DL system for oncology. *Oncoimmunology* (2013) 2:e26097. doi:10.4161/onci.26097
19. Guerra N, Tan YX, Joncker NT, Choy A, Gallardo F, Xiong N, et al. NKG2D-deficient mice are defective in tumor surveillance in models of spontaneous malignancy. *Immunity* (2008) 28:571–80. doi:10.1016/j.immuni.2008.02.016
20. Guerra N, Pestal K, Juarez T, Beck J, Tkach K, Wang L, et al. A selective role of NKG2D in inflammatory and autoimmune diseases. *Clin Immunol* (2013) 149:432–9. doi:10.1016/j.clim.2013.09.003
21. Orange JS. Human natural killer cell deficiencies and susceptibility to infection. *Microbes Infect* (2002) 4:1545–58. doi:10.1016/S1286-4579(02)00038-2
22. Orange JS. Natural killer cell deficiency. *J Allergy Clin Immunol* (2013) 132:515–525; quiz 526. doi:10.1016/j.jaci.2013.07.020
23. Raulet DH. Roles of the NKG2D immunoreceptor and its ligands. *Nat Rev Immunol* (2003) 3:781–90. doi:10.1038/nri1199
24. Nausch N, Cerwenka A. NKG2D ligands in tumor immunity. *Oncogene* (2008) 27:5944–58. doi:10.1038/onc.2008.272
25. Gonzalez S, Lopez-Soto A, Suarez-Alvarez B, Lopez-Vazquez A, Lopez-Larrea C. NKG2D ligands: key targets of the immune response. *Trends Immunol* (2008) 29:397–403. doi:10.1016/j.it.2008.04.007
26. Groh V, Bahram S, Bauer S, Herman A, Beauchamp M, Spies T. Cell stress-regulated human major histocompatibility complex class I gene expressed in gastrointestinal epithelium. *Proc Natl Acad Sci U S A* (1996) 93:12445–50. doi:10.1073/pnas.93.22.12445
27. Lanier LL. NK cell receptors. *Annu Rev Immunol* (1998) 16:359–93. doi:10.1146/annurev.immunol.16.1.359
28. Anfossi N, Andre P, Guia S, Falk CS, Roetynck S, Stewart CA, et al. Human NK cell education by inhibitory receptors for MHC class I. *Immunity* (2006) 25:331–42. doi:10.1016/j.immuni.2006.06.013

29. Eagle RA, Trowsdale J. Promiscuity and the single receptor: NKG2D. *Nat Rev Immunol* (2007) 7:737–44. doi:10.1038/nri2144
30. Fernandez-Messina L, Reyburn HT, Vales-Gomez M. Human NKG2D-ligands: cell biology strategies to ensure immune recognition. *Front Immunol* (2012) 3:299. doi:10.3389/fimmu.2012.00299
31. Choy MK, Phipps ME. MICA polymorphism: biology and importance in immunity and disease. *Trends Mol Med* (2010) 16:97–106. doi:10.1016/j.molmed.2010.01.002
32. Cosman D, Mullberg J, Sutherland CL, Chin W, Armitage R, Fanslow W, et al. ULBPs, novel MHC class I-related molecules, bind to CMV glycoprotein UL16 and stimulate NK cytotoxicity through the NKG2D receptor. *Immunity* (2001) 14:123–33. doi:10.1016/S1074-7613(01)00095-4
33. Li P, Morris DL, Willcox BE, Steinle A, Spies T, Strong RK. Complex structure of the activating immunoreceptor NKG2D and its MHC class I-like ligand MICA. *Nat Immunol* (2001) 2:443–51. doi:10.1038/87757
34. Holmes MA, Li P, Petersdorf EW, Strong RK. Structural studies of allelic diversity of the MHC class I homolog MIC-B, a stress-inducible ligand for the activating immunoreceptor NKG2D. *J Immunol* (2002) 169:1395–400. doi:10.4049/jimmunol.169.3.1395
35. Nachmani D, Gutschner T, Reches A, Diederichs S, Mandelboim O. RNA-binding proteins regulate the expression of the immune activating ligand MICB. *Nat Commun* (2014) 5:4186. doi:10.1038/ncomms5186
36. Gowen BG, Chim B, Marceau CD, Greene TT, Burr P, Gonzalez JR, et al. A forward genetic screen reveals novel independent regulators of ULBP1, an activating ligand for natural killer cells. *Elife* (2015) 4:e08474. doi:10.7554/eLife.08474
37. Schmiedel D, Tai J, Yamin R, Berhani O, Bauman Y, Mandelboim O. The RNA binding protein IMP3 facilitates tumor immune escape by downregulating the stress-induced ligands ULPB2 and MICB. *Elife* (2016) 5:e13426. doi:10.7554/eLife.13426
38. Stern-Ginossar N, Gur C, Biton M, Horwitz E, Elboim M, Stanietsky N, et al. Human microRNAs regulate stress-induced immune responses mediated by the receptor NKG2D. *Nat Immunol* (2008) 9:1065–73. doi:10.1038/ni.1642
39. Tsukerman P, Stern-Ginossar N, Gur C, Glasner A, Nachmani D, Bauman Y, et al. miR-10b downregulates the stress-induced cell surface molecule MICB, a critical ligand for cancer cell recognition by natural killer cells. *Cancer Res* (2012) 72:5463–72. doi:10.1158/0008-5472.CAN-11-2671
40. Whitley RJ, Roizman B. Herpes simplex virus infections. *Lancet* (2001) 357:1513–8. doi:10.1016/S0140-6736(00)04638-9
41. Siegel MA. Diagnosis and management of recurrent herpes simplex infections. *J Am Dent Assoc* (2002) 133:1245–9. doi:10.14219/jada.archive.2002.0366
42. Schepis D, D'Amato M, Studahl M, Bergstrom T, Karre K, Berg L. Herpes simplex virus infection downmodulates NKG2D ligand expression. *Scand J Immunol* (2009) 69:429–36. doi:10.1111/j.1365-3083.2009.02241.x
43. Campbell TM, McSharry BP, Steain M, Slobedman B, Abendroth A. Varicella-zoster virus and herpes simplex virus 1 differentially modulate NKG2D ligand expression during productive infection. *J Virol* (2015) 89:7932–43. doi:10.1128/JVI.00292-15
44. Enk J, Levi A, Weisblum Y, Yamin R, Charpak-Amikam Y, Wolf DG, et al. HSV1 microRNA modulation of GPI anchoring and downstream immune evasion. *Cell Rep* (2016) 17:949–56. doi:10.1016/j.celrep.2016.09.077
45. Ohishi K, Inoue N, Kinoshita T. PIG-S and PIG-T, essential for GPI anchor attachment to proteins, form a complex with GAA1 and GPI8. *EMBO J* (2001) 20:4088–98. doi:10.1093/emboj/20.15.4088
46. Kinoshita T, Fujita M, Maeda Y. Biosynthesis, remodelling and functions of mammalian GPI-anchored proteins: recent progress. *J Biochem* (2008) 144:287–94. doi:10.1093/jb/mvn090
47. Gilden D, Cohrs RJ, Mahalingam R, Nagel MA. Varicella zoster virus vasculopathies: diverse clinical manifestations, laboratory features, pathogenesis, and treatment. *Lancet Neurol* (2009) 8:731–40. doi:10.1016/S1474-4422(09)70134-6
48. Mueller NH, Gilden DH, Cohrs RJ, Mahalingam R, Nagel MA. Varicella zoster virus infection: clinical features, molecular pathogenesis of disease, and latency. *Neurol Clin* (2008) 26:675–697, viii. doi:10.1016/j.ncl.2008.03.011
49. Straus SE, Cohen JI, Tosato G, Meier J. NIH conference. Epstein-Barr virus infections: biology, pathogenesis, and management. *Ann Intern Med* (1993) 118:45–58. doi:10.7326/0003-4819-118-1-199301010-00009
50. Kawa K. Epstein-Barr virus-associated diseases in humans. *Int J Hematol* (2000) 71:108–17.
51. Tao Q, Chan AT. Nasopharyngeal carcinoma: molecular pathogenesis and therapeutic developments. *Expert Rev Mol Med* (2007) 9:1–24. doi:10.1017/S1462399407000312
52. Pappworth IY, Wang EC, Rowe M. The switch from latent to productive infection in Epstein-Barr virus-infected B cells is associated with sensitization to NK cell killing. *J Virol* (2007) 81:474–82. doi:10.1128/JVI.01777-06
53. Nachmani D, Stern-Ginossar N, Sarid R, Mandelboim O. Diverse herpesvirus microRNAs target the stress-induced immune ligand MICB to escape recognition by natural killer cells. *Cell Host Microbe* (2009) 5:376–85. doi:10.1016/j.chom.2009.03.003
54. Song H, Park H, Kim J, Park G, Kim YS, Kim SM, et al. IDO metabolite produced by EBV-transformed B cells inhibits surface expression of NKG2D in NK cells via the c-Jun N-terminal kinase (JNK) pathway. *Immunol Lett* (2011) 136:187–93. doi:10.1016/j.imlet.2011.01.009
55. Griffiths P, Baraniak I, Reeves M. The pathogenesis of human cytomegalovirus. *J Pathol* (2015) 235:288–97. doi:10.1002/path.4437
56. Stern-Ginossar N, Elefant N, Zimmermann A, Wolf DG, Saleh N, Biton M, et al. Host immune system gene targeting by a viral miRNA. *Science* (2007) 317:376–81. doi:10.1126/science.1140956
57. Kubin M, Cassiano L, Chalupny J, Chin W, Cosman D, Fanslow W, et al. ULBP1, 2, 3: novel MHC class I-related molecules that bind to human cytomegalovirus glycoprotein UL16, activate NK cells. *Eur J Immunol* (2001) 31:1428–37. doi:10.1002/1521-4141(200105)31:5<1428:AID-IMMU1428>3.0.CO;2-4
58. Rolle A, Mousavi-Jazi M, Eriksson M, Odeberg J, Soderberg-Naucler C, Cosman D, et al. Effects of human cytomegalovirus infection on ligands for the activating NKG2D receptor of NK cells: up-regulation of UL16-binding protein (ULBP)1 and ULBP2 is counteracted by the viral UL16 protein. *J Immunol* (2003) 171:902–8. doi:10.4049/jimmunol.171.2.902
59. Dunn C, Chalupny NJ, Sutherland CL, Dosch S, Sivakumar PV, Johnson DC, et al. Human cytomegalovirus glycoprotein UL16 causes intracellular sequestration of NKG2D ligands, protecting against natural killer cell cytotoxicity. *J Exp Med* (2003) 197:1427–39. doi:10.1084/jem.20022059
60. Wu J, Chalupny NJ, Manley TJ, Riddell SR, Cosman D, Spies T. Intracellular retention of the MHC class I-related chain B ligand of NKG2D by the human cytomegalovirus UL16 glycoprotein. *J Immunol* (2003) 170:4196–200. doi:10.4049/jimmunol.170.8.4196
61. Eagle RA, Traherne JA, Hair JR, Jafferji I, Trowsdale J. ULBP6/RAET1L is an additional human NKG2D ligand. *Eur J Immunol* (2009) 39:3207–16. doi:10.1002/eji.200939502
62. Ashiru O, Bennett NJ, Boyle LH, Thomas M, Trowsdale J, Wills MR. NKG2D ligand MICA is retained in the cis-Golgi apparatus by human cytomegalovirus protein UL142. *J Virol* (2009) 83:12345–54. doi:10.1128/JVI.01175-09
63. Bennett NJ, Ashiru O, Morgan FJ, Pang Y, Okecha G, Eagle RA, et al. Intracellular sequestration of the NKG2D ligand ULBP3 by human cytomegalovirus. *J Immunol* (2010) 185:1093–102. doi:10.4049/jimmunol.1000789
64. Fielding CA, Aicheler R, Stanton RJ, Wang EC, Han S, Seirafian S, et al. Two novel human cytomegalovirus NK cell evasion functions target MICA for lysosomal degradation. *PLoS Pathog* (2014) 10:e1004058. doi:10.1371/journal.ppat.1004058
65. Zhang Y, Lazaro AM, Lavingia B, Stastny P. Typing for all known MICA alleles by group-specific PCR and SSOP. *Hum Immunol* (2001) 62:620–31. doi:10.1016/S0198-8859(01)00241-5
66. Petersdorf EW, Shuler KB, Longton GM, Spies T, Hansen JA. Population study of allelic diversity in the human MHC class I-related MIC-A gene. *Immunogenetics* (1999) 49:605–12. doi:10.1007/s002510050655
67. Vales-Gomez M. The impact of glycosyl-phosphatidyl-inositol anchored MICA alleles on novel NKG2D-based therapies. *Front Immunol* (2015) 6:193. doi:10.3389/fimmu.2015.00193
68. Seidel E, Le VT, Bar-On Y, Tsukerman P, Enk J, Yamin R, et al. Dynamic co-evolution of host and pathogen: HCMV downregulates the prevalent allele MICA *008 to escape elimination by NK cells. *Cell Rep* (2015) 10:968–82. doi:10.1016/j.celrep.2015.01.029
69. Kondo K, Yamanishi K. HHV-6A, 6B, and 7: molecular basis of latency and reactivation. In: Arvin A, Campadelli-Fiume G, Mocarski E, Moore PS, Roizman B, Whitley R, et al., editors. *Human Herpesviruses: Biology, Therapy, and Immunoprophylaxis*. Cambridge: Cambridge University Press (2007).
70. Ongradi J, Ablashi DV, Yoshikawa T, Stercz B, Ogata M. Roseolovirus-associated encephalitis in immunocompetent and immunocompromised individuals. *J Neurovirol* (2017) 23(1):1–19. doi:10.1007/s13365-016-0473-0

71. Schmiedel D, Tai J, Levi-Schaffer F, Dovrat S, Mandelboim O. Human herpesvirus 6B downregulates expression of activating ligands during lytic infection to escape elimination by natural killer cells. *J Virol* (2016) 90:9608–17. doi:10.1128/JVI.01164-16
72. Schneider CL, Hudson AW. The human herpesvirus-7 (HHV-7) U21 immunoevasin subverts NK-mediated cytoxicity through modulation of MICA and MICB. *PLoS Pathog* (2011) 7:e1002362. doi:10.1371/journal.ppat.1002362
73. Viejo-Borbolla A, Schulz TF. Kaposi's sarcoma-associated herpesvirus (KSHV/HHV8): key aspects of epidemiology and pathogenesis. *AIDS Rev* (2003) 5:222–9.
74. Cesarman E, Chang Y, Moore PS, Said JW, Knowles DM. Kaposi's sarcoma-associated herpesvirus-like DNA sequences in AIDS-related body-cavity-based lymphomas. *N Engl J Med* (1995) 332:1186–91. doi:10.1056/NEJM199505043321802
75. Thomas M, Boname JM, Field S, Nejentsev S, Salio M, Cerundolo V, et al. Down-regulation of NKG2D and NKp80 ligands by Kaposi's sarcoma-associated herpesvirus K5 protects against NK cell cytotoxicity. *Proc Natl Acad Sci U S A* (2008) 105:1656–61. doi:10.1073/pnas.0707883105
76. Stern-Ginossar N, Weisburd B, Michalski A, Le VT, Hein MY, Huang SX, et al. Decoding human cytomegalovirus. *Science* (2012) 338:1088–93. doi:10.1126/science.1227919
77. Arias C, Weisburd B, Stern-Ginossar N, Mercier A, Madrid AS, Bellare P, et al. KSHV 2.0: a comprehensive annotation of the Kaposi's sarcoma-associated herpesvirus genome using next-generation sequencing reveals novel genomic and functional features. *PLoS Pathog* (2014) 10:e1003847. doi:10.1371/journal.ppat.1003847
78. Tycowski KT, Guo YE, Lee N, Moss WN, Vallery TK, Xie M, et al. Viral noncoding RNAs: more surprises. *Genes Dev* (2015) 29:567–84. doi:10.1101/gad.259077.115
79. Zhang Q, Jeang KT. Long non-coding RNAs (lncRNAs) and viral infections. *Biomed Pharmacother* (2013) 3:34–42. doi:10.1016/j.biomed.2013.01.001
80. Samarakoon A, Chu H, Malarkannan S. Murine NKG2D ligands: "double, double toil and trouble". *Mol Immunol* (2009) 46:1011–9. doi:10.1016/j.molimm.2008.09.035
81. Meyer A, Carapito R, Ott L, Radosavljevic M, Georgel P, Adams EJ, et al. High diversity of MIC genes in non-human primates. *Immunogenetics* (2014) 66:581–7. doi:10.1007/s00251-014-0791-4
82. Anzai T, Shiina T, Kimura N, Yanagiya K, Kohara S, Shigenari A, et al. Comparative sequencing of human and chimpanzee MHC class I regions unveils insertions/deletions as the major path to genomic divergence. *Proc Natl Acad Sci U S A* (2003) 100:7708–13. doi:10.1073/pnas.1230533100
83. de Groot NG, Garcia CA, Verschoor EJ, Doxiadis GG, Marsh SG, Otting N, et al. Reduced MIC gene repertoire variation in West African chimpanzees as compared to humans. *Mol Biol Evol* (2005) 22:1375–85. doi:10.1093/molbev/msi127
84. Spreu J, Stehle T, Steinle A. Human cytomegalovirus-encoded UL16 discriminates MIC molecules by their alpha2 domains. *J Immunol* (2006) 177:3143–9. doi:10.4049/jimmunol.177.5.3143
85. Klumkrathok K, Jumnainsong A, Leelayuwat C. Allelic MHC class I chain related B (MICB) molecules affect the binding to the human cytomegalovirus (HCMV) unique long 16 (UL16) protein: implications for immune surveillance. *J Microbiol* (2013) 51:241–6. doi:10.1007/s12275-013-2514-1
86. Renzette N, Pokalyuk C, Gibson L, Bhattacharjee B, Schleiss MR, Hamprecht K, et al. Limits and patterns of cytomegalovirus genomic diversity in humans. *Proc Natl Acad Sci U S A* (2015) 112:E4120–8. doi:10.1073/pnas.1501880112
87. Sijmons S, Thys K, Mbong Ngwese M, Van Damme E, Dvorak J, Van Loock M, et al. High-throughput analysis of human cytomegalovirus genome diversity highlights the widespread occurrence of gene-disrupting mutations and pervasive recombination. *J Virol* (2015) 89:7673–95. doi:10.1128/JVI.00578-15
88. Komatsu-Wakui M, Tokunaga K, Ishikawa Y, Kashiwase K, Moriyama S, Tsuchiya N, et al. MIC-A polymorphism in Japanese and a MIC-A-MIC-B null haplotype. *Immunogenetics* (1999) 49:620–8. doi:10.1007/s002510050658
89. Ota M, Bahram S, Katsuyama Y, Saito S, Nose Y, Sada M, et al. On the MICA deleted-MICB null, HLA-B*4801 haplotype. *Tissue Antigens* (2000) 56:268–71. doi:10.1034/j.1399-0039.2000.560309.x
90. Komatsu-Wakui M, Tokunaga K, Ishikawa Y, Leelayuwat C, Kashiwase K, Tanaka H, et al. Wide distribution of the MICA-MICB null haplotype in East Asians. *Tissue Antigens* (2001) 57:1–8. doi:10.1034/j.1399-0039.2001.057001001.x
91. Aida K, Russomando G, Kikuchi M, Candia N, Franco L, Almiron M, et al. High frequency of MIC null haplotype (HLA-B48-MICA-del-MICB*0107 N) in the Angaite Amerindian community in Paraguay. *Immunogenetics* (2002) 54:439–41. doi:10.1007/s00251-002-0485-1

13

Influence of Irradiated Peripheral Blood Mononuclear Cells on Both *Ex Vivo* Proliferation of Human Natural Killer Cells and Change in Cellular Property

María Delso-Vallejo[1], Jutta Kollet[1], Ulrike Koehl[2] and Volker Huppert[1]*

[1] *Miltenyi Biotec GmbH, Bergisch-Gladbach, Germany,* [2] *Hannover Medical School, Institute for Cellular Therapeutics, IFB-Tx, Hannover, Germany*

Edited by:
Antoine Toubert,
Paris Diderot University, France

Reviewed by:
Philippe Saas,
Etablissement Français du Sang BFC, France
Ralf Dressel,
Universitätsmedizin Göttingen, Germany

****Correspondence:***
María Delso-Vallejo
mar.delso15@gmail.com

Clinical studies with adoptive immunotherapy using allogeneic natural killer (NK) cells showed feasibility, but also limitation regarding the transfused absolute cell numbers. First promising results with peripheral blood mononuclear cells (PBMCs) as feeder cells to improve the final cell number need further optimization and investigation of the unknown controlling mechanism in the cross-talk to NK cells. We investigated the influence of irradiated autologous PBMCs to boost NK cell proliferation in the presence of OKT3 and IL-2. Our findings demonstrate a requirement for receptor–ligand interactions between feeders and NK cells to produce soluble factors that can sustain NK cell proliferation. Thus, both physical contact between feeder and NK cells, and soluble factors produced in consequence, are required to fully enhance NK cell *ex vivo* proliferation. This occurred with an indispensable role of the cross-talk between T cells, monocytes, and NK cells, while B cells had no further influence in supporting NK cell proliferation under these co-culture conditions. Moreover, gene expression analysis of highly proliferating and non-proliferating NK cells revealed important phenotypic changes on 5-day cultured NK cells. Actively proliferating NK cells have reduced Siglec-7 and -9 expression compared with non-proliferating and resting NK cells (day 0), independently of the presence of feeder cells. Interestingly, proliferating NK cells cultured with feeder cells contained increased frequencies of cells expressing RANKL, B7-H3, and HLA class II molecules, particularly HLA-DR, compared with resting NK cells or expanded with IL-2 only. A subset of HLA-DR expressing NK cells, co-expressing RANKL, and B7-H3 corresponded to the most proliferative population under the established co-culture conditions. Our results highlight the importance of the crosstalk between T cells, monocytes, and NK cells in autologous feeder cell-based *ex vivo* NK cell expansion protocols, and reveal the appearance of a highly proliferative subpopulation of NK cells (HLA-DR^{+}RANKL^{+}B7-H3^{+}) with promising characteristics to extend the therapeutic potential of NK cells.

Keywords: natural killer cells, *ex vivo* expansion, immunotherapy, HLA-DR, RANKL, B7-H3

INTRODUCTION

Among the different approaches of immunotherapy to treat cancer, natural killer (NK) cells are very promising cell types with impressive outcomes in clinical studies. NK cells are innate lymphoid cells (1). They are characterized by their potent cytotoxic responses against virus-infected and malignantly transformed cells, without the need of prior immune sensitization, and in a major histocompatibility complex-unrestricted manner (2, 3). In addition, NK cells produce cytokines such as TNF-α and IFN-γ, which enhance immune responses, and engage in reciprocal interactions with other immune cells that contribute to different immune responses including anti-tumor effects (4).

To date, allogeneic NK cells for adoptive immunotherapy have already entered clinical studies successfully for both applications, post stem cell transplantation (5, 6), and in non-transplant settings to treat cancer patients (7–9). However, manufacturing of NK cells directly isolated from apheresis products can result in varying quantity (10, 11) and yield not always sufficient amounts to carry out multiple applications (12–14). An increase in the number of functional NK cells by *ex vivo* expansion methods is therefore of high interest and has recently been summarized (13).

Natural killer cells require multiple signals for their *ex vivo* survival, proliferation, and activation, involving soluble factors and the necessity of physical interactions with other cells. All of these components can be conveniently supplied by feeder cells (14, 15). Different types of feeder cells have been tested for their potential in supporting NK cell *ex vivo* expansion from both, autologous or allogeneic origin. Typically, they are irradiated prior to use and supplemented with survival and activating factors such as the cytokines IL-2 and IL-15 and/or the anti-CD3 monoclonal antibody (mAb) OKT3. Several approaches using autologous peripheral blood mononuclear cells (PBMCs) as feeder cells have demonstrated their utility to generate sufficient NK cell numbers for clinical applications (16–19). In terms of clinical manufacturing, autologous PBMCs are the preferable choice to avoid safety issues that allogeneic feeder cells may rise. Despite these advantages, little is known about the positive effect of autologous feeder cells on NK cell proliferation and activation. A beneficial role of monocytes in promoting NK cell *ex vivo* proliferation has been proposed (20). However, the underlying cellular and molecular changes that NK cells undergo during active proliferation yet need to be unraveled.

In this study, we established a co-culture system with autologous PBMCs to examine which components have a significant influence concerning the enhancement of NK cell proliferation. We further characterized the cellular and molecular changes occurring in actively proliferating NK cells. Our data provide a better understanding of mechanisms influencing and modulating *ex vivo* NK cell proliferation and might be the base to improve harmonized manufacturing protocols for future clinical NK cell studies.

MATERIALS AND METHODS

Cells and Cell Lines

Buffy coats from healthy donors (Klinikum Dortmund) were used for PBMC isolation. Daudi JP cells were a kind gift of Dr. R. Seggewiss-Bernhardt, University Hospital of Wuerzburg, who obtained them from Prof. P. Fisch, University of Freiburg (21), and K562 cells were obtained from the German Collection of Microorganisms and Cell Cultures (DSMZ). Both cell lines were maintained in complete RPMI medium, RPMI 1640 (Biowest) supplemented with 10% fetal bovine serum (Biochrom), and 2 mmol/L L-glutamine (PAA), in a humidified atmosphere with 5% CO_2 at 37°C.

NK Cell Isolation and Co-Culture with Autologous PBMCs

Peripheral blood mononuclear cells were isolated from buffy coats by density gradient centrifugation using Pancoll (PAN-Biotec). To obtain different feeder cell fractions, PBMCs were depleted first of $CD56^+$ cells by MACS sort using anti-CD56 microbeads (Miltenyi Biotec), and when indicated they were further depleted of $CD19^+$ and $CD3^+$ or $CD14^+$ cells using the corresponding microbeads (Miltenyi Biotec). The differently depleted PBMCs fractions were X-ray irradiated (20 Gy) (RS 2000 Biological Research Irradiator, Radsource) and used as autologous feeder cells for co-culture with NK cells within 1–1.5 h post-irradiation. The irradiated autologous CD56-depleted PBMCs ("IAPs") were used as the major feeder cells. NK cells were purified from PBMCs using the human NK cell isolation kit (Miltenyi Biotec), and expanded in 24-well plates, either in co-culture with autologous feeder cells at a 20:1 feeder-NK cell ratio, based on protocol from Ahn et al. (16), or without feeder cells, in complete culture medium; TexMACS medium (Miltenyi Biotec) supplemented with 5% human AB serum (Life Technologies) and 1,000 U/mL of Proleukin S (rhIL-2) (Novartis). Co-cultures were additionally supplemented with 10 ng/mL of anti-CD3 mAb (functional grade OKT3, Miltenyi Biotec). When indicated, NK cells were labeled prior cultivation with CellTrace™ Violet Cell Proliferation Kit [cell trace violet ("CTV") dye] (Life Technologies). Initial total cell densities of cultures were 1×10^6 cells/mL. Culture plates were incubated in a humidified atmosphere with 5% CO_2 at 37°C. To determine NK cell fold expansions, NK cell densities were checked at different indicated time points by volumetric counting and detection of viable $CD3^-CD56^+$ cells using MACSQuant Analyzer 10 (Miltenyi Biotec). For 12-day expansions, NK cells were harvested from 24-well plates on day 7 and transferred to 25 or 75 T-flasks with replenishment of fresh medium without OKT3. Fresh medium was additionally replenished on day 9 or 10.

Co-Cultures in Transwell® Plates

Purified NK cells were cultured for up to 5 days with or without IAPs in 12-well polystyrene plates equipped with Transwell® inserts (Costar). The insert system consisted of a 500 μL upper well (12 mm diameter), separated from the bottom well (1.5 mL) by a 0.4 μm microporous tissue culture-treated polycarbonated membrane. All cells were resuspended in complete culture medium and IAPs suspensions were supplemented with OKT3 as described. Four different culture conditions were established, all of them containing 1 mL of NK cells (5×10^5 cells/mL) at the bottom wells. In two conditions, either 1 mL of complete medium or

1 mL of IAP suspension were added to the NK cells on the bottom wells, and upper wells were left empty. In two other conditions, either 1 mL of IAP suspension (10×10^6 cells/mL) was inoculated through the insert, or a combination of 0.5 mL of IAPs (10×10^6 cells/mL) and 0.5 mL of purified NK cells (5×10^5 cells/mL) was inoculated, remaining the inoculated cells in both situations on the upper well of the insert. Transwell plates were incubated in a humidified atmosphere with 5% CO_2 at 37°C.

Flow Cytometry

Natural killer cell frequencies were determined using the following panel of mAbs from Miltenyi Biotec: CD45-VioGreen (5B1), CD3-FITC (BW264/56), BDCA1-PE (AD5-8E7), CD14-PE (TÜK4), CD19-PEVio770 (LT19), CD16-APC (VEP13), and CD56-APCVio770 (REA196). In experiments with feeder cell fractions containing additional depletions of $CD19^+$ and $CD3^+$ or $CD14^+$ cells, the mAb BDCA-2-PE (AC144) was included instead of CD14-PE, and CD16-APC replaced by CD14-APC. The human FcR Blocking Reagent from Miltenyi was also used to block unspecific antibody binding to Fc receptors on $CD14^+$ enriched feeder cell fractions. Propidium iodide (Miltenyi Biotec) was used at a final concentration of 1 µg/mL to exclude dead cells from the analysis.

To study the phenotypic differences between resting, proliferating, and non-proliferating NK cells, cells were labeled prior cultivation with or without IAPs, with the cell trace proliferation dye eFluor®670 (eBioscience), in order to differentiate proliferating and non-proliferating cells. Different candidate molecules were analyzed using several antibody panels designed based on previous work (22). These panels shared a backbone of mAbs: CD45-VioGreen, CD3-VioBlue, TCRγδ-VioBlue (11F2), CD14-VioBlue, CD19-VioBlue, and CD56-APCVio770, and SYTOX® Blue (Life Technologies) was used to exclude dead cells from the analysis. The backbone was combined with groups of the following mAbs (Miltenyi Biotec unless otherwise indicated): KLRB1-FITC (191B8), CTLA-4-PE (BNI3), Siglec-9-PEVio770 (REA492), Siglec-7-PerCP700 (REA214), LILRB1-FITC (GHI/75), KLRG1-PE (REA261), CD16-PerCP700 (VEP13), RANKL-PE (DN254), B7-H3-PEVio770 (FM276), NKp44-PEVio770 (2.29), NKp80-FITC (4A4.D10), 4-1BB-PE (4b4-1), 2B4-PEVio770 (REA112), NKG2D-PerCPCy5.5 (1D11) (Biolegend), HLA-DP/DQ/DR-FITC (REA332), ALCAM-PEVio770 (REA442), HLA-A/B/C-PerCPVio700 (REA230), and HLA-DR-FITC (AC122). The corresponding mouse immunoglobulin (Ig) G1, IgG2A, IgG2B, IgM, or REAs conjugated with the respective dyes were used as isotype controls. Cells were acquired using MACSQuant Analyzer 10 and analyzed using MACSQuantify 2.8 software (Miltenyi Biotec).

Cytokine Detection

Supernatants from co-cultures of NK cells with IAPs further depleted of $CD19^+$, and $CD3^+$ or $CD14^+$ cells were collected after 5 days and cytokine production was detected using the flow cytometry bead-based array MACSPlex Cytokine 12 kit, human according to manufacturer's instructions (Miltenyi Biotec). The MACSPlex Cytokine 12 kit allows for the detection of human GM-CSF, IFN-α, IFN-γ, IL-2, IL-4, IL-5, IL-6, IL-9, IL-10, IL-12p70, IL-17A, and TNF-α. Samples were acquired using a MACSQuant Analyzer 10 and analyzed using the Express Mode option of the MACSQuantify 2.8 software (Miltenyi Biotec).

Cytotoxicity Assays

Target cell killing of K562 and Daudi JP cell lines was analyzed using a flow cytometry-based assay (23). Briefly, target cells were labeled with CTV dye and seeded in 96 well U-bottom plates at a cell density of 2×10^5 cells/mL, and incubated alone, or with expanded NK cells (2×10^6 cells/mL) at different effector-to-target (E:T) ratios, for 4 h in a humidified atmosphere with 5% CO_2 and 37°C. Antibody-dependent cellular cytotoxicity (ADCC) of expanded NK cells was further analyzed against the $CD20^+$ Daudi cells, by adding 5 µg/mL of the anti-CD20 mAb rituximab (Hoffman-La Roche). After 4 h incubation, plates were transferred to 4°C for at least 30 min to stop cell killing before quantifying viable CTV-positive target cells using the MACSQuant Analyzer 10. The frequency of killed target cells was calculated by the difference between the number of viable target cells in samples with effector NK cells and samples with targets cells containing no effector cells. Representative flow cytometry data are included in Figure S1 in Supplementary Material.

Preparation of NK Cells for Sorting

Natural killer cells from five different donors were labeled with the cell trace proliferation dye eFluor®670 (eBioscience) and co-cultured for 5 days with IAPs. On day 5, cells were harvested and proliferating and non-proliferating NK cells were sorted according to the brightness of the cell trace dye in a FACSAria III cytometer (BD Bioscience, cell sorting facility of the Center for Molecular Medicine of Cologne, Cologne, Germany). 7-AAD (BD Pharmigen™) was used to exclude dead cells. To ease the sorting process, prior to sort, samples were enriched on NK cell content using the human NK cell isolation kit from Miltenyi. Sorted fractions containing the proliferating and non-proliferating ($\approx 0.8 \times 10^5$ cell in each fraction) were lysed in RA1 buffer (Macherey-Nagel) for total RNA isolation, and stored at −20°C until use. For comparison, resting NK cells (1×10^6 cells) were lysed and stored under identical conditions.

RNA Microarrays

Total RNA from sorted samples was isolated using the NucleoSpin® RNA kit (Macherey-Nagel), amplified and labeled using the Agilent Low Input Quick Amp Labeling Kit (Agilent Technologies), prior to hybridization to Agilent Whole Human Genome Oligo Microarrays 8 × 60K V2 chips (Agilent Technologies) according to manufacturer instructions. A detailed description of microarray processing can be found in Supplementary Material.

Pre-Processing of Microarray Data

Raw intensity data from feature extraction output files (FES 10.7.3.1, Agilent Technologies) were analyzed using the Rosetta Resolver® software (Rosetta Biosoftware). The following calculations were performed with software packages within R/Bioconductor (24, 25). Intensity values were corrected with background subtraction and normalized by quantile normalization (26). Reliable signal intensities were considered significant when $p \leq 0.01$, using the Rosetta error model (27). Subsequent statistical analysis was performed on normalized Log2-transformed intensity values. The data set can be found at the NCBI GEO public database with the accession number GSE92512.

Statistical Analysis

Conventional statistics including parametric un-paired Student's *t*-test and one-way ANOVA with Tukey's *post hoc* test, or the non-parametric Wilcoxon test and Kruskal–Wallis test with *post hoc* Dunn's test were performed with Graph Pad Prism 7 software (GraphPad). All statistical analysis were two-sided and $p < 0.05$ considered statistically significant, and indicated as $*p < 0.05$, $**p < 0.01$, and $***p < 0.001$. Only statistical differences are shown.

Description of the statistical analysis of the microarray data can be found in Supplementary Material.

RESULTS

Enhancement of *Ex Vivo* NK Cell Proliferation and Preservation of Function Using Irradiated Autologous CD56-Depleted PBMCs

We first established an NK cell expansion method using irradiated autologous CD56-depleted PBMCs (IAPs) as feeder cells to achieve high expansion rates compared with culture with rhIL-2 only. We compared the proliferation kinetics over 7 days of NK cells cultured with IAPs or with rhIL-2 only. NK cells cultured with IAPs showed significantly higher cell counts visible already at day 5 (mean fold expansion: 3.2, range: 1.4–5.7) compared with cultivation with rhIL-2 only (mean fold expansion: 1.3, range: 0.4–1.9). This differential NK cell expansion rate between the two cultivation methods became much more pronounced after prolonged cultivation, as observed in a set of NK cell expansions over 12 days. Here, cells cultured with IAPs reached particularly high cell numbers (mean fold expansion: 212, range: 80–419) compared with rhIL-2 cultivation only (mean fold expansion: 22.5, range: 6.2–57.3) (**Figure 1A**). The killing capacity of the differently expanded NK cells was tested against two different leukemia tumor cell lines, the K562 cell line to analyze natural cytotoxicity, and the $CD20^+$ Daudi cells to also determine ADCC after opsonization of these cells with anti-CD20 mAb rituximab. The killing capacity of the differently expanded cells was preserved in all tested settings. NK cells displayed comparable natural killing rates toward K562 and Daudi targets (**Figures 1B,C**; Figure S1 in Supplementary Material), as well as comparable ADCC responses upon rituximab opsonization of Daudi cells. *In vitro* ADCC responses can be monitored best at low E:T ratios, where read-out of natural killing activity is lower. Independent of the expansion method, NK cells showed a clearer ADCC effect of rituximab-coated target cells at low E:T ratios (**Figure 1C**; Figure S1 in Supplementary Material). Taken together, co-culture with IAPs significantly enhanced *ex vivo* NK cell proliferation preserving their cytotoxic capacity.

Cross-Talk of NK Cells with Both T Cells and Monocytes Is Necessary to Produce Soluble Factors that Enhance *Ex Vivo* NK Cell Proliferation

Next, we sought to determine whether the enhancement of NK cell *ex vivo* proliferation with our established protocol is a consequence of direct cell-to-cell interactions between NK and feeder cells, or of exposure to soluble factors released by feeder cells. To address this question, we cultured NK cells for 5 days at the bottom of Transwell® plates, alone or with IAPs either in close contact or separated through a permeable membrane, allowing only traffic of soluble factors. NK cell fold expansions were significantly decreased when IAPs were separated through the membrane, reaching similar fold expansions as NK cells cultured alone. On the other hand, the addition of NK cells to IAPs separated by the membrane, rescued the expansion of the NK cells at the bottom of the Transwells (**Figure 2A**). These data indicate that cell-to-cell interactions between IAPs and NK cells are necessary to produce soluble factors beneficial for NK cell proliferation.

IAPs are composed of different immune cells, mainly T cells, B cells, monocytes, and dendritic cells. In this respect, we speculated whether any of these cells could have a predominant role in stimulating NK cell proliferation. NK cell fold expansions were compared after co-culture with a standard IAP feeder cell fraction and three other fractions that were depleted from $CD19^+$ cells (B cells and a small subset of dendritic cells), or additionally further depleted from $CD3^+$ (T cells) or $CD14^+$ cells (monocytes), respectively (**Figure 2B**). Depletion of $CD19^+$ cells from the standard IAP fraction, had no effect on NK cell proliferation, whereas further depletion of T cells or monocytes significantly reduced NK cell expansion, with a more dramatic effect upon the absence of monocytes (**Figure 2C**). Differences in NK cell fold expansions correlated with differences in number of dividing cells as observed with dilution of CTV dye labeling (**Figure 2D**).

We hypothesized that the differences in stimulating NK cell proliferation could be mirrored in the cytokine composition released by the different feeder cell fractions. GM-CSF, IFN-γ, IL-6, TNF-α, IL-5, and IL-9 could be detected at high levels (>100 pg/mL) and were produced at a similar extent in co-cultures of NK cells with IAPs or with CD19-depleted IAPs. A remarkable reduction in cytokine production was observed in co-cultures with further depletion of T cells or monocytes, respectively (**Figure 2E**). This, indeed, supported the differences observed in NK cell proliferation kinetics due to the different cellular composition of the feeder cell fractions. The presence of NK cells in co-culture with the different feeder cell fractions influenced the cytokine production, as observed when feeders were maintained in culture without NK cells (Figure S2 in Supplementary Material). Low detectable levels (<100 pg/mL) of IL-4, IL-10, IL-17A, IFN-α, and IL-12 were also found in all co-culture conditions, without prominent changes upon differential cell subsets depletions from IAPs (data not shown). IL-2 was detected at levels exceeding the maximum standard value (>10.000 pg/mL) in any co-culture condition, as a result of its supplementation at the beginning of the culture (data not shown); which was sufficient to support

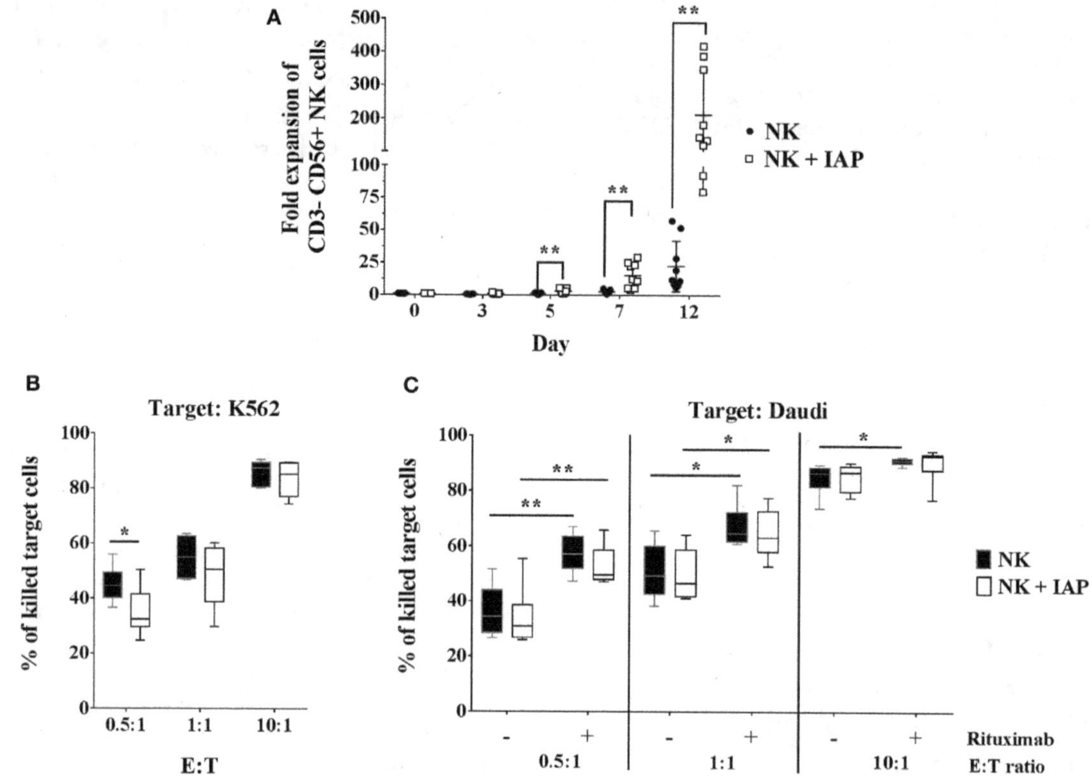

FIGURE 1 | Natural killer (NK) cell expansion using autologous CD56-depleted peripheral blood mononuclear cells (PBMCs) and cytotoxic responses compared with standard culture with rhIL-2 only. **(A)** Proliferation kinetics of NK cells in expansion with irradiated autologous CD56-depleted PBMCs (IAPs) or NK cells from the same donors but cultured with rhIL-2 only were determined by flow cytometry ($n = 9$). A separate set of samples was cultured using both methods to compare expansions up to 12 days (also $n = 9$). Fold expansion of cultured relative to resting CD56+ NK cells are shown for each group per donor including mean ± SD. The non-parametric Wilcoxon test was used for statistical analysis. Natural cytotoxicity of 12-day expanded NK cells with and without IAPs was assessed against K562 **(B)** and Daudi cell lines, including antibody-dependent cellular cytotoxicity triggered by rituximab **(C)**. Frequencies of killed target cells were determined by flow cytometry and shown as mean, minimum to maximum, and SD. Statistical analysis was performed using un-paired Student's *t*-test. Figure S1 in Supplementary Material shows representative raw flow cytometry data of cytotoxicity assays.

NK cell growth independently of the presence or absence of T cells and monocytes. Taken together, our data indicate that the cross-talk between NK cells, T cells, and monocytes, is crucial to enhance *ex vivo* NK cell proliferation when using IAPs, not only due to occurrence of receptor–ligand interactions among these cells but also possibly due to the production of soluble factors as consequence of this cross-talk.

Transcriptional Analysis of Expanded NK Cells Co-Cultured with IAPs Helps to Unravel Characteristics of Highly Proliferating NK Cells

The multitude of signals occurring during the co-culture of NK cells with IAPs helped to boost the *ex vivo* proliferation of NK cells. However, at early time points during culture (day 5) not all NK cells are actively proliferating, but some remain quiescent. We aimed at identifying molecules or pathways responsible for the boost of *ex vivo* NK cell proliferation under the co-culture conditions. For this purpose, we performed a whole genome microarray analysis, with five different donors, to study the transcriptome of resting (R0) NK cells, isolated on day 0, and compare it to proliferating (P), and non-proliferating (NP) NK cells co-cultured for 5 days with IAPs (Figure S3 in Supplementary Material). Principal components analysis (PCA) of the unfiltered transcriptomes separated the samples in three groups corresponding to resting, proliferating, and non-proliferating NK cells, respectively (**Figure 3A**). The expression differences between R0 and 5-day cultured samples appear to contribute most to the variation (principal component 1, 63.1%), while differences between P and NP can be ascribed to the second principal component (principal component 2, 11.5%). A total of 10,299 transcripts were differentially expressed among the three groups, only 2,902 of these transcripts were unique candidates (**Figure 3B**). Hierarchical clustering of the differentially expressed transcripts confirmed the PCA results of the unfiltered data by the formation of two main clusters according to the samples of resting and 5-day expanded NK cells, the latter separated in sub-clusters corresponding to P and NP (**Figure 3C**). A functional annotation analysis of all differentially expressed transcripts revealed to which biological processes and pathways the transcripts were related to. In addition to the anticipated associations with IL-2 pathway, the great

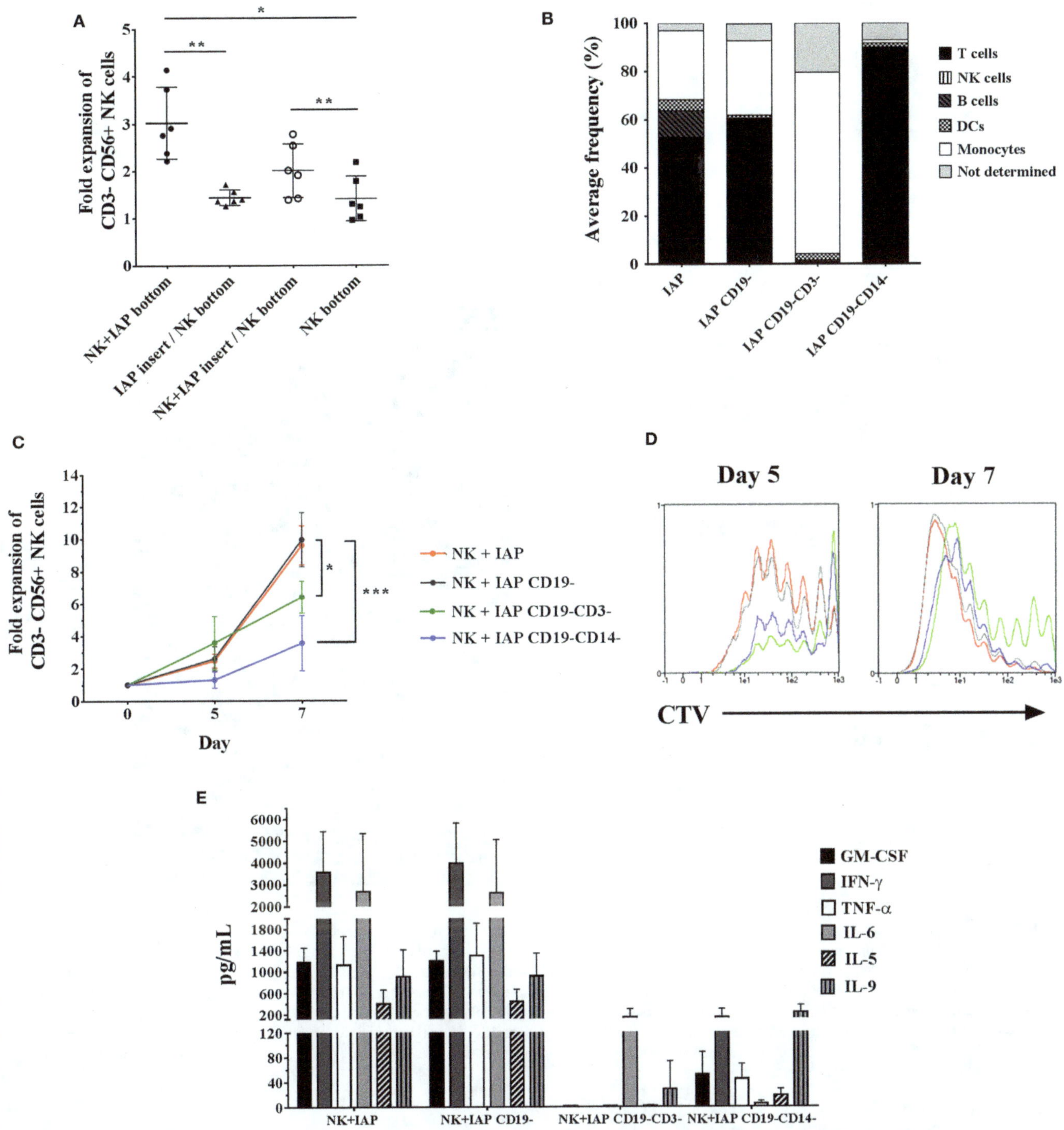

FIGURE 2 | Influence of cell-to-cell interactions from different IAP feeder cell fractions on natural killer (NK) cell expansion and cytokine production upon co-culture. **(A)** NK cell counts were assessed on day 5 of culture in Transwell® plates either alone (NK bottom) or co-cultured with IAPs. For co-cultures, NK cells were either kept in close contact with IAPs (NK + IAP bottom), or IAPs were kept separated from NK cells through an insert with microporous membrane in two conditions: IAPs alone in the insert (IAP insert/NK bottom), or with NK cells (NK + IAP insert/NK bottom). Mean values and SD of NK cell fold expansions from six different donors are depicted. Statistical analysis was performed using one-way ANOVA with Tukey's *post hoc* test and Greenhouse–Geisser adjustment for unequal variances. **(B)** Composition of different IAP fractions analyzed by flow cytometry on day 0 after differential depletions of CD19$^+$ (IAP CD19$^-$) and additional CD3$^+$ (IAP CD19$^-$CD3$^-$) or CD14$^+$ (IAP CD19$^-$CD14$^-$) cells. Average frequencies shown are of four donors. **(C)** NK cell fold expansions analyzed on days 5 and 7 of co-culture with differently depleted IAP fractions ($n = 4$) were determined by flow cytometry. Statistical analysis was performed using one-way ANOVA with Tukey's *post hoc* test. **(D)** Cell trace violet dilution of NK cells expanded with the differently depleted IAP fractions was measured by flow cytometry on days 5 and 7 of co-culture. A representative result of four different donors is shown. **(E)** Cytokine expression levels in supernatants after 5 days of NK cell co-cultures with the differently depleted IAP fractions were determined by cytometric bead array ($n = 4$).

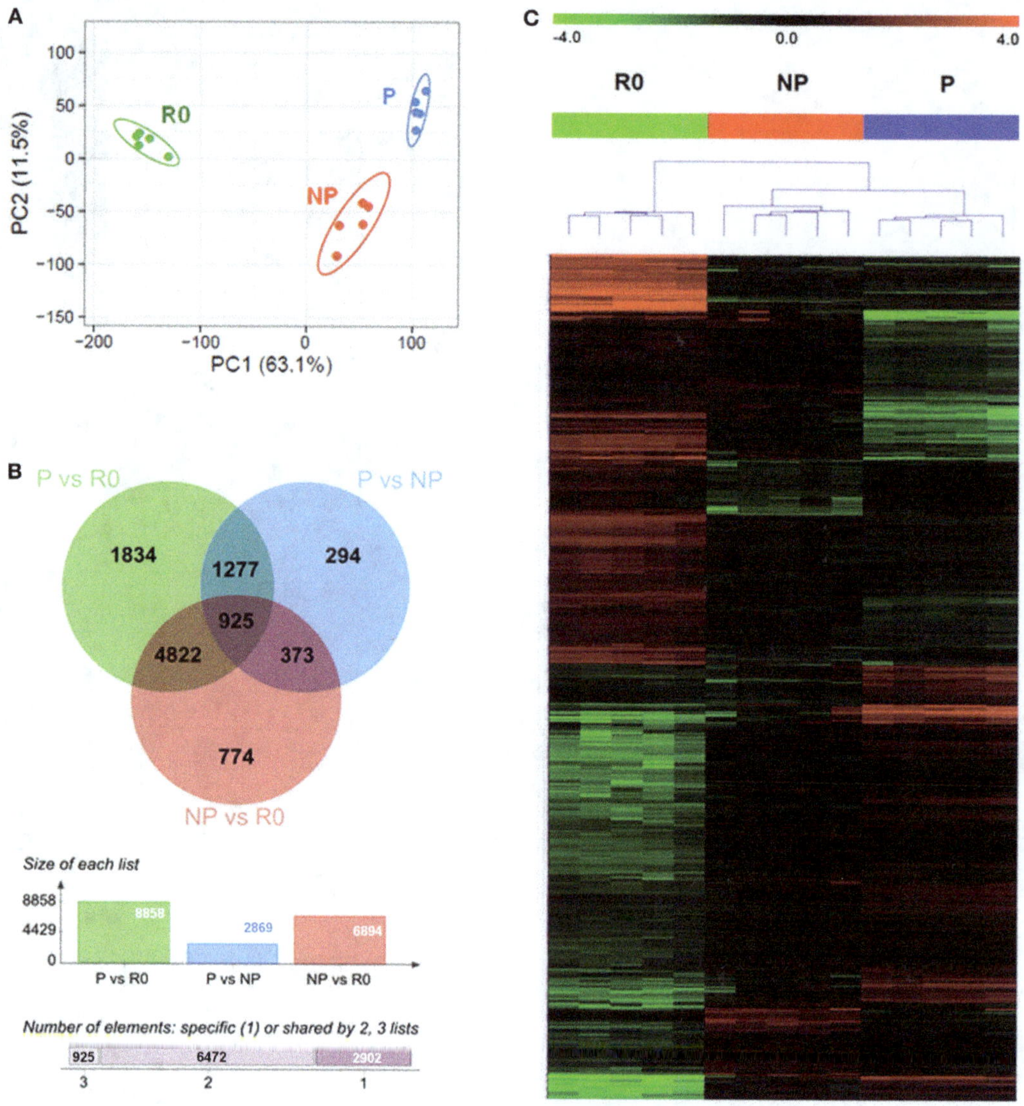

FIGURE 3 | Differential transcriptome signatures between resting, proliferating, and non-proliferating natural killer (NK) cells expanded with IAPs. **(A)** Unsupervised principal component analysis of the transcriptomes of all samples ($n = 5$) for resting (R0); proliferating (P), and non-proliferating (NP) NK cells, displays the samples in a scatterplot of principal component 1 and principal component 2. The prediction ellipses are indicative of that with a 95% probability a new observation from the same group will fall inside the ellipse. **(B)** Venn diagram (top) displaying the overlap between the numbers of differentially expressed transcripts of each pairwise comparison, PvsR0, PvsNP, and NPvsR0. The columns diagram (middle) indicate the total number of differentially expressed reporters per pairwise comparison, and the horizontal bar (bottom) summarizes the number of transcripts specific or shared between two or three groups. **(C)** Heat map of all differentially expressed transcripts after hierarchical clustering (Euclidean distance, complete linkage method). Fold-change differences (row-wise centered to the median) are displayed within color saturation limits −4 (green) to +4 (red).

majority of the identified transcripts were associated with nucleotide metabolism and categories principally related to regulation of cell proliferation and activation (Figure S4 in Supplementary Material). This demonstrates that the sorted fractions indeed represented proliferating and non-proliferating cells.

In this regard, among the top 20 most highly expressed transcripts in proliferating NK cells compared with resting cells, were several positive regulators of the cell cycle and mitosis, such as cyclin A2 (*CCNA2*) and centrosomal protein of 55 kDa (*CEP55*), or the zinc finger proteins *ZBED2* and *ZBTB32* (**Figure 4A**). Zinc finger proteins are structurally a heterogeneous group of molecules which regulate gene expression by binding to DNA and RNA (28). Particularly, Tramtrack bric á brac zinc finger proteins (BTB-ZF) have important roles in controlling development and functional activity of lymphocytes (29, 30). Our data show that in proliferating NK cells, *ZBED2* was the zinc finger protein with highest transcript levels, and the transcription factor *ZBTB32* was the family member of BTB-ZF proteins with highest transcript levels (Figure S5A in Supplementary Material). These results point to a potential role of zinc finger proteins in modulating human *ex vivo* NK cell proliferation.

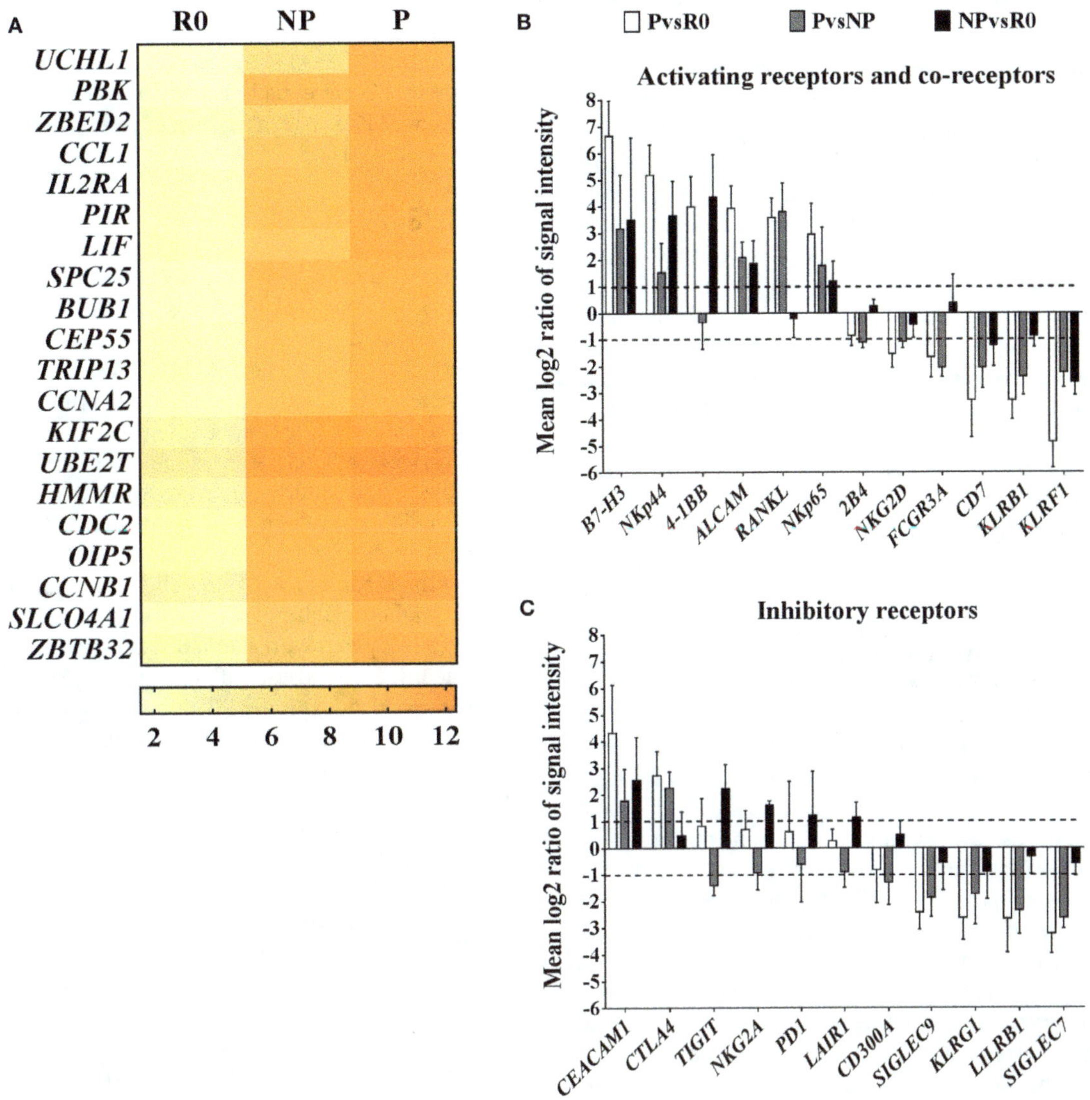

FIGURE 4 | Transcriptional characteristics of proliferating natural killer (NK) cells compared with their non-proliferating counterparts and resting NK cells. **(A)** Top 20 transcripts with highest expression in proliferating NK cells (P) compared with resting cells (R0). Values in the heat map correspond to mean log2 intensities per group ($n = 5$) for each transcript. **(B)** Pairwise analysis (PvsR0, PvsNP, NPvsR0) of differentially expressed transcripts corresponding to activating receptors and co-receptors, or inhibitory receptors **(C)**. Pairwise comparisons are reported as mean values of log2 ratios of signal intensities with SD ($n = 5$). Only transcripts with differential expression levels (mean log2 ratios ≥1 or ≤−1, equivalent to ≥2 or ≤−2-fold difference) in at least one of the pair wise group comparisons are displayed.

Natural killer cell receptors and co-receptors play key roles in controlling NK cell activation. We additionally screened for the most well-known activating and inhibitory receptors and co-receptors differentially expressed among the groups. In general, transcript levels of activating receptors were increased in proliferating cells, meanwhile transcripts of many inhibitory receptors, were decreased (**Figures 4B,C**). Interestingly, among activating receptors and co-receptors, we found that proliferating NK cells compared with resting or non-proliferating contained increased transcript levels of receptor activator of the nuclear factor kappa-B ligand, *RANKL*, and *CD276* molecule known as B7-H3. On the contrary, they had decreased transcript levels of the activating receptors *FCGR3A* (CD16), *KLRF1* (NKp80), or *KLRB1* (CD161) compared with resting and non-proliferating cells (**Figure 4B**). Furthermore, we noticed that transcriptional changes in proliferating NK cells extended also to differences in expression of HLA class I and II molecules. Remarkably, proliferating NK cells expressed high transcript levels of HLA class II, that correlates with their activated status, whereas transcripts corresponding to HLA class I molecules were decreased in comparison with resting and non-proliferating NK cells, representing an unexpected finding (Figure S5B in Supplementary Material). Altogether these results suggest the existence of differential activation responses of NK cells toward the stimuli provided by IAPs, which may translate into a rapid proliferation of certain subsets of NK cells compared with others with delayed or no proliferative response to the same stimuli.

Phenotypic Hallmarks of Proliferating NK Cells

The changes observed in transcript levels of activating and inhibitory receptors and co-receptors are likely to extend at the protein level, and may reveal phenotypic characteristics of proliferating and non-proliferating NK cells. In addition, the detected phenotypic differences may depend on the stimuli used to promote *ex vivo* NK cell expansion. Therefore, we analyzed by flow cytometry, the expression of several selected molecules (Table S1 in Supplementary Material) in resting, proliferating, and non-proliferating NK cells expanded for 5 days with IAPs or rhIL-2 only. Most of the observed changes in protein content confirmed the results obtained by transcriptional profiling, with significant differences especially between resting and proliferating cells (**Figure 5**; Figure S6 in Supplementary Material). Among them, we identified that proliferating NK cells contained significantly decreased frequencies of cells expressing the inhibitory receptors Siglec-7 and -9, independently of the expansion protocol, compared with both non-proliferating and resting cells. Also, independently of the expansion method was a significant reduced expression of the activating receptors KLRB1, NKp80, CD16, and the inhibitory receptor KLRG1 within proliferating cells compared with resting cells (Figure S6 in Supplementary Material). All other analyzed activating receptors and co-receptors were expressed at higher levels in proliferating NK cells, except the NKG2D co-receptor 2B4. Despite almost all resting NK cells expressed 2B4 at the beginning of the culture, this could not be detected on day 5 of expansion with none of the expansion methods used (data not shown). This suggests an oscillating expression of 2B4 on NK cells during expansion, since expression of 2B4 has been reported after longer culture periods (23). Moreover, we could confirm RANKL and B7-H3 expression, absent in resting cells, appeared after 5 days of expansion. Their expression was mainly restricted to proliferating NK cells and reached higher levels when NK cells were expanded with IAPs instead of with rhIL-2 only (**Figures 5A,B**). Similarly, expression of HLA class II molecules and particularly HLA-DR, were increased after 5 days of expansion with IAPs compared with rhIL-2 only. Expression of HLA-DR persisted after longer term culture (data not shown). Interestingly, the reduction observed earlier in the expression of HLA class I molecules after 5 days of culture was verified here too, independently of the expansion protocol used.

Co-Expression of HLA-DR, RANKL, and B7-H3 Define a Subset of Highly Proliferating NK Cells

Given that co-culture with IAPs for 5 days produced an enrichment of HLA-DR expressing cells and induced expression of RANKL and B7-H3, we investigated the correlation of the co-expression of these three molecules with the proliferative status of NK cells by flow cytometry. HLA-DR expressing NK cells contained a subpopulation of double positive RANKL and B7-H3 NK cells that corresponded to the most highly proliferating NK cells (**Figure 6**; Figure S7 in Supplementary Material). This was substantiated by NK cells co-cultured with IAPs in contrast to cultures with rhIL-2 only. The HLA-DR$^+$RANKL$^+$B7-H3$^+$ NK cell population was minimally represented in cultures with rhIL-2 only, but enlarged in co-cultures with IAPs. In summary, we identified a subset of NK cells that preferentially expand in response to the stimuli provided by IAPs and is characterized by expression of HLA-DR, RANKL, and B7-H3.

DISCUSSION

Recent developments in *ex vivo* NK cell expansion protocols for clinical applications have exploited the use of autologous feeder cells in combination with cytokines to boost NK cell proliferation and their activation status (16–19). However, the components responsible for this enhancement remain somewhat elusive, and little is known regarding the phenotypic and intracellular changes occurring when NK cells start to actively proliferate under these conditions. In this study, we established a co-culture protocol using irradiated autologous CD56-depleted PBMCs (IAPs) as feeder cells, to use it as a tool for the analysis of factors influencing *ex vivo* proliferation of NK cells. After ascertaining the achievement of higher numbers of functional NK cells by co-culture with IAPs compared with standard culture with rhIL-2 only, we subsequently analyzed the influence of cell-to-cell interactions, soluble factors, and specific PBMCs subpopulations on NK cell proliferation. The need of homotypic and heterotypic cell-to-cell interactions for NK cell survival, activation, and proliferation has been previously reported (20, 31, 32). Consistent with this, disruption of cell-to-cell interactions between NK cells and IAPs by using Transwell® plates results in lower NK cell proliferation than when feeder cells and NK cells are in close contact. Additionally, here we report that NK cell interactions with IAPs induce changes in production of soluble factors that improve NK cell *ex vivo* proliferation. The role of certain soluble factors, mostly of the cytokines IL-2, IL-15, and IL-21, in inducing mature human NK cell *ex vivo* proliferation has been well acknowledged (33, 34). In our set-up, sufficient IL-2 levels were achieved by supplementation of high concentrations of rhIL-2 at the beginning of the expansions, and also possibly by additional IL-2 production by T cells within the IAPs upon stimulation in culture. Concerning IL-15 and IL-21, we assessed their production using commercial enzyme-linked immunosorbent assays during a preliminary test on supernatants from 5-day expanded NK cells with and without IAPs. Levels of soluble IL-15 were undetected, and only background levels of IL-21 could be detected, without differences upon the presence or absence of IAPs. With this, we concluded that there was no significant production of neither IL-15 nor IL-21 in our co-culture system. Thus, our findings point to additional cytokines and other soluble factors produced during IAPs-NK cell co-cultures that should be further investigated, and may be used to improve *ex vivo* NK cell expansion methods without the use of feeder cells.

Considering the heterogeneous composition of IAPs, we analyzed the influence of subpopulations (mainly T cells, B cells, and monocytes) in promoting NK cell proliferation. In our co-culture method, the presence of both T cells and monocytes is sufficient to completely enhance *ex vivo* NK cell proliferation, whereas the presence of B cells is dispensable. We observed that monocyte depletion substantially reduces the production of cytokines from

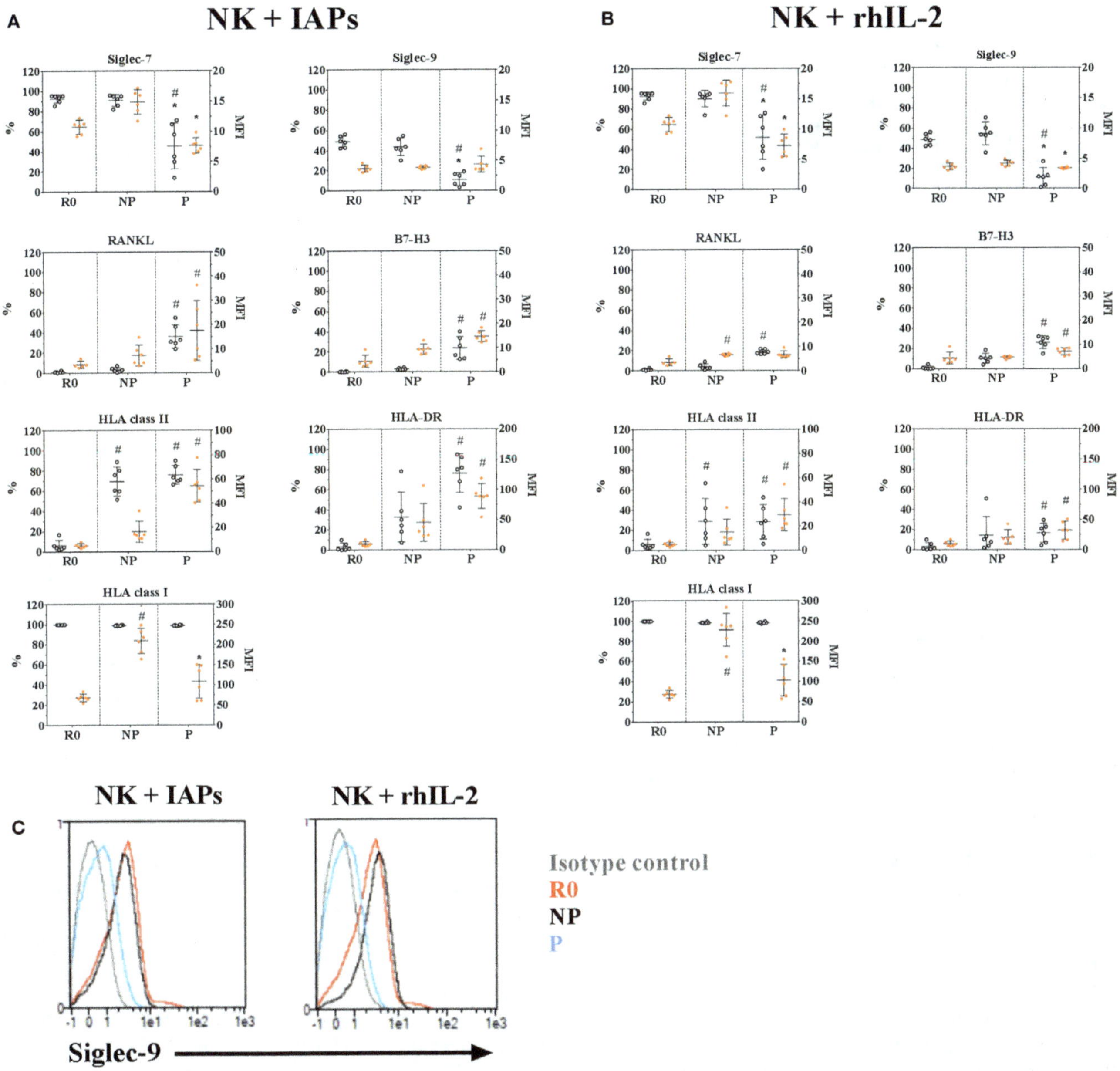

FIGURE 5 | Phenotypic characteristics of proliferating and non-proliferating natural killer (NK) cells compared with resting NK cells. Surface expression of selected inhibitory, activating, and co-stimulatory receptors in 5-days proliferating (P) and non-proliferating (NP) NK cells compared with resting (R0) cells prior culture with IAPs **(A)** or with rhIL-2 only **(B)** were determined by flow cytometry ($n = 6$). Of note, values shown for HLA-DR correspond to a repeated analysis with six other donors. Results are represented for each marker as percentage of positive cells (white circles, referred to the left scale of the plots) and mean fluorescence intensity of positive cells (orange circles, referred to the right scale of the plots) showing mean values and SDs. Statistical analysis was performed using the Kruskal–Wallis test applying Dunn's *post hoc* test. Only significant differences are shown as # for PvsR0 and * for PvsNP. **(C)** Representative histogram of Siglec-9 expression on NK cells at the different time points and proliferative status analyzed R0, P, or NP.

remaining feeder cells, which correlates with the lowest NK cell expansions. On the other hand, the presence of monocytes without T cells almost completely hindered cytokine production, although this caused a less prominent reduction of NK cell expansion. This indicates that the irradiated monocytes besides participating in promoting cytokine production from T cells, may also produce other soluble factors, or express certain surface molecules that result in promoting *ex vivo* NK cell proliferation. To a certain extent, this correlates with previous observations showing the potential of monocytes to boost *ex vivo* NK cell expansion (20, 35), but it is in contrast with the exclusion of a beneficial effect of T cells shown in one of the studies (20). The reason of this difference may be explained by the use of non-irradiated accessory cells and only rhIL-2 supplemented in this former study, while in our set-up the T cells are irradiated and OKT3 is additionally supplemented together rhIL-2. It is important to remark that full

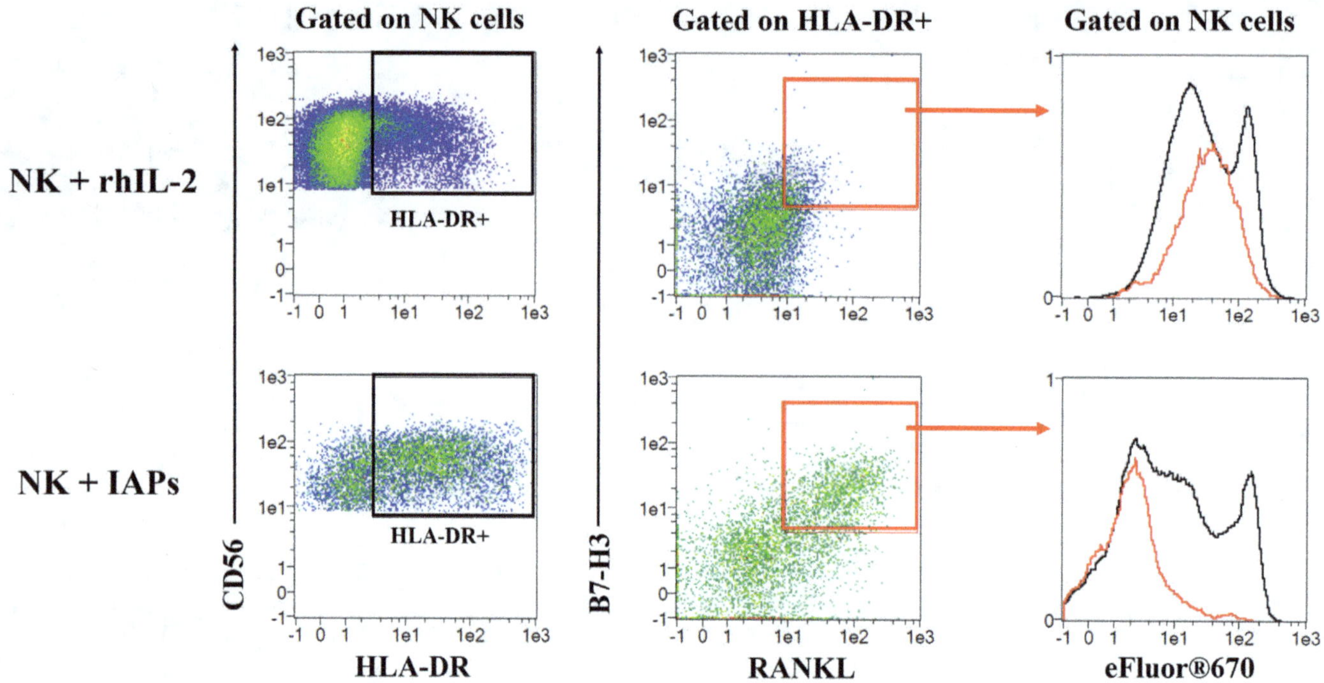

FIGURE 6 | Identification of a subset of HLA-DR expressing natural killer (NK) cells co-expressing RANKL and B7-H3 with high proliferative capacity. eFluor®670 labeled NK cells were expanded in co-culture with IAPs or using rhIL-2 only for 5 days. Co-expression of HLA-DR, RANKL, and B7-H3 was analyzed at this time point by flow cytometry, and compared between the two different culture methods used. First, HLA-DR expressing NK cells were gated and used to assess expression of RANKL and B7-H3. A double positive RANKL and B7-H3 subpopulation of HLA-DR expressing cells was mainly detected in NK cells in co-culture rather than with rhIL-2 only. This subpopulation of HLA-DR+RANKL+B7-H3+ corresponded to the most highly proliferating as observed when depicted (red) against the total pool of expanded NK cells (black) based on the dilution of the eFluor®670 cell trace dye. Plots shown are representative of six different donors. The complete gating strategy is shown in Figure S7 in Supplementary Material.

activation of peripheral blood T cells by OKT3 requires additional signals from accessory cells and the presence of IL-2 (36). These accessory cells are mainly monocytes that through their Fc receptors, bind OKT3, and induce crosslinking of the CD3 molecule on T cells leading to a proper stimulation (36–38). This may explain that upon the absence of monocytes, despite the presence of OKT3 and IL-2, cytokine production by T cells is severely hampered. Moreover, the effect of OKT3, murine IgG2a mAb, on NK cell activation through the Fc receptor CD16 would be minimal, as CD16 has no or low binding affinity to IgG2a isotype (37, 39). In summary, in our co-culture system, the interaction of irradiated T cells and monocytes through bidirectional binding of OKT3, together the presence of IL-2 seems sufficient to induce production of soluble factors at the beginning of the co-culture with clear further beneficial effects on NK cell *ex vivo* growth.

Several studies have performed transcriptional and microRNA expression analysis in human NK cells to determine global changes after short activation or long-term expansion (23, 40, 41). Using also whole genome expression analysis and functional annotation, we focused on transcriptional differences between proliferating NK cells and their remaining non-proliferating counterparts after 5-days in co-culture with IAPs. By choosing this time point for analysis, we assured the emergence of pro-survival and proliferation pathways, absent at earlier time points, or extinguished at later ones. Certainly, we identified among the most highly expressed transcripts in proliferating NK cells many molecules involved in regulation of cell cycle and mitosis. Interestingly also transcripts corresponding to the zinc finger proteins, *ZBED2* and *ZBTB32* were upregulated. While only little is known about ZBED2 (42), the transcription factor ZBTB32 recently has been described to play a role in controlling proliferation of mouse cytomegalovirus (MCMV)-specific NK cells (43). Since in that study ZBTB32 was predominantly involved in the context of infection and inflammation, but was not required for homeostatic proliferation of mouse MCMV-specific NK cells, it was surprising to find increased levels of a transcript corresponding to ZBTB32 in proliferating human NK cells in our co-culture system. Therefore, it will be interesting to analyze which role zinc finger proteins, particularly, ZBED2 and ZBTB32 play in modulating the *ex vivo* expansion of mature human NK cells.

We describe also several phenotypic characteristics of highly proliferating NK cells compared with non-proliferating and resting ones. Actively proliferating NK cells lose expression of the inhibitory receptors Siglec-7 and -9 compared with both resting and non-proliferating counterparts. Several studies have shown that blocking of Siglec-7 and -9 receptors enhances NK cell cytotoxicity against target tumors, allowing for a better activation (44, 45). The reduced expression of Siglec-7 and -9 in proliferating NK cells may reflect a differential activation status related to high proliferation. Also in this regard, we detected reduced expression of the activating receptors NKp80 and CD16 preferentially in the population of proliferating NK cells. Down-regulation of NKp80 and CD16 is known to occur early after NK cell activation (46–48). Therefore, decreased Siglec-7/-9, NKp80, and CD16 expression evidence that differential activation responses within the original pool of NK cells result in either active *ex vivo* proliferation or quiescence. Moreover, the expression of the molecules RANKL and B7-H3 was significantly induced in NK cells particularly proliferating in co-culture with IAPs, as opposed to cultures with rhIL-2 only. Although up-regulation of RANKL has been previously reported to occur in $CD56^{bright}$ NK cells (49) little is known about its function in NK cells. On the other hand, B7-H3, a member of the B7/CD28 superfamily of costimulatory molecules is expressed in many human cancers and induced in several immune cells such as activated monocytes and dendritic cells with a controversial role in modulating T cell activity (50, 51). It would be interesting to determine whether B7-H3 is also a marker of activated NK cells *in vivo* and learn about its functional role.

Co-culture with IAPs also increased expression of particularly HLA-DR molecules in proliferating NK cells compared with culture with rhIL-2 only. Further analysis of the expression of RANKL and B7-H3 in HLA-DR expressing NK cells, revealed that a subpopulation of double positive RANKL and B7-H3 HLA-DR expressing NK cells, corresponds to the most actively expanded upon co-culture with IAPs. Increased frequencies of HLA-DR expressing NK cells have been observed already after few days of culture with IL-2, possibly due to clonal expansion of an original HLA-DR expressing NK cell subpopulation (52). Recently, the expression of HLA-DR molecules was described as a surrogate marker for NK cell clonality in chronic lymphoproliferative disorders (53). Our data support the hypothesis of a preferential clonal expansion of HLA-DR expressing NK cells possibly due to differential activation upon mitogens provided by IAPs as opposed to other NK cell subsets un-responsive to the same stimuli. Identification of the corresponding receptor(s) and pathway(s) responsive to pro-proliferative signaling will be key to optimize NK cell expansion protocols. Additionally, we describe a subset of $HLA\text{-}DR^{+}RANKL^{+}B7\text{-}H3^{+}$ NK cells as the subset with the highest proliferative potential emerging under co-culture with IAPs. Subsequent studies may help to elucidate a functional role of RANKL and B7-H3 molecules in these highly proliferating NK cells. In addition, it will be interesting to verify the occurrence of a preferential expansion of HLA-DR expressing NK cells with other *ex vivo* expansion protocols, and whether this confers antigen presenting characteristics to NK cells, as it has been previously shown (54). Confirmation of antigen presenting properties of expanded NK cells may be further considered to extend the therapeutic potential of NK cells.

In summary, our findings uncover molecular and phenotypic characteristics of *ex vivo* proliferating NK cells and roles of autologous feeder cells that may have an impact in the development of new expansion methods and analysis strategies of expanded NK cells. In addition, we propose that characterization of the newly described subpopulation of $HLA\text{-}DR^{+}RANKL^{+}B7\text{-}H3^{+}$ NK cells will be of interest for the further optimization of NK cell expansion protocols, and improved immunotherapeutic applications.

ETHICS STATEMENT

This study was carried out using samples from healthy donors obtained from Klinikum Dortmund (Dortmund, Germany) with given consent from the Klinikum Dortmund to use those samples for research purposes.

AUTHOR CONTRIBUTIONS

MD-V designed and performed experiments, analyzed data, and wrote the manuscript; JK performed processing and statistical analysis of gene expression data and contributed to write the manuscript; UK provided advice and contributed to write the manuscript; and VH supervised research work, provided advice, and wrote the manuscript. All authors have discussed and revised the manuscript.

ACKNOWLEDGMENTS

We specially thank Silvia Rüberg and Michail Knauel for their assistance with the performance of the microarray and pre-analysis, and Nina Möker for her support in discussions and critical reading of this manuscript.

REFERENCES

1. Vacca P, Montaldo E, Croxatto D, Moretta F, Bertaina A, Vitale C, et al. NK cells and other innate lymphoid cells in hematopoietic stem cell transplantation. *Front Immunol* (2016) 7:188. doi:10.3389/fimmu.2016.00188
2. Kiessling R, Klein E, Pross H, Wigzell H. "Natural" killer cells in the mouse. II. Cytotoxic cells with specificity for mouse Moloney leukemia cells. Characteristics of the killer cell. *Eur J Immunol* (1975) 5:117–21. doi:10.1002/eji.1830050209
3. Raulet DH. Missing self recognition and self tolerance of natural killer (NK) cells. *Semin Immunol* (2006) 18:145–50. doi:10.1016/j.smim.2006.03.003
4. Vivier E, Tomasello E, Baratin M, Walzer T, Ugolini S. Functions of natural killer cells. *Nat Immunol* (2008) 9:503–10. doi:10.1038/ni1582
5. Ruggeri L, Capanni M, Urbani E, Perruccio K, Shlomchik WD, Tosti A, et al. Effectiveness of donor natural killer cell alloreactivity in mismatched hematopoietic transplants. *Science* (2002) 295:2097–100. doi:10.1126/science.1068440
6. Locatelli F, Pende D, Mingari MC, Bertaina A, Falco M, Moretta A, et al. Cellular and molecular basis of haploidentical hematopoietic stem cell transplantation in the successful treatment of high-risk leukemias: role of alloreactive NK cells. *Front Immunol* (2013) 4:15. doi:10.3389/fimmu.2013.00015
7. Miller JS, Soignier Y, Panoskaltsis-Mortari A, McNearney SA, Yun GH, Fautsch SK, et al. Successful adoptive transfer and in vivo expansion of human haploidentical NK cells in patients with cancer. *Blood* (2005) 105:3051–7. doi:10.1182/blood-2004-07-2974
8. Rubnitz JE, Inaba H, Ribeiro RC, Pounds S, Rooney B, Bell T, et al. NKAML: a pilot study to determine the safety and feasibility of haploidentical natural killer cell transplantation in childhood acute myeloid leukemia. *J Clin Oncol* (2010) 28:955–9. doi:10.1200/JCO.2009.24.4590
9. Yang Y, Lim O, Kim TM, Ahn Y-O, Choi H, Chung H, et al. Phase I study of random healthy donor-derived allogeneic natural killer cell therapy in patients

with malignant lymphoma or advanced solid tumors. *Cancer Immunol Res* (2016) 4:215–24. doi:10.1158/2326-6066.CIR-15-0118
10. Koepsell SA, Miller JS, McKenna DH. Natural killer cells: a review of manufacturing and clinical utility. *Transfusion* (2013) 53:404–10. doi:10.1111/j.1537-2995.2012.03724.x
11. Koehl U, Brehm C, Huenecke S, Zimmermann S-Y, Kloess S, Bremm M, et al. Clinical grade purification and expansion of NK cell products for an optimized manufacturing protocol. *Front Oncol* (2013) 3:118. doi:10.3389/fonc.2013.00118
12. Klingemann H. Challenges of cancer therapy with natural killer cells. *Cytotherapy* (2015) 17:245–9. doi:10.1016/j.jcyt.2014.09.007
13. Koehl U, Kalberer C, Spanholtz J, Lee DA, Miller JS, Cooley S, et al. Advances in clinical NK cell studies: donor selection, manufacturing and quality control. *Oncoimmunology* (2016) 5:e1115178. doi:10.1080/2162402X.2015.1115178
14. Chabannon C, Mfarrej B, Guia S, Ugolini S, Devillier R, Blaise D, et al. Manufacturing natural killer cells as medicinal products. *Front Immunol* (2016) 7:504. doi:10.3389/fimmu.2016.00504
15. Becker PSA, Suck G, Nowakowska P, Ullrich E, Seifried E, Bader P, et al. Selection and expansion of natural killer cells for NK cell-based immunotherapy. *Cancer Immunol Immunother* (2016) 65:477–84. doi:10.1007/s00262-016-1792-y
16. Ahn Y, Kim S, Kim TM, Song EY, Park MH, Heo DS. Irradiated and activated autologous PBMCs induce expansion of highly cytotoxic human NK cells in vitro. *J Immunother* (2013) 36:373–81. doi:10.1097/CJI.0b013e3182a3430f
17. Lim O, Lee Y, Chung H, Her JH, Kang SM, Jung M, et al. GMP-compliant, large-scale expanded allogeneic natural killer cells have potent cytolytic activity against cancer cells in vitro and in vivo. *PLoS One* (2013) 8:e53611. doi:10.1371/journal.pone.0053611
18. Rujkijyanont P, Chan WK, Eldridge PW, Lockey T, Holladay M, Rooney B, et al. Ex vivo activation of CD56(+) immune cells that eradicate neuroblastoma. *Cancer Res* (2013) 73:2608–18. doi:10.1158/0008-5472.CAN-12-3322
19. Sakamoto N, Ishikawa T, Kokura S, Okayama T, Oka K, Ideno M, et al. Phase I clinical trial of autologous NK cell therapy using novel expansion method in patients with advanced digestive cancer. *J Transl Med* (2015) 13:277. doi:10.1186/s12967-015-0632-8
20. Miller JS, Oelkers S, Verfaillie C, McGlave P. Role of monocytes in the expansion of human activated natural killer cells. *Blood* (1992) 80:2221–9.
21. Hassold N, Seystahl K, Kempf K, Urlaub D, Zekl M, Einsele H, et al. Enhancement of natural killer cell effector functions against selected lymphoma and leukemia cell lines by dasatinib. *Int J Cancer* (2012) 131:E916–27. doi:10.1002/ijc.27537
22. Veluchamy JP, Delso-Vallejo M, Kok N, Bohme F, Seggewiss-Bernhardt R, van der Vliet HJ, et al. Standardized and flexible eight colour flow cytometry panels harmonized between different laboratories to study human NK cell phenotype and function. *Sci Rep* (2017) 7:43873. doi:10.1038/srep43873
23. Granzin M, Soltenborn S, Müller S, Kollet J, Berg M, Cerwenka A, et al. Fully automated expansion and activation of clinical-grade natural killer cells for adoptive immunotherapy. *Cytotherapy* (2015) 17:621–32. doi:10.1016/j.jcyt.2015.03.611
24. Gentleman R, Carey V, Bates D, Bolstad B, Dettling M, Dudoit S, et al. Bioconductor: open software development for computational biology and bioinformatics. *Genome Biol* (2004) 5:R80. doi:10.1186/gb-2004-5-10-r80
25. R Development Core Team. *R: A Language and Environment for Statistical Computing*. Vienna Austria: R Found Stat Comput (2016). ISBN 3-900051-07-0.
26. Bolstad BM, Irizarry RA, Astrand M, Speed TP. A comparison of normalization methods for high density oligonucleotide array data based on variance and bias. *Bioinformatics* (2003) 19:185–93. doi:10.1093/bioinformatics/19.2.185
27. Weng L, Dai H, Zhan Y, He Y, Stepaniants SB, Bassett DE. Rosetta error model for gene expression analysis. *Bioinformatics* (2006) 22:1111–21. doi:10.1093/bioinformatics/btl045
28. Klug A. The discovery of zinc fingers and their development for practical applications in gene regulation and genome manipulation. *Q Rev Biophys* (2010) 43:1–21. doi:10.1017/S0033583510000089
29. Siggs OM, Beutler B. The BTB-ZF transcription factors. *Cell Cycle* (2012) 11:3358–69. doi:10.4161/cc.21277
30. Beaulieu AM, Sant'Angelo DB. The BTB-ZF family of transcription factors: key regulators of lineage commitment and effector function development in the immune system. *J Immunol* (2011) 187:2841–7. doi:10.4049/jimmunol.1004006
31. Choi I, Yoon SR, Park S-Y, Kim H, Jung S-J, Jang YJ, et al. Donor-derived natural killer cells infused after human leukocyte antigen-haploidentical hematopoietic cell transplantation: a dose-escalation study. *Biol Blood Marrow Transplant* (2014) 20:696–704. doi:10.1016/j.bbmt.2014.01.031
32. Cho D, Campana D. Expansion and activation of natural killer cells for cancer immunotherapy. *Korean J Lab Med* (2009) 29:89–96. doi:10.3343/kjlm.2009.29.2.89
33. Becknell B, Caligiuri MA. Interleukin-2, interleukin-15, and their roles in human natural killer cells. *Adv Immunol* (2005) 86:209–39. doi:10.1016/S0065-2776(04)86006-1
34. Parrish-novak J, Dillon SR, Nelson A, Hammond A, Sprecher C, Gross JA, et al. Interleukin 21 and its receptor are involved in NK cell expansion and regulation of lymphocyte function. *Nature* (2000) 408:57–63. doi:10.1038/35040504
35. Kloss M, Decker P, Baltz KM, Baessler T, Jung G, Rammensee H-G, et al. Interaction of monocytes with NK cells upon toll-like receptor-induced expression of the NKG2D ligand MICA. *J Immunol* (2008) 181:6711–9. doi:10.4049/jimmunol.181.10.6711
36. Schwab R, Crow MK, Russo C, Weksler ME. Requirements for T cell activation by OKT3 monoclonal antibody: role of modulation of T3 molecules and interleukin 1. *J Immunol* (1985) 135:1714–8.
37. Itoh K, Balch CM, Trial J, Platsoucas CD. CD8+ T cells lyse autologous monocytes in the presence of anti-CD3 monoclonal antibody: association with interleukin-1 production. *Cell Immunol* (1988) 114:257–71. doi:10.1016/0008-8749(88)90320-6
38. Rinnooy Kan EA, Platzer E, Welte K, Wang CY. Modulation induction of the T3 antigen by OKT3 antibody is monocyte dependent. *J Immunol* (1984) 133:2979–85.
39. Li P, Jiang N, Nagarajan S, Wohlhueter R, Selvaraj P, Zhu C. Affinity and kinetic analysis of Fcgamma receptor IIIa (CD16a) binding to IgG ligands. *J Biol Chem* (2007) 282:6210–21. doi:10.1074/jbc.M609064200
40. Dybkaer K, Iqbal J, Zhou G, Geng H, Xiao L, Schmitz A, et al. Genome wide transcriptional analysis of resting and IL2 activated human natural killer cells: gene expression signatures indicative of novel molecular signaling pathways. *BMC Genomics* (2007) 8:230. doi:10.1186/1471-2164-8-230
41. Park KU, Jin P, Sabatino M, Feng J, Civini S, Khuu H, et al. Gene expression analysis of ex vivo expanded and freshly isolated NK cells from cancer patients. *J Immunother* (2011) 33:945–55. doi:10.1097/CJI.0b013e3181f71b81
42. Hayward A, Ghazal A, Andersson G, Andersson L, Jern P. ZBED evolution: repeated utilization of DNA transposons as regulators of diverse host functions. *PLoS One* (2013) 8:e59940. doi:10.1371/journal.pone.0059940
43. Beaulieu AM, Zawislak CL, Nakayama T, Sun JC. The transcription factor Zbtb32 controls the proliferative burst of virus-specific natural killer cells responding to infection. *Nat Immunol* (2014) 15:546–53. doi:10.1038/ni.2876
44. Jandus C, Boligan KF, Chijioke O, Liu H, Dahlhaus M, Démoulins T, et al. Interactions between Siglec-7/9 receptors and ligands influence NK cell-dependent tumor immunosurveillance. *J Clin Invest* (2014) 124:1810–20. doi:10.1172/JCI65899DS1
45. Xiao H, Woods EC, Vukojicic P, Bertozzi CR. Precision glycocalyx editing as a strategy for cancer immunotherapy. *Proc Natl Acad Sci USA* (2016) 113:10304–9. doi:10.1073/pnas.1608069113
46. Klimosch SN, Bartel Y, Wiemann S, Steinle A. Genetically coupled receptor-ligand pair NKp80-AICL enables autonomous control of human NK cell responses. *Blood* (2013) 122:2380–9. doi:10.1182/blood-2013-01-479790
47. Huenecke S, Zimmermann SY, Kloess S, Esser R, Brinkmann A, Tramsen L, et al. IL-2-driven regulation of NK cell receptors with regard to the distribution of CD16+ and CD16- subpopulations and in vivo influence after haploidentical NK cell infusion. *J Immunother* (2010) 33:200–10. doi:10.1097/CJI.0b013e3181bb46f7
48. Romee R, Foley B, Lenvik T, Wang Y, Zhang B, Ankarlo D, et al. NK cell CD16 surface expression and function is regulated by a disintegrin and metalloprotease-17 (ADAM17). *Blood* (2013) 121:3599–608. doi:10.1182/blood-2012-04-425397
49. Wilk E, Kalippke K, Buyny S, Schmidt RE, Jacobs R. New aspects of NK cell subset identification and inference of NK cells' regulatory capacity by assessing functional and genomic profiles. *Immunobiology* (2008) 213:271–83. doi:10.1016/j.imbio.2007.10.012
50. Hashiguchi M, Inamochi Y, Nagai S, Otsuki N, Piao J, Kobori H, et al. Human B7-H3 binds to Triggering receptor expressed on myeloid cells-like transcript

2 (TLT-2) and enhances T cell responses. *Open J Immunol* (2012) 2:9–16. doi:10.4236/oji.2012.21002

51. Loos M, Hedderich DM, Friess H, Kleeff J. B7-h3 and its role in antitumor immunity. *Clin Dev Immunol* (2010) 2010:683875. doi:10.1155/2010/683875
52. Evans JH, Horowitz A, Mehrabi M, Wise EL, Pease JE, Riley EM, et al. A distinct subset of human NK cells expressing HLA-DR expand in response to IL-2 and can aid immune responses to BCG. *Eur J Immunol* (2011) 41:1924–33. doi:10.1002/eji.201041180
53. Bárcena P, Jara-Acevedo M, Tabernero MD, López A, Sánchez ML, García-Montero AC, et al. Phenotypic profile of expanded NK cells in chronic lymphoproliferative disorders: a surrogate marker for NK-cell clonality. *Oncotarget* (2015) 6:42938–51. doi:10.18632/oncotarget.5480
54. Roncarolo MG, Bigler M, Haanen JB, Yssel H, Bacchetta R, de Vries JE, et al. Natural killer cell clones can efficiently process and present protein antigens. *J Immunol* (1991) 147:781–7.

14

MICA and NKG2D: Is There an Impact on Kidney Transplant Outcome?

Matilde Risti and Maria da Graça Bicalho*

LIGH – Immunogenetics and Histocompatibility Laboratory, Federal University of Paraná, Curitiba, Brazil

Edited by:
Gianfranco Pittari,
Hamad Medical Corporation, Qatar

Reviewed by:
Nicolas Dulphy,
Paris Diderot University, France
Myra Coppage,
University of Rochester Medical Center, USA

***Correspondence:**
Maria da Graça Bicalho
ligh@ufpr.br

This paper aims to present an overview of MICA and natural killer group 2 member D (NKG2D) genetic and functional interactions and their impact on kidney transplant outcome. Organ transplantation has gone from what can accurately be called a “clinical experiment” to a routine and reliable practice, which has proven to be clinically relevant, life-saving and cost-effective when compared with non-transplantation management strategies of both chronic and acute end-stage organ failures. The kidney is the most frequently transplanted organ in the world (transplant-observatory[1]). The two treatment options for end-stage renal disease (ESRD) are dialysis and/or transplantation. Compared with dialysis, transplantation is associated with significant improvements in quality of life and overall longevity. A strong relationship exists between allograft loss and human leukocyte antigens (HLA) antibodies (Abs). HLA Abs are not the only factor involved in graft loss, as multiple studies have shown that non-HLA antigens are also involved, even when a patient has a good HLA matche and receives standard immunosuppressive therapy. A deeper understanding of other biomarkers is therefore important, as it is likely to lead to better monitoring (and consequent success) of organ transplants. The objective is to fill the void left by extensive reviews that do not often dive this deep into the importance of MICA and NKG2D in allograft acceptance and their partnership in the immune response. There are few papers that explore the relationship between these two protagonists when it comes to kidney transplantation. This is especially true for the role of NKG2D in kidney transplantation. These reasons give a special importance to this review, which aims to be a helpful tool in the hands of researchers in this field.

Keywords: transplantation, kidney, allograft, MICA, MICA-129, NKG2D, LNK1, HNK1

INTRODUCTION

Genetic diversity is the hallmark of MHC genes (1). The main antigenic barrier to transplantation is molecules, which are polypeptide products of a cluster of genes known, in humans, as human leukocyte antigens (HLA). In addition, a family of highly glycosylated MHC-encoded molecules, the *MHC class I chain-related (MIC)* genes, has been identified (2) as a second lineage of mammalian MHC I genes, which could constitute an antigenic barrier to transplantation as well (3). The MIC molecules possess a low degree of homology to other *MHC class I* encoded genes and interact with

[1] http://www.transplant-observatory.org.

both T-cell and natural killer (NK)-cell receptors (2). MIC proteins act as ligands for NK cells, γδ T cells, and αβ CD8+ T cells, which express natural killer group 2 member D (NKG2D) ligand (4). The importance of the MICA protein in kidney transplantation has been acknowledged in recent years, and the role they play in graft rejection has been intensely pursued.

MICA GENE: STRUCTURE, POLYMORPHISMS, AND FUNCTION

The *MIC* gene family consists of seven members (*MICA–MICG*) (**Figure 1**), five of which are pseudogenes, and two, *MICA* and *MICB*, of which are functional (5, 6). *MICA* and *MICB* are the most divergent members of the human MHC-encoded class I genes identified to date, having an average of 19, 25, and 35% similarity in the extracellular α1, α2, and α3 domains, respectively, to those of other MHC α-polypeptides (7).

The *MICA* gene is located 46.4 kb centromeric to *HLA-B* on the short (p) arm of chromosome 6 at position 21.33 (3).

MICA and *MICB* have been shown to differ in the transcriptional control regions from common HLA class I genes. *MICA/B* genes lack the prototypic MHC class I gene promoter regulatory elements, the SXY module [heterotrimeric X-box-binding factor—regulatory factor X; X2-box-binding factor—cyclic-AMP-responsive-element-binding protein; Y-box-binding factor—nuclear transcription factor Y (NF-Y); and an as-yet-unidentified S-box-binding factor]. In contrast, the regulatory promoter module of *MICA/B* contains heat shock elements resembling those of HSP70 genes, a CCAAT box that binds to nuclear transcription factor Y (NF-Y), and a GC box that binds to Sp1, Sp3, and Sp4 transcription factors (8).

There are 12 known possible haplotypes of *MICA* 5′ promoter regions, including a null haplotype due to a deletion of the entire *MICA* gene (*MICA*-P12), which are more densely distributed in both ends compared to the central portion of 5′ promoter (8, 9).

MICA has six exons separated by five introns (**Figure 2**): exon 1 encodes the leader peptide, exons 2–4 encode three extracellular globular domains, exon 5 encodes the transmembrane domain, and exon 6 encodes the cytoplasmatic tail (6, 10). An intron of 6,840 bp follows exon 1 and is unusually large for a class I gene. The remainder of the *MICA* gene has a quite similar organization to classical class I genes, except for the presence of a relatively long intron 5 and the fusion of the cytoplasmic tail and 3′ UTR sequence in a single last exon (11).

It is considered that *MICA* gene has a codominant expression, and the presence of heat shock elements within the promoter suggests that MICA transcription is indued under stress conditions, and that therefore the MICA protein functions as an indicator of cell stress (11–13). The first intron of the *MICA* gene contains an NFkB-binding site that binds p65 (RelA)/p50 heterodimers and p50/p50 homodimers of the NFkB transcription factor family. The role of the proximal −130 bp NFkB site was reported as necessary and sufficient for transcriptional transactivation of *MICA* in response to TNFα in primary endothelial cells (ECs) (14).

Gene transcription isoforms are mRNAs transcribed from the same locus that differs in their transcription start sites and/or untranslated regions or protein coding DNA sequences (CDSs) also producing different protein isoforms. The alternative splicing of *MICA* leads to the formation of four isoforms. Two of them were described by Zou and Stastny (15) (MICA isoforms 1 and 2), and they did not appear to be tissue specific.

MICA isoform 1 (*1*001*) is the longest isoform, derived from the *MICA*001* allele. *MICA* isoform 2 (*1*008:01*) is a variant isoform derived from the *MICA*008:01* allele that contains a four-nucleotide insertion (rs9279200), which causes a frameshift mutation and subsequent truncation of the CDS, compared to isoform 1 (allele *MICA*001*) (15). The other two isoforms of *MICA*, isoforms 3 and 4, are described only in the *ncbi.nih.gov/gene*[2] website. *MICA* isoform 3 is, like isoform 2, encoded by the *MICA*008:01* allele; however, it is shorter than isoform 2 at the N-terminus, containing an alternate 5′ exon, differences in the 5′ UTR, and lacking a portion of the 5′ coding region, with translation being initiated from a downstream in frame start codon. *MICA* isoform 4 contains an alternate 5′ exon and uses an alternate splice site in an internal exon. It differs in the 5′ UTR, lacks a portion of the 5′ coding region, and initiates translation from an alternate start codon, compared to variant 1 (*MICA*008:01 allele*). Isoform 4 has a distinct and shorter N-terminus, compared to isoform 2.

The *MIC* genes are transcribed in keratinocytes, ECs, fibroblasts, monocytes, epithelial cell lines and epithelial tissues of cell lines, and freshly isolated cells (2, 16) and are not usually transcribed in CD4+ T cells, CD8+ T cells, and CD19+ cells (17). MIC protein is only expressed on the cell surface of freshly isolated ECs, fibroblasts (17), and gastric epithelium (12). MIC protein acts as a ligand for NK cells, γδ T cells, and αβ CD8+ T cells, which express NKG2D ligand (NKG2DL) (4).

[2]https://www.ncbi.nlm.nih.gov/gene?Db=gene&Cmd=DetailsSearch&Term=100507436#.

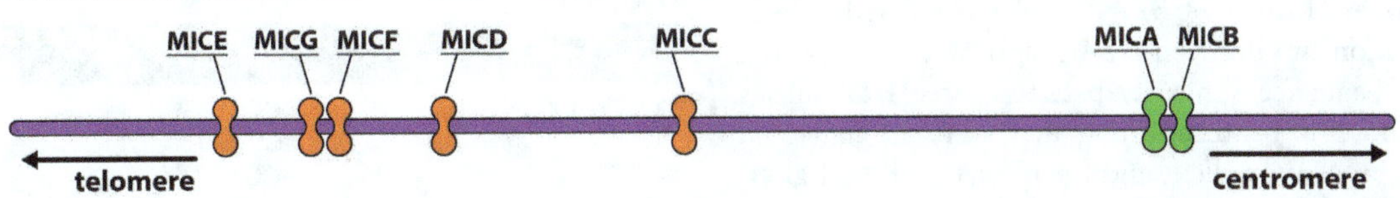

FIGURE 1 | **Representation of *MHC class I chain-related (MIC)* genes**. The functional genes are represented in green and the pseudogenes are in orange (image by Matilde Risti).

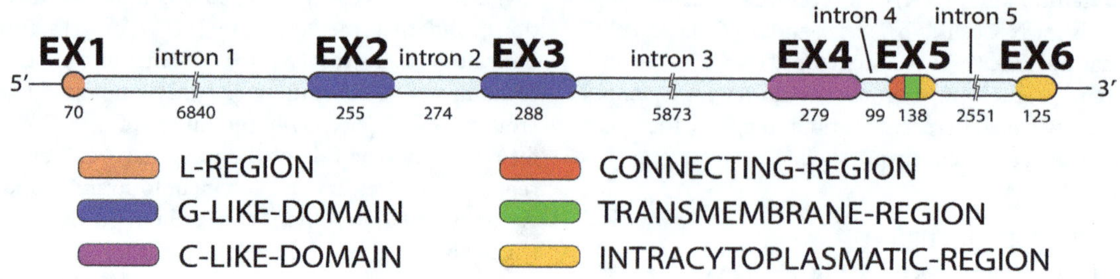

FIGURE 2 | *MICA* gene exon–intron organization. The *MICA* gene has five introns and six exons (image by Matilde Risti).

MICA Polymorphism and -129Met/Val Dimorphism

Bahram et al. (3) first described *MICA*01* to *MICA*05* alleles with a total of 18 nucleotide substitutions resulting in 14 amino acid changes in the final polypeptide. Fodil et al. (7) described the alleles, *MICA*06* to *MICA*16*, with nine nucleotide substitutions and eight amino acids changes. One year later, Mizuki et al. (18) showed a variable number of trinucleotide GCT repeats that encode 4, 5, 6, 7, 9, or 10 alanine (A, Ala). The short tandem repeats or microsatellite alleles were labeled as *A4*, *A5*, *A6*, *A7*, *A8*, *A9*, and *A10*. There is also an *A5.1* allele that contains five triplet repeats of GCT plus an additional guanine nucleotide insertion (GGCT). This insertion causes a frameshift mutation leading to a premature intradomain stop codon within the transmembrane region, which deletes the MICA cytoplasmic tail. The A4, A5, A6, A7, A8, A9, A10, and A5.1 sizes are, respectively, 179, 182, 185, 194, and 183 bp (18–20). At the time of writing (October 2016) hla.alleles.org[3] reports 105 *MICA* alleles, 2 of which considered null, result in 82 different MICA proteins. All *MICA* alleles from **001* to **087* producing different proteins and their nucleotides variations on exons 2–6 are shown in **Table 1**.

Several studies have documented *MICA* allele frequencies within different populations (**Figure 3**), and the frequency distribution varies between them. For example, the same group of three alleles (*MICA*008*, *MICA*002*, and *MICA*004*) accounts for more than 50% of the allele frequencies commonly found in several Caucasoid populations (21–24) but at the same time *MICA*027*'s frequency is extremely different in a comparison between the South American Indian and Caucasoid populations (25). Single high-frequency *MICA* alleles are each associated with more than one different *HLA*-B allele, but this pattern is not reciprocal. Most specific *HLA-B* alleles, including *B*07:02* and *B*08:01* variations, are usually linked to a single *MICA* allele. This pattern suggests that the *MICA* alleles had an earlier origin than major branches of *HLA-B* alleles (26).

The evolutionary history of *HLA-B* alleles is recognizable in the linkage relationship between *HLA-B* and *MICA* genes. The high degree of sequence similarity between three *HLA* alleles (*B*35*, *B*53*, and *B*58*) indicates that they were all generated from the same progenitor allele, and the observation that globally they are all linked to the *MICA*002* allele further supports

[3]http://hla.alleles.org/nomenclature/stats.html.

TABLE 1 | Nucleotide variations on exons 2–6 for *MICA* alleles from **001* to **087*.

	EXON 2 α1	
CODON 6	CTG (LEU)	CGC (ARG)
		CTC (PRO)
CODON 14	TGG (TRP)	GGG (GLY)
CODON 23	CTC (LEU)	GTT (LEU)
CODON 24	ACT (THR)	GCT (VAL)
CODON 26	GTA (VAL)	GGA (GLY)
CODON 36	TGT (CYS)	TAT (TYR)
CODON 38	AGG (ARG)	AGC (SER)
CODON 39	CAG (GLN)	TAG (Stop)
CODON 55	GGA (GLY)	GGC (GLY)
CODON 56	AAT (ASN)	AAC (ASN)
CODON 64	AGA (ARG)	AAG (ARG)
CODON 69	AAC (ASN)	AAT (ASN)
	EXON 3 α2	
CODON 90	CTC (LEU)	TTC (PHE)
CODON 91	CAG (GLN)	CGG (ARG)
CODON 93	ATT (ILE)	ATG (MET)
CODON 102	AAC (ASN)	AGC (SER)
CODON 105	AAG (ARG)	AAG (LYS)
CODON 112	TAC (TYR)	TAT (TYR)
CODON 114	GGG (GLY)	AGG (ARG)
CODON 122	CTG (LEU)	GTG (VAL)
CODON 124	ACT (THR)	TCT (SER)
CODON 125	AAG (LYS)	GAG (GLU)
CODON 129	ATG (MET)	GTG (VAL)
CODON 130	CCC (PRO)	TCC (SER)
CODON 139	GCC (ALA)	GCA (ALA)
CODON 142	GTC (VAL)	ATC (ILE)
CODON 151	ATG (MET)	GTG (VAL)
CODON 156	CAC (HIS)	CTC (LEU)
		CGC (ARG)
CODON 169	CGG (ARG)	TGG (TRP)
CODON 173	AAA (LYS)	GAA (GLU)
CODON 174	TCC (SER)	TCT (SER)
CODON 175	GGC (GLY)	AGC (SER)
		GGT (GLY)
CODON 176	GTA (VAL)	ATA (ILE)
CODON 181	ACA (THR)	AGA (ARG)
	EXON 4 α3	
CODON 190	CGC (ARG)	TGC (CYS)
CODON 191	AGC (SER)	AGT (SER)
CODON 193	GCC (ALA)	GCA (ALA)
CODON 198	ATT (ILE)	ATC (ILE)
CODON 205	TCT (SER)	TCC (SER)
CODON 206	GGC (GLY)	AGC (SER)

(Continued)

TABLE 1 | Continued

	EXON 4 α3	
CODON 208	TAT (TYR)	TGT (CYS)
CODON 210	TGG (TRP)	CGG (ARG)
CODON 213	ACA (THR)	ATA (ILE)
CODON 215	AGC (SER)	ACC (THR)
CODON 221	GTA (VAL)	CTA (LEU)
CODON 230	TGG (TRP)	TCG (SER)
CODON 244	TGG (TRP)	TGA (Stop)
CODON 247	AAC (THR)	ACT (THR)
CODON 250	TGC (CYS)	CGC (ARG)
CODON 251	CAA (GLN)	CGA (ARG)
CODON 253	GAG (GLU)	AAG (LYS)
CODON 254	GAG (GLU)	AAG (LYS)
CODON 256	AAG (ARG)	AGT (SER)
		AAG (LYS)
CODON 265	GGG (GLY)	AGG (ARG)
CODON 268	AGC (SER)	GGC (GLY)
CODON 269	ACT (THR)	ATT (ILE)
CODON 271	CCT (PRO)	GCT (ALA)
	EXON 5 TM	
CODON 295	CGT (ALA)	GCGT
CODON 304	TAT (TYR)	TAC (TYR)
CODON 306	CGT (ARG)	TGT (CYS)
	EXON 6	
CODON 350	GAT (ASP)	GCT (ALA)
CODON 354	ACT (THR)	GCT (ALA)
CODON 359	GGC (GLY)	GGT (GLY)
CODON 360	GCC (ALA)	ACC(THR)

*Codons are shown in the first column. The second column shows the triplets and their corresponding amino acids in the consensus sequence (MICA*001). The third column lists that triplet's possible variations in other alleles compared with the consensus sequence. Amino acid substitutions in MICA on the three external protein domains (exons 2–4), on the transmembrane domain TM (exon 5) and carboxy-terminal cytoplasmic tail (exon 6). The G nucleotide insertion is represented in red in the exon 5 TM.*

this conclusion. Specific *MICA* alleles also tend to associate with serological HLA-B groups. A rare exception can be found in *B*44*, whose two subgroups *B*44:02* and *B*44:03* have exclusive associations with *MICA*008* and *MICA*004* (26).

The *MICA-129Val/Met* dimorphism, caused by an SNP (rs1051792) at nucleotide position 454 (G>A) of the *MICA* gene is of particular interest. The substitution of valine (Val) for methionine (Met) at position 129 in the α2 domain of the MICA protein has been reported to affect NKG2D binding avidity (36–40). This dimorphism divides the *MICA* alleles into two groups (**Table 2**). In 2015, it has been observed that *MICA-129Met* alelles increased the risk of experiencing acute graft-versus-host disease. This effect could be the consequence on NKG2D signaling by MICA-129Met variant (40). In addition to this, it has been shown that the MICA-129 dimorphism may directly affect plasma membrane expression and shedding of MICA, and these functional effects might contribute to the numerous disease associations (41).

MICA Molecule

MICA is a highly glycosylated membrane-anchored cell surface protein composed of 383 amino acids (12). Unglycosylated MICA appeared less stable than those incorporating glycosylated MICA (36). Its expression has been reported on the surface of different cells and resembles the domain organization (**Figure 4**) of the α chain of MHC class I molecules (16, 42). MICA α chain does not bind β2-microglobulin and is independent of any transporter-associated protein. Attempts to identify peptides bound to MICA have been unsuccessful (10, 12). The crystal structure of MICA shows four distinct α helices arranged in an eight-stranded antiparallel β sheet. These helices in MICA roughly correspond to the two helices that define the peptide-binding groove in peptide-binding MHC class I proteins and homologs (42).

MICA is generally concentrated in lipid rafts and is *S*-acylated, similar to other lipid rafts-associated proteins. *In vitro* mutation of the S-acylation site, replacing a cysteine residue with a stop codon at aminoacid position 39, yields a truncated form of MICA, unable to activate NK cells (43).

The MICA molecule interacts with NK cells, γδ T cells, and αβ CD8+ T cells, which express NKG2D, a common activating NK cell receptor (4, 10, 44). NKG2D recognizes the human MICA protein in conjunction with a transmembrane signaling adaptor protein, DNAX-activation protein (DAP10) (4, 10).

It is noteworthy that the MICA molecule can also be recognized by γδ T cells with the TCR variable region $V_{\delta}1$ (4, 45–47).

Both types of receptors, $V_{\delta}1$TCR and NKG2D, can simultaneously recognize and bind to MICA on a $V_{\delta}1$ cell surface. There is close association between the tissue distribution of $V_{\delta}1$ cells and the physiological expression of MICA, as MICA affects $V_{\delta}1$ cell lineage development (46). In $V_{\delta}1$ $_{\gamma\delta}$T cells, the strength of the binding between TCR and MICA is weaker than that between NKG2D and MICA. Although weak, TCR:MICA complexes show unusual stability after they are formed, with long half-lives. TCR and NKG2D receptors compete for binding to MIC ligands, and it has been suggested that initial interactions at the point of contact may be dominated by NKG2D:MIC binding events, which then give way to longer-lived γδ TCR:MIC complexes (47).

Conclusions on MICA

The *MICA* gene is polymorphic, and it is in linkage disequilibrium with *HLA-B* genes. The MICA protein is expressed on the cell surface, and it is possibly the proteolytic cleavage of the α_3 domain which in turn releases soluble MICA (sMICA). The MICA molecule does not present a peptide in its groove and can interact with the NKG2D receptor, which is the focus of the following paragraphs.

NKG2D OR *KILLER CELL LECTIN-LIKE RECEPTOR K1 (KLRK1)* GENE: STRUCTURE, POLYMORPHISMS, AND FUNCTION

NKG2D gene, also known as KLRK1, is located in the natural killer complex (NKC) on chromosome 12 (42, 48, 49). Human *NKG2D* (**Figure 5**) has 10 exons (50). Exons 2–4 encode the intracellular/transmembrane domain; exons 5–8 encode the ligand-binding ectodomain, which is a membrane-bound domain protruding into extracellular space (50, 51). *NKG2D* has a low number of nucleotide variations (48). *NKG2D* appears to be conserved

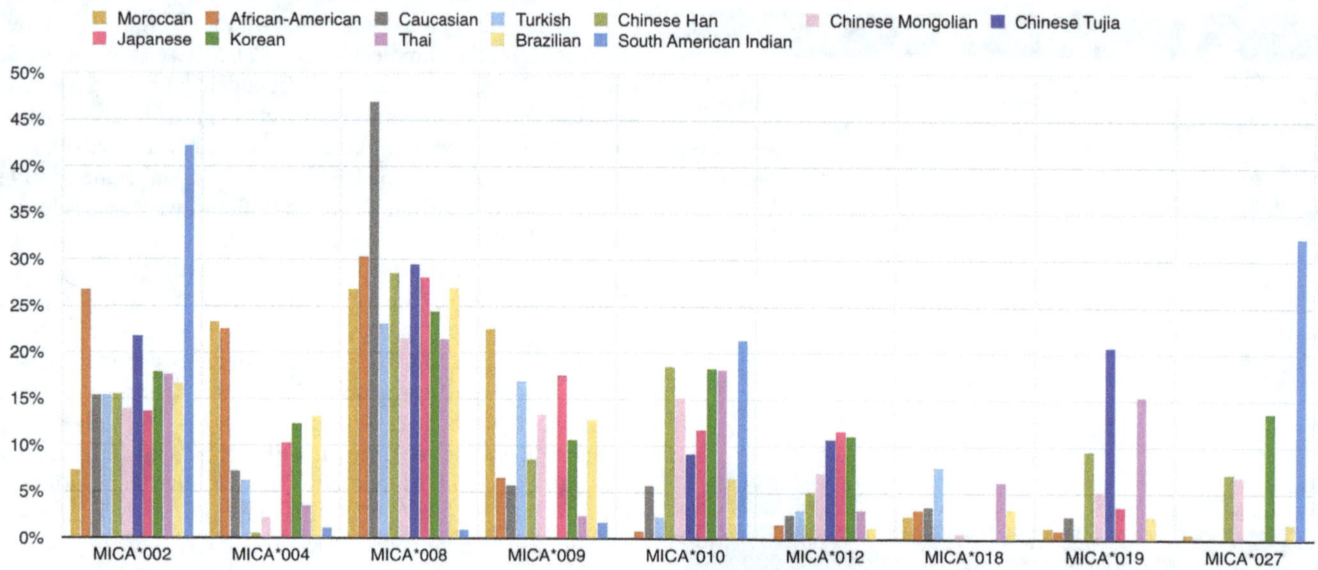

FIGURE 3 | Frequencies (%) of common *MICA* alleles within 12 populations. The allele frequencies of nine *MICA* alleles are shown for 12 populations: Caucasoid (21–23), Korean (19), North-Eastern Thai (27), Japanese (13, 28), African-American (21, 29), South American Indian (25), Moroccan (30), Turkish (31), Brazilian (24, 32), Chinese Mongolian (33), and Chinese Tujia (34, 35) (image by Matilde Risti).

TABLE 2 | Dimorphism 129 Val/Met divides *MICA* alleles into two groups.

Dimorphism 129 val/met divides *MICA* alleles in two groups	
ATG (Met)	**GTG (Val)**
*MICA*001, ***002**, *007, *011, ***012**, *014, *015, *017, ***018**, *020, *023, *025, *026, *029, *030, *031, *032, *034, *035, *036, *037, *038, *039, *040, *041, *042, *043, *045, *046, *047, *050, *051, *052, *055, *059, *060, *061, *068, *072, *075, *078, *079, *081, *083, *084, *086*	*MICA***004**, *005, *006, ***008**, ***009**, ***010**, *013, *016, ***019**, *022, *024, ***027**, *028, *033, *044, *048, *049, *053, *054, *056, *057, *058, *062, *063, *064, *065, *066, *067, *069, *070, *073, *074, *076, *077, *080, *082, *085, *087*

The most frequent alleles present in **Figure 3** *are shown in bold. The MICA alleles shown are from MICA*001 to MICA*087.*[4] *"MICA*003:01" label has never been assigned to any sequence. MICA*021 sequence was renamed MICA*012:03 in August 2007. The sequence originally labeled MICA*071 was proven to contain errors and to be identical to MICA*017 (March 2013) (see text footnote 3).*

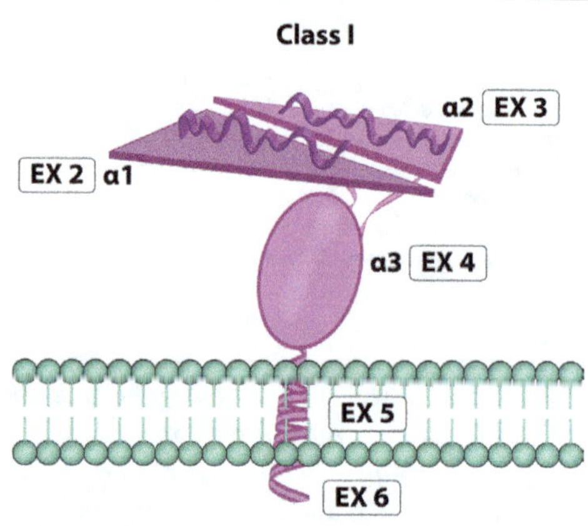

FIGURE 4 | MICA molecule. Exon 1 encodes one leader peptide, exons 2–4 encode three extracellular globular domains, exon 5 encodes one transmembrane domain, and exon 6 encodes a cytoplasmic tail (image by Matilde Risti).

during evolution, with orthologs of *KLRK1* are present in the genome of all mammals, as well as in marsupials (4, 52).

Human *NKG2D* is expressed from at least three distinct alleles, and several gene transcription isoforms have been described, including an alternatively spliced variant that introduces a nonsense mutation resulting in a protein isoform that lacks the entire extracellular ligand-binding domain (53).

Hayashi et al. (54) evaluated the SNPs in the NKC gene region. They selected 20 SNPs with a >10% higher frequency in Caucasoid or Japanese populations (**Table 3**); these SNPs covered *CD94*, *NKG2D*, *NKG2F*, *NKG2E*, *NKG2A*, and *Ly49* genes. They selected 8 out of the 20 SNPs that were closely associated with natural cytotoxic activity, having *P* values <0.001. All these SNPs are located in the NKG2D gene region, except for rs1983526 that is located in the promoter region of the NKG2A gene. These eight SNPs were split into two groups: group 1 (rs1049174, rs2617160, rs2617170, rs2617171, and rs1983526) and group 2 (rs2255336, rs2246809, and rs2617169). All the SNP combinations of group 1/group 1 and group 2/group 2 revealed a strong linkage disequilibrium, with r^2 values >0.9, whereas group 1/group 2 combinations showed much weaker linkage disequilibrium, with r^2 values <0.5. This indicates that the five group 1 and three group 2 SNPs belong to two different haplotype blocks (NKG2D hb-1 and hb-2), each of which generates two major haplotypes associated with low

[4]http://hla.alleles.org/alleles/classo.html.

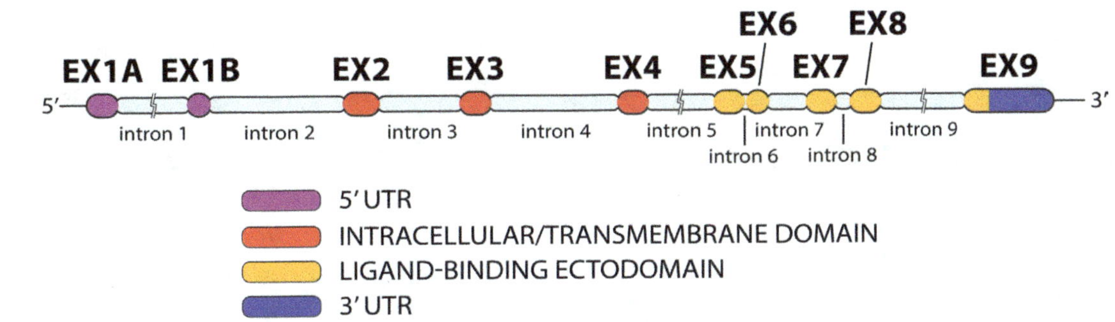

FIGURE 5 | *NKG2D* gene exon–intron organization. The *NKG2D* gene has 10 exons and 9 introns (image by Matilde Risti).

TABLE 3 | 20 SNPs selected by Hayashi et al. in their study (54).

SNP ID	Variation	SNP ID	Variation	NKG2D hb-1	Low	High
rs3759272	G>T	rs2617170	T>C	**rs1049174**	C	G
rs2537752	T>A	rs2617171	C>G	rs2617170	C	T
rs1049174	G>C	rs1971939	C>G	rs2617171	C	G
rs2255336	A>G	rs1915319	A>G	rs1983526	C	G
rs2294148	G>A	rs4763525	G>A	rs2617160	T	A
rs2049796	A>C	rs3003	C>T	**NKG2D hb-2**	**Low**	**High**
rs2617160	A>T	rs1983526	C>G	rs2255336	G	A
rs7972757	A>G	rs10772285	G>C	rs2246809	G	A
rs2246809	A>G	rs1915325	G>A	rs2617169	T	A
rs2617169	T>A	rs2607893	C>T			

Blue fields belong to group 1 and green ones represent group 2. Each of the different haplotype blocks (NKG2D hb-1 and hb-2) is split in low and high natural cytotoxic activity haplotypes. hb-1 and hb-2 may be successfully predicted knowing only rs1049174 (in bold).

(LNK) and high (HNK) natural cytotoxic activity phenotypes (**Table 3**) (54).

A separate study on a European population confirmed that the NKG2D region haplotype associated with increased cancer susceptibility in the Japanese population also exists in Europeans at similar frequency. Therefore, the conclusions of the original study may also be applicable to this population (55).

NKG2D: HNK1 and LNK1 Haploblocks

Several studies have demonstrated that high and low natural cytotoxic activity haplotype alleles (HNK1 or LNK1) belonging to NKG2D haplotype blocks 1 (hb-1) may be successfully predicted by only a single SNP (dbSNP: rs1049174) (54, 56, 57).

A study on Japanese individuals demonstrated that the *HNK1* haplotype is associated with a greater activity of NK cells in the peripheral blood and a lower prevalence of cancers originating from epithelial cells (58). Espinoza et al. showed an association between the *NKG2D-HNK1* haplotype (haplotype frequency, 61%) in bone marrow donors and a significantly reduced transplant-related mortality and better overall survival for unrelated donors of HLA-matched myeloablative bone marrow recipients with standard-risk disease (58).

The rs1049174 distribution for 25 populations (**Figure 6**) is reported on the 1,000 genomes website.[5] HNK is reported to be associated with the rs1049174 (G) allele, and LNK with rs1049174 (C) (54, 56).

NKG2D Protein

The NKG2D is a member of a C-type lectin-like family receptor called CD94/NKG2 (42). Despite its inclusion in the NKG2 family, NKG2D displays only limited sequence similarity to other members of the NKG2 family of NK cell surface receptors (NCRs) and CD94 and forms homodimers, rather than heterodimers, with CD94, as do other NKG2 NCRs (42).

Natural killer group 2 member D is a transmembrane-anchored receptor expressed as a disulfide-linked homodimer on the cell surface, with a molecular weight of ~42 kDa (42).

In humans, each NKG2D homodimer (**Figure 7**) associates with two DAP10 homodimers to form a hexameric structure (59), which can signal by recruitment of phosphatidylinositol 3-kinase (36).

Human NKG2DLs are MICA and MICB, and a group of glycosylphosphatidylinositol-bound surface molecules including UL16 binding protein(ULBP)-1, -2, -3, and -4 (6), RAET1G (or ULBP5), and RAET1L (or ULBP6) (60), which share about 25% identical amino acids in their $\alpha_1\alpha_2$ domains that are variably scattered throughout the aligned sequences without discernible patterns of sequence conservation (36).

Signals triggered by the NKG2D receptor are transmitted through the associated DAP10 dimer (**Figure 7**) (59) because NKG2D lacks a tyrosine-based inhibitory motif in its cytoplasmic tail (4, 61).

Natural killer group 2 member D is expressed by all human NK cells, γδ T lymphocytes, αβ CD8$^+$ T lymphocytes (6), interferon-producing killer DC (62), invariant NKT cells cells, and a small subset of effector or memory CD4$^+$ T cells (4, 52, 63). Expression of NKG2D on NK cells and CD8$^+$ T cells can be modulated by cytokines due to their effects on transcription and posttranscriptional processing of *NKG2D* and *DAP10*. In

[5] http://browser.1000genomes.org/Homo_sapiens/Variation/Population?db=core;r=12:10524865-10525865;v=rs1049174;vdb=variation;vf=750969.

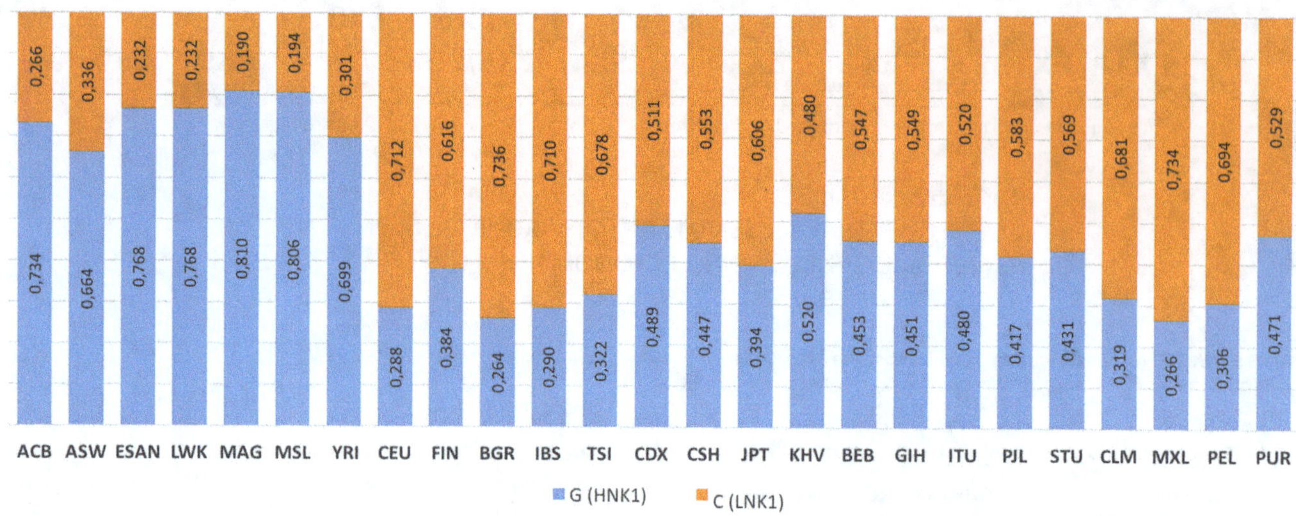

FIGURE 6 | 1000 Genomes frequency for the G>C alleles (NKG2D hb-1) (see Footnote 5). The population represented are African Caribbean in Barbados (ACB), African Ancestry in Southwest US (ASW), Esan in Nigeria (ESAN), Luhya in Webuyu Kenya (LWK), Mandinka in Degambia (MAG), Mende in Sierra Leone (MSL), Yoruba in Ibadan Nigeria (YRI), Utah Residence with Northern and Western European Ancestry (CEU), Finnish in Finland (FIN), British in England and Scotland (BGR), Iberian populations in Spain (IBS), Toscani in Italy (TSI), Chinese Dai in Xishuangbanna, China (CDX), Han Chinese in Bejing, China (CSH), Japanese in Tokyo, Japan (JPT), Kinh in Ho Chi Minh City Vietnam (KHV), Bengali in Bangladesh (BEB), Gujarat Indian in Houston Texas (GIH), Indian Telegu in the UK (ITU), Punjabi in Lahore Pakistan (PJL), Srilankan Tamil in the UK (STU), Colombian in Medellin Colombia (CLM), Mexican Ancestry in Los Angeles, California, USA (MXL), Peruvian in Lima Peru (PEL), and Puerto Rican in Puerto Rico (PUR).

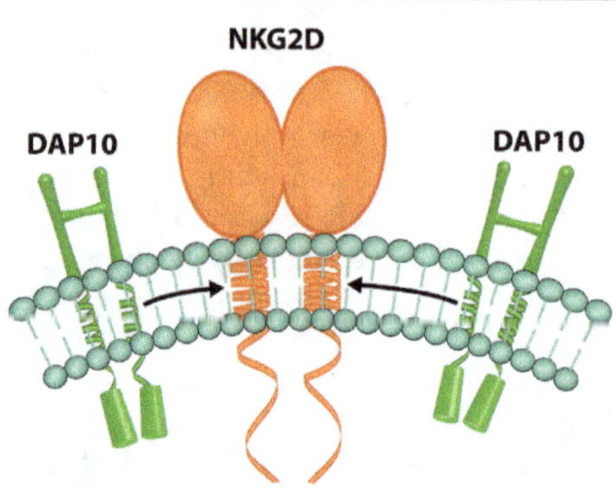

FIGURE 7 | NKG2D and DAP10. Representation of the hexametric structure formed by one NKG2D and two DAP10 homodimers (image by Matilde Risti).

humans, IL-2, IL-7, IL-12 (64), IL-15, and IFN-α (65) upregulate *NKG2D* expression, whereas TGFβ (65–67), IFNβ1 (68), and IL-21 (69), IL-4, IL-12, and IFNγ (65) downmodulate *NKG2D*. This downregulation can also be attributed to the overexposure to soluble or membrane-bound NKG2DLs, which promote the internalization and subsequent degradation of the receptor or catabolites produced on macrophage activation [reactive oxygen species (ROS) and L-kynurenine] (65). This is a possible explanation of the mechanism of oxidative stress, which is a common feature of chronic renal failure. ROS trigger the upregulation of MICA and downregulation of NKG2D in NK cells *in vitro* (70). DAP10 availability is also a decisive factor in NKG2D surface expression, and miRNAs can downregulate NKG2D expression in NK cells, reducing its cytotoxic effect (65).

Fernandez-Sanchez et al. (65) have shown for the first time that epigenetic mechanisms are involved in the regulation of NKG2D expression. They analyzed the region around the translation initiation site of the *NKG2D* gene (which included 11 CpG sites between −992 and +263 positions), and they found the greatest differences in DNA methylation patterns between the positions −992 and −255. These CpGs were highly methylated in Jurkat, HUT78 cell lines and CD4$^+$ T cells, partially methylated in CD8$^+$ T lymphocytes and NK cells, and fully demethylated in NK cells lines. They discovered that the acetylation of histone H3 lysine 9 (H3K9) is important for correct NKG2D expression in NK and CD8$^+$ T cells, while DNA demethylation may be associated with an increased expression of NKG2D in CD4$^+$ T cells. The DNA methylation profile of *DAP10* gene was also analyzed, but no differences were found. CD4$^+$ T lymphocytes and T cell lines (Jurkat and HUT78) had a DNA methylation; instead NKG2D-positive cells (CD8$^+$ T lymphocytes, NK cells, and NKL cell line) had an unmethylated *NKG2D* gene and high levels of histone H3 lysine 9 acetylation (H3K9Ac). It was observed that the histone acetyltransferase inhibitor, curcumin, reduced H3K9Ac levels in the *NKG2D* gene, downregulated NKG2D transcription, and led to a marked reduction in the NKG2D-mediated lytic capacity of NK cell lines (65).

Another interesting study by Karimi et al. (71) of human primary NK and CD8$^+$ T cells discovered a novel splice variant of human NKG2D that encodes a truncated receptor lacking the ligand-binding ectodomain (NKG2D^{TR}). Overexpression of this truncated isoform severely attenuated cell killing and

IFNγ release mediated by full-length NKG2D ($NKG2D^{FL}$). A specific knockdown of an $NKG2D^{TR}$ isoform enhanced NKG2D-mediated cytotoxicity, suggesting that $NKG2D^{TR}$ is a negative regulator of $NKG2D^{FL}$. At the biochemical level, it was demonstrated that $NKG2D^{TR}$ bound to DAP10 and interfered with the DAP10–$NKG2D^{FL}$ interaction. In addition, $NKG2D^{TR}$ formed heterodimers with $NKG2D^{FL}$ and negatively modulated $NKG2D^{FL}$ preventing its surface expression. Therefore, $NKG2D^{TR}$ constitutes a mechanism for regulation of NKG2D-mediated function in human $CD8^+$ T cells and NK cells (71).

Unlike $CD8^+$ T cells, TCR-mediated activation is not sufficient to induce NKG2D expression on $CD4^+$ T cells, and the factors responsible for induction of NKG2D on $CD4^+$ T cells are still unknown (71).

Saez-Borderias et al. (63) provided the first evidence that a subset of human cytomegalovirus (HCMV)-specific $CD4^+$ T cells displays NKG2D. Their data suggest that $CD4^+NKG2D^+$ cells expanding in HCMV-stimulated cultures correspond to virus-specific memory T cells that have acquired NKG2D while losing CD28 (63).

Conclusions on NKG2D

The *NKG2D* gene can be split into two haploblocks: *HNK1* and *LNK1* (high and low cytotoxic activity related). The NKG2D protein is a homodimer associated with two DAP10 molecules and can interact with MICA. In NK cells, the NKG2D protein is an activation receptor which is able by itself to trigger cytotoxicity. This is the main reason why it is interesting to study the relationship between MICA and NKG2D in depth in the following paragraphs.

MICA LIGAND AND ITS RECEPTOR NKG2D: FUNCTIONAL INTERACTIONS

The crystal structure of the MICA–NKG2D complex shows that NKG2D binds to one MICA molecule as a homodimer. One of the NKG2D molecules binds mostly to the α1 domain of MICA, while the other binds mostly to the α2 domain (6). The contact between these two molecules creates a small pocket (roughly 6 Å wide × 6 Å thick × 14 Å long) (42).

The NKG2D homodimer overlays MICA diagonally in way that resembles αβTCR overlaying MHC I molecules. The central section of the MICA α2 domain is disordered when MICA is crystallized in isolated form, but it becomes ordered when MICA is bound to NKG2D and forms part of the interface between the two molecules (6).

MICA glycosylation was not essential, but it enhanced complex formation with NKG2D. Likewise, the glycosylation state of NKG2D had no substantial effect on complex formation (36).

MICA–NKG2D is considered a versatile ligand–receptor pair. As a matter of fact, NKG2D can act as a primary receptor or costimulatory molecule during infections, autoimmunity, or antitumor immune responses (6). For example, it has been shown that endothelial MICA triggeres an activating signal in allogeneic polyclonal NK cells through the immunoreceptor NKG2D, which may have account for a significant part in EC lysis by allogeneic NK cells. *In vitro* coculture assays show that interaction of endothelial MICA with NKG2D provides an immune suppressive pathway by downregulating NKG2D on the NK cell surface (14).

Boukouaci et al. (72) suggested that endocytosis of the NKG2D receptor, upon binding to sMICA, is considerably more rapid than the replenishment of cell surface NKG2D by *de novo* synthesis. The same authors also found that sMICA down regulates NKG2D receptor expression on $CD8^+$ T cells. sMICA upregulates the IFNγ production only by cytokines-activated NK cells, while it has no effect on non-activated cells. The researchers demonstrated that sMICA upregulates IFNγ expression by IL-12/IL-18-activated CD3 $CD56^+$ NK cells, demonstrating the pro-inflammatory effect of sMICA (72). A study with a mouse model found that Lewis rat hearts transplanted into BALB/c mice developed typical acute rejection (AR) in 6 days. The severity of xenograft rejection increased with time, from 2 to 6 days. Also increasing over time, the MICA protein and MICA mRNA reached their highest value after 6 h. The prevalence of anti-MICA was significantly higher among mice with severe AR. However, sMICA was significantly increased during AR at 2 h, then gradually decreased, and reaching its lowest value after 6 h (73).

MICA–NKG2D AND KIDNEY TRANSPLANT

In the last few decades, the role of MICA and NKG2D in kidney transplants has emerged (**Table 4**). The involvement of NK cells was discovered in 1995 when some indirect evidence was reported during rejection of kidney transplants. Accumulation of $CD56^+$ NK cells expressing granzyme in kidney biopsies of patients undergoing AR suggested a role of their cytolytic activity in kidney-allograft rejection (74). Over the years, the association between NK cells and the mechanisms of microcirculation injury during antibody-mediated rejection (AMR) in kidney transplants has become increasingly evident. The researchers proposed that donor-specific antibodies (DSA) were able to bind to the endothelium and to recruit NK cells that produce IFNγ and trigger antibody (Ab)-dependent cellular cytotoxicity (75).

NK Cells and Kidney Damage in Mice and Cell Lines

Natural killer group 2 member D-ligand engagement delivers a strong dominant activating signal that overrides the inhibitory signal delivered by self-MHC class I, thus activating NKG2D-expressing cells, resulting in innate and adaptive immunity activation (113).

Zhang et al. (116) reported a study on ischemia/reperfusion injury (IRI) on mice and discovered the capacity of NK cells to injure renal tubular epithelial cells *in vitro*. *In vivo* data supported the hypothesis that NK cells interact with tubular epithelia through NKG2D/Rae-1 interaction to mediate kidney damage following IRI.

Luo et al. (89) performed an *in vitro* study on human renal proximal tubular epithelial cell line (HK-2). They discovered that hypoxia-inducible factor-1-α (HIF-1α) plays a very important role in upregulating MICA expression and enhancing NK cell cytotoxicity toward target cells during hypoxia/reoxygenation in HK-2 cells. HIF is a heterodimer consisting of an α-subunit (HIF-1α) and a β-subunit (HIF-1β), the HIF-1β protein is constitutively

TABLE 4 | Relevant published work regarding NKG2D, MICA, and kidney transplants.

Reference	Summary	MICA biomarker
Relevant published works regarding MICA and transplants		
Zwirner et al. (76)	Several patients had specific antibodies (Abs) against MICA. Most of them were detected in serum samples collected at different times after organ rejection	Yes
Hankey et al. (77)	MHC class I chain-related expression was documented in allografted kidneys and pancreas. Expression of MICB was observed in epithelial cells in allografted kidney and pancreas that showed histologic evidence of rejection and/or cellular injury	Yes
Opelz (78)	This work showed that non-HLA immunity contributed substantially to long-term kidney transplant failure. The targets for Abs causing late rejections could be called minor histocompatibility antigens	Yes
Mizutani et al. (79)	Patients who rejected transplants had anti-HLA and anti-MICA Abs more frequently than those with functioning grafts. These Abs found in the peripheral circulation were not necessarily donor-specific, but their association with failure was consistent with a causality hypothesis	Yes
Amezaga et al. (80)	Anti-MICA Abs were not detected pretransplant nor posttransplant in patients receiving a compatible graft. Anti-MICA Abs were detected posttransplant acute antibody-mediated rejection in patients receiving an incompatible graft	Yes
Mizutani et al. (81)	Anti-HLA and anti-MICA Abs were present independently on a more frequent basis in patients with failed grafts than those with functioning grafts	Yes
Panigrahi et al. (82)	Patients who developed both anti-HLA and anti-MICA Abs rejected their grafts more frequently than those having either of these Abs	Yes
Zou et al. (83)	Pre-sensitization of kidney transplant recipients against MICA antigens had been associated with an increased frequency of graft loss and might contribute to allograft loss among recipients who were well matched for HLA	Yes
Seiler et al. (62)	Unlike previous reports, in this work the researchers could not detect elevated MICA mRNA levels in kidney biopsies derived from patients undergoing acute rejection (AR) or chronic allograft nephropathy. In contrast, they observed a strong NKG2D mRNA induction during renal-allograft rejection, which was verified by immunohistology in kidney biopsies	No
Suarez-Alvarez et al. (84)	Anti-MICA Abs were detected in 17.6% of the patients and correlated with the development of AR. The presence of anti-MICA Abs could be an important marker for diagnosis because of their contribution to the outcome of the graft, regardless of presence of anti-HLA Abs	Yes
Alvarez-Marquez et al. (85)	At the time of the biopsy, 21% patients had only anti-HLA I Abs, 15.8% had anti-GSTT1 Abs, 10.5% had anti-HLA II Abs, and 10.5% had anti-MICA Abs. Besides anti-HLA Abs, donor-specific Abs against MICA and GSTT1 antigens could be responsible for the occurrence of Ab-mediated kidney graft rejection	Yes
Racca et al. (86)	This work did not show a correlation between MICA expression and renal graft state. The state of kidney allograft could be measured by using HLA-G1 isoforms, but not MICA mRNA levels, as markers	No
Lemy et al. (87)	The comparison between anti-MICA Abs$^+$ and anti-MICA Abs$^-$ patients showed that the incidence of AR episodes during the first year was similar in both groups. MICA Abs did not adversely affect renal graft outcomes	No
Li et al. (88)	Anti-MICA Abs were detected in 11 of the 15 transplant patients, irrespective of interval acute graft rejection. Also, integrative genomics predicted localization of the MICA antigen on the glomerulus in the kidney. MICA localization may explain both immunoregulatory and pathogenic roles for MICA after transplantation	Yes
Luo et al. (89)	HIF-1α plays a very important role in upregulating MICA expression and enhancing natural killer (NK) cell cytotoxicity toward target cells during hypoxia/reoxygenation in HK-2 cells. Their results demonstrated that hypoxia/reoxygenation-promoted MICA expression on HK-2 cells is through a HIF-1 pathway	Yes
Cox et al. (90)	Anti-MICA and anti-HLA Abs significantly associated with AR and anti-MICA donor-specific antibodies (DSA) and anti-HLA DSA correlated with decreased graft function by univariate and multivariate analysis. The researchers concluded that mismatching for MICA epitopes in renal transplantation is a mechanism leading to production of MICA Abs that associate with AR and graft dysfunction	Yes
Narayan et al. (91)	Case report: this case demonstrated that donor-specific anti-MICA Abs could be associated with both acute antibody-mediated rejection (AMR) and type IIA acute cellular rejection and emphasized the necessity of treating both humoral and cellular components of the rejection	Yes
Yao et al. (92)	The authors proved that Anti-MICA Abs$^+$ rate was significantly higher in sensitized recipients and it had significant effect on the recovery of allograft function in early postoperative period. Protein A immunoadsorption plays an important role in decreasing preexisting Abs, especially the anti-MICA Abs	Yes

(Continued)

TABLE 4 | **Continued**

Reference	Summary	MICA biomarker
Zhang et al. (93)	Anti-MICA Abs were present in 28.9% of patients and they were associated with renal-allograft deterioration. The researchers concluded that, besides anti-HLA Abs, the presence of posttransplant anti-MICA Abs was associated with poor graft outcome and increased the risk of graft failure	Yes
Lemy et al. (94)	Anti-MICA Abs$^+$ patients were more frequently anti-HLA Abs sensitized and regrafted. Four-year death-censored graft survival was not different between MICA$^+$ and MICA$^-$ patients. These data did not support an independent pathogenic role for MICA in long-term renal graft injury	No
Li et al. (95)	The levels of the peak mean fluorescence intensity of MICA Abs in patients with impaired renal function were significantly higher than those in normal renal function controls. They also concluded that some MICA Abs might be more important than others in mediating graft rejection	Yes
Seyhun et al. (96)	Anti-HLA class II and anti-MICA Abs$^+$ were only important predictors of graft failure when present together with anti-HLA I Abs$^+$. Patients who developed anti-HLA Abs alone or both anti-HLA Abs and anti-MICA Abs rejected their grafts more frequently than Abs$^-$ recipients	Yes
Rodriguez Ferrero et al. (97)	They compared patients with versus without preformed circulating antibodies (circulating anti-MICA Abs and anti-HLA Abs), and they did not observe a significant difference in graft survival or renal function at 3-month follow-up	No
Solgi et al. (98)	This research supported the idea that monitoring of anti-HLA and anti-MICA Abs as well as soluble CD30 levels early after transplant had predictive value for early and late allograft dysfunctions and the presence of these factors was detrimental to graft function and survival	Yes
Akgul et al. (99)	In this study, the scientist observed the role of anti-HLA II Abs in the development of chronic active AMR and in long-term allograft survival. It is observed that anti-MICA and anti-GSTT1 Abs showed no effect on rejection mechanisms	No
Chaudhuri et al. (100)	Anti-MICA and anti-HLA Abs appeared in approximately 25% of unsensitized pediatric patients, placing them at greater risk for acute and chronic rejection with accelerated loss of graft function	Yes
Ding et al. (101)	When comparing patients with acute graft rejection against recipients with stable renal functions, the researchers highlighted a significantly higher positivity rate of anti-MICA Abs. The status of anti-MICA Abs can predict the occurrence and treatment outcomes of AR, and affect the long-term survival of the renal grafts	Yes
He et al. (102)	By following transplantation recipients during follow-ups, anti-HLA and anti-MICA Abs expression was proven to have a predictive value for early and late allograft dysfunction. The presence of donor-specific Ab is detrimental to graft function and graft survival	Yes
Jin et al. (103)	They observed the prevalence of panel-reactive antibody (PRA) and anti-MICA Abs to be increased among Ptc, albeit not significantly different from C4d AR. These results implied that Ptc could be an early indicator of AR	Yes
Li et al. (104)	CD19$^+$ B cells and CD19$^+$CD27$^+$ memory B-cell subsets were detected from peripheral blood mononuclear cells obtained from six anti-MICA-sensitized kidney recipients. Kidney recipients had a higher percentage of CD19$^+$CD27$^+$ B cells compared with healthy controls. This study thus showed that B cells may be stimulated to secrete Abs	Yes
Sanchez-Zapardiel et al. (105)	The researchers detected that pretransplantation sensitization against anti-MICA and anti-HLA Abs were independent events. Preformed anti-MICA Abs independently increase risk for kidney rejection and enhance the deleterious effect of PRA$^+$ status early after transplantation	Yes
Tonnerre et al. (106)	The researcher found that individual carrying *MICA A5.1/MICA A5.1* had 10-fold higher levels of *MICA* mRNA and MICA proteins at the endothelial cell surface. They also demonstrate a significant association between D/R MICA A5.1 mismatch and anti-MICA alloimmunization, particularly when donors carry the A5.1 mutation. They concluded that A5.1 mutation is an immunodominant factor and a potential risk factor for transplant survival	Yes
Zhang et al. (107)	5 years after transplantation, the frequencies of *de novo* anti-HLA and anti-MICA Abs were 25.8 and 12%, while 26.5% of patients had proteinuria. All of these factors have been associated with poor graft survival	Yes
Sapak et al. (108)	The researchers did not prove a complete correlation between the recipient anti-MICA Abs specificities and MICA antigens of the donor. They assumed that anti-MICA Ab induction occurred not only due to the allogeneic stimulation itself but also due to other factors that needed to be elucidated	No
Ming et al. (109)	Case report: the patient's HLA alloantibodies were not specific to the first kidney donor, but the MICA alloantibodies were. This indicates the importance of MICA virtual crossmatch in the process of selection for the kidney donor if the recipient is sensitized.	Yes
Xu et al. (110)	Serum anti-HLA II Abs, anti-MICA Abs, and anti-HLA plus MICA Abs all statistically increased in renal-transplanted recipients	Yes

(*Continued*)

TABLE 4 | Continued

Reference	Summary	MICA biomarker
Cai et al. (111)	Transplant recipients had Abs against denatured HLA class I, II, and MICA antigens. However, only C1q-fixing Abs were associated with graft failure, which was related to AMR	Yes (only for c1q-fixing denaturated MICA Abs)
Sanchez-Zapardiel et al. (112)	Occasionally, preformed anti-MICA Abs may be cytotoxic by activating and fixing complement. This could lead to a reduced function in early kidney grafts	Yes
Relevant published works regarding NKG2D and transplant		
Feng et al. (113)	Ischemia/reperfusion injury (IRI) caused mRNA expression of Rae-1 and protein expression of Rae-1 in ischemic kidneys. This study suggested that the expression of the NKG2D ligand, Rae-1, may play a potential role in innate immunity associated with IRI	
Zheng et al. (114)	The absence of enhancement of NKG2D expression in the kidney in AN in immunodeficient mice suggested that the populations expressing NKG2D were likely to be CD8 or γδ T cells, which were not present in the immunodeficient mice, rather than macrophages, which were present and activated in both models of AN	
Seiler et al. (62)	Unlike previous reports, in this paper, the researchers could not detect elevated MICA mRNA levels in kidney biopsies derived from patients undergoing AR or chronic allograft nephropathy. In contrast, they observed a strong mRNA induction of NKG2D during renal-allograft rejection, which could be verified by immunohistology in kidney biopsies	
Hadaya et al. (115)	The results of this paper have shown an expansion of the NKG2D$^+$ NK cell population during acute cytomegalovirus (CMV) infection (after kidney transplantation), which decreased over time to a level very similar to that of the control group. This suggests that the NKG2D receptor could play a similar role in NK and CD4$^+$ T cells	
Zhang et al. (116)	In this study, the researchers demonstrated for the first time that NK cells could induce kidney TEC death *in vitro* and that NKG2D and Rae-1 interactions played a critical role in this killing in mice	
Shabir et al. (117)	Cytotoxic CD4$^+$ CD28^{null} cell is an important biomarker for and potential mediator of adverse events after kidney transplantation. NKG2D represents an integral component of CMV immunosurveillance and immunoevasion and was upregulated on CD4$^+$ CD27$^-$ CD28^{null} cells isolated from patients of this study. The researchers proposed it as an important component of the cytotoxic effects (either protective or pathogenic) of these cells	

"Yes" and "No" labels have been used if, in the studies analyzed, MICA has been valued as a possible biomarker ("Yes") or not ("No").

present, while HIF-1α has a unique O_2-dependent degradation domain, which leads to its degradation under normoxia conditions. The authors speculate that HIF-1α upregulates the surface expression of MICA on grafts during renal IRI, causing NK cells cytotoxicity against the organ (89).

Possible Causes of End-stage Renal Disease

A 2009 study of the possible causes of end-stage renal disease (ESRD) (70), while note directly related to kidney transplants, inevitably reported findings of consequences for kidney transplantation. Peraldi et al. evaluated seven patients with ESRD that were treated with peritoneal dialysis, and not with the hemodialysis procedure; NKG2D expression on NK cells was significantly decreased in these patients compared to healthy donors, indicating that reduction in NKG2D expression was independent of the dialysis procedure and linked with chronic renal failure. The authors also discovered that oxidative stress in presence of increased ROS production is one of the most significant consequences of chronic renal failure, alone or in concert with other mediators, and it seems to decrease the NKG2D levels on NK cells in ESRD and to favor the upregulation of MICA expression (70).

Anti-MICA Abs and Rejection

Some mechanisms have been proposed for MIC-mediated organ rejection. MICA antigens expressed in the allograft could induce the generation of anti-MICA Abs, which in turn might injure cells in the presence of complement.

This section contains no works that focus solely on NKG2D since most of the manuscripts are almost exclusively conserved with anti-MICA Abs: NKG2D is often just a side note; its presence and the link with MICA are given.

MICA-Sensitized Kidney Recipients and Higher Percentage of CD19$^+$CD27$^+$B Cells

CD19$^+$CD27$^+$ B cells are the subset of memory B cells that have the potential ability to secrete Abs. Li et al. (104) assessed the serum from 68 long-term survival kidney recipients and found 11 subjects who were MICA positive. They analyzed six MICA-sensitized kidney transplant recipients and six healthy volunteers who did not receive a transplant (control group). Healthy controls had a higher percentage of CD19$^+$CD27$^-$ in PBMCs than transplant patients, while the percentage of CD19$^+$CD27$^+$ in B cells was higher in transplant patients. The MICA-sensitized transplant patients had a significantly lower average percentage of CD19$^+$ B cells in PBMC than healthy controls (3.58 ± 0.80 versus $8.53 \pm 1.04\%$; $P < 0.01$). These results suggest that CD19$^+$CD27$^+$ B cells from sensitized patients have the potential ability to secrete Abs. In the same study, PBMC cells were isolated and cultured and stimulated with different molecules [toll-like receptor-9 ligand ODN-2006 CpG, PMA, B-cell activating factor (BAFF), CD40 ligand (CD40L), human recombinant IL-2 (rhuIL-2), rhuIL-10, rhuIL-4, rhuIL-21, CD40L, and BAFF] including MICA antigens. After stimulation, B cells from healthy controls and transplant patients had a lower percentage of apoptosis than non-stimulated cells. The average percentage of apoptosis cells from transplant patients was significantly higher than from healthy controls, and the IgM production (the first Ab produced by B cells after antigen stimulation) was higher in stimulated B cells from transplant patients than from healthy controls. The authors speculate that the B-cell population may be compromised by the transplant because patients are under immunosuppressive regimens, which may alter the apoptosis of B stimulated cells compared with healthy controls. The same study also performed an *in vitro* study with drugs and found that bortezomib and mycophenolic acid could inhibit B-cell Ab secretion (104).

MICA Abs

Hankey et al. (77) first reported that MICA and MICB expression on epithelial cells in transplanted kidneys and pancreases with histological evidence of rejection and cellular injury played a role in allograft rejection. The study showed that in a healthy kidney there was no immunochemical evidence of MIC expression. In contrast, the majority of biopsies with histologic proof of rejection or acute tubular necrosis (ATN) showed MICA positive staining of the tubular epithelium in the proximal and distal tubules. For this reason, it was concluded that alloantibodies against MICA might play a role in allograft rejection.

Zwirner et al. (76) found that several patients who had undergone a kidney transplant had specific Abs against MICA, and most of them were detected in serum samples collected at different times after organ rejection. However, these Abs were not directed against the alleles expressed by the patients, and it was speculated that if the presence of MICA Abs was probably caused by multiple blood transfusions received by the patients while awaiting a transplant, or resulting from a pregnancy or a previous transplant (76).

Lemy et al. (87) analyzed the MICA Abs from 494 controls and 597 patients with chronic kidney disease. They found a three times higher prevalence of MICA Abs in patients with chronic kidney disease when compared with controls (14.9 versus 4.7%). Nevertheless, they speculated that even if the increase in MICA Abs prevalence among patients affected by chronic kidney disease was probably related to previous renal transplantation and transfusions. Logistic regression analysis and analysis of chronic kidney disease patients who have not been subjected to transfusions and renal transplantations suggest that the increase of urea (and other nitrogenous waste) in the blood is connected to an increase of MICA immunization. The authors also reported that MICA Abs were more frequent in men than in women, despite pregnancy being an independent risk factor for the development of MICA Abs (87). This finding is in sharp contrast with other published work. The fact that nearly one-third of MICA chronic kidney disease stage V patients have never experienced any identifiable immunizing event indicates that there must be other causes for MICA sensitization. At the same time, one-fifth of the same patients showed the presence of autoreactive MICA IgG Abs, distinctly rare with respect to HLA Abs. The authors showed that patients with MICA Abs had a somewhat better overall graft survival than MICA Abs^{-} patients. Finally, Lemy et al. found in MICA Abs^{+} and MICA Abs^{-} patients a similar incidence of AR episodes during the first year (10.2 versus 12.8%), as well as similar levels of proteinuria and creatinine (87).

Another study of MICA Abs screened 147 recipients with end-stage renal disease; 82 of these patients were Abs^{+} (55.8%). Forty patients had both anti-HLA and anti-MICA, 33 had only anti-HLA, and 9 only anti-MICA Abs in the posttransplant period. The authors found that patients who developed HLA alone, or both HLA and MICA Abs, rejected their grafts more frequently than Abs^{-} recipients. The rates of HLA class I, class II, or both Abs^{+} were greater in the rejection patients than the non-rejection patients ($P = 0.011$, 0.037, and 0.0275, respectively). So the authors speculated that HLA class II and MICA Abs^{+} were the only important predictors of graft failure when both of them were present with HLA class I Abs^{+} (96).

In a retrospective study, Solgi et al. (98) analyzed sera samples of 40 living unrelated donor kidney recipients, looking at anti-HLA and anti-MICA Abs and the levels of soluble CD30 (sCD30) and sMICA. They found that patients with pre- and posttransplant HLA Abs had a higher incidence of AR episodes ($P = 0.01$ and $P = 0.02$), more graft loss ($P = 0.001$), and lower graft survival during a mean follow-up of 3 years. This group of patients also had higher levels of sCD30 and serum creatinine and decreased contents of sMICA early after transplantation, as compared to the patients without HLA Abs. Anti-MICA Abs were observed in 8/40 (20%) and 5/40 (12.5%) of all patients pre- and posttransplant, respectively. HLA and MICA Abs were both found in two out of four cases with graft loss. In a comparison of transplant rejecting to functioning graft groups, sCD30 levels increased at day 14 ($P = 0.001$), while sMICA levels were insignificantly lower in the first group (98).

Chaudhuri et al. (100) studied the evolution of humoral immunity in low-risk pediatric patients during the first 2 years after renal transplantation. They correlated the presence of serum anti-HLA DSA and serum MICA Abs with clinical outcomes and histology (the biopsies were performed at 0, 6, 12, and 24 months). They found anti-HLA Abs in 22% of patients, 6% of which were donor-specific, while 6% developed anti-MICA Abs. Three percent of patients developed *de novo* Abs to both HLA and MICA. The presence of *de novo* Abs was associated with significantly higher risks for AR ($P = 0.02$), chronic graft injury ($P = 0.02$), and decline in graft function ($P = 0.02$). Graft function was monitored by the difference between creatinine clearances. Anti-MICA and -HLA Abs were found in 25% of unsensitized pediatric patients. This was correlated with a greater risk of acute and chronic rejection (100).

Zhang et al. (107) associated the presence of *de novo* MICA Abs and proteinuria with graft failure, after renal transplantation. They investigated 275 patients without preexisting anti-HLA and -MICA Abs. Five years after renal transplantation, 25.8% showed *de novo* anti-HLA Abs, 12% showed *de novo* anti-MICA Abs, and 26.5% proteinuria. *De novo* anti-HLA Abs were associated with increased proteinuria after transplantation (relative risk, 3.12). Anti-HLA Abs and proteinuria were both associated with poor 5-year graft survival ($P = 0.027$ and $P = 0.006$, respectively). Patients with *de novo* anti-MICA Abs were also apt to have proteinuria. The authors concluded that *de novo* anti-HLA and -MICA Abs and proteinuria are all associated with poor graft survival (107).

Pretransplant Panel-Reactive Abs and Preexistent Circulating Abs

Opelz (78) studied the influence of pretransplant panel-reactive antibody (PRA) status on the long-term outcome of kidney grafts from HLA-A, -B, and -DR, identical sibling donors. In over 10 years of follow-up, he discovered that non-HLA-directed immunity and Abs against HLA had a similar influence for the long-term results for kidney recipients with PRA. Opelz suggested that the targets for Abs causing late rejections could be the so-called minor histocompatibility antigens (78).

Sanchez-Zapardiel et al. (105) studied 727 transplanted patients and showed that the effect of anti-MICA Abs occurs independently of the presence of anti-HLA Abs. Pacients were categorized into four groups according to the presence (+) or absence (−) of anti-HLA and anti-MICA Abs: HLA^+MICA^+ (n = 27); HLA^-MICA^- (n = 510); HLA^+MICA^- (n = 165), and HLA^-MICA^+ (n = 25). A notable difference was observed 3 months after transplantation, when HLA^-MICA^+ patients had a graft rejection rate of 8% compared with 2% in HLA^-MICA^- patients. The patients were also grouped according to the presence of preexisting anti-HLA Abs, as measured by % PRA (PRA^+ or PRA^-): PRA^+MICA^+ (n = 7), PRA^-MICA^- (n = 610), PRA^+MICA^- (n = 65), and PRA^-MICA^+ (n = 45). The incidence of rejection was found to be superior in PRA^+MICA^- cohort versus PRA^-MICA^- patients (24 months after transplantation), but allograft rejection rate was the highest when comparing PRA^+MICA^+ patients with PRA^-MICA^- patients 3 months after transplantation, a finding which was repeated at 6 months (105). This work is of interest because it performed a comparative study on the effects of anti-MICA and anti-HLA Abs on kidney transplants.

The Rodriguez Ferrero et al.'s (97) study included 22 recipients of kidney transplantations from deceased donors, and no differences between patients that showed preexistent circulating antibodies (CA) and those that did not were reported. In regards to the incidence of AR episodes, the only factor associated with CA was re-transplantation. So the authors concluded that CA monitoring is important for highly sensitized renal transplants, but they did not observe a difference in graft survival or renal function in the first 3-month follow-up (97).

Cd4 Deposition and C1q-Fixing Abs

A study of patients with acute antibody-mediated rejection (AAMR), who had MICA*008 Ab, showed that the presence of anti-MICA Abs and the deposition of C4d in biopsies performed at the time of AAMR was associated with the detection of DSA or Abs against HLA (80). The observation that the control group of 30 patients with long-term functioning grafts did not have anti-MICA*008 Abs provided indirect evidence of the importance of anti-MICA Abs in chronic rejection. Furthermore, all patients receiving an allograft fully matched at MICA had functioning grafts (80). It is also important to mention that MICA Abs are able to activate complement in *in vitro* experiments (80).

Alvarez-Marquez et al. (85) selected 58 patients that underwent a kidney biopsy because of primary non-function, delayed graft function or acute dysfunction of a previously functional graft, suspected by oliguria, increase of serum creatinine levels, or proteinuria. At the time of the transplant, all patients showed negative complement-dependent cytotoxicity crossmatches. Researchers demonstrated that 80% of a group of 19 patients with clinically evident graft dysfunction and with C4d deposition in kidney biopsies had Abs directed against donor-specific HLA class I, class II, MICA, or GSTT1 (glutathione-*S*-transferase T1) antigens (85).

In the Li et al.'s (88) study, a human ProtoArray platform was used to study 37 serum samples from 15 renal transplant patients (pediatric and young adult) with (n = 10) and without (n = 5) AR, and seven normal controls. To test serum Abs, they used a ProtoArray containing 5,056 non-redundant human proteins expressed in a baculovirus system, purified from insect cells and printed in duplicate onto a nitrocellulose-coated glass slide. Moreover, all patients were primary transplant recipients, and the biopsies were graded by the Banff classification. The authors found that the mean immune response signal in posttransplant patient serum showed an increase in anti-MICA Abs when compared with healthy normal controls (n = 7), but anti-MICA Abs signal intensity was unrelated to the sampling time interval posttransplantation. Mean MICA Abs signal intensity was higher in transplant patients with $C4d^+AR$ (121.4) versus $C4d^-AR$ (4.3), so a correlation between high MICA Abs levels and $C4d^+$ graft rejection $r = 0.54$ ($P = 0.039$) was observed. On ProtoArray, each gene on the cDNA platform was compared between a specific kidney compartment versus all other compartments, by a two-unpaired class comparison and a multi-class comparison. The signal intensity of anti-MICA Abs ranked in the top 15 for glomerulus, so the MICA antigen was found to have a 2.7-fold higher expression in the glomerulus when compared to the other 6 normal kidney compartments. Cytoplasmic granular staining for MICA in normal and stable transplanted kidneys was observed solely in podocytes within glomeruli. In AR, in addition to the persisting glomerular staining, the infiltrating mononuclear lymphocytes also showed strong positive staining for MICA. So the authors demonstrated that Ab responses in patients are modulated by MICA after transplantation in patients, irrespective of graft rejection (88).

Another study correlates Cd4 deposition and creatinine levels. Ding et al. (101) evaluated serum anti-MICA Abs before and after kidney transplant, and they also examined PRA, serum creatinine, urine, graft ultrasound, lymphocyte subsets, and the pathology of graft biopsy. The study was split into two parts. In the first part, patients with AR were grouped into $MICA^+$, $MICA^-$ ($P < 0.05$) and control groups. There were a significantly higher number of anti-MICA Abs positive patients with acute graft rejection compared with stable renal functions patients (control group).

Two to three days after the occurrence of AR, the anti-MICA Abs level increased gradually. Anti-rejection treatment had no effect on anti-MICA Abs but lowered serum creatinine to a normal level. In the second part, the authors analyzed chronic graft rejection patients. The number of anti-MICA Ab positive patients was significantly higher than those with stable renal function ($P < 0.05$), and the serum creatinine levels were significantly higher in MICA$^+$ than in MICA$^-$ cases ($P < 0.05$). The authors also found that graft biopsy of all MICA$^+$ cases showed C4d deposition (101).

Jin et al. (103) studied 53 cases of AR that showed C4d deposition in the peritubular capillaries, 50 cases of ARs without C4d deposition, 30 with peritubular capillaries alone, 28 with ATN, and 78 patients with surveillance biopsies (control group). The authors observed that the prevalence of PRA and anti-MICA Abs was increased among the peritubular capillaries alone group (30.0 and 43.3%, respectively), albeit not significantly different from the group with C4d$^+$ AR (49.1 and 39.6%, respectively). They also observed that the immunophenotype of infiltrating T lymphocytes and serum Abs (85.9% of control biopsies presented) had a regulatory phenotype while in the peritubular capillaries cohort, 93.3% of biopsies showed the cytotoxic phenotype. These results showed that peritubular capillaries in biopsy specimens from patients with early renal-allograft dysfunction could be an indicator of AR, especially acute humoral rejection (103).

Cai et al. (111) collected samples from 975 kidney transplant recipients, and they tested for C1q-fixing Abs against denatured HLA class I, class II, and MICA antigens. Among 169 patients who lost renal grafts, 44% had c1q-fixing Abs against denatured HLA/MICA antigens, which was significantly higher in patients with functioning renal transplants (25%). They concluded that C1q-fixing Abs were significantly associated with graft failure caused by AMR (72.73%) and they affirmed that only c1q-fixing Abs were associated with graft failure and AMR (111).

MICA Allele Epitopes and Eplets

Regarding the anti-MICA Abs, Duquesnoy et al. (118) developed an eplet-based version of the HLA-Matchmaker algorithm as a tool to assess the epitope specificity of these Abs. A repertoire of 38 potentially immunogenic *MICA* eplets was selected (based on MICA structure molecular viewing and the amino acid sequence differences between *MICA* alleles). These eplets are based on a functional epitope structure (a configuration of amino acids within a 3 Å radius of an Ab accessible polymorphic residue on the molecular surface). In this study, the eplet frequencies were calculated from *MICA* allele frequencies in 1,245 European-Americans and 605 African-Americans. Many eplets are shared by very similar groups of *MICA* alleles. For instance, the combination of eplets called CMGWS "supereplet" is composed by 36C, 129M, 206GW, and 215S epitopes and shared by the same group of *MICA* alleles (*A*001, A*002, A*007, A*011, A*012, A*015, A*017, A*018, A*021, A*030, A*041, A*043, A*045, A*046, A*047, A*014, A*020, A*023, A*026, A*029, A*036, A*040, A*050, A*052*, and *A*055*). The random chance that these eplets are a mismatch is 20.1% in African-Americans and 24.0% in European-Americans. Alternatively, the combination of eplets named AYVE "supereplet" is composed by 25AY, 129V, and 173E and was shared by another group of *MICA* alleles (*A*004, A*006, A*008, A*009, A*010, A*016, A*019, A*024*, and *A*044*). The random chance of their being a mismatch is 28.2% in African-Americans and 20.1% in European-Americans (118).

Panigrahi et al. (82, 119) analyzed the presence of Abs against MICA*001, MICA*002, MICA*004, MICA*008, and MICA*009 in serum samples of 185 patients transplanted with live related donor kidneys. Sixteen percent of all recipients developed anti-MICA Abs during the posttransplant period, 83% of the patients whose grafts eventually failed had both anti-HLA and anti-MICA Abs as compared to 29% patients who had only anti-MICA Abs, and 11% of those without any of the Abs (HLA or MICA) (82, 119).

Analysis of anti-MICA*001, MICA*002, MICA*004, MICA*008, and MICA*009 Abs in serum samples from 1,910 kidney recipients showed that a correlation between the presence of anti-MICA Abs and the reduced in kidney-allograft survival was not influenced by the simultaneous presence of Abs against HLA (120). In this study, decreased renal-allograft survival is associated with anti-MICA Abs formed before transplantation. It was also found that patients with Abs against MICA before transplantation did not received more transfusions than patients without such Abs, in contrast with the Zwirner et al.'s study (76). So the authors speculate that cross-reactivity with substances from the environment may play a role in priming the immune system, facilitating anti-MICA Ab production (120).

Suarez-Alvarez et al. (84) screened 284 kidney transplant sera for anti-MICA Abs and mapped the epitopes of MICA by screening a library of synthetic overlapping peptides from the extracellular domains of the protein against the sera from kidney transplant patients with anti-MICA Abs. Anti-MICA Abs were detected in 50 of 284 patients (17.6%), and they correlated with the development of AR. The authors found that nine regions were reactive with anti-MICA Abs. Five epitopes were located in constant regions (II, III, IV, VI, and IX) and were present in all *MICA* alleles, while the other four regions (I, V, VII, and VIII) mapped to variable sites of polymorphic amino acids among the different alleles products of MICA. In particular, regions V, VII, and VIII were the regions with the highest amino acid variability. Three polymorphic residues, 173 (E/K), 175 (S/G), and 181 (R/T), had determined allele-specific epitopes. The aminoacid 208Y and 213T, instead, contributed in the cross-reactivity among alleles (84).

Cox et al. (90) identified MICA IgG Abs directed against MICA*001, *002, *004, *007, *008, *009, *012, *017, *018, *019, and *027. Analysis of 116 healthy control subjects revealed only one subject with anti-MICA Abs (0.9%) and five subjects (4%) with anti-HLA class II Abs, while in a subgroup of 227 transplant recipients and their donors the coproduction of Abs to HLA and MICA significantly associated with acute cellular rejection (ACR). Analysis of patients with AAMR established strong associations with the presence of Abs against HLA class I and II, but not anti-MICA. By aligning *MICA* allele profiles present in the subgroup of 227 renal graft recipients and their respective donors, it was possible to establish the precise position of amino acid mismatches that correlate strongly with MICA Ab production. Mismatching at residues 36, 129, 173, 175, 213, and 251 showed the strongest association with anti-MICA Ab production in transplant recipients, while 91, 125, 156, and 221 residues were also mismatched between recipients and donors, but were not significantly associated with anti-MICA Ab production. There are two immunodominant motifs: MICA-G1 is characterized by residues 36 cysteine (C), 129 methionine (M), 173 lysine (K), 206 glycine (G), 210 tryptophan (W), and 215 serine (S). Alternatively MICA-G2 epitopes share residues 36 tyrosine (Y), 129 valine (V), 173 glutamic acid (E), 206 serine (S), 210 arginine (R), and 215 threonine (T). The majority of these recipients (10 out of 17 individuals, 59%) developed *de novo* donor-specific anri-MICA Abs posttransplantation, and there was a significant association of graft dysfunction with the presence of anti-MICA DSA alone after 2 years. In conclusion, it was discovered that mismatching *MICA* alleles lead to the development of anti-MICA Abs in some renal graft recipients, and the presence of anti-MICA DSA was independently associated with decreased glomerular filtration rate (eGFR) and poorer graft outcome (90).

Tonnerre et al. (106) went beyond the usual studies of anti-MICA Abs and focused on searching for a specific allele that could lead to a poorer outcome. The authors performed a study that showed that the MICA*008 (A5.1) molecule is a major antigenic determinant and target for recipient sensitization of kidney transplant patients. *MICA A5.1* is associated with four alleles: **023*, **028*, **053*, and **008*. The authors divided primary EC cultures from transplant donors in *MICA A5.1* homozygous, heterozygous, and control. The MICA surface expression was significantly higher on ECs from *A5.1/A5.1* donors than from controls. The *MICA A5.1* allele also leads to a reduction of sMICA and an increase in the MICA level in exosomes in ECs. Anti-MICA (A5.1) Abs intensities in the sera of recipients with anti-MICA Abs were not higher than intensities observed for other anti-MICA (control) Abs. However, when tested on EC cultures expressing physiologic levels of membrane-bound MICA, the sera only bound to ECs from *MICA A5.1* donors. This seemed to show that anti-MICA Abs bind ECs' targets in an allele-specific manner.

In fact, the combination of the donor carrying *MICA A5.1* and the recipient having a non-*MICA A5.1* allele was overrepresented in the group of MICA-sensitized patients compared with the group of non-immunized recipients (106).

Sapak et al. (108) concluded that anti-MICA Abs could not be responsible for the rejection if they were not directly detected in the transplanted graft. In the sera of 124 renal recipients, the authors found only 22 patients positive for anti-MICA Abs. The most frequent anti-MICA Abs were directed against *MICA*018* and *MICA*001*. *MICA*008* had the highest gene frequency (31%), followed by *MICA*002* (14%). Comparing *MICA* allele profiles of donors and anti-MICA Ab epitopes of their respective recipients, Sapak et al. found a match in only in 9 donor–recipient pairs (41%) while the sera of the other 13 patients was negative for Abs against graft MICA molecules, but positive for Abs against other MICA antigens. The majority (59%) of anti-MICA Abs in patients were not donor-specific, so the authors suggested that anti-MICA Ab induction was not caused by renal graft allogeneic stimulation but was also probably stimulated by other still unknown immune mechanisms (108).

Sanchez-Zapardiel et al. (112) studied 727 kidney recipients. They found that $PRA^{+}MICA^{+}$ recipients exhibited a longer time to reach optimal serum creatinine level after transplantation ($P = 0.005$) had the lowest eGFR at 3 months and $PRA^{+}MICA^{+}$ status independently increased the risk for chronic kidney disease stage 5 at month 3. Pretransplant anti-MICA Abs were poly-specific; anti-AYVE supereplet reactivity was higher in $HLA^{+}MICA^{+}$ versus $HLA^{-}MICA^{+}$ patients and superior than anti-CMGWS supereplet within $HLA^{+}MICA^{+}$ patients. The authors also found that some preformed anti-MICA Abs might bind complement, using the C1q Luminex assay. Sanchez-Zapardiel et al. analyzed 13 anti-$MICA^{+}$ pretransplant sera that were positive for the C1q binding assay and one of them (serum 3) exclusively recognized the AYVE supereplet with a strong reactivity against MICA*027 antigen. The authors concluded that these preformed anti-MICA Abs are able to mediate cell death by fixing and activating the complement cascade. So they speculated that the anti-MICA Abs might contribute to worse early kidney graft function (112).

Correlation between Anti-MICA Abs and Creatinine Levels or Estimated Glomerular Filtration Rate (eGRF) or Death-Censored Graft Survival (DCGSs)

Yao et al. (92) included 29 sensitized recipient patients who had undergone living-related donor renal transplantation between 2007 and 2009. They found a statistical difference in postoperative serum creatinine levels within 1 week between anti-MICA Ab-positive (135.4 ± 21.4 mol/L) and anti-MICA Ab-negative groups (108.6 ± 31.6 mol/L), but no significant difference between the two groups at discharge. To decrease the preexisting Abs (mainly IgG, IgM, and IgE), all recipients were treated with protein A immunoadsorptions, and this therapy was effective in decreasing anti-MICA Abs (92).

Zhang et al. (93) studied patients receiving primary kidney transplants (all from deceased donors) between 2004 and 2007. No significant association was found between the presence of anti-MICA and -HLA Abs, nor between the presence of anti-MICA Abs and 1-year graft survival rate. However, during the follow-up period, eGFR decreased 24.0 ± 3.4% in the anti-MICA Abs positive group, while it decreased only 8.4 ± 3.0% in anti-MICA Abs negative patients. A strong correlation between the production of anti-MICA Abs and renal impairment was also found. For these reasons, the authors concluded that patients with anti-MICA Abs had a more rapid deterioration of graft function, compared to those without anti-MICA Abs (93).

In another study that did not recognize MICA as a biomarker, sera from 779 kidney transplant recipients was tested with two single-antigen flow bead assays 1 year after transplantation. Thirteen of the 779 patients were lost to follow-up, 50 had lost their graft, and 33 died with a functioning graft. The prevalence of anti-MICA Abs was 5.3% at 1-year posttransplantation, and that $MICA^{+}$ patients were more frequently HLA sensitized and regrafted. However, 4-year DCGSs were not different between $MICA^{+}$ and $MICA^{-}$ patients (97 versus 94%, $P = 0.28$), and 4- and 8-year survival rates were similar in $MICA^{+}$ and $MICA^{-}$ patients. Thus, the hypothesis of an independent pathogenic role for MICA in long-term renal graft injury was not supported, and the authors questioned the utility of monitoring anti-MICA Abs posttransplant with single-antigen flow bead assays (94).

MICA Abs in Case Study

Narayan et al.'s (91) case study focused on a 14-year-old girl with branchiooto renal syndrome who underwent re-transplantation with an HLA crossmatch-negative deceased donor kidney. She lost her first kidney transplant to chronic rejection at the age of 10 and underwent allograft nephrectomy. She was highly sensitized, and to improve her chances for transplantation, she underwent desensitization with high-dose IVIG and rituximab. When she received a deceased donor renal transplant, the pretransplant anti-HLA Ab testing showed no anti-donor HLA Abs. The patient maintained good allograft function until postoperative day 10 when she presented with fever and anuric renal failure. The only Ab found was donor-specific anti-MICA Ab, specifically directed against MICA*012 protein. Evaluation of the pretransplant serum revealed preformed anti-MICA*012 Abs with levels that were elevated both before transplant and at the time of rejection. Anti-MICA Abs levels declined with the initiation of plasmapheresis and IVIG and correlated well with normalization of renal function and resolution of ACR and AMR. The authors speculated that the sensitization to the MICA*012 protein was caused by prior sensitization from the first renal transplant or previous infections or transfusions. The conclusion of their research is that donor-specific anti-MICA Abs can be associated with both AMR and Banff type IIA ACR and may require treatment with plasmapheresis (91).

Ming et al. (109) studied a patient who suffered early aggressive AMR in the presence of DSA against MICA after her first renal transplant. The researchers found that anti-MICA–DSA in recipient serum could bind MICA-G1 antigens expressed in the cultured human umbilical cord vein endothelial cells (HUVECs). The recipient serum was cytotoxic to these HUVECs, but not against HUVECs that did not express MICA-G1 antigens in the presence of complement. The researchers discovered that the patient had been sensitized to MICA antigens and HLA, before transplantation, and the HLA alloantibodies were not specific to the first kidney donor, but the MICA alloantibodies were. In light of this discovery, the second renal transplant was with a negative MICA virtual crossmatch, and it was successful (109).

microRNA and mRNA's Analysis

Seiler et al. (62) showed that an elevated NKG2D mRNA expression in biopsy material was correlated with the severity of AR and detected NKG2D$^+$ cells located in clusters around tubules in biopsies derived from patients diagnosed with acute and chronic rejection. The expression of *NKG2D* mRNA was also detected in urinary sediments obtained 2–3 days before the AR episode. However, significant levels of MICA mRNA were not detected in the patient groups analyzed (62). For the first time, the focus was on the importance of the role of the NKG2D molecule, which is responsible for MICA signal transduction.

Another controversial paper regarding the role of MICA is the Racca et al.'s study (86), in which the authors obtained peripheral blood samples from 29 renal-transplanted patients (19 men). They classified patients it into three groups: AR group (9 patients with acute grade I/II allograft rejection), chronic rejection group (10 patients with chronic allograft rejection), and stable evolution group (10 patients with clinically stable allograft evolution). The authors observed that MICA mRNA levels in peripheral blood mononuclear cells showed similar expression levels in all groups evaluated and in the control group. They also found similar levels of MICA expression in a comparison of biopsy specimens from AR and nephrotoxic ATN patients. They did not find a correlation between MICA expression and renal graft state (86). It is interesting to note that the MICA expression in biopsies did not have a healthy control group, while expression of MICA mRNA may be a posttranscriptional control that modules MICA expression on the cell surface. The Racca et al.'s (86) study still represents an interesting opportunity to discuss the role of MICA as a biomarker.

Xu et al. (110) studied miR-338-5p, a microRNA downregulated in AMR renal allografts, and negatively correlated with BAFF. This molecule plays an important role in the differentiation, development, and proliferation of B lymphocytes. BAFF could be released in a soluble form (sBAFF) after cleavage and would bind to BAFF receptor. The receptor-associated factor 3 is a sort of adaptor for the BAFF–BAFF-R connection, it is implicated in a signal transduction, and it appeared to be a candidate target for miR-338-5p. In the study, 49 follow-up renal-transplanted recipients and a healthy control group were examined, and it was found that anti-HLA II Ab, anti-MICA Ab, and anti-HLA + MICA mixed Abs were all statistically increased in recipients. Serum miR-338-5p was significantly downregulated in renal-transplanted recipients compared with healthy volunteers and was inversely correlated with sBAFF. The authors speculate that miR-338-5p may regulate the BAFF signal, and they suggested that sBAFF was significantly negatively correlated with anti-MICA Abs (110).

Cytomegalovirus (CMV) and Polyomavirus and Transplantation

Cytomegalovirus infection is the most common viral complication after renal transplantation and solid organ transplantation in general. One hundred ninety-six recipients who underwent kidney transplantation during the past 6 years were assessed with at follow-up of at least 12 months. In this study, it was shown that the activating receptor NKG2D was expressed in a significantly higher number of NK cells at day 0 and day 20 compared to day 180 ($P = 0.01$ and $P = 0.003$, respectively) and compared to the control group ($P = 0.0003$ and $P = 0.0004$, respectively) (121). This finding suggests a possible mechanism for the activation of NKG2D that goes beyond organ rejection, but it is closely related. In fact, in the Hadaya et al. (121) study, it was shown that an expansion of the NKG2D$^+$ NK cell population occurred during acute CMV infection which decreased over time to a level very similar to that of the control group.

An interesting study that involved NKG2D, performed by Shabir et al. (117), demonstrated that CD4$^+$CD28^{null} T cell expansion is driven by latent CMV infection inflammation. The immune surveillance of CMV may have an unwanted consequence in the development of endothelial injury, which was proven to be mediated by CD4$^+$CD27$^-$CD28^{null} cells in *in vitro* experiments. NKG2D was upregulated on CD4$^+$CD27$^-$CD28^{null} cells isolated from patients in this study and might have an important component of the cytotoxic effects of these cells. In fact, CD4$^+$CD28^{null} cells were found predominantly in CMV-seropositive patients, and expanded in the posttransplantation period, and expressed markers of cytotoxicity (NKG2D and perforin) and endothelial homing (CX3CR1). Isolated CD4$^+$CD27$^-$CD28^{null} cells previously exposed only to CMV-derived antigens showed signs of endothelial damage and apoptosis, and this effect was mitigated by NKG2D-blocking Ab. They concluded that the increase in $CD4^{+}CD28^{null}$ cell frequencies was associated with delayed graft function and lower eGRF at end follow-up, and this could be mediated by NKG2D (117).

Another study by Tonnerre et al. (122) investigated the implication of MICA in BK polyomavirus (BKPyV) reactivation in a cohort of 144 transplant donor/recipient pairs including recipients with no reactivation (control). BKPyV is frequently reactivated in kidney transplant recipients receiving an immunosuppressive regimen and is associated with nephropathy (BKPyVAN) and graft rejection. They investigated the impact of the *MICA A5.1* mutation on recipient BKPyV reactivation, and they found that recipients carrying a *non-MICA A5.1 (nA5.1)* genotype transplanted with a kidney from a donor carrying the *A5.1 MICA* variant had a lower risk of BKPyV reactivation ($P = 0.0148$). So they speculated that *MICA A5.1* could be a protective allele toward BKPyV infection (122). Interestingly, these researchers also found that the donor (A5.1)–recipient (nA5.1) combination was overrepresented in the group of MICA-sensitized patients, but in the latter, MICA A5.1 seemed to be a protective factor for a virus related to graft rejection (106).

CONCLUSION

Since the *MICA* gene was first described, it has been the subject of many studies aiming to comprehend its immunobiology and the role it plays in fine-tuning the innate and adaptive immune response. MICA appears to be involved in transplant rejection, immune response against viruses and intracellular bacteria, inflammation, homeostasis of epithelia, immune response against tumors, and tumor immune escape mechanisms. However, there remain a number of open issues to be addressed surrounding MICA's functions and roles. Developing and implementing typing strategies for *MICA* alleles may increase the chance for positive outcomes in solid organ transplantation by allowing better matching. MICA's biological function is achieved through its interaction with the NKG2D receptor. This activating receptor and its ligands are deeply involved in the outcomes of transplanted grafts, in fact, the overexpression of NKG2DLs could be involved in rejection episodes and can contribute to graft loss (44).

Various studies have shown that anti-MICA Abs, binding to MICA molecules expressed at the endothelial allograft cell surface, may have relevance to kidney transplantation outcome (81, 106). However, it is important to note that some studies, such as that of Lemy et al. (87), where the presence of anti-MICA Abs do not show adverse effects in renal graft outcomes (87). Also, *MICA* mRNA level analysis in blood mononuclear cells did not show a correlation between *MICA* expression and renal graft state (86). Seiler et al. (62) did not find an enhancement of mRNA expression levels of MICA in kidney biopsies from patients undergoing AR or chronic allograft nephropathy, but they observed increased *NKG2D* expression. In an interesting study performed by Sapak et al. (108), 41% of the detected anti-MICA Abs were donor-specific, but an astonishing 36% were anti-MICA Abs against self-MICA antigens and several patients (27%) produced both (108).

Regarding NKG2D, there are studies that report that it is possible to prolong graft survival and to prevent CD28-independent rejection of cardiac allografts after blocking NKG2D (123).

We can conclude that the role of MICA and NKG2D in transplant outcome is not yet clear; MICA-mediated rejection probably is not just a reaction to the MICA non-self protein. The stress condition following a transplant causes a general inflammatory status in the recipient. This could increase MICA production, thus activating the response *via* the NKG2D receptor. The clinical impact of these interactions will remain unclear until further studies are performed.

AUTHOR CONTRIBUTIONS

MR: planning and organizing structure of the review; research and analysis of the papers; wrote the review; and planning and creation of figures and tables. MB: planning and organizing structure of the review and contributions to the sections writing/critical review of the manuscript.

REFERENCES

1. Suthanthiran M, Strom TB. Renal transplantation. *N Engl J Med* (1994) 331(6):365–76. doi:10.1056/nejm199408113310606
2. Bahram S. MIC genes: from genetics to biology. *Adv Immunol* (2000) 76:1–60. doi:10.1016/S0065-2776(01)76018-X
3. Bahram S, Bresnahan M, Geraghty DE, Spies T. A second lineage of mammalian major histocompatibility complex class I genes. *Proc Natl Acad Sci U S A* (1994) 91(14):6259–63.
4. Bauer S, Groh V, Wu J, Steinle A, Phillips JH, Lanier LL, et al. Activation of NK cells and T cells by NKG2D, a receptor for stress-inducible MICA. *Science* (1999) 285(5428):727–9.
5. Collins RW. Human MHC class I chain related (MIC) genes: their biological function and relevance to disease and transplantation. *Eur J Immunogenet* (2004) 31(3):105–14. doi:10.1111/j.1365-2370.2004.00457.x
6. Zwirner NW, Fuertes MB, Girart Ma V, Domaica CI, Rossi LE. Immunobiology of the human MHC class I chain-related gene A (MICA): from transplantation immunology to tumor immune escape. *Immunologia* (2006) 25(1):25–38.
7. Fodil N, Laloux L, Wanner V, Pellet P, Hauptmann G, Mizuki N, et al. Allelic repertoire of the human MHC class I MICA gene. *Immunogenetics* (1996) 44(5):351–7.
8. Luo J, Tian W, Pan F, Liu X, Li L. Allelic and haplotypic diversity of 5'promoter region of the MICA gene. *Hum Immunol* (2014) 75(4):383–8. doi:10.1016/j.humimm.2013.12.010
9. Cox ST, Madrigal JA, Saudemont A. Diversity and characterization of polymorphic 5' promoter haplotypes of MICA and MICB genes. *Tissue Antigens* (2014) 84(3):293–303. doi:10.1111/tan.12400
10. Choy MK, Phipps ME. MICA polymorphism: biology and importance in immunity and disease. *Trends Mol Med* (2010) 16(3):97–106. doi:10.1016/j.molmed.2010.01.002

11. Bahram S, Mizuki N, Inoko H, Spies T. Nucleotide sequence of the human MHC class I MICA gene. *Immunogenetics* (1996) 44(1):80–1.
12. Groh V, Bahram S, Bauer S, Herman A, Beauchamp M, Spies T. Cell stress-regulated human major histocompatibility complex class I gene expressed in gastrointestinal epithelium. *Proc Natl Acad Sci U S A* (1996) 93(22):12445–50.
13. Katsuyama Y, Ota M, Ando H, Saito S, Mizuki N, Kera J, et al. Sequencing based typing for genetic polymorphisms in exons, 2, 3 and 4 of the MICA gene. *Tissue Antigens* (1999) 54(2):178–84.
14. Chauveau A, Tonnerre P, Pabois A, Gavlovsky PJ, Chatelais M, Coupel S, et al. Endothelial cell activation and proliferation modulate NKG2D activity by regulating MICA expression and shedding. *J Innate Immun* (2014) 6(1):89–104. doi:10.1159/000351605
15. Zou Y, Stastny P. Alternatively spliced forms of MICA and MICB lacking exon 3 in a human cell line and evidence of presence of similar RNA in human peripheral blood mononuclear cells. *Immunogenetics* (2002) 54(9):671–4. doi:10.1007/s00251-002-0496-y
16. Zwirner NW, Fernández-Viña MA, Stastny P. MICA, a new polymorphic HLA-related antigen, is expressed mainly by keratinocytes, endothelial cells, and monocytes. *Immunogenetics* (1998) 47(2):139–48.
17. Zwirner NW, Dole K, Stastny P. Differential surface expression of MICA by endothelial cells, fibroblasts, keratinocytes, and monocytes. *Hum Immunol* (1999) 60(4):323–30.
18. Mizuki N, Ota M, Kimura M, Ohno S, Ando H, Katsuyama Y, et al. Triplet repeat polymorphism in the transmembrane region of the MICA gene: a strong association of six GCT repetitions with Behcet disease. *Proc Natl Acad Sci U S A* (1997) 94(4):1298–303.
19. Pyo CW, Hur SS, Kim YK, Choi HB, Kim TY, Kim TG. Distribution of MICA alleles and haplotypes associated with HLA in the Korean population. *Hum Immunol* (2003) 64(3):378–84. doi:10.1016/S0198-8859(02)00826-1
20. Gambelunghe G, Brozzetti A, Ghaderi M, Candeloro P, Tortoioli C, Falorni A. MICA gene polymorphism in the pathogenesis of type 1 diabetes. *Ann N Y Acad Sci* (2007) 1110:92–8. doi:10.1196/annals.1423.011
21. Zhang Y, Han M, Vorhaben R, Giang C, Lavingia B, Stastny P. Study of MICA alleles in 201 African Americans by multiplexed single nucleotide extension (MSNE) typing. *Hum Immunol* (2003) 64(1):130–6. doi:10.1016/S0198-8859(02)00743-7
22. Petersdorf EW, Shuler KB, Longton GM, Spies T, Hansen JA. Population study of allelic diversity in the human MHC class I-related MIC-A gene. *Immunogenetics* (1999) 49(7–8):605–12.
23. Norris S, Kondeatis E, Collins R, Satsangi J, Clare M, Chapman R, et al. Mapping MHC-encoded susceptibility and resistance in primary sclerosing cholangitis: the role of MICA polymorphism. *Gastroenterology* (2001) 120(6):1475–82. doi:10.1053/gast.2001.24041
24. Marin ML, Savioli CR, Yamamoto JH, Kalil J, Goldberg AC. MICA polymorphism in a sample of the São Paulo population, Brazil. *Eur J Immunogenet* (2004) 31(2):63–71. doi:10.1111/j.1365-2370.2004.00446.x
25. Zhang Y, Lazaro AM, Zou Y, Lavingia B, Moraes EM, Moraes RJ, et al. MICA polymorphism in South American Indians. *Immunogenetics* (2002) 53(10–11):900–6. doi:10.1007/s00251-001-0426-4
26. Gao X, Single RM, Karacki P, Marti D, O'Brien SJ, Carrington M. Diversity of MICA and linkage disequilibrium with HLA-B in two North American populations. *Hum Immunol* (2006) 67(3):152–8. doi:10.1016/j.humimm.2006.02.009
27. Romphruk AV, Naruse TK, Romphruk A, Kawata H, Puapairoj C, Kulski JK, et al. Diversity of MICA (PERB11.1) and HLA haplotypes in Northeastern Thais. *Tissue Antigens* (2001) 58(2):83–9. doi:10.1034/j.1399-0039.2001.580203.x
28. Komatsu-Wakui M, Tokunaga K, Ishikawa Y, Kashiwase K, Moriyama S, Tsuchiya N, et al. MIC-A polymorphism in Japanese and a MIC-A-MIC-B null haplotype. *Immunogenetics* (1999) 49(7–8):620–8.
29. Tian W, Boggs DA, Ding WZ, Chen DF, Fraser PA. MICA genetic polymorphism and linkage disequilibrium with HLA-B in 29 African-American families. *Immunogenetics* (2001) 53(9):724–8. doi:10.1007/s00251-001-0392-x
30. Piancatelli D, Del Beato T, Oumhani K, El Aouad R, Adorno D. MICA polymorphism in a population from north Morocco, Metalsa Berbers, using sequence-based typing. *Hum Immunol* (2005) 66(8):931–6. doi:10.1016/j.humimm.2005.06.008
31. Mizuki N, Meguro A, Tohnai I, Gul A, Ohno S, Mizuki N. Association of major histocompatibility complex class I chain-related gene A and HLA-B alleles with Behcet's disease in Turkey. *Jpn J Ophthalmol* (2007) 51(6):431–6. doi:10.1007/s10384-007-0473-y
32. Ribas F, Oliveira LA, Petzl-Erler ML, Bicalho MG. Major histocompatibility complex class I chain-related gene A polymorphism and linkage disequilibrium with HLA-B alleles in Euro-Brazilians. *Tissue Antigens* (2008) 72(6):532–8. doi:10.1111/j.1399-0039.2008.01142.x
33. Wang W, Tian W, Zhu F, Liu X, Li L, Wang F. MICA, MICB polymorphisms and linkage disequilibrium with HLA-B in a Chinese Mongolian population. *Scand J Immunol* (2016) 83(6):456–62. doi:10.1111/sji.12437
34. Wang YJ, Zhang NJ, Chen E, Chen CJ, Bu YH, Yu P. MICA/B genotyping of Tujias from Zhangjiajie, Hunan Province, China. *Hum Immunol* (2016) 77(4):340–1. doi:10.1016/j.humimm.2016.01.014
35. Wang YJ, Zhang NJ, Chen E, Chen CJ, Bu YH, Yu P. Allele polymorphism and haplotype diversity of MICA/B in Tujia nationality of Zhangjiajie, Hunan Province, China. *Hum Immunol* (2016) 77(5):411–7. doi:10.1016/j.humimm.2016.03.005
36. Steinle A, Li P, Morris DL, Groh V, Lanier LL, Strong RK, et al. Interactions of human NKG2D with its ligands MICA, MICB, and homologs of the mouse RAE-1 protein family. *Immunogenetics* (2001) 53(4):279–87. doi:10.1007/s002510100325
37. Boukouaci W, Busson M, Peffault de Latour R, Rocha V, Suberbielle C, Bengoufa D, et al. MICA-129 genotype, soluble MICA, and anti-MICA antibodies as biomarkers of chronic graft-versus-host disease. *Blood* (2009) 114(25):5216–24. doi:10.1182/blood-2009-04-217430
38. Lopez-Hernandez R, Valdes M, Lucas D, Campillo JA, Martinez-Garcia P, Salama H, et al. Association analysis of MICA gene polymorphism and MICA-129 dimorphism with inflammatory bowel disease susceptibility in a Spanish population. *Hum Immunol* (2010) 71(5):512–4. doi:10.1016/j.humimm.2010.02.003
39. Zhao J, Jiang Y, Lei Y, Zou K, Wang C, Huang S, et al. Functional MICA-129 polymorphism and serum levels of soluble MICA are correlated with ulcerative colitis in Chinese patients. *J Gastroenterol Hepatol* (2011) 26(3):593–8. doi:10.1111/j.1440-1746.2010.06524.x
40. Isernhagen A, Malzahn D, Viktorova E, Elsner L, Monecke S, von Bonin F, et al. The MICA-129 dimorphism affects NKG2D signaling and outcome of hematopoietic stem cell transplantation. *EMBO Mol Med* (2015) 7(11):1480–502. doi:10.15252/emmm.201505246
41. Isernhagen A, Schilling D, Monecke S, Shah P, Elsner L, Walter L, et al. The MICA-129Met/Val dimorphism affects plasma membrane expression and shedding of the NKG2D ligand MICA. *Immunogenetics* (2016) 68(2):109–23. doi:10.1007/s00251-015-0884-8
42. Li P, Morris DL, Willcox BE, Steinle A, Spies T, Strong RK. Complex structure of the activating immunoreceptor NKG2D and its MHC class I-like ligand MICA. *Nat Immunol* (2001) 2(5):443–51. doi:10.1038/87757
43. Eleme K, Taner SB, Onfelt B, Collinson LM, McCann FE, Chalupny NJ, et al. Cell surface organization of stress-inducible proteins ULBP and MICA that stimulate human NK cells and T cells via NKG2D. *J Exp Med* (2004) 199(7):1005–10. doi:10.1084/jem.20032194
44. Suarez-Alvarez B, Lopez-Vazquez A, Baltar JM, Ortega F, Lopez-Larrea C. Potential role of NKG2D and its ligands in organ transplantation: new target for immunointervention. *Am J Transplant* (2009) 9(2):251–7. doi:10.1111/j.1600-6143.2008.02526.x
45. Wu J, Groh V, Spies T. T cell antigen receptor engagement and specificity in the recognition of stress-inducible MHC class I-related chains by human epithelial gamma delta T cells. *J Immunol* (2002) 169(3):1236–40. doi:10.4049/jimmunol.169.3.1236
46. Li J, Cui L, He W. Distinct pattern of human Vdelta1 gammadelta T cells recognizing MICA. *Cell Mol Immunol* (2005) 2(4):253–8.
47. Xu B, Pizarro JC, Holmes MA, McBeth C, Groh V, Spies T, et al. Crystal structure of a gammadelta T-cell receptor specific for the human MHC class I homolog MICA. *Proc Natl Acad Sci U S A* (2011) 108(6):2414–9. doi:10.1073/pnas.1015433108
48. Ho EL, Heusel JW, Brown MG, Matsumoto K, Scalzo AA, Yokoyama WM. Murine Nkg2d and Cd94 are clustered within the natural killer complex and are expressed independently in natural killer cells. *Proc Natl Acad Sci U S A* (1998) 95(11):6320–5.

49. Vance RE, Tanamachi DM, Hanke T, Raulet DH. Cloning of a mouse homolog of CD94 extends the family of C-type lectins on murine natural killer cells. *Eur J Immunol* (1997) 27(12):3236–41. doi:10.1002/eji.1830271222
50. Glienke J, Sobanov Y, Brostjan C, Steffens C, Nguyen C, Lehrach H, et al. The genomic organization of NKG2C, E, F, and D receptor genes in the human natural killer gene complex. *Immunogenetics* (1998) 48(3):163–73.
51. Uhrberg M, Valiante NM, Shum BP, Shilling HG, Lienert-Weidenbach K, Corliss B, et al. Human diversity in killer cell inhibitory receptor genes. *Immunity* (1997) 7(6):753–63.
52. Jamieson AM, Diefenbach A, McMahon CW, Xiong N, Carlyle JR, Raulet DH. The role of the NKG2D immunoreceptor in immune cell activation and natural killing. *Immunity* (2002) 17(1):19–29. doi:10.1016/S1074-7613(02)00333-3
53. Shum BP, Flodin LR, Muir DG, Rajalingam R, Khakoo SI, Cleland S, et al. Conservation and variation in human and common chimpanzee CD94 and NKG2 genes. *J Immunol* (2002) 168(1):240–52. doi:10.4049/jimmunol.168.1.240
54. Hayashi T, Imai K, Morishita Y, Hayashi I, Kusunoki Y, Nakachi K. Identification of the NKG2D haplotypes associated with natural cytotoxic activity of peripheral blood lymphocytes and cancer immunosurveillance. *Cancer Res* (2006) 66(1):563–70. doi:10.1158/0008-5472.CAN-05-2776
55. Ucisik-Akkaya E, Dorak MT. A study of natural killer cell lectin-like receptor K1 gene (KLRK1/NKG2D) region polymorphisms in a European population sample. *Tissue Antigens* (2009) 73(2):177–83. doi:10.1111/j.1399-0039.2008.01181.x
56. Furue H, Matsuo K, Kumimoto H, Hiraki A, Suzuki T, Yatabe Y, et al. Decreased risk of colorectal cancer with the high natural killer cell activity NKG2D genotype in Japanese. *Carcinogenesis* (2008) 29(2):316–20. doi:10.1093/carcin/bgm260
57. Furue H, Kumimoto H, Matsuo K, Suzuki T, Hasegawa Y, Shinoda M, et al. Opposite impact of NKG2D genotype by lifestyle exposure to risk of aerodigestive tract cancer among Japanese. *Int J Cancer* (2008) 123(1):181–6. doi:10.1002/ijc.23456
58. Espinoza JL, Takami A, Onizuka M, Sao H, Akiyama H, Miyamura K, et al. NKG2D gene polymorphism has a significant impact on transplant outcomes after HLA-fully-matched unrelated bone marrow transplantation for standard risk hematologic malignancies. *Haematologica* (2009) 94(10):1427–34. doi:10.3324/haematol.2009.008318
59. Garrity D, Call ME, Feng J, Wucherpfennig KW. The activating NKG2D receptor assembles in the membrane with two signaling dimers into a hexameric structure. *Proc Natl Acad Sci U S A* (2005) 102(21):7641–6. doi:10.1073/pnas.0502439102
60. Bacon L, Eagle RA, Meyer M, Easom N, Young NT, Trowsdale J. Two human ULBP/RAET1 molecules with transmembrane regions are ligands for NKG2D. *J Immunol* (2004) 173(2):1078–84. doi:10.4049/jimmunol.173.2.1078
61. Groh V, Rhinehart R, Randolph-Habecker J, Topp MS, Riddell SR, Spies T. Costimulation of CD8alphabeta T cells by NKG2D via engagement by MIC induced on virus-infected cells. *Nat Immunol* (2001) 2(3):255–60. doi:10.1038/85321
62. Seiler M, Brabcova I, Viklicky O, Hribova P, Rosenberger C, Pratschke J, et al. Heightened expression of the cytotoxicity receptor NKG2D correlates with acute and chronic nephropathy after kidney transplantation. *Am J Transplant* (2007) 7(2):423–33. doi:10.1111/j.1600-6143.2006.01625.x
63. Saez-Borderias A, Guma M, Angulo A, Bellosillo B, Pende D, Lopez-Botet M. Expression and function of NKG2D in CD4+ T cells specific for human cytomegalovirus. *Eur J Immunol* (2006) 36(12):3198–206. doi:10.1002/eji.200636682
64. Park YP, Choi SC, Kiesler P, Gil-Krzewska A, Borrego F, Weck J, et al. Complex regulation of human NKG2D-DAP10 cell surface expression: opposing roles of the gammac cytokines and TGF-beta1. *Blood* (2011) 118(11):3019–27. doi:10.1182/blood-2011-04-346825
65. Fernandez-Sanchez A, Baragano Raneros A, Carvajal Palao R, Sanz AB, Ortiz A, Ortega F, et al. DNA demethylation and histone H3K9 acetylation determine the active transcription of the NKG2D gene in human CD8+ T and NK cells. *Epigenetics* (2013) 8(1):66–78. doi:10.4161/epi.23115
66. Crane CA, Han SJ, Barry JJ, Ahn BJ, Lanier LL, Parsa AT. TGF-beta downregulates the activating receptor NKG2D on NK cells and CD8+ T cells in glioma patients. *Neuro Oncol* (2010) 12(1):7–13. doi:10.1093/neuonc/nop009
67. Codo P, Weller M, Meister G, Szabo E, Steinle A, Wolter M, et al. MicroRNA-mediated down-regulation of NKG2D ligands contributes to glioma immune escape. *Oncotarget* (2014) 5(17):7651–62. doi:10.18632/oncotarget.2287
68. Muntasell A, Magri G, Pende D, Angulo A, Lopez-Botet M. Inhibition of NKG2D expression in NK cells by cytokines secreted in response to human cytomegalovirus infection. *Blood* (2010) 115(25):5170–9. doi:10.1182/blood-2009-11-256479
69. Burgess SJ, Marusina AI, Pathmanathan I, Borrego F, Coligan JE. IL-21 down-regulates NKG2D/DAP10 expression on human NK and CD8+ T cells. *J Immunol* (2006) 176(3):1490–7. doi:10.4049/jimmunol.176.3.1490
70. Peraldi MN, Berrou J, Dulphy N, Seidowsky A, Haas P, Boissel N, et al. Oxidative stress mediates a reduced expression of the activating receptor NKG2D in NK cells from end-stage renal disease patients. *J Immunol* (2009) 182(3):1696–705. doi:10.4049/jimmunol.182.3.1696
71. Karimi MA, Aguilar OA, Zou B, Bachmann MH, Carlyle JR, Baldwin CL, et al. A truncated human NKG2D splice isoform negatively regulates NKG2D-mediated function. *J Immunol* (2014) 193(6):2764–71. doi:10.4049/jimmunol.1400920
72. Boukouaci W, Al-Daccak R, Dulphy N, Lauden L, Amokrane K, Fortier C, et al. Soluble MICA-NKG2D interaction upregulates IFN-gamma production by activated CD3-CD56+ NK cells: potential impact on chronic graft versus host disease. *Hum Immunol* (2013) 74(12):1536–41. doi:10.1016/j.humimm.2013.08.281
73. Yu R, Xu S, Wang Y, Cai H, Xu P. Role of MICA expression, anti-MICA antibodies and serum MICA during acute rejection in a rat-to-mouse cardiac transplantation model. *Int J Clin Exp Pathol* (2015) 8(11):14514–20.
74. Kummer JA, Wever PC, Kamp AM, ten Berge IJ, Hack CE, Weening JJ. Expression of granzyme A and B proteins by cytotoxic lymphocytes involved in acute renal allograft rejection. *Kidney Int* (1995) 47(1):70–7.
75. Hidalgo LG, Sis B, Sellares J, Campbell PM, Mengel M, Einecke G, et al. NK cell transcripts and NK cells in kidney biopsies from patients with donor-specific antibodies: evidence for NK cell involvement in antibody-mediated rejection. *Am J Transplant* (2010) 10(8):1812–22. doi:10.1111/j.1600-6143.2010.03201.x
76. Zwirner NW, Marcos CY, Mirbaha F, Zou Y, Stastny P. Identification of MICA as a new polymorphic alloantigen recognized by antibodies in sera of organ transplant recipients. *Hum Immunol* (2000) 61(9):917–24. doi:10.1016/S0198-8859(00)00162-2
77. Hankey KG, Drachenberg CB, Papadimitriou JC, Klassen DK, Philosophe B, Bartlett ST, et al. MIC expression in renal and pancreatic allografts. *Transplantation* (2002) 73(2):304–6. doi:10.1097/00007890-200201270-00029
78. Opelz G. Non-HLA transplantation immunity revealed by lymphocytotoxic antibodies. *Lancet* (2005) 365(9470):1570–6. doi:10.1016/S0140-6736(05)66458-6
79. Mizutani K, Terasaki P, Rosen A, Esquenazi V, Miller J, Shih RN, et al. Serial ten-year follow-up of HLA and MICA antibody production prior to kidney graft failure. *Am J Transplant* (2005) 5(9):2265–72. doi:10.1111/j.1600-6143.2005.01016.x
80. Amezaga N, Crespo M, Lopez-Cobos M, Millan MA, Vinas O, Sole M, et al. Relevance of MICA antibodies in acute humoral rejection in renal transplant patients. *Transpl Immunol* (2006) 17(1):39–42. doi:10.1016/j.trim.2006.09.016
81. Mizutani K, Terasaki P, Bignon JD, Hourmant M, Cesbron-Gautier A, Shih RN, et al. Association of kidney transplant failure and antibodies against MICA. *Hum Immunol* (2006) 67(9):683–91. doi:10.1016/j.humimm.2006.06.002
82. Panigrahi A, Gupta N, Siddiqui JA, Margoob A, Bhowmik D, Guleria S, et al. Post transplant development of MICA and anti-HLA antibodies is associated with acute rejection episodes and renal allograft loss. *Hum Immunol* (2007) 68(5):362–7. doi:10.1016/j.humimm.2007.01.006
83. Zou Y, Stastny P, Susal C, Dohler B, Opelz G. Antibodies against MICA antigens and kidney-transplant rejection. *N Engl J Med* (2007) 357(13):1293–300. doi:10.1056/NEJMoa067160
84. Suarez-Alvarez B, Alonso-Arias R, Bravo-Mendoza C, Lopez-Vazquez A, Ortega T, Baltar JM, et al. Identification of epitopes and immunodominant regions on the MICA protein defined by alloantibodies from kidney

transplant patients. *Transplantation* (2009) 88(3 Suppl):S68–77. doi:10.1097/TP.0b013e3181afeb7a

85. Alvarez-Marquez A, Aguilera I, Gentil MA, Caro JL, Bernal G, Fernandez Alonso J, et al. Donor-specific antibodies against HLA, MICA, and GSTT1 in patients with allograft rejection and C4d deposition in renal biopsies. *Transplantation* (2009) 87(1):94–9. doi:10.1097/TP.0b013e31818bd790
86. Racca AL, Veaute CM, Bailat AS, Gaite L, Arriola M, Hajos SE, et al. Expression of HLA-G and MICA mRNA in renal allograft. *Transpl Immunol* (2009) 21(1):10–2. doi:10.1016/j.trim.2009.01.002
87. Lemy A, Andrien M, Wissing KM, Ryhahi K, Vandersarren A, Racape J, et al. Major histocompatibility complex class 1 chain-related antigen a antibodies: sensitizing events and impact on renal graft outcomes. *Transplantation* (2010) 90(2):168–74. doi:10.1097/TP.0b013e3181e228f7
88. Li L, Chen A, Chaudhuri A, Kambham N, Sigdel T, Chen R, et al. Compartmental localization and clinical relevance of MICA antibodies after renal transplantation. *Transplantation* (2010) 89(3):312–9. doi:10.1097/TP.0b013e3181bbbe4c
89. Luo L, Lu J, Wei L, Long D, Guo JY, Shan J, et al. The role of HIF-1 in up-regulating MICA expression on human renal proximal tubular epithelial cells during hypoxia/reoxygenation. *BMC Cell Biol* (2010) 11:91. doi:10.1186/1471-2121-11-91
90. Cox ST, Stephens HA, Fernando R, Karasu A, Harber M, Howie AJ, et al. Major histocompatibility complex class I-related chain A allele mismatching, antibodies, and rejection in renal transplantation. *Hum Immunol* (2011) 72(10):827–34. doi:10.1016/j.humimm.2011.05.004
91. Narayan S, Tsai EW, Zhang Q, Wallace WD, Reed EF, Ettenger RB. Acute rejection associated with donor-specific anti-MICA antibody in a highly sensitized pediatric renal transplant recipient. *Pediatr Transplant* (2011) 15(1):E1–7. doi:10.1111/j.1399-3046.2010.01407.x
92. Yao QC, Wang W, Li XB, Yin H, Zhang XD. Expression characteristics of major histocompatibility complex class I-related chain A antibodies and immunoadsorption effect in sensitized recipients of kidney transplantation. *Chin Med J (Engl)* (2011) 124(5):669–73. doi:10.3760/cma.j.issn.0366-6999.2011.05.006
93. Zhang M, Lu FM, Qu LX, He J, Yuan XN, Gu Y. Serum major-histocompatibility-complex class I-related chain A antibody detection for the evaluation of graft dysfunction in renal allograft recipients. *Chin Med J (Engl)* (2011) 124(14):2127–31. doi:10.3760/cma.j.issn.0366-6999.2011.14.009
94. Lemy A, Andrien M, Lionet A, Labalette M, Noel C, Hiesse C, et al. Posttransplant major histocompatibility complex class I chain-related gene A antibodies and long-term graft outcomes in a multicenter cohort of 779 kidney transplant recipients. *Transplantation* (2012) 93(12):1258–64. doi:10.1097/TP.0b013e31824fd8f1
95. Li Z, Luo M, Qiu J, Liu Y, Fan Y, Jahr FM, et al. Detection of antibodies against major histocompatibility complex class I-related chain A in long-term renal graft recipients. *Exp Clin Transplant* (2012) 10(3):239–42. doi:10.6002/ect.2011.0103
96. Seyhun Y, Ozdilli K, Oguz F, Karahan G, Onal E, Turkmen A, et al. Human leukocyte antigen and major histocompatibility complex class I-related chain A antibodies after kidney transplantation in Turkish renal transplant recipients. *Transplant Proc* (2012) 44(6):1660–6. doi:10.1016/j.transproceed.2012.04.018
97. Rodriguez Ferrero ML, Arroyo D, Panizo N, Vicario JL, Balas A, Anaya F. Monitoring of circulating antibodies in a renal transplantation population: preliminary results. *Transplant Proc* (2012) 44(9):2548–50. doi:10.1016/j.transproceed.2012.09.072
98. Solgi G, Furst D, Mytilineos J, Pourmand G, Amirzargar AA. Clinical relevance of pre and post-transplant immune markers in kidney allograft recipients: anti-HLA and MICA antibodies and serum levels of sCD30 and sMICA. *Transpl Immunol* (2012) 26(2–3):81–7. doi:10.1016/j.trim.2011.12.002
99. Akgul SU, Oguz FS, Caliskan Y, Kekik C, Cagatay P, Turkmen A, et al. The effect of anti-human leukocyte antigen, anti-major histocompatibility complex class 1 chain-related antigen a, and anti-glutathione transferase-T1 antibodies on the long-term survival of renal allograft. *Transplant Proc* (2013) 45(3):890–4. doi:10.1016/j.transproceed.2013.02.097
100. Chaudhuri A, Ozawa M, Everly MJ, Ettenger R, Dharnidharka V, Benfield M, et al. The clinical impact of humoral immunity in pediatric renal transplantation. *J Am Soc Nephrol* (2013) 24(4):655–64. doi:10.1681/ASN.2012070663
101. Ding X, Jiao F, Wang X, Xue W, Tian P, Li Y, et al. [Significance of MICA antibody monitoring in management of acute and chronic rejection after renal transplantation]. *Nan Fang Yi Ke Da Xue Xue Bao* (2013) 33(10):1427–31. doi:10.3969/j.issn.1673-4254.2013.10.04
102. He J, Li C, Yuan XN, Zhang JL, Li Y, Wei XD, et al. Anti-human leukocyte antigens and anti-major histocompatibility complex class I-related chain A antibody expression in kidney transplantation during a four-year follow-up. *Chin Med J (Engl)* (2013) 126(15):2815–20. doi:10.3760/cma.j.issn.0366-6999.20131143
103. Jin J, Xu Y, Wang H, Huang H, He Q, Wu P, et al. Peritubular capillaritis in early renal allograft dysfunction is an indicator of acute rejection. *Transplant Proc* (2013) 45(1):163–71. doi:10.1016/j.transproceed.2012.08.018
104. Li Z, Wen X, Wang Y, Luo M, Qiu J. Effect of mycophenolic acid and bortezomib on purified human B cells: an in vitro study of long-term functionally stable MICA-sensitized renal recipients. *Exp Clin Transplant* (2013) 11(6):482–8. doi:10.6002/ect.2012.0281
105. Sanchez-Zapardiel E, Castro-Panete MJ, Castillo-Rama M, Morales P, Lora-Pablos D, Valero-Hervas D, et al. Harmful effect of preformed anti-MICA antibodies on renal allograft evolution in early posttransplantation period. *Transplantation* (2013) 96(1):70–8. doi:10.1097/TP.0b013e3182943506
106. Tonnerre P, Gerard N, Chatelais M, Poli C, Allard S, Cury S, et al. MICA variant promotes allosensitization after kidney transplantation. *J Am Soc Nephrol* (2013) 24(6):954–66. doi:10.1681/ASN.2012080814
107. Zhang LW, Peng ZG, Xian WH, Cui XQ, Sun HB, Li EG, et al. Association of de novo human leukocyte antigen and major histocompatibility complex class I chain-related gene-A antibodies and proteinuria with graft survival 5 years after renal transplantation. *Transplant Proc* (2013) 45(9):3249–53. doi:10.1016/j.transproceed.2013.08.029
108. Sapak M, Chrenova S, Tirpakova J, Zilinska Z, Durmanova V, Shawkatova I, et al. Donor non-specific MICA antibodies in renal transplant recipients. *Immunobiology* (2014) 219(2):109–12. doi:10.1016/j.imbio.2013.08.006
109. Ming Y, Hu J, Luo Q, Ding X, Luo W, Zhuang Q, et al. Acute antibody-mediated rejection in presence of MICA-DSA and successful renal re-transplant with negative-MICA virtual crossmatch. *PLoS One* (2015) 10(5):e0127861. doi:10.1371/journal.pone.0127861
110. Xu H, Ma X, He X, Dong P, Xue D, Zhang Y, et al. Serum miR-338-5p, soluble B-cell-activating factor, allo-antibodies, and renal transplantation. *Transplant Proc* (2015) 47(2):337–42. doi:10.1016/j.transproceed.2014.11.030
111. Cai J, Terasaki PI, Zhu D, Lachmann N, Schonemann C, Everly MJ, et al. Complement-fixing antibodies against denatured HLA and MICA antigens are associated with antibody mediated rejection. *Exp Mol Pathol* (2016) 100(1):45–50. doi:10.1016/j.yexmp.2015.11.023
112. Sanchez-Zapardiel E, Castro-Panete MJ, Mancebo E, Morales P, Laguna-Goya R, Morales JM, et al. Early renal graft function deterioration in recipients with preformed anti-MICA antibodies: partial contribution of complement-dependent cytotoxicity. *Nephrol Dial Transplant* (2016) 31(1):150–60. doi:10.1093/ndt/gfv308
113. Feng L, Cheng F, Ye Z, Li S, He Y, Yao X, et al. The effect of renal ischemia-reperfusion injury on expression of RAE-1 and H60 in mice kidney. *Transplant Proc* (2006) 38(7):2195–8. doi:10.1016/j.transproceed.2006.06.013
114. Zheng G, Zheng L, Wang Y, Wu H, Kairaitis L, Zhang C, et al. NK cells do not mediate renal injury in murine adriamycin nephropathy. *Kidney Int* (2006) 69(7):1159–65. doi:10.1038/sj.ki.5000244
115. Hadaya K, de Rham C, Bandelier C, Bandelier C, Ferrari-Lacraz S, Jendly S, et al. Natural killer cell receptor repertoire and their ligands, and the risk of CMV infection after kidney transplantation. *Am J Transplant* (2008) 8(12):2674–83. doi:10.1111/j.1600-6143.2008.02431.x
116. Zhang ZX, Wang S, Huang X, Min WP, Sun H, Liu W, et al. NK cells induce apoptosis in tubular epithelial cells and contribute to renal ischemia-reperfusion injury. *J Immunol* (2008) 181(11):7489–98. doi:10.4049/jimmunol.181.11.7489
117. Shabir S, Smith H, Kaul B, Pachnio A, Jham S, Kuravi S, et al. Cytomegalovirus-associated CD4(+) CD28(null) cells in NKG2D-dependent glomerular

endothelial injury and kidney allograft dysfunction. *Am J Transplant* (2016) 16(4):1113–28. doi:10.1111/ajt.13614

118. Duquesnoy RJ, Mostecki J, Hariharan J, Balazs I. Structurally based epitope analysis of major histocompatibility complex class I-related chain A (MICA) antibody specificity patterns. *Hum Immunol* (2008) 69(12): 826–32. doi:10.1016/j.humimm.2008.10.003
119. Panigrahi A, Gupta N, Siddiqui JA, Margoob A, Bhowmik D, Guleria S, et al. Monitoring of anti-HLA and anti-major histocompatibility complex class I related chain A antibodies in living related renal donor transplantation. *Transplant Proc* (2007) 39(3):759–60. doi:10.1016/j.transproceed.2007.02.009
120. Zou Y, Stastny P, Süsal C, Döhler B, Opelz G. Antibodies against MICA antigens and kidney-transplant rejection. *N Engl J Med* (2007) 357(13):1293–300. doi:10.1056/NEJMoa067160
121. Hadaya K, de Rham C, Bandelier C, Ferrari-Lacraz S, Jendly S, Berney T, et al. Natural killer cell receptor repertoire and their ligands, and the risk of CMV infection after kidney transplantation. *Am J Transplant* (2008) 8(12):2674–83. doi:10.1111/j.1600-6143.2008.02431.x
122. Tonnerre P, Gerard N, Gavlovsky PJ, Mazalrey S, Hourmant M, Cheneau ML, et al. MICA mutant A5.1 influences BK polyomavirus reactivation and associated nephropathy after kidney transplantation. *J Infect Dis* (2016) 214(5):807–16. doi:10.1093/infdis/jiw168
123. McNerney ME, Lee KM, Zhou P, Molinero L, Mashayekhi M, Guzior D, et al. Role of natural killer cell subsets in cardiac allograft rejection. *Am J Transplant* (2006) 6(3):505–13. doi:10.1111/j.1600-6143.2005.01226.x

15

Shaping of Natural Killer Cell Antitumor Activity by *Ex Vivo* Cultivation

Markus Granzin[1], Juliane Wagner[2,3], Ulrike Köhl[4], Adelheid Cerwenka[5,6], Volker Huppert[7] and Evelyn Ullrich[2,3]**

[1] Clinical Research, Miltenyi Biotec Inc., Gaithersburg, MD, USA, [2] Division for Stem Cell Transplantation and Immunology, Department for Children and Adolescents Medicine, Hospital of the Goethe University, Frankfurt, Germany, [3] LOEWE Center for Cell and Gene Therapy, Cellular Immunology, Goethe University, Frankfurt, Germany, [4] Institute of Cellular Therapeutics, Integrated Research and Treatment Center Transplantation, Hannover Medical School, Hannover, Germany, [5] Innate Immunity Group, German Cancer Research Center, Heidelberg, Germany, [6] Division of Immunbiochemistry, Medical Faculty Mannheim, Heidelberg University, Heidelberg, Germany, [7] R&D Reagents, Miltenyi Biotec GmbH, Bergisch Gladbach, Germany

Edited by:
Hermann Einsele,
University of Würzburg, Germany

Reviewed by:
Reem Al-Daccak,
Institut national de la santé et de la recherche médicale (INSERM), France
Stanislaw Stepkowski,
University of Toledo, USA

***Correspondence:**
Markus Granzin
markus.granzin@miltenyibiotec.com;
Evelyn Ullrich
evelyn.ullrich@kgu.de

Natural killer (NK) cells are a promising tool for the use in adoptive immunotherapy, since they efficiently recognize and kill tumor cells. In this context, *ex vivo* cultivation is an attractive option to increase NK cells in numbers and to improve their antitumor potential prior to clinical applications. Consequently, various strategies to generate NK cells for adoptive immunotherapy have been developed. Here, we give an overview of different NK cell cultivation approaches and their impact on shaping the NK cell antitumor activity. So far, the cytokines interleukin (IL)-2, IL-12, IL-15, IL-18, and IL-21 are used to culture and expand NK cells. The selection of the respective cytokine combination is an important factor that directly affects NK cell maturation, proliferation, survival, distribution of NK cell subpopulations, activation, and function in terms of cytokine production and cytotoxic potential. Importantly, cytokines can upregulate the expression of certain activating receptors on NK cells, thereby increasing their responsiveness against tumor cells that express the corresponding ligands. Apart from using cytokines, cocultivation with autologous accessory non-NK cells or addition of growth-inactivated feeder cells are approaches for NK cell cultivation with pronounced effects on NK cell activation and expansion. Furthermore, *ex vivo* cultivation was reported to prime NK cells for the killing of tumor cells that were previously resistant to NK cell attack. In general, NK cells become frequently dysfunctional in cancer patients, for instance, by downregulation of NK cell activating receptors, disabling them in their antitumor response. In such scenario, *ex vivo* cultivation can be helpful to arm NK cells with enhanced antitumor properties to overcome immunosuppression. In this review, we summarize the current knowledge on NK cell modulation by different *ex vivo* cultivation strategies focused on increasing NK cytotoxicity for clinical application in malignant diseases. Moreover, we critically discuss the technical and regulatory aspects and challenges underlying NK cell based therapeutic approaches in the clinics.

Keywords: natural killer cells, natural killer cell cultivation, natural killer cell expansion, natural killer cell therapy, natural killer cell cytotoxicity, *ex vivo* stimulation

INTRODUCTION

As an important part of the innate immune system, natural killer (NK) cells are deployed as first line of defense against aberrant cells caused by viral infections or malignancies. Human NK cells can be identified *via* their morphology as large granular lymphocytes, and *via* their surface marker profile, as they express by definition CD56, but not CD3. The NK cell compartment can be further divided into subpopulations. There are two main NK cell subsets that can be distinguished, the $CD56^{high}CD16^{neg}$ subpopulation, which has mostly immune modulatory function, mainly accomplished by interferon (IFN)-γ secretion, and the $CD56^{low}CD16^{pos}$ fraction with direct cytotoxic capacity (1–3). NK cell activation is based on a balanced system integrating signals from activating and inhibitory receptors. Inhibitory signals derive mainly from germ-line encoded inhibitory killer cell immunoglobulin-like receptors (KIRs). Ligands for inhibitory KIRs, in humans major histocompatibility complex (MHC) class I molecules, are highly expressed by healthy cells and thereby prevent NK cell activation. Malignant cells often downregulate MHC class I molecules on their surface to evade T cell attack (4). However, these so-called "missing-self" cells are recognized by NK cells through inhibitory receptors, and as signals from activating receptors prevail, NK cells become active and react against the encountered targets. Alternatively, NK cells can be activated by overexpression of stress-induced surface ligands on infected or abnormal cells, which are recognized by activating receptors, such as the natural cytotoxicity receptors (NCRs) NKp30, NKp44, and NKp46, and the so-called C-type lectin-like receptors, such as NKG2D (1, 5–9). In this case, activating signals outbalance inhibitory self-signals and lead to NK cell activation. Furthermore, NK cells become activated upon encounter of antibody-coated targets by CD16, which binds to the Fc portion of the antibody and mediates a strong activating signal. By means of activating and inhibitory receptors, NK cells, unlike T and B-lymphocytes, can react immediately without prior priming or antigen presentation.

Activated NK cells execute effector functions through different mechanisms. NK cells mediate direct cytotoxicity *via* the exocytosis pathway with release of cytotoxic granules, which contain granzymes and perforin, resulting in lysis of the target cell (10). In addition, NK cells induce apoptosis of target cells by expression of death receptor ligands, such as Fas ligand or tumor necrosis factor-related apoptosis-inducing ligand (TRAIL) (11). Production and release of IFN-γ by NK cells after activation also has multiple functional consequences, with particular relevance in tumor surveillance, as IFN-γ inhibits tumor angiogenesis, has antimetastatic activity, and acts pro-apoptotic (12, 13).

The ability of tumor cells to bypass the immune response is a basic prerequisite for cancer formation and progression. Within immune editing, tumors undergo genetic, epigenetic, and phenotypic changes, thereby becoming a heterogeneous cell population that is hardly visible to or assailable by immune cells due to downregulation of tumor antigens and NCR ligands (14). Additionally, malignant cells suppress NK cells by blocking the NKG2D receptor *via* shedding of NKG2D ligands (15–17) or upregulation of inhibitory MHC class I molecules (18, 19). Immunosuppressive cytokines such as transforming growth factor-β, interleukin (IL)-10, or immunosuppressive enzymes, such as indoleamin 2,3-dioxigenase, further impair antitumor NK cell responses of cancer patients (20–22).

Ex vivo modulation of NK cell receptor expression is therefore an important tool to overcome immune response inhibition. A number of studies reported an upregulation of DNAM-1, NKG2D, and other NK cell-activating receptors under certain culture conditions, mostly involving stimulation by IL-2 (23–26). In addition, other ILs such as IL-12, IL15, IL-18, or IL-21 and Type I IFNs shape the NK cell receptor expression profile (27–31).

Natural killer cells can play an important role for cellular immunotherapy and the adoptive transfer of NK cells represents an attractive strategy to treat cancer patients (32, 33). In this context, *ex vivo* expansion of NK cells prior to their clinical application is not only required to increase the applicable cell doses but it is also reasonable to pre-activate and modify their antitumor features. For *ex vivo* cultivation, NK cells from different sources can be stimulated with different cytokines, and, to reach efficient expansion rates, NK cells are cultured among autologous accessory cells or together with different types of growth-inactivated autologous or allogeneic feeder cells (**Figure 1**). Of note, it is possible to genetically engineer NK cells *ex vivo* to further augment their antitumor activity, for example, to integrate chimeric antigen receptors against distinct tumor antigens (34, 35). In this review, we focus on the cultivation of NK cells without genetic modifications. Many different protocols exist for *ex vivo* expansion of NK cells, all with different features and capacities. Here, we give a comprehensive overview of strategies to obtain appropriate amounts of functional NK cells. We will discuss starting material and culture systems as well as the use of cytokines, feeder cells, and other additives.

STARTING MATERIAL FOR NK CELL EXPANSION AND ROLE OF NK CELL PURITY

Until recent, 92% of clinical studies used NK cells from peripheral blood, either donor (79% of recruiting trials) or patient derived (13% of recruiting trials) (36). Alternatives are the use of NK cell lines, or the differentiation of NK cells from umbilical cord blood or pluripotent stem cells (37–39). NK cell lines, such as NK-92, avoid the need for donor selection and enable the production of large cell doses to treat patients on a flexible schedule (40). Nevertheless, NK cell lines require growth inactivation mainly achieved by irradiation, possibly reducing their antitumor potential due to short *in vivo* persistence. Differentiation of NK cells from cord blood $CD34^+$ cells is attractive because of the "off-the-shelf" availability from a cord blood bank. Similarly, NK cells from pluripotent stem cells are a promising concept for the future but still in early development (39, 41). In this overview, we focus on peripheral blood-derived NK cells, currently the main source for NK cells for clinical use.

The NK cell purity, meaning the frequency of NK cells among other cells, is an important factor for the intended therapeutic

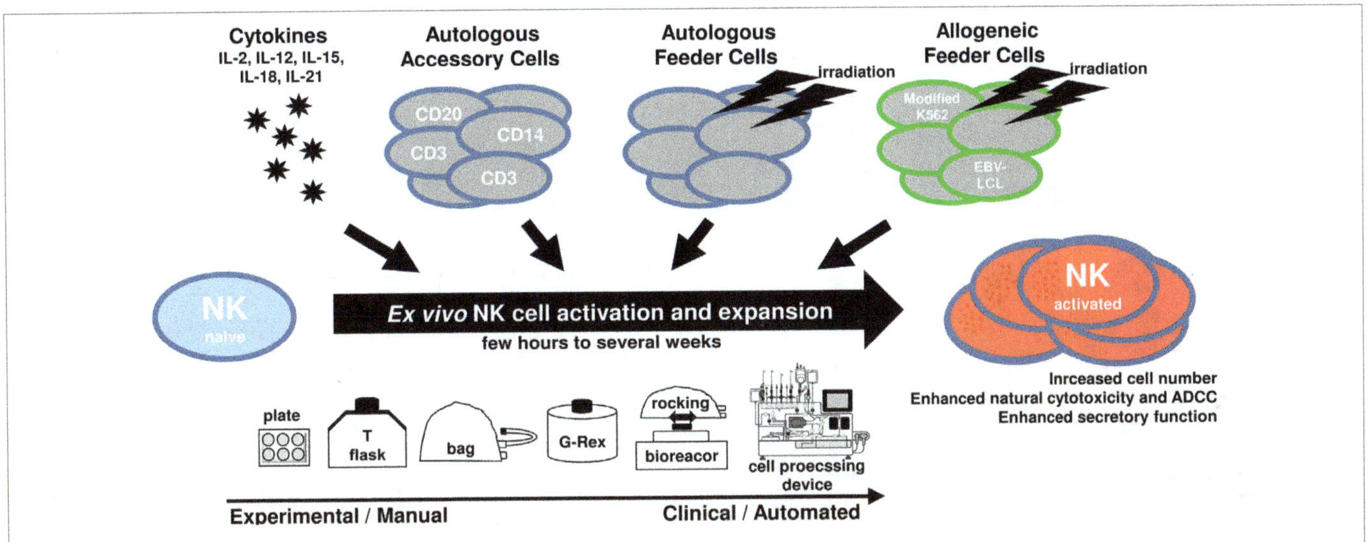

FIGURE 1 | **Scheme showing main components utilized for *ex vivo* natural killer (NK) cell activation and expansion procedures.**

application. For *ex vivo* expansion, NK cells are often cultured within a mixture of cells, such as PBMC, thereby avoiding further purification. Whereas the cultivation of NK cells among other accessory cells is a practical strategy for autologous therapeutic settings, it may be critical for allogeneic applications, since non-NK cells may induce unwanted side effects. Alloreactive T cells are a major risk factor for the patient, as they mediate "graft-versus-host disease" (GvHD), a severe complication following allogeneic hematopoietic stem cell transplantation (HSCT) (42). Furthermore, donor-derived B cells can lead to B cell lymphoproliferative disorder after reactivation of an Epstein–Barr virus (EBV) infection (43, 44), and they can cause the passenger lymphocyte syndrome (45), both critical side effects for the patient. Therefore, purification of NK cells might be required and is realized so far in most clinical settings by magnetic cell separation, for instance, by depletion of CD3-expressing cells and subsequent enrichment for CD56-expressing cells (26, 46–49). In addition, a first proof of concept is shown for good manufacturing practice (GMP)-compliant fluorescence-activated cell sorting to purify for NK cell subsets, such as NK cells expressing a single KIR (50).

CYTOKINE-INDUCED NK CELL EXPANSION

Aims of adoptive transfer of *ex vivo* expanded NK cells are the enhancement of natural cytotoxicity and homing to tumor sites under maintenance of "self" protection. Studies performed with cytokine-stimulated NK cells or PBMC have shown the safety of this approach and indicated some clinical responses upon adoptive NK cell transfer following HSCT. In the next paragraph, we summarize *ex vivo* NK cell expansion protocols starting with purified NK cells (**Table 1**) or PBMC (**Table 2**). Concepts administering cytokines in the presence of growth-inactivated feeder cells will be discussed in later sections of this article.

THE ROLE OF IL-2

Interleukin-2 plays an important role in activation of NK cells *via* binding to the IL-2 receptor (IL-2R), a heterotrimeric protein expressed on NK cells and other immune cells. This led to the interest in both (i) using IL-2 for stimulation of autologous NK cells in cancer patients and (ii) *ex vivo* activation and expansion of allogeneic donor NK cells for adaptive immunotherapy. At the beginning of the 1980s, researchers around Rosenberg and colleagues showed that IL-2 exposed lymphokine-activated killer (LAK) cells were able to attack autologous fresh tumor cells and that this effect could mainly be ascribed to NK cells (71, 72). Nevertheless, in first clinical trials using adoptive transfer of LAK cells and IL-2 therapy, the clinical response did not exceed the efficacy of IL-2 monotherapy (73).

Importantly, during the last 20 years, it has been elaborated that NK cells play a major role in the regulation of the balance between GvL and GvHD after allogeneic HSCT, especially haploidentical HSCT (33, 74–78), demonstrating improved anticancer activity while avoiding GvHD. In order to make haploidentical NK cells available for clinical use, large-scale GMP-conform manufacturing protocols were established. After starting with a leukapheresis product that was depleted for $CD3^+$ cells and enriched for $CD56^+$ cells, cultivation in medium containing IL-2 (1,000 U/mL) for up to 2 weeks yielded $0.1–3 \times 10^9$ $CD56^+CD3^-$ NK cells (**Table 1**), sometimes sufficient for multiple infusions to patients with hematological malignancies (26). Median NK cell expansion was fivefold and median NK cell purity was >94 with <0.1% T cell contamination (26). *Ex vivo* stimulation with IL-2 induced elevated cytokine secretion by NK cells, enhanced intracellular STAT3/AKT signaling, and upregulation of various NCRs and NKG2D receptors (52). Depletion of CD3 cells from leukapheresis products without subsequent CD56 enrichment and short-term activation with IL-2 overnight led to a final product containing 40% NK cells (**Table 2**). In all cases, IL-2-activated NK cells demonstrated a much higher cytotoxic activity against

TABLE 1 | ***Ex vivo*** **cultivation of pure natural killer (NK) cells with cytokines only.**

Protocol features	Starting material/culture system	NK cell expansion rate	NK cell purity	NK cell phenotype	NK cell function	Setting	Reference
IL-2 + IL-15	PBMC, CD3 depleted and CD56 enriched or PBMC, CD3/CD19 depleted in flasks	>1 (5 days)	75–100% NK ≤0.1% T cells	*Upregulated*: CD69, NKp30, and NKp44	Cytolysis of leukemia cell lines and primary acute leukemic blasts	*In vitro*	(51)
IL-2	PBMC, CD3 depleted, and CD56 enriched in bags and flasks	4–5 (12–14 days)	~92–95% NK <0.1% T cells	increased CD56^{+}CD16^{-} frequency; increased p-STAT3 and p-AKT; increased lytic activity, upregulation of CD69, NKG2D, and natural cytotoxicity receptors (NCRs); increasing amount of NK cells without killer cell immunoglobulin-likereceptors	Improved cytotoxic activity against leukemia and tumors	Clinical	(26, 52, 53)
IL-15	Isolated CD3^{-}CD56^{+} cells	N/A (2–5 weeks)	94–99% NK	Enhanced killing *via* NCRs, DNAM-1 and NKG2D	Enhanced cytolysis of lymphoma and rhabdomyosarcoma cell lines *via* NCRs	*In vitro*	(54)
IL-2 + IL-21	PBMC, sorted for CD3^{-}CD56^{+}	None with IL-21 only; strongly with IL-2 + IL-21	N/A	*Upregulated*: CD69, CD25; activation of STAT3	Enhanced cytotoxicity against K562	*In vitro*	(55)
IL-12 + IL-15 + IL-18	PBMC, sorted for CD3^{-}CD56^{+} cells PBMC, CD3 depleted and CD56 enriched	N/A (12–16 h)	≥90% NK	*Upregulated*: CD94, NKG2A, NKp30, NKp44, NKG2D, NKp46, CD69, and CD25 *Downregulated*: NKp80	Memory: increased IFN-γ production upon stimulation that is preserved during cell division Responsive to picomolar concentrations of IL-2	*In vitro* Clinical	(56–59)

K562 target cells compared to unstimulated NK cells (26, 43, 52, 53, 60). In addition, after cryopreservation and thawing, NK cells showed a moderate to high viability when activated with IL-2, whereas the viability of unstimulated NK cells was low (26).

Transfer into the clinic in 2004 and 2005 with first patient studies using those IL-2-activated donor NK cells were performed in parallel in Europe and the US, for both, haploidentical HSCT (53), and in the non-transplant setting (43). In the latter one, Miller and coworkers used IL-2 expanded haploidentical NK to treat 43 patients with advanced cancer (43), with 19 of them suffering from acute myeloid leukemia, followed by studies in patient with ovarian and breast cancer and B-cell non-Hodgkin lymphoma (60, 79). Importantly, the authors reported *in vivo* persistence and even expansion of the alloreactive donor NK cells in patients pretreated with high dose preparative regimen, consisting of 5 days of 60 mg/kg intravenous cyclophosphamide and 25 mg/m^2 intravenous fludarabine (43). Of note, successful NK cell engraftment was dependent on the patients' pretreatment regimen, which was also responsible for the patients' elevated IL-15 plasma concentrations (43). In addition, it was demonstrated that *in vivo* persistence of donor NK cells at day 7 after infusion and successful *in vivo* expansion (more than 100 donor-derived NK cells per microliter of patient blood 14 days after transfer) correlated with leukemia clearance (60). Expansion of host regulatory T cells was associated with low numbers of NK cells (60). In parallel, Koehl et al. reported on three pediatric patients with multiply relapsed leukemia (still in blast persistence at HSCT) treated with repeated transfusions of IL-2-activated donor NK cells post-haploidentical HSCT (53), which led to complete remission remaining for several weeks up to some months. In the following clinical study, they also demonstrated a small clinical benefit in patients with various malignancies receiving IL-2-activated compared to patients receiving resting NK cells only (80). Interestingly, IL-2-stimulated NK cells but not unstimulated NK cells promoted NK cell trafficking and changes in the distribution of leukocyte subpopulations in the peripheral blood. In the meanwhile, safety and feasibility using IL-2-activated and -expanded NK cells for adaptive immunotherapy has been demonstrated in various clinical studies as summarized in a recent review by Koehl and others (33).

IMPACT OF IL-15 ON NK CELL EXPANSION

Carson et al. postulated that NK cells might be dependent on other cytokines than IL-2 such as IL-15 (81, 82). The trimeric IL-15 receptor on NK cells shares two subunits with the IL-2R, but not CD25 forming the high affinity IL-2R. Therefore, they also share some functions, e.g., maintenance of NK cell survival (82). Similarities and differences between IL-2 and IL-15 effects on NK cells have been extensively reviewed elsewhere, and IL-15 might be the preferable cytokine for cancer therapy as it inhibits activation-induced cell death and it is considered safe (83–85). In addition, compared to IL-2, IL-15 leads to more sustained antitumor capacity of NK cells *via* signaling through mammalian target of rapamycin and stress-activated gene expression (86). However,

TABLE 2 | *Ex vivo* cultivation of natural killer (NK) cells with accessory cells.

Protocol features	Starting material/ culture system	NK cell expansion rate	NK cell purity	NK cell phenotype	NK cell function	Setting	Reference
IL-2	PBMC, CD3 depleted in bags and flasks	N/A (overnight)	33% NK 0.1% T cells	N/A	N/A	Clinical	(60)
		N/A (14–16 h)	26.7%		Enhanced cytotoxicity *In vitro*		(61, 62)
		N/A (overnight)	40% NK 0.9% T cells		Enhanced cytotoxicity *In vitro*		(43)
IL-15	PBMC, CD56 enriched	23 (20 days)	98% NK	Expression of NKp30, NKp44, NKp46, NKG2D, and 2B4	Cytotoxic *in vitro*	Clinical	(63)
IL-15 + IL-21	PBMC, CD3 depleted	3.7 $CD56^+/CD122^+$ (2–3 weeks)	>90% $CD56^+/CD122^+$ <0.3% $CD3^+/CD56^-$ <3% $CD3^+/CD56^+$	67% $CD56^+CD16^+$	Cytotoxic against K562 and patient bone marrow blasts	Clinical	(64)
OKT-3 + IL-2	PBMC in plates	193 (21 days)	~55% NK ~22% T cells	N/A	Substantial cytotoxicity against K562	*In vitro*	(65)
	PBMC in flasks	1,625 (20 days)	~65% NK ~22% T cells	*Upregulated*: 2B4, CD8, CD16, CD27, CD226, NKG2C, NKG2D, NKp30, NKp44, NKp46, LIR-1, KIR2DL3, and CXCR3 *Downregulated*: CCR7	Increased cytotoxicity against tumor cell lines and primary MM cells *In vitro*	*In vitro*	(25)
	PBMC	1,036 (total cells) (19 days)	~30% NK ~40% T cells	*Upregulated*: NKG2A, LILR-B1, NKG2D, NKp30, NKp44, and NKp46	*In vitro* cytotoxicity increases during culture	Clinical	(66)
	PBMC in a bioreactor, flasks, and plates	77—bioreactor 530—bags 770—flasks (20 days)	38%—bioreactor 31%—bags 44%—flasks	Bioreactor compared to flasks: higher expression of CD11b, NKG2D, and NKp44	Bioreactor compared to flasks: higher cytotoxicity	*In vitro*	(67)
OKT-3 + IL-2 + Alemtuzumab	PBMC in plates, flasks, and bags	646 (14 days) 1,537 (18 days)	60% NK 37% T cells <0.1% B cells	*Upregulated*: 2B4, NKG2D, NKp30, NKp44, KIR2DL1, LIR-1, and CD16 *Downregulated*: CCR7	Increased cytotoxicity *In vitro* and *in vivo*	Clinical	(68)
OKT-3 + IL-2 + IL-15	PBMC or $CD56^+$ + $CD56^-$ (1:1) in flasks and bioreactor (Cellbag)	PBMC: 112 1:1 Mix: 89 (21 days)	With PBMC: 34% With "1:1 Mix": 92%	*Upregulated*: NKp30, NKp44, DNAM-1, NKG2D, and CD11a	Increased activity against neuroblastoma cell lines *in vitro* and *in vivo*	Preclinical model	(69)
aCD16 mAb + OK432 + IL-2	PBMC in flasks and bags	637–5,712 (day 21)	79% NK 8.4% T cells (day 21)	*Upregulated*: NKG2D, NKp44, and CD69 *Downregulated*: CD16 (transient)	Increased cytotoxicity against tumor cell lines and primary cancer cells *in vitro* ADCC activity	*In vitro*	(70)

recent data revealed that continuous IL-15 signaling causes functional exhaustion of NK cells by decreased fatty acid oxidation, resulting in lower cytotoxicity *in vitro* and decreased tumor control *in vivo* (87). Thus, optimal dosing and timing of IL-15 is critical for *ex vivo* NK cell activation. Purified NK cells expanded using IL-15 exhibit upregulation of NCRs and CD69 and cytolysis of leukemia and primary ALL blasts (51). Enhanced cytotoxicity of IL-15-stimulated NK cells against leukemia and rhabdomyosarcoma cell lines could be attributed to NCRs, DNAM-1 and NKG2D (54). Using IL-15 to expand NK cells from CD56-enriched PBMC for 20 days resulted in a 23-fold expansion of $CD3^-CD56^+$ NK cells with a final purity of about 98% (63). NK cells generated with the latter protocol were transferred to 15 non-small lung cancer patients in a phase I clinical trial in two to four doses of $0.2–29 \times 10^6$ NK cells/kg, showing the safety of the approach (63).

IL-21 ENHANCES NK CELL EFFECTOR FUNCTIONS

The cytokine IL-21, in combination with IL-2 or IL-15, is utilized in some protocols for NK cell stimulation (55, 64). IL-21 belongs to the IL-2 family and signals through a heterodimer consisting of the common γ-chain and the IL-21 receptor α-chain. Activated $CD4^+$ T cells are the main producers of IL-21 and IL-21 affects many different cell types expressing the IL-21 receptor, including NK cells (88). IL-21 plays a role in the development of NK cells from bone marrow progenitors (89), and, in mice, it dampens the expansion of NK cells but is required for functional NK cell maturation (90, 91). Recently, expansion of "memory-like" NK cells has been shown to be IL-21 dependent in the context of tuberculosis infection (92). Wendt et al. observed increased proliferation of $CD56^{bright}$ human NK cells (55), but another group reported no effect of IL-21 on the proliferation of NK cells from healthy human donors and from HIV patients (93). Moreover, IL-21 is known to trigger apoptosis, resulting in a shorter lifespan of NK cells *in vitro* (90, 94). Thus, the time span NK cells are exposed to IL-21 appears critical (95, 96). Besides its effect on NK cell proliferation, IL-21 enhances the effector functions of NK cells, including secretory and cytotoxic functions as well as enhanced ADCC responses (93, 97, 98). Culturing CD3-depleted PBMC for 13–20 days with IL-21 and IL-15 without additional feeder cells yields activated NK cells with a purity of >90%, which were applied in a clinical trial with 41 leukemia patients receiving infusions of donor-derived NK cells 2–3 weeks after HSCT (64). Although the NK cells expanded weakly under this condition (3.7-fold), they possessed potent cytotoxic activity against primary bone marrow blasts prior to transplantation, and infusions with a median dose of 2×10^8 NK cells/kg were well tolerated and correlated with a reduction in leukemia progression compared to historical controls (64).

IL-12/15/18 INDUCED MEMORY NK CELLS

Interleukin-12 was originally discovered as NK cell-stimulating factor, inducing proliferation, enhanced cytotoxicity, and production of IFN-γ by NK cells when added to PBMC (99, 100). IL-12 is produced by DCs, macrophages, and B cells, and its receptor consists of two subunits (α and β), which mediate signaling through members of the JAK–STAT family (101). IL-2 enhances the response of NK cells to IL-12 by increasing the expression of the IL-12 receptor and STAT4, a relevant factor for IL-12 signaling (102). Furthermore, it was revealed that IL-12-mediated IFN-γ production of NK cells requires priming with IL-18, a cytokine also known to enhance IL-15-induced NK cell proliferation (103, 104). Due to the synergistic effects, it seems reasonable to combine the different cytokines for *ex vivo* stimulation of NK cells. In this context, the combination of IL-12, IL-15, and IL-18 raised special interest, as it leads to the so-called "cytokine-induced memory-like" (CIML) NK cells in mice and humans, which exhibit an increased capacity to produce IFN-γ upon re-stimulation at later time points (56, 105). Importantly, this memory response is a cell intrinsic effect that is passed on to offspring cells and is maintained up to several months (56). In mice, the intrinsic ability for mediated IFN-γ production coincided with demethylation of the conserved non-coding sequence 1 in the IFN-γ locus (106). Furthermore, adoptive transfer of CIML NK cells had a clear antitumor activity against established melanoma or lymphoma *in vivo*, which required IL-2 from $CD4^+$ T cells (57, 106). For both, murine and human NK cells, IL-12, IL-15, and IL-18 together induce an increased expression of CD25, making CIML NK cells responsive to low concentrations of IL-2 *in vitro* and *in vivo* (57, 58). Thus, there is a clear rationale to apply adoptive transfer of *ex vivo*-generated CIML NK cells together with IL-2 injections as a combination therapy. Recently, CIML NK cells together with low dose IL-2 therapy were evaluated in a first-in-human phase I clinical trial with promising results, as clinical response was observed in five of nine treated patients (107).

AUTOLOGOUS ACCESSORY CELLS AND AUTOLOGOUS FEEDER CELLS FOR NK CELL EXPANSION

Although cytokines efficiently activate NK cells and result in cell products with advanced effector functions, cytokines alone do not allow pronounced *ex vivo* expansion (**Table 1**). Consequently, in addition to the activation with cytokines, stimuli from autologous accessory cells can be used to further enhance the expansion of NK cells to overcome the hurdle of limited NK cell doses for adoptive NK cell therapy (**Table 2**). Outgrowth of NK cells from the whole PBMC fraction is more effective than cultivation of pure NK cells, because other cell types provide additional factors for NK cell proliferation. $CD14^+$ cells, for instance, enhance the *ex vivo* NK cell proliferation *via* direct cell contact and soluble factors (108, 109). After activation, for instance by concanavalin A, T cells also trigger NK cell proliferation (110).

Stimulation of PBMC with IL-2 and the clinically approved anti-CD3 antibody OKT-3 leads to a profound outgrowth of NK cells (25, 65–67, 111), probably by activation of T cells and this is utilized by several clinical protocols for NK cell cultivation. Nevertheless, starting the culture from PBMC goes along with extensive coexpansion of unwanted $CD3^+/CD56^-$ T cells and $CD3^+/CD56^+$ NK-like T (NKT) cells, accounting for the majority of cells in the final cellular product. Surprisingly, infusion of this heterogeneous cell product without removal of potentially alloreactive T cells did not cause side effects, such as GvHD, in a safety trial with five cancer patients, evaluating the cultivated cellular product in an allogeneic setting (66). This can be explained by the fact that T cells may lose their alloreactivity during extended *ex vivo* expansion (112). Thus, low NK cell purities may be less critical for long-term cultivated cellular products compared to NK cells directly obtained from a donor, but more clinical data are required to prove this hypothesis. Of note, the approach also allows efficient expansion of functional patient-derived NK cells, as shown for B cell chronic lymphocytic leukemia and multiple myeloma patients, enabling therapy with autologous NK cells and further circumventing possible safety risks of therapy with donor-derived cells (25, 111).

Starting with PBMC enriched for CD56 cells together with the corresponding non-CD56 PBMC in a 1:1 mixture favors a 89-fold NK cell expansion with a final product consisting of 92% NK cells after 21 days (69). Alternatively, adding irradiated autologous PBMC to the culture is a strategy to benefit from these "feeder cells" for NK cell activation and expansion but to avoid their coexpansion (**Table 3**). Of note, to make a clear difference, we use the term "feeder cells" for all inactivated cells that are added to the culture, whereas cocultured non-NK cells that are not inactivated are defined as "accessory cells." Besides its growth inactivating function, irradiation can induce upregulation of stress-regulated surface molecules on PBMC, such as ULBP1–3, that further trigger NK cell activation, e.g., through NKG2D (113). Still, irradiated autologous PBMC induce only weak NK cell proliferation without additional activation of the feeder cells (e.g., only 16-fold expansion within 2 weeks) (24). Whereas irradiated autologous PBMC previously activated with IL-2, OKT-3 and RetroNectin allow a median 4,720-fold NK cell expansion after 3 weeks with a NK cell purity of 91% starting from PBMC (114). To obtain a more pure final product with 98% NK cells, it is possible to start the culture with already CD3-depleted PBMC and add irradiated autologous PBMC as feeder cells together with IL-2 and OKT-3 (23). The highest purity can be achieved by cell sorting, representing also the method of choice to expand defined NK cell subpopulations. As demonstrated by Siegler et al., GMP-sorted and highly pure single KIR^+ NK cells can be expanded 160- to 390-fold in 19 days with IL-2, IL-15, OKT-3, and irradiated autologous PBMC (50).

NK CELL EXPANSION WITH ALLOGENEIC FEEDER CELLS

Using irradiated allogeneic cells as feeder cells is another option to stimulate NK cell expansion *ex vivo* (118) (**Table 4**). Compared to autologous PBMC, allogeneic PBMC may be even more efficient as feeder cells for NK stimulation. Accordingly, in a study testing the expansion of NK cells from patients with advanced lymphomas or terminal solid tumors, 300-fold NK expansion was obtained with irradiated allogeneic PBMC feeder cells from healthy donors, whereas only 169-fold expansion was achieved with irradiated autologous PBMC feeder cells from the patients (115). Furthermore, whereas the availability of autologous feeder cells is limited, as they have to be obtained directly from the patient, for allogeneic feeder cells it is possible to utilize established cell lines. Cell lines can be grown easily to sufficient numbers and different cell lines in fact trigger NK cell proliferation, such as HFWT, K562, RPMI 1866, Daudi, KL-1, MM-170, and different EBV-transformed lymphoblastoid cell lines (EBV-LCL) (99, 119–122).

Culturing PBMC together with the Wilms tumor cell line HFWT and IL-2 leads to significant NK cell expansion (124, 145), and interestingly under this condition NK cells not only arise from mature $CD3^-CD56^+$ NK cells but also from $CD3^-CD14^-C D19^-CD56^-$ NK cell precursors expressing CD122 (146). In 2004, early clinical data showed that adoptive transfer of autologous NK cells generated by coculture with irradiated HFWT is safe and patients with recurrent malignant glioma partially responded to the treatment (125).

Another advantage of cell lines is that it is relatively easy to genetically modify them and to integrate additional factors for NK cell stimulation. In recent years, modified K562 cells have been utilized, such as K562 expressing membrane-bound IL-15 and 41BBL (K562-mb15-41BBL) (126). While unmodified K562 only induce a weak NK cell proliferation (2.5-fold NK cell expansion in 1 week), with K562-mb15-41BBL the NK cell number can be significantly increased by 20- or 1,000-fold in 1 or 3 weeks (126). In addition, stimulation of NK cells with K562-mb15-41BBL demonstrated that NK cells actually have a substantial proliferative potential *ex vivo*, with up to 30 population doublings and 5.9×10^4-fold NK cell expansion (147). NK cells expanded with K562-mb15-41BBL exhibit enhanced natural cytotoxicity against several allogeneic and autologous tumors *in vitro*, efficiently mediate ADCC and showed antitumor efficacy in mouse xenograft models for the treatment of sarcoma and myeloma (128, 148, 149). Of note, in a clinical trial assessing adoptive transfer of K562-mb15-41BBL following HSCT, acute GvHD occurred in five of nine patients, although the donors were completely HLA matched and the doses of injected NK cells and cotransferred T cells were low ($1–10 \times 10^5$ and $\leq 2 \times 10^4$/kg) (131). These observations suggested that the acute GvHD was T cell mediated, but NK cells apparently may promote this severe side effect indirectly (150). Importantly, another group utilized NK cells expanded with a similar K562 variant expressing 41BBL and IL-15 in another treatment setting and did not observe GvHD, although up to 1×10^8 NK cells/kg were administered (129).

Furthermore, Denman and colleagues revealed that K562 expressing 41BBL and membrane-bound IL-21 instead of IL-15 are even more effective for *ex vivo* expansion of NK cells, and weekly restimulation with this cell line supports a sustained NK cell proliferation over several weeks (134). In coculture with K562 expressing membrane-bound IL-21 and 41BBL, NK cells show an increased telomere length and enhanced activation of the STAT-3 signaling pathway, explaining the positive effect for sustained expansion of NK cells over long time (134, 151). Adoptive transfer of NK cells expanded with K562 expressing membrane-bound IL-21 and 41BBL into tumor-bearing mice improved the survival of the animals, indicating a therapeutic effect of these NK cells (135).

TABLE 3 | *Ex vivo* cultivation of natural killer (NK) cells with autologous feeder cells.

Protocol features	Starting material/culture system	NK cell expansion rate	NK cell purity	NK cell phenotype	NK cell function	Setting	Reference
Irr. autologous PBMC (depleted for CD3$^-$/CD56$^+$ cells) + IL-2 + IL-15	PBMC, CD3 depleted, and CD56 enriched in flasks	16 (14 days)	97% NK 0.2% T cells	*Upregulated*: NKG2D, DNAM-1, NKp30, NKp44, CD158a, and CD158e	Efficient degranulation and lysis of K562 *In vitro*	*In vitro*	(24)
Irr. autologous PBMC activated with OK432, FN-CH296 and OKT-3 + IL-2	PBMC in flasks and bags	4,720 (21–22 days)	91% NK ~12% NK-like T and $^-$	Strong expression of NKG2D and CD16	Elevated cytotoxicity that is maintained for up to 4 weeks after infusion to patients	Clinical	(114)
Irr. autologous PBMC + OKT-3 + IL-2	PBMC, CD3 depleted, and CD56 enriched in plates	169 (14 days)	84% NK	*Upregulated*: CD16, CD56, NKG2D, NKp30, and NKp44	Increased cytotoxicity against tumor cell lines *in vitro*	*In vitro*	(115)
	PBMC, CD3 depleted in flasks and bags	278–1,097 (21–26 days)	91–98% NK	Most cells express NKG2D, CD16, CD94, NKp46, KIR2DL1, KIR3DL1, and KIR2DL2/3	Efficient lysis of tumor cell lines *in vitro*; persistence in patients up to several months; cytotoxic potential is lost *in vivo*, while ability for ADCC is maintained	Clinical	(116)
	PBMC, CD3 depleted in bags	691 (14 days)	98% NK 0.06% T cells	*Upregulated*: NKG2C, NKp30, NK44, CXCR4, CD25, CD62L, and CD69	Increased cytotoxicity against tumor cell lines *in vitro*; antitumor effect and ADCC activity in a leukemia xenograft mouse model; up to 4 days persistence in patients	Preclinical model	(23)
		758 (14 days)	98% NK 0.4% T cells			Clinical	(117)
Irr. autologous PBMC (depleted for CD3$^-$/CD56$^+$ cells) + OKT-3 + IL-2	PBMC, CD3 depleted, and CD56 enriched in plates and flasks	546 (14 days)	94.9% NK 2.2% T cells	*Upregulated*: NKG2D, NKp30, NKp44, tumor necrosis factor-related apoptosis-inducing ligand, and DNAM-1 *Downregulated*: NKp80	Increased cytotoxicity against tumor cell lines *in vitro*	*In vitro*	(113)
Irr. autologous PBMC + OKT-3 + IL-2 ± IL-15	PBMC, CD3 depleted, and CD56 enriched in plates and bags	117/63 in bags (±IL-15) 993 in plates (19 days)	Bags: 45% NK 0.6% T cells	*Upregulated*: NKG2D, NKp44	High cytotoxicity against K562 and high productivity of IFN-γ	*In vitro*	(50)
	Good manufacturing practice killer cell immunoglobulin-like receptor (KIR) sorted NK cells in bags	160–390	~100% NK >0.01% T cells	Single KIR + NK cells	Anti-leukemic activity against primary acute myeloid leukemia cells *in vitro* and *in vivo*	Preclinical model	(50)

TABLE 4 | *Ex vivo* cultivation of natural killer (NK) cells with allogeneic feeder cells.

Protocol features	Starting material/culture system	NK cell expansion rate	NK cell purity	NK cell phenotype	NK cell function	Setting	Reference
Irr. allogeneic PBMC activated with ConA + IL-2	*In vivo* IL-2 primed PBMC depleted for non-NK cells in flasks	1–148 (14 days)	64–98% NK	N/A	Cytotoxic activity against leukemic cell lines	Clinical	(123)
Irr. allogeneic PBMC activated with ConA, PHA and ionomycin + IL-2 + IL-15	PBMC, depleted for CD3, CD4, CD19, and CD33 in bags	80–200 (15 days)	91% CD56 0.3% CD3 (day 12)	*Upregulated*: CD16, CD25	Increased cytotoxicity against tumor cell lines *in vitro*; decreased frequency of INF-g producing cells	*In vitro*	(118)
Irr. allogeneic PBMC + OKT-3 + IL-2	PBMC, CD3 depleted, and CD56 enriched in plates	300 (14 days)	94% NK	*Upregulated*: CD16, CD56, NKG2D, NKp30, and NKp44	Increased cytotoxicity against tumor cell lines *in vitro*	*In vitro*	(115)
Irr. HFWT + IL-2	PBMC in flasks	113 (2 weeks)	86% CD56⁺/CD16⁺	N/A	Cytotoxic against tumor cell lines *in vitro*	Clinical	(124, 125)
Irr. Jurkat/KL-1 + IL-2	PBMC in flasks	~130 (2 weeks)	40–90% NK	*Upregulated*: CD54, CD11a, CD48, CD2, CD49d, CD58, NKp30, NKp44, 2B4, DNAM-1, NKG2D, CD25, and CD69 *Downregulated*: CD16	Increased cytotoxicity against tumor cell lines *in vitro* and antitumor activity *in vivo*	Preclinical model	(121)
Irr. K562 expressing membrane-bound IL-15 and 41BBL + IL-2	PBMC in plates	1,089 (3 weeks)	"Virtually pure"	N/A	N/A	*In vitro*	(126)
	PBMC in bags	23, 152, and 277 after 7, 14, and 21 days	96.8% NK 3.1% T cells (day 21)	Marked differences of gene expression profile compared to unstimulated or IL-2-stimulated NK cells	Increased cytotoxicity against tumor cell lines *in vitro* and antitumor activity *in vivo*	Preclinical model	(127)
	PBMC	447 (days 10–14)	88% NK 2.2% T cells (day 14)	Upregulated genes for cytolytic activity, cytokines, chemokines, activating receptors, adhesion molecules, cell cycle regulators, and multiple pathways	Increased cytotoxicity against primary MM cells *in vitro* and *in vivo*; high productivity of IFN-γ	Preclinical model	(128)
	PBMC in G-Rex, bags	442—G-Rex 227—bags (10 days)	70% NK 5–35% T cells	*Upregulated*: NKp30, NKp44, NKG2D, CD26, CD70, and CXCR3 *Downregulated*: CD16, CD62L	Increased cytotoxicity and ADCC against primary tumor cells *in vitro*; robust *in vivo* proliferation post-infusion	Clinical	(129, 130)
Irr. K562 expressing membrane-bound IL-15 and 41BBL + IL-15	PBMC, CD3 depleted, and CD56 enriched	1,000 (21 days)	N/A	*Upregulated*: CD56, NKG2D, tumor necrosis factor-related apoptosis-inducing ligand (TRAIL), CD158a, CD158b, and CD158e1	Increased cytotoxicity *in vitro* independent of killer cell immunoglobulin-like receptor mismatch; NK infusion contributed to acute graft-versus-host disease in first clinical trial	Clinical	(131, 132)
Plasma membrane particles of K562 expressing IL-15 and 41BBL + IL-2	PBMC in plates and flasks	1,265 (17 days)	86% NK cells 9% T cells 2% NK-like T	*Upregulated*: NKp30, NKp44, NKp46, NKG2D, 2B4, NKG2A, TRAIL, and Fas ligand (FasL) *Downregulated*: CD16	Increased cytotoxicity against leukemic cell lines and primary acute myeloid leukemia (AML) cells *in vitro*	*In vitro*	(133)
Irr. K562 expressing membrane-bound IL-21, 41BBL, CD64, CD86, and CD19 + IL-2	PBMC in flasks	4.8×10^4 (21 days)	21.7% T cells	High expression of natural cytotoxicity receptors, CD16, and NKG2D	Cytotoxic against tumor cell lines *in vitro*; capable of ADCC; increased telomere length	*In vitro*	(134)
		2,363 (14 days)	83% NK 9.1% T cells	*Upregulated*: DNAM-1, NKG2D, CD16, and CD56	Cytotoxic and capable of ADCC against neuroblastoma cell lines *in vitro* and *in vivo*	Preclinical model	(135)

(*Continued*)

TABLE 4 | **Continued**

Protocol features	Starting material/culture system	NK cell expansion rate	NK cell purity	NK cell phenotype	NK cell function	Setting	Reference
Plasma membrane particles of K562 expressing membrane-bound IL-21 and 41BBL + IL-2	PBMC	825 (14 days) >10^5 (28 days)	>90% NK (day 14)	N/A	Increased cytotoxicity against leukemic cell lines and primary AML cells *in vitro*; enhanced proliferation *in vivo*	Preclinical model	(136)
Irr. allogeneic PBMC; irr. EBV transformed lymphoblastoid cell lines (EBV-LCL) (LAZ 388 cells) + PHA + IL-2	PBMC depleted for CD3 and monocytes in bags and plates	~43 (31–21 days)	90% NK <5% T cells	N/A	Increased cytotoxicity against tumor cell lines *in vitro*	Clinical	(137, 138)
Irr. EBV-LCL (TM-LCL) + IL-2	PBMC, CD3 depleted, and CD56 enriched in bags	800–1,000 (2 weeks)	98% NK	*Upregulated*: TRAIL, FasL, NKG2D, NKp30, NKp44, NKp46, CD48, CD25, LTB, MX1, and BAX	Increased cytotoxicity against tumor cell lines *in vitro*	*In vitro*	(139, 140)
Irr. EBV-LCL (SMI-LCL) + IL-2	PBMC, CD3 depleted, and CD56 enriched in bags	3,637 (24–27 days)	99.7% NK			Clinical	(141)
	PBMC, CD3 depleted, and CD56 enriched in CliniMACS Prodigy	850 (14 days)	>99% NK	*Upregulated*: TRAIL, FasL, NKG2D, NKp30, NKp44, and DNAM-1	Increased cytotoxicity and ADCC against tumor cell lines *in vitro*	*In vitro*	(142)
Irr. EBV-LCL (SMI-LCL) + IL-2 + IL-21	PBMC depleted for non-NK cells (research kit) in plates and flasks	2,900 (14 days) 2.7×10^{11} (46 days)	>99% NK	*Upregulated*: TRAIL, NKG2D, and DNAM-1	Cytotoxic against tumor cell lines *in vitro* and *in vivo*; enhanced and sustained production of IFN-γ and TNF-α	Preclinical model	(96)
Lysate of CTV-1	PBMC, CD3 depleted, and CD56 enriched	N/A (overnight)	97–98% NK	*Upregulated*: CD69 *Downregulated*: CD16	Cytotoxic against NK-resistant leukemia cell lines and primary tumors *in vitro*	Clinical	(143, 144)

The stimulatory effect of EBV-LCL on NK cell proliferation was discovered more than 30 years ago (152). In 1994, an early clinical trial already evaluated the adoptive transfer of autologous NK cells expanded with the LAZ 388 cell line to treat 10 patients with metastatic renal cell adenocarcinoma (137). More recently, the cell lines TM-LCL and SMI-LCL were reported for NK cell expansion, allowing around 800-fold expansion of highly pure NK cells within 2 weeks (139–142). NK cells generated with these EBV-LCL feeder cells are currently applied in a study testing them for adoptive transfer in an autologous setting with intended doses up to 1×10^9 NK cells/kg (141). Recently, it was reported that repeated stimulation with SMI-LCL in IL-2-containing medium and adding IL-21 only at start of cultivation enables 10^{11}-fold NK cell expansion after 6 weeks, to our knowledge representing the most efficient protocol to expand NK cells at the moment (96). NK cells generated with the latter method are highly cytotoxic *in vitro*, show a sustained high productivity of IFN-γ and TNF-α, similar to CIML NK cells, and they efficiently controlled melanoma in a xenograft mouse model (96).

Although feeder cells, and allogeneic feeder cell lines in particular, make it possible to generate substantial numbers of NK cells for adoptive therapy, from a regulatory point of view this strategy has drawbacks as feeder cell lines must be qualified as safe for human use. The cell line qualification of modified K562 cells, for instance, includes costly viral testing and assays to prove absence of bacterial and *Mycoplasma* contamination (153). In this context, lysates from cell lines containing the NK cell-stimulating factors could be an alternative to the intact feeder cells to minimize regulatory concerns. It was demonstrated that short cultivation of NK cells with lysate of the leukemia cell line CTV-1 primes NK cells to specifically lyse cell lines that are resistant to resting NK cells (143). Interestingly, the priming effect of CTV-1 on NK cells is KIR independent and does not require supplementation of cytokines, such as IL-2 or IL-15, making this an unique approach for NK cell activation (154). NK cells primed with CTV-1 were evaluated in the first UK clinical trial of a cell therapy regulated as a medicine, with an anti-leukemia effect in four of seven treated patients and no evidence of NK cell infusion-related toxicities (144). Another step forward from a regulatory standpoint could be to add only specific fragments of feeder cells to the culture that are responsible for the desired NK cell activation, instead of using intact feeder cells or their lysates. Of note, NK cells can be expanded *ex vivo* with IL-2 and plasma membrane particles prepared from K562-expressing membrane-bound IL-15 and 41BBL with a rate of expansion that is comparable to stimulation with intact feeder cells and far better than stimulation with soluble IL-15, 41BBL, and IL-2 (133). Plasma membrane particles from K562 expressing membrane-bound IL-21 and 41BBL work for *ex vivo* NK cell expansion as well and may be an option for *in vivo* NK cell expansion, as demonstrated in a first proof of concept using a mouse model (136).

TECHNICAL ASPECTS OF NK CELL EXPANSION

In general, one encounters technical challenges and opportunities when manufacturing NK cells as medicinal products, as reviewed recently (155). In this section, we focus on technical options for NK cell culture, ranging from simple cell culture plates for small scale experiments to highly standardized and automated systems for clinical scale. The selection of the adequate culture system is based on the intended application of the cells. Most preclinical experimental studies grow NK cells in cell culture plates or tissue culture (T) flasks. These are commonly used and very convenient to test and compare different culture additives in parallel, e.g., different cytokine concentrations. However, for clinical applications in large scale, cultivation in plates and flasks is rather inappropriate for different reasons. First, due to the small volume of T flasks, numerous T flasks have to be handled at the same time, with for instance 51 T flasks for the treatment of a single patient (116). In addition, T flasks have to be opened from time to time for medium exchange or harvesting of cells, bearing the risk of contaminating the cellular product. Although the likelihood of contamination for each T flask is reduced to a minimum by sterile workflows in safety cabinets, the remaining risk potentates by the number of flasks.

To overcome the drawbacks of small cell culture vessels, clinical NK cell cultivation is often done in cell culture bags, which make it possible to culture high volumes in a closed system, as all required steps can be done by sterile welding of tubing connections for the transfer of media, harvesting of cells, etc. Unfortunately, different reports describe that the NK cell expansion performance is reduced after transition of a protocol from T flasks to larger scale in cell culture bags (50, 67). In addition, bag systems still require several labor-intensive interventions during the culture, especially when different cultures are set up in parallel.

The G-Rex vessel is another system avoiding frequent processing steps for exchange of medium during the culture. In contrast to normal cell culture flasks, the bottom of the G-Rex is highly gas permeable, ensuring optimal CO_2 exchange and O_2 supply for the cells. Thus, by its design, G-Rex flasks can be filled directly with a high level of cell culture medium and exchange of medium is not necessary for long time. For NK cell culture, G-Rex were used for example for 10 days of culture without any cell manipulation or feeding, and resulted in higher fold expansion of NK cells compared to cell culture bags (130). Unfortunately, although G-Rex are scalable in general, multiple G-Rex flasks are still required to achieve high cell numbers for clinical trials, which can be cumbersome and costly, and G-Rex flasks are still an open system and may require adaption to a closed system (156).

Automated systems combine the need for reduced interventions during the culture with a closed system. Automation of the cell manufacturing ensures constant product quality without the need for highly skilled experts, is finally cost saving, and may be required for cellular therapy to become available beyond specialized academic centers (157). Although early integration of automation is associated with higher capital costs in the development phase, it allows a smooth transition at later stages of clinical development (158). A first feasibility study of automated NK cell cultivation with a stirred bioreactor was already published in 1996, showing advantages of the bioreactor culture over manually handled controls (159). More recently, different investigators report automated NK cell expansion procedures with a rocking motion bioreactor (67, 69, 156, 160), yielding $2–10 \times 10^9$ NK cells under GMP-compliant conditions. However, the latter system still needs preceding manual cultivation, because relatively high cell numbers are required as inoculum for the automated culture (67, 69, 156, 160). Alternatively, fully automated NK cell expansion with an automated cell processing device can be performed for clinical use, with as little as 10^6 NK cells being sufficient to initiate the automated culture that can yield up to 2.7×10^9 NK cells after stimulation with clinical grade feeder cells (142). Of note, in addition to the culture process, the cell processing device is designed for GMP-compliant cell separation, concentration, and washing applications, so that combined NK cell purification, cultivation, and final formulation of the cellular product is possible fully automated (161). Thus, the whole processing, from the starting material, such as a leukapheresis product, to the finally expanded NK cells, readily prepared for infusion, can be covered by a single instrument.

Centralized processing of NK cell products probably will be carried out mainly in specialized centers for manufacturing of cellular products. Consequently, after *ex vivo* cultivation, storage of the NK cell product and shipment to the location of use will be needed. Compared to naive NK cells, IL-2-activated NK cells are less sensitive to freezing, as they show higher recovery and viability after thawing (26). Still, different groups state that cryopreservation of cultivated NK cells goes along with a drop in cell viability and cytotoxicity, whereas the latter can be restored by a short re-stimulation, e.g., by a short resting in IL-2-containing medium (139, 156). Poor survival of the NK cells can be an issue during further *in vitro* culture post thawing, so that shipping of freshly formulated cells for direct infusion may be advantageous (129). Interestingly, some groups recently claim that freezing and thawing does not influence the cytotoxicity or the proliferative ability of cultivated NK cells in their hands (24, 68). These divergent observations possibly result from different cultivation methods and different protocols for freezing and thawing, which should be investigated further. Without freezing, transport of the readily prepared cells in an appropriate time frame is challenging, and any delay during the shipment affects the quality of the cellular product with critical consequences for the patient. Alternatively, automated and closed systems for cell processing open the way for scale out strategies and de-centralized NK cell manufacturing directly at the location of intended use, avoiding the freezing and shipment process (142). But, although de-centralized manufacturing in the clinics seems promising, cellular therapeutics are very complex and still in early development, so that manufacturing by well-trained specialists in specific facilities is reasonable at that state.

REGULATORY ASPECTS OF NK CELL CULTIVATION FOR CLINICAL USE

Apart from technical difficulties, one has to consider regulatory aspects for the use of *ex vivo*-generated NK cells with regulations varying in time and geographical policies (153). In Europe, for instance, cytokine-activated and -expanded NK cells are currently classified as advanced therapy medicinal products and will be regulated accordingly either centralized or under the hospital exemption by the member states [Regulation (EC) No 1394/2007; Directive 2001/83/EC and Regulation (EC) No 726/2004]. Quality aspects related to somatic cell therapy medicinal product as defined in guidelines (CPMP/BWP/3088/99; EMEA/CHMP/410869/2006; Ph. Eur. 0784: Ph. Eur. 5.14) will apply to the identity, potency, and activity. The establishment of correspondingly adequate in process and quality controls as well as of process target values and product specifications will have to take into account the variability of the primary effector cell as the starting material (162).

CONCLUSION AND OUTLOOK

Comparing different protocols for NK cell cultivation in detail is challenging as these are extremely heterogeneous. The duration of *ex vivo* NK cell cultivation ranges from a few hours for short NK cell activation up to several weeks for long-term expansion, different starting materials are in use with varying NK cell purities, different cytokines are combined at different doses, and NK cells often are cocultured with different feeder cells at different NK-to-feeder ratios. Nevertheless, overall differently *ex vivo* expanded NK cells exhibit some common characteristics.

In general, *ex vivo* cultivated NK cells show an increased cytotoxicity and may become even responsive against tumor targets previously appearing resistant to NK cell lysis. This explains the use of IL-2 or IL-15 in virtually every protocol, as it is known since a long time that both cytokines amplify NK cell activity (81, 163). However, upon NK cell activation with different stimuli, including IL-2 and IL-15, downregulation of CD16 surface levels occurs by metalloproteases-mediated shedding of CD16 (164–166). The Fc receptor CD16 is crucial for NK cells to perform ADCC and would be of particular importance for potential combination therapies using NK cells together with therapeutic antibodies. Of note, although reduced levels of CD16 on NK cells are observed for several NK cell cultivation protocols the NK cells still mediate ADCC (70, 129, 142). Nevertheless, inhibition of the relevant metalloproteases to maintain CD16 on NK cells could be an option to further increase the ADCC function of *ex vivo* activated NK cells (164, 167).

Another clinically highly relevant aspect is the tumor-induced immunosuppression as important challenge for all cell therapeutic strategies. Remarkably, it ruled out from most preclinical and clinical NK cell studies that NK cells may gain the capability to overcome tumor immunosuppression. Different research groups have reported signs of NK cell suppression in cancer patients such as a lower expression of NK cell receptors, e.g., NCRs, NKG2D, DNAM-1, and 2B4 (22, 25, 168–170), the shedding of tumor cell ligands, such as NKp30 and NKG2D (171–174), or the release of blocking NKG2D ligands, such as MICA and ULBP3, *via* tumor-derived exosomes (175, 176). Notably, *ex vivo* cultivation of patient-derived NK cells is often possible with same efficacy as for donor-derived NK cells (25, 111) and can normalize the NK cell phenotype and activation (25). Additionally, elevated levels of NKG2D on *ex vivo*-activated NK cells can scavenge shed NKG2D ligands and counter their inhibitory effect (177). Furthermore, the high cytotoxicity of *ex vivo* expanded NK cells has been shown to be independent of KIR inhibition for some protocols (107, 132).

In comparison to other cell therapeutic approaches using, e.g., T cells, donor-derived allogeneic NK cells mediate GVL without an elevated risk for GVHD or even with a GVHD-reducing effect, as reported in mice and men (74, 75, 77, 78). However, contradictory results regarding GVHD induction have been reported in clinical trials assessing adoptive transfer of NK cells expanded with K562 feeder cell variants expressing 41BBL and IL-15 (129, 131). These reports show that there are still open questions that have to be unraveled to better understand the complex role of NK cells and their specific subsets in the bidirectional regulation of GVL and GVHD.

In conclusion, many different protocols are in use to expand NK cells *in vitro*, each with its specific advantages and disadvantages in regard of cell numbers, function, and handling efforts. The data summarized in this review underline the complexity related to the design of an optimal NK cell therapeutic protocol that should be not only reliable and safe in use but also highly efficient in targeting different forms of malignancies. With this in mind, additional studies need to be envisioned that not only further address *ex vivo* NK cell purification, expansion, and activation strategies but also the final clinical setting including pre-conditioning, dosing, and timing of the NK cell application. Efforts for harmonization of protocols at the European and worldwide level should be undertaken to ensure highest quality and efficacy of the NK cell product for clinical application. Finally, with regard to the possible tumor-mediated immunosuppression, therapeutic concepts have to be developed that either directly strengthen NK cells to deal with the hostile tumor environment and/or specifically counteract tumor-induced immunosuppressive mechanisms.

AUTHOR CONTRIBUTIONS

MG, JW, and EU extensively reviewed the current literature on *ex vivo* NK cell cultivation and expansion and prepared a comprehensive overview that is listed in the tables. MG, JW, UK, AC, VH, and EU wrote and critically reviewed the manuscript.

ACKNOWLEDGMENTS

The authors apologize to all investigators whose works were not cited in this article due to space limitations. EU and JW have been supported by the LOEWE Center for Cell and Gene Therapy, Frankfurt, funded by the Hessian Ministry of Higher Education, Research and the Arts, Germany (III L 4-518/17.004). AC's work on cytokine-induced NK cells is supported by the Grant of German Cancer Aid 111455 and the GRK 2099.

REFERENCES

1. Moretta A, Bottino C, Vitale M, Pende D, Cantoni C, Mingari MC, et al. Activating receptors and coreceptors involved in human natural killer cell-mediated cytolysis. *Annu Rev Immunol* (2001) 19:197–223. doi:10.1146/annurev.immunol.19.1.197
2. Voskoboinik I, Smyth MJ, Trapani JA. Perforin-mediated target-cell death and immune homeostasis. *Nat Rev Immunol* (2006) 6:940–52. doi:10.1038/nri1983
3. Vivier E, Tomasello E, Baratin M, Walzer T, Ugolini S. Functions of natural killer cells. *Nat Immunol* (2008) 9:503–10. doi:10.1038/ni1582
4. Garrido F, Ruiz-Cabello F, Cabrera T, Pérez-Villar JJ, López-Botet M, Duggan-Keen M, et al. Implications for immunosurveillance of altered HLA class I phenotypes in human tumours. *Immunol Today* (1997) 18:89–95. doi:10.1016/S0167-5699(96)10075-X
5. Bottino C, Castriconi R, Moretta L, Moretta A. Cellular ligands of activating NK receptors. *Trends Immunol* (2005) 26:221–6. doi:10.1016/j.it.2005.02.007
6. Pende D, Cantoni C, Rivera P, Vitale M, Castriconi R, Marcenaro S, et al. Role of NKG2D in tumor cell lysis mediated by human NK cells: cooperation with natural cytotoxicity receptors and capability of recognizing tumors of nonepithelial origin. *Eur J Immunol* (2001) 31:1076–86. doi:10.1002/1521-4141(200104)31:4<1076:AID-IMMU1076>3.0.CO;2-Y
7. Pende D, Spaggiari GM, Marcenaro S, Martini S, Rivera P, Capobianco A, et al. Analysis of the receptor-ligand interactions in the natural killer-mediated lysis of freshly isolated myeloid or lymphoblastic leukemias: evidence for the involvement of the Poliovirus receptor (CD155) and Nectin-2 (CD112). *Blood* (2005) 105:2066–73. doi:10.1182/blood-2004-09-3548
8. Bauer S, Groh V, Wu J, Steinle A, Phillips JH, Lanier LL, et al. Activation of NK cells and T cells by NKG2D, a receptor for stress-inducible MICA. *Science* (1999) 285:727–9. doi:10.1126/science.285.5428.727
9. Ullrich E, Koch J, Cerwenka A, Steinle A. New prospects on the NKG2D/NKG2DL system for oncology. *Oncoimmunology* (2013) 2:e26097. doi:10.4161/onci.26097
10. Smyth MJ, Cretney E, Kelly JM, Westwood JA, Street SEA, Yagita H, et al. Activation of NK cell cytotoxicity. *Mol Immunol* (2005) 42:501–10. doi:10.1016/j.molimm.2004.07.034
11. Smyth MJ, Cretney E, Takeda K, Wiltrout RH, Sedger LM, Kayagaki N, et al. Tumor necrosis factor-related apoptosis-inducing ligand (TRAIL) contributes to interferon gamma-dependent natural killer cell protection from tumor metastasis. *J Exp Med* (2001) 193:661–70. doi:10.1084/jem.193.6.661
12. Smyth MJ, Crowe NY, Pellicci DG, Kyparissoudis K, Kelly JM, Takeda K, et al. Sequential production of interferon-gamma by NK1.1(+) T cells and natural killer cells is essential for the antimetastatic effect of alpha-galactosylceramide. *Blood* (2002) 99:1259–66. doi:10.1182/blood.V99.4.1259
13. Chawla-Sarkar M, Lindner DJ, Liu Y-F, Williams BR, Sen GC, Silverman RH, et al. Apoptosis and interferons: role of interferon-stimulated genes as mediators of apoptosis. *Apoptosis* (2003) 8:237–49. doi:10.1023/A:1023668705040
14. Costello RT, Sivori S, Marcenaro E, Lafage-Pochitaloff M, Mozziconacci M-J, Reviron D, et al. Defective expression and function of natural killer cell-triggering receptors in patients with acute myeloid leukemia. *Blood* (2002) 99:3661–7. doi:10.1182/blood.V99.10.3661
15. Salih HR, Rammensee H-G, Steinle A. Cutting edge: down-regulation of MICA on human tumors by proteolytic shedding. *J Immunol* (2002) 169:4098–102. doi:10.4049/jimmunol.169.8.4098
16. Salih HR, Goehlsdorf D, Steinle A. Release of MICB molecules by tumor cells: mechanism and soluble MICB in sera of cancer patients. *Hum Immunol* (2006) 67:188–95. doi:10.1016/j.humimm.2006.02.008
17. Kaiser BK, Yim D, Chow I-T, Gonzalez S, Dai Z, Mann HH, et al. Disulphide-isomerase-enabled shedding of tumour-associated NKG2D ligands. *Nature* (2007) 447:482–6. doi:10.1038/nature05768
18. Rouas-Freiss N, Moreau P, Ferrone S, Carosella ED. HLA-G proteins in cancer: do they provide tumor cells with an escape mechanism? *Cancer Res* (2005) 65:10139–44. doi:10.1158/0008-5472.CAN-05-0097
19. Urosevic M, Dummer R. Human leukocyte antigen-G and cancer immunoediting. *Cancer Res* (2008) 68:627–30. doi:10.1158/0008-5472.CAN-07-2704
20. Castriconi R, Dondero A, Bellora F, Moretta L, Castellano A, Locatelli F, et al. Neuroblastoma-derived TGF-β1 modulates the chemokine receptor repertoire of human resting NK cells. *J Immunol* (2013) 190:5321–8. doi:10.4049/jimmunol.1202693
21. Stringaris K, Sekine T, Khoder A, Alsuliman A, Razzaghi B, Sargeant R, et al. Leukemia-induced phenotypic and functional defects in natural killer cells predict failure to achieve remission in acute myeloid leukemia. *Haematologica* (2014) 99:836–47. doi:10.3324/haematol.2013.087536
22. Pietra G, Manzini C, Rivara S, Vitale M, Cantoni C, Petretto A, et al. Melanoma cells inhibit natural killer cell function by modulating the

expression of activating receptors and cytolytic activity. *Cancer Res* (2012) 72:1407–15. doi:10.1158/0008-5472.CAN-11-2544
23. Lim O, Lee Y, Chung H, Her JH, Kang SM, Jung M-Y, et al. GMP-compliant, large-scale expanded allogeneic natural killer cells have potent cytolytic activity against cancer cells in vitro and in vivo. *PLoS One* (2013) 8:e53611. doi:10.1371/journal.pone.0053611
24. Torelli GF, Rozera C, Santodonato L, Peragine N, D'Agostino G, Montefiore E, et al. A good manufacturing practice method to ex vivo expand natural killer cells for clinical use. *Blood Transfus* (2015) 13:464–71. doi:10.2450/2015.0231-14
25. Alici E, Sutlu T, Björkstrand B, Gilljam M, Stellan B, Nahi H, et al. Autologous antitumor activity by NK cells expanded from myeloma patients using GMP-compliant components. *Blood* (2008) 111:3155–62. doi:10.1182/blood-2007-09-110312
26. Koehl U, Brehm C, Huenecke S, Zimmermann S-Y, Kloess S, Bremm M, et al. Clinical grade purification and expansion of NK cell products for an optimized manufacturing protocol. *Front Oncol* (2013) 3:118. doi:10.3389/fonc.2013.00118
27. Colombo MP, Trinchieri G. Interleukin-12 in anti-tumor immunity and immunotherapy. *Cytokine Growth Factor Rev* (2002) 13:155–68. doi:10.1016/S1359-6101(01)00032-6
28. Becknell B, Caligiuri MA. Interleukin-2, interleukin-15, and their roles in human natural killer cells. *Adv Immunol* (2005) 86:209–39. doi:10.1016/S0065-2776(04)86006-1
29. Theofilopoulos AN, Baccala R, Beutler B, Kono DH. Type I interferons (alpha/beta) in immunity and autoimmunity. *Annu Rev Immunol* (2005) 23:307–36. doi:10.1146/annurev.immunol.23.021704.115843
30. Burgess SJ, Marusina AI, Pathmanathan I, Borrego F, Coligan JE. IL-21 down-regulates NKG2D/DAP10 expression on human NK and CD8+ T cells. *J Immunol* (2006) 176:1490–7. doi:10.4049/jimmunol.176.3.1490
31. Ortaldo JR, Winkler-Pickett R, Wigginton J, Horner M, Bere EW, Mason AT, et al. Regulation of ITAM-positive receptors: role of IL-12 and IL-18. *Blood* (2006) 107:1468–75. doi:10.1182/blood-2005-04-1579
32. Terme M, Ullrich E, Delahaye NF, Chaput N, Zitvogel L. Natural killer cell-directed therapies: moving from unexpected results to successful strategies. *Nat Immunol* (2008) 9:486–94. doi:10.1038/ni1580
33. Koehl U, Kalberer C, Spanholtz J, Lee D, Miller J, Cooley S, et al. Advances in clinical NK cell studies: donor selection, manufacturing and quality control. *Oncoimmunology* (2015) 5(4):e1115178. doi:10.1080/2162402X.2015.1115178
34. Glienke W, Esser R, Priesner C, Suerth JD, Schambach A, Wels WS, et al. Advantages and applications of CAR-expressing natural killer cells. *Front Pharmacol* (2015) 6:21. doi:10.3389/fphar.2015.00021
35. Klingemann H. Challenges of cancer therapy with natural killer cells. *Cytotherapy* (2015) 17:245–9. doi:10.1016/j.jcyt.2014.09.007
36. Cany J, Dolstra H, Shah N. Umbilical cord blood-derived cellular products for cancer immunotherapy. *Cytotherapy* (2015) 17:739–48. doi:10.1016/j.jcyt.2015.03.005
37. Spanholtz J, Tordoir M, Eissens D, Preijers F, van der Meer A, Joosten I, et al. High log-scale expansion of functional human natural killer cells from umbilical cord blood CD34-positive cells for adoptive cancer immunotherapy. *PLoS One* (2010) 5:e9221. doi:10.1371/journal.pone.0009221
38. Suck G, Odendahl M, Nowakowska P, Seidl C, Wels WS, Klingemann HG, et al. NK-92: an "off-the-shelf therapeutic" for adoptive natural killer cell-based cancer immunotherapy. *Cancer Immunol Immunother* (2015) 65:485–92. doi:10.1007/s00262-015-1761-x
39. Knorr DA, Kaufman DS. Pluripotent stem cell-derived natural killer cells for cancer therapy. *Transl Res* (2010) 156:147–54. doi:10.1016/j.trsl.2010.07.008
40. Klingemann H, Boissel L, Toneguzzo F. Natural killer cells for immunotherapy – advantages of the NK-92 cell line over blood NK cells. *Front Immunol* (2016) 7:91. doi:10.3389/fimmu.2016.00091
41. Eguizabal C, Zenarruzabeitia O, Monge J, Santos S, Vesga MA, Maruri N, et al. Natural killer cells for cancer immunotherapy: pluripotent stem cells-derived NK cells as an immunotherapeutic perspective. *Front Immunol* (2014) 5:439. doi:10.3389/fimmu.2014.00439
42. Ferrara JLM, Levine JE, Reddy P, Holler E. Graft-versus-host disease. *Lancet* (2009) 373:1550–61. doi:10.1016/S0140-6736(09)60237-3
43. Miller JS, Soignier Y, Panoskaltsis-Mortari A, McNearney SA, Yun GH, Fautsch SK, et al. Successful adoptive transfer and in vivo expansion of human haploidentical NK cells in patients with cancer. *Blood* (2005) 105:3051–7. doi:10.1182/blood-2004-07-2974
44. Ho VT, Soiffer RJ. The history and future of T-cell depletion as graft-versus-host disease prophylaxis for allogeneic hematopoietic stem cell transplantation. *Blood* (2001) 98:3192–204. doi:10.1182/blood.V98.12.3192
45. Skeate R, Singh C, Cooley S, Geller M, Northouse J, Welbig J, et al. Hemolytic anemia due to passenger lymphocyte syndrome in solid malignancy patients treated with allogeneic natural killer cell products. *Transfusion* (2013) 53:419–23. doi:10.1111/j.1537-2995.2012.03942.x
46. McKenna DH, Sumstad D, Bostrom N, Kadidlo DM, Fautsch S, McNearney S, et al. Good manufacturing practices production of natural killer cells for immunotherapy: a six-year single-institution experience. *Transfusion* (2007) 47:520–8. doi:10.1111/j.1537-2995.2006.01145.x
47. Iyengar R, Handgretinger R, Babarin-Dorner A, Leimig T, Otto M, Geiger TL, et al. Purification of human natural killer cells using a clinical-scale immunomagnetic method. *Cytotherapy* (2003) 5:479–84. doi:10.1080/14653240310003558
48. Passweg JR, Tichelli A, Meyer-Monard S, Heim D, Stern M, Kuhne T, et al. Purified donor NK-lymphocyte infusion to consolidate engraftment after haploidentical stem cell transplantation. *Leukemia* (2004) 18:1835–8. doi:10.1038/sj.leu.2403524
49. Meyer-Monard S, Passweg J, Siegler U, Kalberer C, Koehl U, Rovó A, et al. Clinical-grade purification of natural killer cells in haploidentical hematopoietic stem cell transplantation. *Transfusion* (2009) 49:362–71. doi:10.1111/j.1537-2995.2008.01969.x
50. Siegler U, Meyer-Monard S, Jorger S, Stern M, Tichelli A, Gratwohl A, et al. Good manufacturing practice-compliant cell sorting and large-scale expansion of single KIR-positive alloreactive human natural killer cells for multiple infusions to leukemia patients. *Cytotherapy* (2010) 12:750–63. doi:10.3109/14653241003786155
51. van Ostaijen-ten Dam MM, Prins H-J, Boerman GH, Vervat C, Pende D, Putter H, et al. Preparation of cytokine-activated NK cells for use in adoptive cell therapy in cancer patients: protocol optimization and therapeutic potential. *J Immunother* (2016) 39:90–100. doi:10.1097/CJI.0000000000000110
52. Huenecke S, Zimmermann SY, Kloess S, Esser R, Brinkmann A, Tramsen L, et al. IL-2-driven regulation of NK cell receptors with regard to the distribution of CD16+ and CD16- subpopulations and in vivo influence after haploidentical NK cell infusion. *J Immunother* (2010) 33:200–10. doi:10.1097/CJI.0b013e3181bb46f7
53. Koehl U, Sörensen J, Esser R, Zimmermann S, Grüttner HP, Tonn T, et al. IL-2 activated NK cell immunotherapy of three children after haploidentical stem cell transplantation. *Blood Cells Mol Dis* (2004) 33:261–6. doi:10.1016/j.bcmd.2004.08.013
54. Boerman GH, van Ostaijen-ten Dam MM, Kraal KCJM, Santos SJ, Ball LM, Lankester AC, et al. Role of NKG2D, DNAM-1 and natural cytotoxicity receptors in cytotoxicity toward rhabdomyosarcoma cell lines mediated by resting and IL-15-activated human natural killer cells. *Cancer Immunol Immunother* (2015) 64:573–83. doi:10.1007/s00262-015-1657-9
55. Wendt K, Wilk E, Buyny S, Schmidt RE, Jacobs R. Interleukin-21 differentially affects human natural killer cell subsets. *Immunology* (2007) 122:486–95. doi:10.1111/j.1365-2567.2007.02675.x
56. Romee R, Schneider SE, Leong JW, Chase JM, Keppel CR, Sullivan RP, et al. Cytokine activation induces human memory-like NK cells. *Blood* (2012) 120:4751–60. doi:10.1182/blood-2012-04-419283
57. Ni J, Miller M, Stojanovic A, Garbi N, Cerwenka A. Sustained effector function of IL-12/15/18-preactivated NK cells against established tumors. *J Exp Med* (2012) 209:2351–65. doi:10.1084/jem.20120944
58. Leong JW, Chase JM, Romee R, Schneider SE, Sullivan RP, Cooper MA, et al. Preactivation with IL-12, IL-15, and IL-18 induces CD25 and a functional high-affinity IL-2 receptor on human cytokine-induced memory-like natural killer cells. *Biol Blood Marrow Transplant* (2014) 20:463–73. doi:10.1016/j.bbmt.2014.01.006
59. Berrien-Elliott MM, Wagner JA, Fehniger TA. Human cytokine-induced memory-like natural killer cells. *J Innate Immun* (2015) 7:563–71. doi:10.1159/000382019
60. Geller MA, Cooley S, Judson PL, Ghebre R, Carson LF, Argenta PA, et al. A phase II study of allogeneic natural killer cell therapy to treat patients with recurrent ovarian and breast cancer. *Cytotherapy* (2011) 13:98–107. doi:10.3109/14653249.2010.515582.A

61. Klingemann H, Grodman C, Cutler E, Duque M, Kadidlo D, Klein AK, et al. Autologous stem cell transplant recipients tolerate haploidentical related-donor natural killer cell-enriched infusions. *Transfusion* (2013) 53:412–8. doi:10.1111/j.1537-2995.2012.03764.x
62. Koepsell SA, Kadidlo DM, Fautsch S, McCullough J, Klingemann H, Wagner JE, et al. Successful “in-flight” activation of natural killer cells during long-distance shipping. *Transfusion* (2013) 53:398–403. doi:10.1111/j.1537-2995.2012.03695.x
63. Iliopoulou EG, Kountourakis P, Karamouzis MV, Doufexis D, Ardavanis A, Baxevanis CN, et al. A phase I trial of adoptive transfer of allogeneic natural killer cells in patients with advanced non-small cell lung cancer. *Cancer Immunol Immunother* (2010) 59:1781–9. doi:10.1007/s00262-010-0904-3
64. Choi I, Yoon SR, Park S-Y, Kim H, Jung S-J, Jang YJ, et al. Donor-derived natural killer cells infused after human leukocyte antigen-haploidentical hematopoietic cell transplantation: a dose-escalation study. *Biol Blood Marrow Transplant* (2014) 20:696–704. doi:10.1016/j.bbmt.2014.01.031
65. Carlens S, Gilljam M, Chambers BJ, Aschan J, Guven H, Ljunggren HG, et al. A new method for in vitro expansion of cytotoxic human CD3-CD56+ natural killer cells. *Hum Immunol* (2001) 62:1092–8. doi:10.1016/S0198-8859(01)00313-5
66. Barkholt L, Alici E, Conrad R, Sutlu T, Gilljam M, Stellan B, et al. Safety analysis of ex vivo-expanded NK and NK-like T cells administered to cancer patients: a phase I clinical study. *Immunotherapy* (2009) 1:753–64. doi:10.2217/imt.09.47
67. Sutlu T, Stellan B, Gilljam M, Quezada HC, Nahi H, Gahrton G, et al. Clinical-grade, large-scale, feeder-free expansion of highly active human natural killer cells for adoptive immunotherapy using an automated bioreactor. *Cytotherapy* (2010) 12:1044–55. doi:10.3109/14653249.2010.504770
68. Masuyama J, Murakami T, Iwamoto S, Fujita S. Ex vivo expansion of natural killer cells from human peripheral blood mononuclear cells co-stimulated with anti-CD3 and anti-CD52 monoclonal antibodies. *Cytotherapy* (2015) 18:1–6. doi:10.1016/j.jcyt.2015.09.011
69. Rujkijyanont P, Chan WK, Eldridge PW, Lockey T, Holladay M, Rooney B, et al. Ex vivo activation of CD56+ immune cells that eradicate neuroblastoma. *Cancer Res* (2013) 73:2608–18. doi:10.1158/0008-5472.CAN-12-3322
70. Deng X, Terunuma H, Nieda M, Xiao W, Nicol A. Synergistic cytotoxicity of ex vivo expanded natural killer cells in combination with monoclonal antibody drugs against cancer cells. *Int Immunopharmacol* (2012) 14:593–605. doi:10.1016/j.intimp.2012.09.014
71. Lotze MT, Grimm EA, Mazumder A, Strausser JL, Rosenberg SA. Lysis of fresh and cultured autologous tumor by human lymphocytes cultured in T-cell growth factor. *Cancer Res* (1981) 41:4420–5.
72. Grimm EA, Mazumder A, Zhang HZ, Rosenberg SA. Lymphokine-activated killer cell phenomenon. Lysis of natural killer-resistant fresh solid tumor cells by interleukin 2-activated autologous human peripheral blood lymphocytes. *J Exp Med* (1982) 155:1823–41. doi:10.1084/jem.155.6.1823
73. Burns LJ, Weisdorf DJ, DeFor TE, Vesole DH, Repka TL, Blazar BR, et al. IL-2-based immunotherapy after autologous transplantation for lymphoma and breast cancer induces immune activation and cytokine release: a phase I/II trial. *Bone Marrow Transplant* (2003) 32:177–86. doi:10.1038/sj.bmt.1704086
74. Asai O, Longo DL, Tian ZG, Hornung RL, Taub DD, Ruscetti FW, et al. Suppression of graft-versus-host disease and amplification of graft-versus-tumor effects by activated natural killer cells after allogeneic bone marrow transplantation. *J Clin Invest* (1998) 101:1835–42. doi:10.1172/JCI1268
75. Olson JA, Leveson-Gower DB, Gill S, Baker J, Beilhack A, Negrin RS. NK cells mediate reduction of GVHD by inhibiting activated, alloreactive T cells while retaining GVT effects. *Blood* (2010) 115:4293–301. doi:10.1182/blood-2009-05-222190
76. Ruggeri L, Capanni M, Urbani E, Perruccio K, Shlomchik WD, Tosti A, et al. Effectiveness of donor natural killer cell alloreactivity in mismatched hematopoietic transplants. *Science* (2002) 295:2097–100. doi:10.1126/science.1068440
77. Velardi A, Ruggeri L, Mancusi A, Aversa F, Christiansen FT. Natural killer cell allorecognition of missing self in allogeneic hematopoietic transplantation: a tool for immunotherapy of leukemia. *Curr Opin Immunol* (2009) 21:525–30. doi:10.1016/j.coi.2009.07.015
78. Meinhardt K, Kroeger I, Bauer R, Ganss F, Ovsiy I, Rothamer J, et al. Identification and characterization of the specific murine NK cell subset supporting graft-versus-leukemia- and reducing graft-versus-host-effects. *Oncoimmunology* (2015) 4:e981483. doi:10.4161/2162402X.2014.981483
79. Bachanova V, Burns LJ, McKenna DH, Curtsinger J, Panoskaltsis-Mortari A, Lindgren BR, et al. Allogeneic natural killer cells for refractory lymphoma. *Cancer Immunol Immunother* (2010) 59:1739–44. doi:10.1007/s00262-010-0896-z
80. Brehm C, Huenecke S, Quaiser A, Esser R, Bremm M, Kloess S, et al. IL-2 stimulated but not unstimulated NK cells induce selective disappearance of peripheral blood cells: concomitant results to a phase I/II study. *PLoS One* (2011) 6:e27351. doi:10.1371/journal.pone.0027351
81. Carson WE, Giri JG, Lindemann MJ, Linett ML, Ahdieh M, Paxton R, et al. Interleukin (IL) 15 is a novel cytokine that activates human natural killer cells via components of the IL-2 receptor. *J Exp Med* (1994) 180:1395–403. doi:10.1084/jem.180.4.1395
82. Carson WE, Fehniger TA, Haldar S, Eckhert K, Lindemann MJ, Lai CF, et al. A potential role for interleukin-15 in the regulation of human natural killer cell survival. *J Clin Invest* (1997) 99:937–43. doi:10.1172/JCI119258
83. Waldmann TA. The biology of interleukin-2 and interleukin-15: implications for cancer therapy and vaccine design. *Nat Rev Immunol* (2006) 6:595–601. doi:10.1038/nri1901
84. Fehniger TA, Cooper MA, Caligiuri MA. Interleukin-2 and interleukin-15: immunotherapy for cancer. *Cytokine Growth Factor Rev* (2002) 13:169–83. doi:10.1016/S1359-6101(01)00021-1
85. Waldmann TA, Lugli E, Roederer M, Perera LP, Smedley JV, Macallister RP, et al. Safety (toxicity), pharmacokinetics, immunogenicity, and impact on elements of the normal immune system of recombinant human IL-15 in rhesus macaques. *Blood* (2011) 117:4787–95. doi:10.1182/blood-2010-10-311456
86. Mao Y, van Hoef V, Zhang X, Wennerberg E, Lorent J, Witt K, et al. IL-15 activates mTOR and primes stress-activated gene expression leading to prolonged antitumor capacity of NK cells. *Blood* (2016) 128:1475–89. doi:10.1182/blood-2016-02-698027
87. Felices M, Lenvik A, Chu S, McElmurry R, Cooley S, Tolar J, et al. Continuous IL-15 signaling leads to functional exhaustion of human natural killer cells through metabolic changes that alters their in vivo anti-tumor activity. *Blood* (2016) 128:551.
88. Davis MR, Zhu Z, Hansen DM, Bai Q, Fang Y. The role of IL-21 in immunity and cancer. *Cancer Lett* (2015) 358:107–14. doi:10.1016/j.canlet.2014.12.047
89. Parrish-novak J, Dillon SR, Nelson A, Hammond A, Sprecher C, Gross JA, et al. Interleukin 21 and its receptor are involved in NK cell expansion and regulation of lymphocyte function. *Nature* (2000) 408:57–63. doi:10.1038/35040504
90. Kasaian MT, Whitters MJ, Carter LL, Lowe LD, Jussif JM, Deng B, et al. IL-21 limits NK cell responses and promotes antigen-specific T cell activation: a mediator of the transition from innate to adaptive immunity. *Immunity* (2002) 16:559–69. doi:10.1016/S1074-7613(02)00295-9
91. Brady J, Hayakawa Y, Smyth MJ, Nutt SL. IL-21 induces the functional maturation of murine NK cells. *J Immunol* (2004) 172:2048–58. doi:10.4049/jimmunol.172.4.2048
92. Venkatasubramanian S, Cheekatla S, Paidipally P, Tripathi D, Welch E, Tvinnereim AR, et al. IL-21-dependent expansion of memory-like NK cells enhances protective immune responses against *Mycobacterium tuberculosis*. *Mucosal Immunol* (2016):1–12. doi:10.1038/mi.2016.105
93. Strbo N, de Armas L, Liu H, Kolber MA, Lichtenheld M, Pahwa S. IL-21 augments natural killer effector functions in chronically HIV-infected individuals. *AIDS* (2008) 22:1551–60. doi:10.1097/QAD.0b013e3283089367
94. Li Q, Ye L-J, Ren H-L, Huyan T, Li J, Shi J-L, et al. Multiple effects of IL-21 on human NK cells in ex vivo expansion. *Immunobiology* (2015) 220:876–88. doi:10.1016/j.imbio.2015.01.009
95. Lim D-P, Jang Y-Y, Kim S, Koh SS, Lee J-J, Kim J-S, et al. Effect of exposure to interleukin-21 at various time points on human natural killer cell culture. *Cytotherapy* (2014) 16:1419–30. doi:10.1016/j.jcyt.2014.04.008
96. Granzin M, Stojanovic A, Miller M, Childs R, Huppert V, Cerwenka A. Highly efficient IL-21 and feeder cell-driven ex vivo expansion of human NK cells with therapeutic activity in a xenograft mouse model of melanoma. *Oncoimmunology* (2016) 5:e1219007. doi:10.1080/2162402X.2016.1219007
97. McMichael EL, Jaime-Ramirez AC, Guenterberg KD, Luedke E, Atwal LS, Campbell AR, et al. IL-21 enhances natural killer cell response to

cetuximab-coated pancreatic tumor cells. *Clin Cancer Res* (2017) 23:489–502. doi:10.1158/1078-0432.CCR-16-0004
98. Iannello A, Boulassel M-R, Samarani S, Tremblay C, Toma E, Routy J-P, et al. IL-21 enhances NK cell functions and survival in healthy and HIV-infected patients with minimal stimulation of viral replication. *J Leukoc Biol* (2010) 87:857–67. doi:10.1189/jlb.1009701
99. Kobayashi M, Fitz L, Ryan M, Hewick RM, Clark SC, Chan S, et al. Identification and purification of natural killer cell stimulatory factor (NKSF), a cytokine with multiple biologic effects on human lymphocytes. *J Exp Med* (1989) 170:827–45. doi:10.1084/jem.170.3.827
100. D'Andrea A, Rengaraju M, Valiante NM, Chehimi J, Kubin M, Aste M, et al. Production of natural killer cell stimulatory factor (interleukin 12) by peripheral blood mononuclear cells. *J Exp Med* (1992) 176:1387–98. doi:10.1084/jem.176.5.1387
101. Vignali DAA, Kuchroo VK. IL-12 family cytokines: immunological playmakers. *Nat Immunol* (2012) 13:722–8. doi:10.1038/ni.2366
102. Wang KS, Frank DA, Ritz J. Interleukin-2 enhances the response of natural killer cells to interleukin-12 through up-regulation of the interleukin-12 receptor and STAT4. *Blood* (2000) 95:3183–90.
103. Chaix J, Tessmer MS, Hoebe K, Fuseri N, Ryffel B, Dalod M, et al. Cutting edge: priming of NK cells by IL-18. *J Immunol* (2008) 181:1627–31. doi:10.4049/jimmunol.181.3.1627
104. French AR, Holroyd EB, Yang L, Kim S, Yokoyama WM. IL-18 acts synergistically with IL-15 in stimulating natural killer cell proliferation. *Cytokine* (2006) 35:229–34. doi:10.1016/j.cyto.2006.08.006
105. Cooper MA, Elliott JM, Keyel PA, Yang L, Carrero JA, Yokoyama WM. Cytokine-induced memory-like natural killer cells. *Proc Natl Acad Sci U S A* (2009) 106:1915–9. doi:10.1073/pnas.0813192106
106. Ni J, Hölsken O, Miller M, Hammer Q, Luetke-Eversloh M, Romagnani C, et al. Adoptively transferred natural killer cells maintain long-term antitumor activity by epigenetic imprinting and CD4(+) T cell help. *Oncoimmunology* (2016) 5:e1219009. doi:10.1080/2162402X.2016.1219009
107. Romee R, Rosario M, Berrien-Elliott MM, Wagner JA, Jewell BA, Schappe T, et al. Cytokine-induced memory-like natural killer cells exhibit enhanced responses against myeloid leukemia. *Sci Transl Med* (2016) 8:357ra123. doi:10.1126/scitranslmed.aaf2341
108. Miller JS, Oelkers S, Verfaillie C, McGlave P. Role of monocytes in the expansion of human activated natural killer cells. *Blood* (1992) 80:2221–9.
109. Rölle A, Pollmann J, Ewen E, Le VTK, Halenius A, Hengel H, et al. IL-12-producing monocytes and HLA-E control HCMV-driven NKG2C+ NK cell expansion. *J Clin Invest* (2014) 124:5305–16. doi:10.1172/JCI77440
110. Rabinowich H, Sedlmayr P, Herberman RB, Whiteside TL. Increased proliferation, lytic activity, and purity of human natural killer cells cocultured with mitogen-activated feeder cells. *Cell Immunol* (1991) 135:454–70. doi:10.1016/0008-8749(91)90290-R
111. Guven H, Gilljam M, Chambers BJ, Ljunggren HG, Christensson B, Kimby E, et al. Expansion of natural killer (NK) and natural killer-like T (NKT)-cell populations derived from patients with B-chronic lymphocytic leukemia (B-CLL): a potential source for cellular immunotherapy. *Leukemia* (2003) 17:1973–80. doi:10.1038/sj.leu.2403083
112. Sutlu T, Alici E. Ex vivo expansion of natural killer cells: a question of function. *Cytotherapy* (2011) 13:767–8. doi:10.3109/14653249.2011.563295
113. Ahn Y-O, Kim S, Kim TM, Song EY, Park MH, Heo DS. Irradiated and activated autologous PBMCs induce expansion of highly cytotoxic human NK cells in vitro. *J Immunother* (2013) 36:373–81. doi:10.1097/CJI.0b013e3182a3430f
114. Sakamoto N, Ishikawa T, Kokura S, Okayama T, Oka K, Ideno M, et al. Phase I clinical trial of autologous NK cell therapy using novel expansion method in patients with advanced digestive cancer. *J Transl Med* (2015) 13:277. doi:10.1186/s12967-015-0632-8
115. Kim E-K, Ahn Y-O, Kim S, Kim TM, Keam B, Heo DS. Ex vivo activation and expansion of natural killer cells from patients with advanced cancer with feeder cells from healthy volunteers. *Cytotherapy* (2013) 15:231–41.e1. doi:10.1016/j.jcyt.2012.10.019
116. Parkhurst MR, Riley JP, Dudley ME, Rosenberg SA. Adoptive transfer of autologous natural killer cells leads to high levels of circulating natural killer cells but does not mediate tumor regression. *Clin Cancer Res* (2011) 17:6287–97. doi:10.1158/1078-0432.CCR-11-1347
117. Yang Y, Lim O, Kim TM, Ahn Y-O, Choi H, Chung H, et al. Phase I study of random healthy donor-derived allogeneic natural killer cell therapy in patients with malignant lymphoma or advanced solid tumors. *Cancer Immunol Res* (2016) 4:215–24. doi:10.1158/2326-6066.CIR-15-0118
118. Luhm J, Brand JM, Koritke P, Hoppner M, Kirchner H, Frohn C. Large-scale generation of natural killer lymphocytes for clinical application. *J Hematother Stem Cell Res* (2002) 11:651–7. doi:10.1089/15258160260194794
119. Bae DS, Lee JK. Development of NK cell expansion methods using feeder cells from human myelogenous leukemia cell line. *Blood Res* (2014) 49:154. doi:10.5045/br.2014.49.3.154
120. Fuchshuber PR, Lotzova E. Feeder cells enhance oncolytic and proliferative activity of long-term human bone marrow interleukin-2 cultures and induce different lymphocyte subsets. *Cancer Immunol Immunother* (1991) 33:15–20. doi:10.1007/BF01742522
121. Lim SA, Kim T-J, Lee JE, Sonn CH, Kim K, Kim J, et al. Ex vivo expansion of highly cytotoxic human nk cells by cocultivation with irradiated tumor cells for adoptive immunotherapy. *Cancer Res* (2013) 73(8):2598–607. doi:10.1158/0008-5472.CAN-12-2893
122. Warren HS, Kinnear BF, Skipsey LJ. Human natural killer (NK) cells: requirements for cell proliferation and expansion of phenotypically novel subpopulations. *Immunol Cell Biol* (1993) 71(Pt 2):87–97. doi:10.1038/icb.1993.9
123. Lister J, Rybka WB, Donnenberg AD, DeMagalhaes-Silverman M, Pincus SM, Bloom EJ, et al. Autologous peripheral blood stem cell transplantation and adoptive immunotherapy with activated natural killer cells in the immediate posttransplant period. *Clin Cancer Res* (1995) 1:607–14.
124. Harada H, Saijo K, Watanabe S, Tsuboi K, Nose T, Ishiwata I, et al. Selective expansion of human natural killer cells from peripheral blood mononuclear cells by the cell line, HFWT. *Jpn J Cancer Res* (2002) 93:313–9. doi:10.1111/j.1349-7006.2002.tb02174.x
125. Ishikawa E, Tsuboi K, Saijo K, Harada H, Takano S, Nose T, et al. Autologous natural killer cell therapy for human recurrent malignant glioma. *Anticancer Res* (2004) 24:1861–71.
126. Imai C, Iwamoto S, Campana D. Genetic modification of primary natural killer cells overcomes inhibitory signals and induces specific killing of leukemic cells. *Blood* (2005) 106:376–83. doi:10.1182/blood-2004-12-4797
127. Fujisaki H, Kakuda H, Shimasaki N, Imai C, Ma J, Lockey T, et al. Expansion of highly cytotoxic human natural killer cells for cancer cell therapy. *Cancer Res* (2009) 69:4010–7. doi:10.1158/0008-5472.CAN-08-3712
128. Garg TK, Szmania SM, Khan JA, Hoering A, Malbrough PA, Moreno-Bost A, et al. Highly activated and expanded natural killer cells for multiple myeloma immunotherapy. *Haematologica* (2012) 97:1348–56. doi:10.3324/haematol.2011.056747
129. Szmania S, Lapteva N, Garg T, Greenway A, Lingo J, Nair B, et al. Ex vivo-expanded natural killer cells demonstrate robust proliferation in vivo in high-risk relapsed multiple myeloma patients. *J Immunother* (2015) 38:24–36. doi:10.1097/CJI.0000000000000059
130. Lapteva N, Durett AG, Sun J, Rollins LA, Huye LL, Fang J, et al. Large-scale ex vivo expansion and characterization of natural killer cells for clinical applications. *Cytotherapy* (2012) 14:1131–43. doi:10.3109/14653249.2012.700767
131. Shah NN, Baird K, Delbrook CP, Fleisher TA, Kohler ME, Rampertaap S, et al. Acute GVHD in patients receiving IL-15/4-1BBL activated NK cells following T cell depleted stem cell transplantation. *Blood* (2014) 125:784–92. doi:10.1182/blood-2014-07-592881
132. Zhang H, Cui Y, Voong N, Sabatino M, Stroncek DF, Morisot S, et al. Activating signals dominate inhibitory signals in CD137L/IL-15 activated natural killer cells. *J Immunother* (2011) 34:187–95. doi:10.1097/CJI.0b013e31820d2a21
133. Oyer JL, Igarashi RY, Kulikowski AR, Colosimo DA, Solh MM, Zakari A, et al. Generation of highly cytotoxic natural killer cells for treatment of AML using feeder-free, particle based approach. *Biol Blood Marrow Transplant* (2015) 21:632–39. doi:10.1016/j.bbmt.2014.12.037
134. Denman CJ, Senyukov VV, Somanchi SS, Phatarpekar PV, Kopp LM, Johnson JL, et al. Membrane-bound IL-21 promotes sustained ex vivo proliferation of human natural killer cells. *PLoS One* (2012) 7:e30264. doi:10.1371/journal.pone.0030264
135. Liu Y, Wu H-W, Sheard MA, Sposto R, Somanchi SS, Cooper LJN, et al. Growth and activation of natural killer cells ex vivo from children with neuroblastoma for adoptive cell therapy. *Clin Cancer Res* (2013) 19:2132–43. doi:10.1158/1078-0432.CCR-12-1243
136. Oyer JL, Pandey V, Igarashi RY, Somanchi SS, Zakari A, Solh M, et al. Natural killer cells stimulated with PM21 particles expand and biodistribute in vivo:

clinical implications for cancer treatment. *Cytotherapy* (2016) 18:653–63. doi:10.1016/j.jcyt.2016.02.006
137. Escudier B, Farace F, Angevin E, Charpentier F, Nitenberg G, Triebel F, et al. Immunotherapy with interleukin-2 (IL2) and lymphokine-activated natural killer cells: improvement of clinical responses in metastatic renal cell carcinoma patients previously treated with IL2. *Eur J Cancer* (1994) 30A:1078–83. doi:10.1016/0959-8049(94)90460-X
138. Hercend T, Farace F, Baume D, Charpentier F, Droz JP, Triebel F, et al. Immunotherapy with lymphokine-activated natural killer cells and recombinant interleukin-2: a feasibility trial in metastatic renal cell carcinoma. *J Biol Response Mod* (1990) 9:546–55.
139. Berg M, Lundqvist A, McCoy P Jr, Samsel L, Fan Y, Tawab A, et al. Clinical-grade ex vivo-expanded human natural killer cells up-regulate activating receptors and death receptor ligands and have enhanced cytolytic activity against tumor cells. *Cytotherapy* (2009) 11:341–55. doi:10.1080/14653240902807034
140. Park KU, Jin P, Sabatino M, Feng J, Civini S, Khuu H, et al. Gene expression analysis of ex vivo expanded and freshly isolated NK cells from cancer patients. *J Immunother* (2010) 33:945–55. doi:10.1097/CJI.0b013e3181f71b81
141. Childs RW, Berg M. Bringing natural killer cells to the clinic: ex vivo manipulation. *Hematology Am Soc Hematol Educ Program* (2013) 2013:234–46. doi:10.1182/asheducation-2013.1.234
142. Granzin M, Soltenborn S, Müller S, Kollet J, Berg M, Cerwenka A, et al. Fully automated expansion and activation of clinical-grade natural killer cells for adoptive immunotherapy. *Cytotherapy* (2015) 17:621–32. doi:10.1016/j.jcyt.2015.03.611
143. North J, Bakhsh I, Marden C, Pittman H, Addison E, Navarrete C, et al. Tumor-primed human natural killer cells lyse NK-resistant tumor targets: evidence of a two-stage process in resting NK cell activation. *J Immunol* (2007) 178:85–94. doi:10.4049/jimmunol.178.1.85
144. Kottaridis PD, North J, Tsirogianni M, Marden C, Samuel ER, Jide-Banwo S, et al. Two-stage priming of allogeneic natural killer cells for the treatment of patients with acute myeloid leukemia: a phase I trial. *PLoS One* (2015) 10:e0123416. doi:10.1371/journal.pone.0123416
145. Peng B-G, Liang L-J, He Q, Huang J-F, Lu M-D. Expansion and activation of natural killer cells from PBMC for immunotherapy of hepatocellular carcinoma. *World J Gastroenterol* (2004) 10:2119–23. doi:10.3748/wjg.v10.i14.2119
146. Harada H, Watanabe S, Saijo K, Ishiwata I, Ohno T. A Wilms tumor cell line, HFWT, can greatly stimulate proliferation of CD56+ human natural killer cells and their novel precursors in blood mononuclear cells. *Exp Hematol* (2004) 32:614–21. doi:10.1016/j.exphem.2004.03.011
147. Fujisaki H, Kakuda H, Imai C, Mullighan CG, Campana D. Replicative potential of human natural killer cells. *Br J Haematol* (2009) 145:606–13. doi:10.1111/j.1365-2141.2009.07667.x
148. Cho D, Shook DR, Shimasaki N, Chang Y-H, Fujisaki H, Campana D. Cytotoxicity of activated natural killer cells against pediatric solid tumors. *Clin Cancer Res* (2010) 16:3901–9. doi:10.1158/1078-0432.CCR-10-0735
149. Voskens CJ, Watanabe R, Rollins S, Campana D, Hasumi K, Mann DL. Ex-vivo expanded human NK cells express activating receptors that mediate cytotoxicity of allogeneic and autologous cancer cell lines by direct recognition and antibody directed cellular cytotoxicity. *J Exp Clin Cancer Res* (2010) 29:134. doi:10.1186/1756-9966-29-134
150. Lee DA. The off-target effects of nonspecific NK cells. *Blood* (2015) 125:744–5. doi:10.1182/blood-2014-12-616359
151. Wang X, Lee DA, Wang Y, Wang L, Yao Y, Lin Z, et al. Membrane-bound interleukin-21 and CD137 ligand induce functional human natural killer cells from peripheral blood mononuclear cells through STAT-3 activation. *Clin Exp Immunol* (2013) 172:104–12. doi:10.1111/cei.12034
152. Hercend T, Meuer S, Reinherz EL, Schlossman SF, Ritz J. Generation of a cloned NK cell line derived from the "null cell" fraction of human peripheral blood. *J Immunol* (1982) 129:1299–305.
153. Lee DA. Regulatory considerations for NK cells used in human immunotherapy applications. *Methods Mol Biol* (2016) 1441:347–61. doi:10.1007/978-1-4939-3684-7_29
154. Sabry M, Tsirogianni M, Bakhsh IA, North J, Sivakumaran J, Giannopoulos K, et al. Leukemic priming of resting NK cells is killer Ig-like receptor independent but requires CD15-mediated CD2 ligation and natural cytotoxicity receptors. *J Immunol* (2011) 187:6227–34. doi:10.4049/jimmunol.1101640
155. Chabannon C, Mfarrej B, Guia S, Ugolini S, Devillier R, Blaise D, et al. Manufacturing natural killer cells as medicinal products. *Front Immunol* (2016) 7:504. doi:10.3389/fimmu.2016.00504
156. Lapteva N, Szmania SM, van Rhee F, Rooney CM. Clinical grade purification and expansion of natural killer cells. *Crit Rev Oncog* (2014) 19:121–32. doi:10.1615/CritRevOncog.2014010931
157. Kalos M, June CH. Adoptive T cell transfer for cancer immunotherapy in the era of synthetic biology. *Immunity* (2013) 39:49–60. doi:10.1016/j.immuni.2013.07.002
158. Rafiq QA, Thomas RJ. The evolving role of automation in process development and manufacture of cell and gene-based therapies. *Cell Gene Ther Insights* (2016) 2:473–9. doi:10.18609/cgti.2016.058
159. Pierson BA, Europa AF, Hu WS, Miller JS. Production of human natural killer cells for adoptive immunotherapy using a computer-controlled stirred-tank bioreactor. *J Hematother* (1996) 5:475–83. doi:10.1089/scd.1.1996.5.475
160. Spanholtz J, Preijers F, Tordoir M, Trilsbeek C, Paardekooper J, de Witte T, et al. Clinical-grade generation of active NK cells from cord blood hematopoietic progenitor cells for immunotherapy using a closed-system culture process. *PLoS One* (2011) 6:e20740. doi:10.1371/journal.pone.0020740
161. Apel M, Brüning M, Granzin M, Essl M, Stuth J, Blaschke J, et al. Integrated clinical scale manufacturing system for cellular products derived by magnetic cell separation, centrifugation and cell culture. *Chemie-Ingenieur-Technik* (2013) 85:103–10. doi:10.1002/cite.201200175
162. Schüle S, Renner M, Longhurst S, Narayanan G. Regulatory requirements for clinical trial and marketing authorisation application for gene therapy medicinal products. *Bundesgesundheitsblatt Gesundheitsforschung Gesundheitsschutz* (2010) 53:30–7. doi:10.1007/s00103-009-0988-0
163. Caligiuri MA, Zmuidzinas A, Manley TJ, Levine H, Smith KA, Ritz J. Functional consequences of interleukin 2 receptor expression on resting human lymphocytes. Identification of a novel natural killer cell subset with high affinity receptors. *J Exp Med* (1990) 171:1509–26. doi:10.1084/jem.171.5.1509
164. Romee R, Foley B, Lenvik T, Wang Y, Zhang B, Ankarlo D, et al. NK cell CD16 surface expression and function is regulated by a disintegrin and metalloprotease-17 (ADAM17). *Blood* (2013) 121:3599–608. doi:10.1182/blood-2012-04-425397
165. Peruzzi G, Femnou L, Gil-Krzewska A, Borrego F, Weck J, Krzewski K, et al. Membrane-type 6 matrix metalloproteinase regulates the activation-induced downmodulation of CD16 in human primary NK cells. *J Immunol* (2013) 191:1883–94. doi:10.4049/jimmunol.1300313
166. Lajoie L, Congy-Jolivet N, Bolzec A, Gouilleux-Gruart V, Sicard E, Sung HC, et al. ADAM17-mediated shedding of FcγRIIIA on human NK cells: identification of the cleavage site and relationship with activation. *J Immunol* (2014) 192:741–51. doi:10.4049/jimmunol.1301024
167. Wiernik A, Foley B, Zhang B, Verneris MR, Warlick E, Gleason MK, et al. Targeting natural killer cells to acute myeloid leukemia in vitro with a CD16 x 33 bispecific killer cell engager and ADAM17 inhibition. *Clin Cancer Res* (2013) 19:3844–55. doi:10.1158/1078-0432.CCR-13-0505
168. Fauriat C, Just-Landi S, Mallet F, Arnoulet C, Sainty D, Olive D, et al. Deficient expression of NCR in NK cells from acute myeloid leukemia: evolution during leukemia treatment and impact of leukemia cells in NCRdull phenotype induction. *Blood* (2007) 109:323–30. doi:10.1182/blood-2005-08-027979
169. Le Maux Chansac B, Moretta A, Vergnon I, Opolon P, Lécluse Y, Grunenwald D, et al. NK cells infiltrating a MHC class I-deficient lung adenocarcinoma display impaired cytotoxic activity toward autologous tumor cells associated with altered NK cell-triggering receptors. *J Immunol* (2005) 175:5790–8. doi:10.1182/blood-2005-08-027979
170. Carlsten M, Norell H, Bryceson YT, Poschke I, Schedvins K, Ljunggren H-G, et al. Primary human tumor cells expressing CD155 impair tumor targeting by down-regulating DNAM-1 on NK cells. *J Immunol* (2009) 183:4921–30. doi:10.4049/jimmunol.0901226
171. Doubrovina ES, Doubrovin MM, Vider E, Sisson RB, O'Reilly RJ, Dupont B, et al. Evasion from NK cell immunity by MHC class I chain-related molecules expressing colon adenocarcinoma. *J Immunol* (2003) 171:6891–9. doi:10.4049/jimmunol.171.12.6891

172. Hilpert J, Grosse-Hovest L, Grünebach F, Buechele C, Nuebling T, Raum T, et al. Comprehensive analysis of NKG2D ligand expression and release in leukemia: implications for NKG2D-mediated NK cell responses. *J Immunol* (2012) 189:1360–71. doi:10.4049/jimmunol.1200796
173. Reiners KS, Topolar D, Henke A, Simhadri VR, Kessler J, Sauer M, et al. Soluble ligands for NK cell receptors promote evasion of chronic lymphocytic leukemia cells from NK cell anti-tumor activity. *Blood* (2013) 121:3658–65. doi:10.1182/blood-2013-01-476606
174. Schlecker E, Fiegler N, Arnold A, Altevogt P, Rose-John S, Moldenhauer G, et al. Metalloprotease-mediated tumor cell shedding of B7-H6, the ligand of the natural killer cell-activating receptor NKp30. *Cancer Res* (2014) 74:3429–40. doi:10.1158/0008-5472.CAN-13-3017
175. Ashiru O, Boutet P, Fernández-Messina L, Agüera-González S, Skepper JN, Valés-Gómez M, et al. Natural killer cell cytotoxicity is suppressed by exposure to the human NKG2D ligand MICA*008 that is shed by tumor cells in exosomes. *Cancer Res* (2010) 70:481–9. doi:10.1158/0008-5472.CAN-09-1688
176. Fernández-Messina L, Ashiru O, Boutet P, Agüera-González S, Skepper JN, Reyburn HT, et al. Differential mechanisms of shedding of the glycosylphosphatidylinositol (GPI)-anchored NKG2D ligands. *J Biol Chem* (2010) 285:8543–51. doi:10.1074/jbc.M109.045906
177. Kloess S, Huenecke S, Piechulek D, Esser R, Koch J, Brehm C, et al. IL-2-activated haploidentical NK cells restore NKG2D-mediated NK-cell cytotoxicity in neuroblastoma patients by scavenging of plasma MICA. *Eur J Immunol* (2010) 40:3255–67. doi:10.1002/eji.201040568

16

Elusive Role of the CD94/NKG2C NK Cell Receptor in the Response to Cytomegalovirus: Novel Experimental Observations in a Reporter Cell System

Aldi Pupuleku[1], Marcel Costa-García[1], Domènec Farré[2], Hartmut Hengel[3,4], Ana Angulo[2,5], Aura Muntasell[6†] and Miguel López-Botet[1,6†]*

[1] Department of Experimental and Health Sciences, University Pompeu Fabra, Barcelona, Spain, [2] Immunology Unit, Department of Biomedical Sciences, Medical School, University of Barcelona, Barcelona, Spain, [3] Institute of Virology, Albert Ludwigs University of Freiburg, Freiburg, Germany, [4] Faculty of Medicine, Albert Ludwigs University of Freiburg, Freiburg, Germany, [5] Institut d'Investigacions Biomèdiques August Pi i Sunyer, Barcelona, Spain, [6] Hospital del Mar Medical Research Institute (IMIM), Barcelona, Spain

Edited by:
Ulrike Koehl,
Hannover Medical School,
Germany

Reviewed by:
Christine Susanne Falk,
Hannover Medical School,
Germany
Federico Simonetta,
Stanford University,
United States

Correspondence:
Miguel López-Botet
lbotet@imim.es,
miguel.lopez-botet@upf.edu

†These authors share credit for senior authorship.

Human cytomegalovirus (HCMV) infection promotes the differentiation and persistent expansion of a mature NK cell subset, which displays high surface levels of the activating CD94/NKG2C NK cell receptor, together with additional distinctive phenotypic and functional features. The mechanisms underlying the development of adaptive NK cells remain uncertain but some observations support the involvement of a cognate interaction of CD94/NKG2C with ligand(s) displayed by HCMV-infected cells. To approach this issue, the heterodimer and its adaptor (DAP12) were expressed in the human Jurkat leukemia T cell line; signaling was detected by transfection of a reporter plasmid encoding for Luciferase (Luc) under NFAT/AP1-dependent control. Engagement of the receptor by solid-phase bound CD94- or NKG2C-specific monoclonal antibodies (mAbs) triggered Luc expression. Moreover, reporter activation was detectable upon interaction with HLA-E+ 721.221 (.221-AEH) cells, as well as with 721.221 cells incubated with synthetic peptides, which stabilized surface expression of endogenous HLA-E; the response was specifically antagonized by soluble NKG2C- and HLA-E-specific mAbs. By contrast, activation of Jurkat-NKG2C+ was undetectable upon interaction with Human Fetal Foreskin Fibroblasts (HFFF) infected with HCMV laboratory strains (i.e., AD169, Towne), regardless of their differential ability to preserve surface HLA-E expression. On the other hand, infection with two clinical isolates or with the endotheliotropic TB40/E strain triggered Jurkat-NKG2C+ activation; yet, this response was not inhibited by blocking mAbs and was independent of CD94/NKG2C expression. The results are discussed in the framework of previous observations supporting the hypothetical existence of specific ligand(s) for CD94/NKG2C in HCMV-infected cells.

Keywords: human, natural killer cell, cytomegalovirus, CD94, NKG2C, HLA-E, UL40

INTRODUCTION

Inhibitory receptors specific for MHC class I molecules with immunoreceptor tyrosine-based inhibitory motifs play a key role in preventing NK cell responses against normal autologous cells. This function is mainly fulfilled by members of the human killer-cell immunoglobulin-like receptor (KIR) family, which recognize sets of classical HLA class I (HLA-I) molecules, and by the CD94/NKG2A lectin-like heterodimer specific for HLA-E. Conversely, other KIRs and CD94/NKG2C, which display a lower affinity for HLA-I ligands trigger protein tyrosine kinase pathways through DAP12, an adaptor with immunoreceptor tyrosine-based activation motifs. Similar inhibitory and activating NK cell receptors (NKR) have been identified among the murine Ly49 and NKG2 lectin-like receptor families (1, 2). The hypothesis that MHC-specific activating NKR may contribute to the innate response against pathogens was supported by the evidence that Ly49H specifically interacts with the MHC class I-related murine cytomegalovirus glycoprotein m157, triggering NK cell effector functions and the development of a memory-like response that confers resistance against the viral infection in some mice strains (3–5).

With this remarkable exception, no formal proof has been thus far obtained supporting the involvement of other activating KIR, NKG2, or Ly49 receptors in direct recognition of pathogen molecules (6). In this regard, human cytomegalovirus (HCMV) infection has been shown to promote the differentiation and persistent expansion of a mature NK cell subset, which displays high surface levels of the activating CD94/NKG2C NKR ($NKG2C^{bright}$), together with additional distinctive phenotypic and functional features (7–12). The magnitude of such adaptive NK cell subset redistribution appears variable in healthy blood donors, being undetectable in some HCMV+ individuals. This NK cell response pattern has been as well observed following active HCMV infection in newborns and immunocompromised patients (13–17). Although expansions of NKG2C+ NK cells have been reported in the context of other infections (18–21), the effect appears restricted to individuals coinfected by HCMV, thus suggesting that it is specifically induced by this herpes virus, being potentially amplified along the immune response to other pathogens.

The mechanisms underlying differentiation and expansion of $NKG2C^{bright}$ NK cells remain uncertain. Engagement of CD94/NKG2C by specific monoclonal antibodies (mAbs) or HLA-E, expressed in the 721.221 (.221) HLA-I defective cell line, triggered NKG2C+ NK-cell effector functions and proliferation in response to IL-2 or IL-15, strongly suggesting that the receptor might play a direct role in the response to HCMV infection (22, 23). *In vitro* proliferation of NKG2C+ cells was observed coculturing PBMCs or purified NK cells from some HCMV+ donors with HCMV-infected fibroblasts. The response required the participation of cytokines (i.e., IL-12, IL-15) and was antagonized by anti-CD94 (22), -NKG2C, or -HLA-E mAbs (23). These observations supported the hypothesis of an instructive process driven by a cognate interaction of the CD94/NKG2C receptor with ligand(s) displayed by HCMV-infected cells (24). Paradoxically, no formal evidence has been obtained supporting an active role of the CD94/NKG2C receptor in triggering *in vitro* NK cell effector functions against HCMV-infected cells, suggesting that NKG2C-mediated NK cell activation might be hampered by viral immune evasion mechanism(s) (25). By contrast, antibody-dependent stimulation *via* CD16 (FcγR-IIIA) efficiently activates adaptive NKG2C+ NK cells to mediate specific cytotoxicity, cytokine production, and proliferation in response to HCMV- and other virus-infected cells (26–29). CD2 has been shown to play an important co-stimulatory role in antibody-dependent activation of NKG2C+ cells (30, 31). Recently, increased baseline proportions of adaptive NKG2C+ NK cells in kidney transplant recipients have been directly related with a reduced incidence of posttransplant HCMV infection (32), suggesting that they may play a role in antiviral defense, involving CD94/NKG2C and/or CD16-dependent activation (33).

Previous reports revealed that binding of HLA-E to a peptide from the HCMV UL40 leader sequence preserves its expression in infected cells, engaging the CD94/NKG2A inhibitory receptor (34, 35). On the other hand, viral MHC class I-modulating molecules (i.e., US2-US11) were shown to play a prevalent role in governing the response of NK cells against infected targets (36).

In the present study, we approached the identification of putative ligand(s) for CD94/NKG2C in HCMV-infected cells, reducing the complexity of NK cell-infected target interactions. To this end, both receptor subunits and DAP12 were stably expressed in the human Jurkat leukemia T cell line. Signaling was detected by transient transfection of a reporter plasmid encoding for Luciferase (Luc) under NFAT/AP1-dependent control. Our results are discussed in the hypothetical framework on the development of adaptive NKG2C+ cells in response to HCMV.

MATERIALS AND METHODS

mAbs and Flow Cytometry Analysis

Flow cytometry was performed using mAbs specific for the following surface molecules: anti-NKG2C-PE (clone 134591) R&D Systems (Minneapolis, MN, USA), anti-HLA-I-APC (clone HP-1F7) generated in our laboratory and conjugated by Immunostep (Salamanca, Spain). The following indirect antibodies were used as purified or culture supernatants: anti-HLA-E (clone 3D12) provided by Dr. D. E. Geraghty (Fred Hutchinson Cancer Research Centre, Seattle, WA, USA), anti-CD3 (clone SpvT3B); anti-NKG2A (clone Z199), anti-NKG2D (clone BAT221), anti-NKp46 (clone Bab281), anti-NKp30 (clone AZ20), anti-DNAM1 (clone F22), anti-CD16 (KD1) provided by Dr. A. Moretta (University of Genova), and Dr. D. Pende (National Institute for Cancer Research, Genova); anti-LFA1 (clone TS/18), anti-ICAM1 (clone HU5/3) provided by Dr. F. Sánchez-Madrid (Hospital Univ. de la Princesa, Madrid); anti-KIR3DL1 (clone DX9) provided by Dr. L. Lanier (University of California San Francisco, CA, USA); anti-KIR2DL2/S2/L3 (clone CH-L) provided by Dr. S. Ferrini (National Institute for Cancer Research, Genova, Italy); anti-KIR3DL1/3DL2/2DS4/2DS5/2DS2/3DS1 (clone 5.133), provided by Dr. M. Colonna (University of Saint

Louis, MO, USA). Anti-CD94 (clone HP-3B1), anti-ILT2 (LILRB1, LIR1) (clone HP-F1), anti-CD2 (clone MAR206), anti-KIR2DL1 (clone HP-DM1), anti-KIR2DL1/2DS1/2DS3/2DS5 (clone HP-MA4), anti-KIR2DL5 (clone UP-R1), and anti-KIR2DL1/S1/S4 (clone HP-3E4) were produced in our laboratory.

Briefly, cells were pretreated with human IgG (10 μg/ml) to block Fc receptors, incubated with individual NKR-specific mAbs for 30 min, washed, and further incubated with a secondary PE-tagged F(ab')2 rabbit anti-mouse Ig (The Jackson Immunoresearch, West Grove PA, USA); anti-myc mAb (9E10, IgG1) was used as negative control. Data were acquired on FACSCalibur flow cytometer (BD Biosciences) and processed using FlowJo software (TreeStar, OR, USA).

Cell Lines and Culture Conditions

The Jurkat leukemia T cell line and its transfectants were grown in RPMI-1640 medium (Gibco, Grand Island, New York, NY, USA) supplemented with 10% heat-inactivated Fetal Bovine Serum (Gibco), 100 U/ml penicillin, and 100 μg/ml streptomycin (Gibco), termed as complete medium. Jurkat-CD94+ cells were kindly provided by Dr. Lewis Lanier, obtained as previously described (37) and cultured in complete medium with G418 (1 mg/ml) (InvivoGen, San Diego, CA, USA).

The 721.221 (.221) HLA-I-deficient EBV-transformed B lymphoblastoid cell line and its transfectant .221-AEH (kindly provided by Dr. D. E. Geraghty, Fred Hutchinson Cancer Research Centre, Seattle, WA, USA) were cultured in RPMI-1640 complete medium .221-AEH cells were generated by stable transfection of .221 cells with a construct in which the leader sequence of the HLA-E*0101 allele was replaced by that of HLA-A2 and were selected in the presence of 300 μg/ml hygromycin B (Invitrogen, Carlsbad, CA, USA) (38).

Synthetic leader sequence peptides from HLA-G (VMAPRTLFL) or the AD169 UL40 viral protein (VMAPRTLIL) were purchased from CRG-UPF proteomic core facility (Parc de Recerca Biomèdica de Barcelona, Spain). As described (39), to stabilize HLA-E surface expression, HLA-Ia-defective 721.221 cells were incubated overnight with peptides (10 mM) at 26°C; HLA-E surface expression was monitored before and after incubation with peptides by flow cytometry.

Human Fetal Foreskin Fibroblast (HFFF) cells provided by Prof. John Trowsdale (University of Cambridge, UK), and the human lung fibroblast cell line MRC-5 provided by Dr. A. Angulo, were maintained in Dulbecco modified essential medium (DMEM) (Gibco) supplemented with 10% FBS, penicillin, and streptomycin.

Generation of Jurkat-NKG2C+ Reporter Cells

Jurkat-NKG2C+ cells were established using a retroviral expression system to stably express DAP12 and NKG2C proteins in Jurkat-CD94+ cells. DAP12 and NKG2C cDNA constructs were subcloned from pJFE14 expression vector to pBABE-puro retroviral vector using XbaI-EcoRI (DAP12) and BamHI-EcoRI (NKG2C) restriction sites. As described (40), the retroviral constructs were individually transfected using the non-modified polyethyleneimine reagent (PEI, Sigma-Aldrich, St. Louis, MO, USA) into the helper-virus free amphotropic producer cell line Phoenix-A, a derivative of the human embryonic kidney cell line 293T (provided by Dr. Ramon Gimeno, IMIM, Barcelona, Spain). At 48, 72, and 96 h post-transfection, supernatants containing retroviral particles of DAP12 and NKG2C were collected, filtered with 45 μm filter (Millipore, Billerica, MA, USA), and centrifuged with Beckman SW28 rotor at 25,000 rpm for 90 min at 4°C. Pelleted virus were resuspended in 1 ml of RPMI medium and used to transduce Jurkat-CD94+ cells. To this end, 500,000 cells were plated (48 well/plates) in 1 ml mixed concentrated retroviral medium (0.5 ml DAP12 and 0.5 ml NKG2C) in the presence of 8 μg/ml polybrene (Sigma-Aldrich) and spinned for 90 min at 930 *g*. Cells were supplemented with fresh medium at 6 h post-transduction and selected with 1.5 μg/ml puromycin (Sigma-Aldrich) for 48 h; subsequently, cells positive for NKG2C surface expression were sorted (Influx Cell sorter, BD Bioscience), cloned by limiting dilution and expanded.

HCMV Preparations and Infection of Human Fetal Foreskin Fibroblasts (HFFF)

This work was carried out in an authorized UPF p2-level biohazard facility, in compliance with the official requirements for CMV manipulation. Stocks of concentrated HCMV strains AD169 and Towne, provided by Dr. A. Angulo, TB40/E, provided by Dr. C. Sinzger (Institute for Medical Virology, University of Tübingen), and two HCMV clinical isolates: UL1271 (41) and #119, provided by Dr. H. Hengel, were prepared as follows. Almost confluent MRC-5 fibroblasts were infected at low multiplicity of infection (MOI) and supernatants were recovered when maximum cytopathic effect was reached (7–10 days) followed by clearing of cellular debris by centrifugation at 1,750 *g* for 10 min (42). Thereafter, the virus was concentrated for 3 h by centrifugation at 29,000 *g* at 15°C. Pelleted virus was resuspended in serum-free Dulbecco medium, stored at −80°C, and titrated by standard plaque assays.

Human fetal foreskin fibroblasts were seeded in 48-well plates 2 days prior infection at 4×10^4 cells/well. Confluent cells were incubated alone (mock) or with different viral strains at MOI of 10. After 2 h of absorption at 37°C, cells were washed twice with PBS and then fresh DMEM medium was added. Depending on the experimental design, HFFF cells were washed again at 24, 48, or 72 h postinfection. The infection rate was assessed by monitoring expression of the IE1 protein by indirect immunofluorescence with mAb MAB810R (clone 8B1.2) (Millipore) and Alexa Fluor 488-Labeled F(ab')$_2$ goat anti-mouse secondary (Invitrogen, Carlsbad, CA, USA). Alternatively, infected cells were indirectly identified assessing down-modulationof surface HLA-I expression at 72 h postinfection.

Preparation and Activation of CD94-NKG2C+ Reporter Cells

Human cytomegalovirus-infected HFFFs were cocultured with Jurkat-NKG2C+ cells previously transfected with a reporter plasmid encoding Luciferase (Luc) under the control of NFAT/AP1 promoter (3X NFAT/AP1-Luc) generated as described (43) and provided by Dr. Jose Aramburu (Universitat Pompeu Fabra, Barcelona, Spain). Transfection of Jurkat cells was carried out using the Neon Transfection System (Thermo Fisher Scientific, Waltham, MA, USA) following the protocol provided by the manufacturer. Luc-transfected Jurkat cells were cocultured with the different targets for 18–24 h. After coculture, cells were collected, lysed, and Luc activity was measured using Promega Luciferase Assay system (Promega, Madison, WI, USA). The data were normalized referring the specific luminescence counts to those of non-treated (NT) Jurkat-NKG2C+ cells and are represented as fold-change induction. Stimulation of Jurkat with plate-bound anti-CD3, -CD94, -NKG2C, or co-culture with 721.221 and .221-AEH cells were used as controls. To verify the involvement of NFAT/AP1 in the reporter activation, experiments were carried out in the presence of 1 μM FK506 calcineurin inhibitor (Sigma-Aldrich), pretreating Jurkat-NKG2C+ cells with the drug for 2 h. For antibody-blocking assays, Jurkat-NKG2C+ cells or infected HFFF cells were, respectively, preincubated for 2 h with anti-NKG2C (clone 134522, R&D Systems) or anti-HLA-E (clone 3D12) mAbs (5 μg/ml) prior to coculture; anti-CD94 F(ab')$_2$ fragments obtained from the HP-3B1 clone (37) were employed in some experiments.

Statistical Analysis

Jurkat-NKG2C+ cell activation, assessed by induction of Luc activity in response to control stimuli (i.e., anti-NKG2C or .221-AEH cells) in different experiments ($n = 15$), was verified to follow a normal distribution applying the conventional Shapiro–Wilk test. Statistical analysis of the results was carried out applying the Student's *t*-test.

RESULTS

Generation and Phenotypic Characterization of a Human CD94/NKG2C+ Reporter T Cell Line

To study the role of the CD94/NKG2C NKR in recognition of HCMV-infected cells, a reporter cell system was developed expressing the receptor segregated from other NKR. For this purpose, NKG2C and DAP12 were stably transduced in an available CD94-transfected human Jurkat leukemia T cell line. Jurkat-NKG2C+ cells were sorted and cloned by limiting dilution assessing the expression of adhesion molecules and NKR. A clone (97), which expressed CD94/NKG2C was selected (**Figure 1A**). These cells had downregulated CD3, displayed adhesion/co-stimulatory molecules (i.e., LFA-1, CD2, ICAM-1, and DNAM-1), but lacked activating (i.e., NKG2D, NKp46, NKp44, NKp30, CD16, aKIR) and inhibitory (i.e., iKIR, NKG2A, TIGIT, ILT2) NKR (**Figure 1B**). The marginal expression of ILT2 was particularly important as this HLA-I specific inhibitory receptor, which interacts with the UL18 HCMV molecules, was detectable in the parental Jurkat-CD94+ cells (not shown).

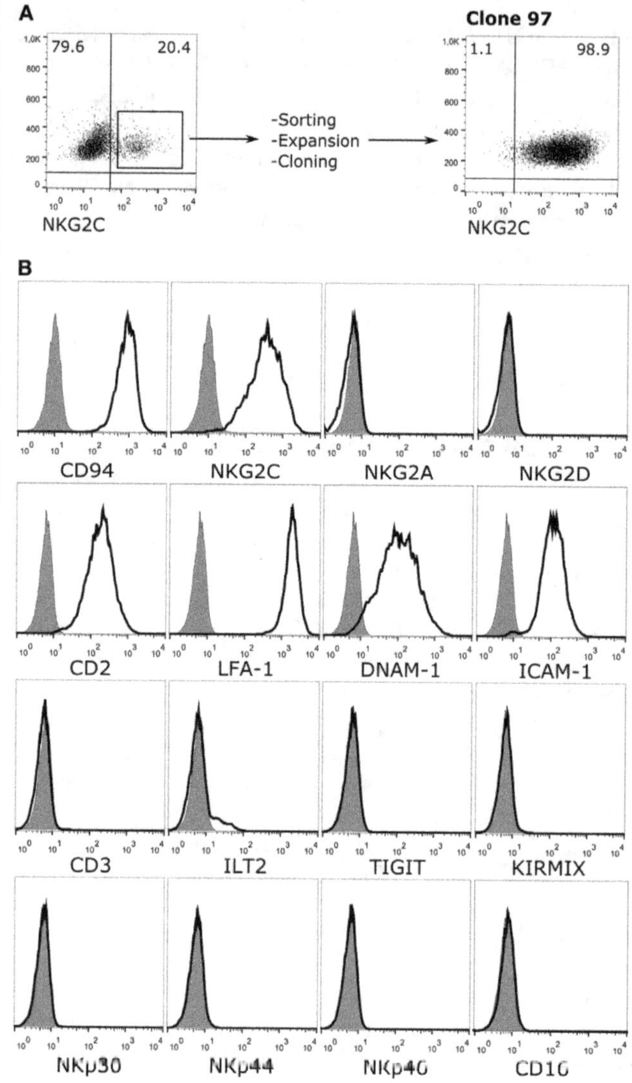

FIGURE 1 | Phenotypic characterization of Jurkat-NKG2C+ reporter cells. **(A)** Jurkat-CD94+ cells were transduced with retroviral particles carrying the sequences encoding for NKG2C and DAP12 as described in Section "Materials and Methods." After sorting, expansion, and limiting dilution, clone 97, which expressed high levels of the CD94/NKG2C heterodimer was selected. **(B)** Jurkat-NKG2C+ cells were stained for the indicated receptors and analyzed by flow cytometry. KIRMIX corresponds to a mixture of the following monoclonal antibodies: HP-MA4 (KIR2DL1/2DS1/2DS3/2DS5), CHL (KIR2DL2/2DL3/2DS2), 5.133 (KIR3DL1/3DL2/2DS4/2DS5/2DS2/3DS1), DX9 (KIR3DL1), UP-R1 (KIR2DL5). Gray histograms correspond to the isotype control.

Specific Recognition of HLA-E by Jurkat-NKG2C+ Reporter Cells

In order to detect signaling by the CD94/NKG2C-DAP12 complex, Jurkat-NKG2C+ cells were transiently transfected with a plasmid encoding for Luciferase (Luc) under the control of NFAT/AP1-dependent promoter. Engagement of the receptor by solid-phase bound CD94- or NKG2C-specific mAbs triggered Luc expression (**Figure 2A**). Of note, a slightly higher background in Jurkat-NKG2C+ cells compared to the parental cell line suggested that the low constitutive HLA-E expression in Jurkat-NKG2C+ cells promoted a limited self-activation of the reporter (Figure S1 in Supplementary material). However, coculture of Jurkat-NKG2C+ cells with the .221-AEH cell line,

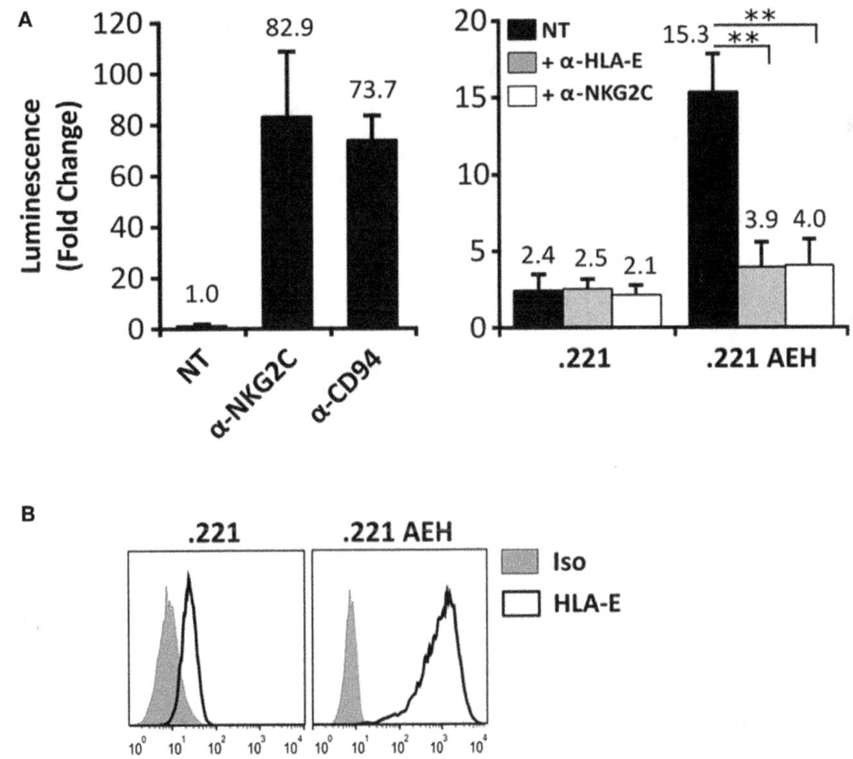

FIGURE 2 | Specific engagement of the CD94/NKG2C receptor by HLA-E activates the Jurkat-NKG2C+ reporter. **(A)** Jurkat-NKG2C+ cells were electroporated with 3x NFAT/AP1-Luc plasmid, followed by stimulation with: (a) anti-NKG2C or anti-CD94 monoclonal antibodies (mAbs) pre-adsorbed to culture plates or (b) with .221 or .221-AEH cells. After 18–24 h, cells were collected, lysed, and Luc activity was measured. Blocking experiments were carried out pretreating .221 and .221-AEH cells with anti-HLA-E 3D12 mAb (gray bars) or Jurkat-NKG2C+ cells with anti-NKG2C mAb (white bars); NT, not treated. Data correspond to three independent experiments (mean ± SD). **(B)** HLA-E expression by .221 and .221-AEH was assessed prior to coculture with Jurkat-NKG2C+ cells (gray histograms represent the isotype control). Statistically significant differences are indicated (**$p < 0.01$).

which displays HLA-E in the absence of classical HLA-I, induced Luc expression and the response was specifically antagonized by soluble NKG2C- and HLA-E-specific mAbs (**Figures 2A,B**). Similar results were obtained stimulating Jurkat-NKG2C+ cells with the .221 cell line preincubated with synthetic HLA-I leader sequence peptides, known to stabilize surface HLA-E expression (**Figures 3A,B**). These results validated the sensitivity and specificity of the reporter system to detect CD94/NKG2C receptor–ligand interaction.

Response of Jurkat-NKG2C+ Reporter Cells to Fibroblasts Infected by HCMV Laboratory Strains

On that ground, experiments were carried out incubating Jurkat-NKG2C+ cells with HFFF, infected at different time-points (24–72 h) with HCMV laboratory strains (AD169 and Towne). As compared to positive controls stimulated with anti-NKG2C mAb or .221-AEH cells, no differences were perceived comparing Luc expression in Jurkat-NKG2C+ cocultured either with mock- or HCMV-infected HFFF (**Figure 4A**); similar results were obtained incubating the reporter with MRC5 fibroblasts (not shown). Total HLA-I expression was downregulated in cells infected by both HCMV strains. By contrast, HLA-E was preserved by AD169, which displays a canonical UL40 leader peptide (VMAPRTLIL) binding to the HLA class Ib molecule, but was lost in Towne-infected cells (**Figure 4B**). As reported (44) and according to the annotated GenBank sequences, this HCMV strain contains a deletion spanning the UL40 leader sequence. These results point out that the undetectable response of Jurkat-NKG2C+ to fibroblasts infected by laboratory strains was unrelated to their ability to sustain surface HLA-E expression.

Response of Jurkat-NKG2C+ Reporter Cells to Fibroblasts Infected by HCMV Clinical Isolates

TB40/E HCMV strain retains the ability to replicate in endothelial and myelomonocytic cells, differing from AD169 and Towne (45). Moreover, it is well established that HCMV clinical isolates substantially diverge from laboratory strains, rapidly undergoing important genomic changes upon *in vitro* passage, as reviewed in Ref. (46). Thus, additional experiments were carried out coculturing Jurkat-NKG2C+ cells with HFFF infected with TB40/E, which encodes for a UL40 leader peptide variant unable to preserve HLA-E surface expression (VVAPRTLIL) (25), or with two HCMV clinical isolates

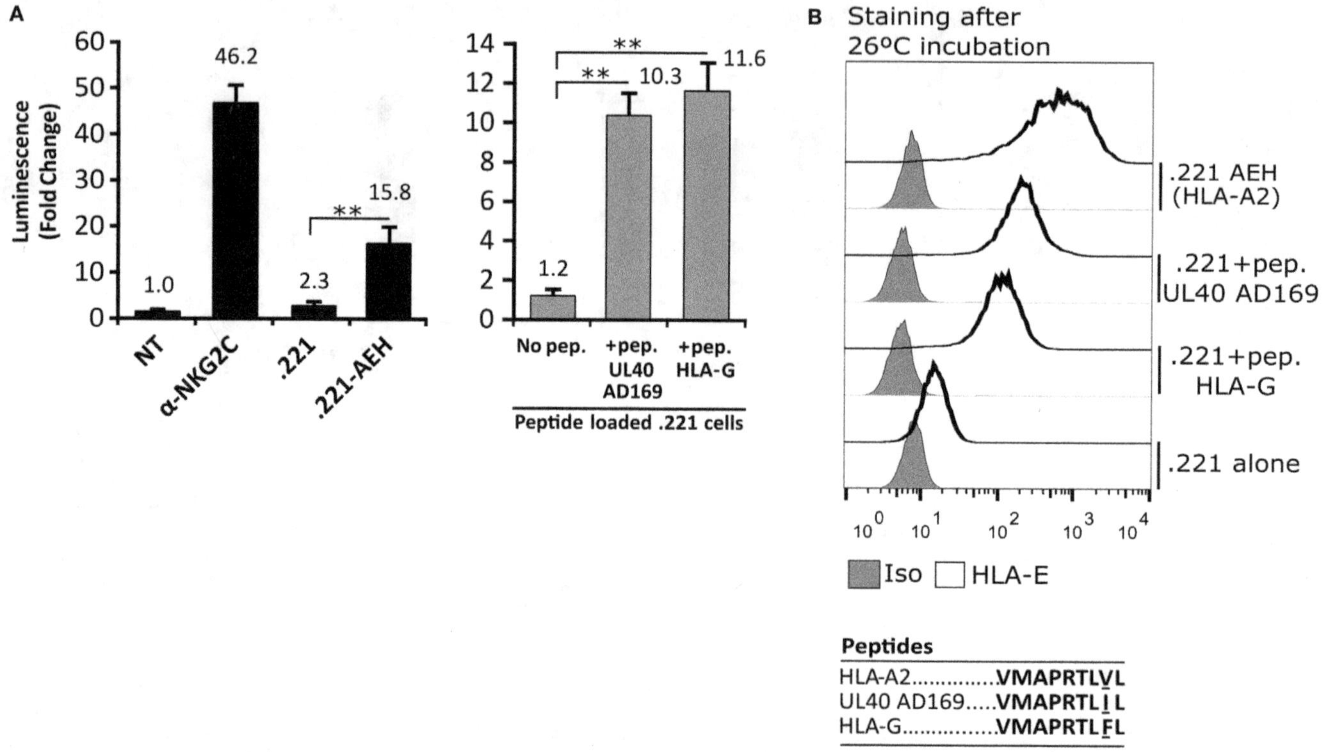

FIGURE 3 | Activation of Jurkat-NKG2C+ cells by HLA-E loaded with the canonical human cytomegalovirus UL40 leader sequence-derived synthetic peptide on 721.221 cells. **(A)** As described for **Figure 2**, Jurkat-NKG2C+ cells were: (a) untreated (NT), (b) stimulated with plate coated anti-NKG2C, or (c) cocultured with the .221 or .221-AEH cells (black bars). In parallel, the response of Jurkat-NKG2C+ cells was assessed coculturing them with .221 cells preincubated at 26°C overnight with synthetic peptides corresponding to HLA-G (VMAPRTLFL) and AD169 UL40 (VMAPRTLIL) leader sequences, reported to stabilize HLA-E expression (gray bars). Data correspond to three independent experiments (mean ± SD). **(B)** HLA-E expression by .221-AEH and .221 cells preincubated with peptides prior to coculture with Jurkat-NKG2C+ cells; gray histograms represent the isotype control. Statistically significant differences are indicated (**$p < 0.01$).

(#UL1271 and #119), which share the canonical UL40-nonamer and preserved HLA-E expression (**Figures 5A,B**). Remarkably, a specific response of the reporter was detected upon infection with these viruses, which appeared greater for #119 and gradually increased along postinfection time. Yet, controls indicated that Luc expression in response to HFFF infected with HCMV clinical isolates was not inhibited by pretreatment with anti-NKG2C or -HLA-E mAbs (**Figures 6A,B**); similar results were obtained with anti-CD94 F(ab′)$_2$ fragments (data not shown), ruling out the potential influence of an overlapping agonistic effect mediated by anti-NKG2C bound to virus-encoded FcR. Moreover, the effect was also induced in CD94+ and wild-type Jurkat cells, which lacked NKG2C, unequivocally revealing that reporter activation was independent of this NKR (**Figure 6C**); similar results were obtained with TB40/E-infected cells (data not shown). As this response was dominant, the possibility that it might eventually mask a subtle involvement of CD94/NKG2C cannot be entirely ruled out.

DISCUSSION

A number of observations indirectly support the hypothesis that a specific and direct interaction of CD94/NKG2C with infected cells contributes to drive the adaptive NK cell response; yet, molecular evidence remains thus far elusive (10). To explore the presence of putative ligand(s) for CD94/NKG2C in HCMV-infected cells, the receptor was stably expressed along with DAP12 in the human Jurkat leukemia cell line, transiently transfected with an NFAT/AP1-dependent Luc-encoding reporter. The rationale for using a human parental cell line was based on our previous experience with heterologous CD94/NKG2C+ rat basophilic leukemia cells (RBL), which failed to respond to receptor engagement by HLA-E+ .221-AEH cells (not shown). A Jurkat-NKG2C+ clone that shared with NK cells key adhesion receptors (i.e., LFA-1, CD2, and DNAM1) but lacked inhibitory NKR, was selected.

The possibility that constitutive HLA-E expression by Jurkat-NKG2C+ cells might promote self-activation of the reporter, potentially impairing its responsiveness to external stimuli, was beforehand considered a potential drawback. Yet, Luc expression in Jurkat-NKG2C+ cells was induced following stimulation with CD94 or NKG2C-specific mAbs. Importantly, reporter activation was detectable upon interaction with HLA-E+ .221-AEH cells, as well as with .221 cells incubated with synthetic peptides which stabilized surface expression of endogenous HLA-E; in these settings, inhibition by anti-NKG2C and -HLA-E mAbs supported specific receptor-ligand engagement. By contrast, reporter activation was undetectable upon interaction of

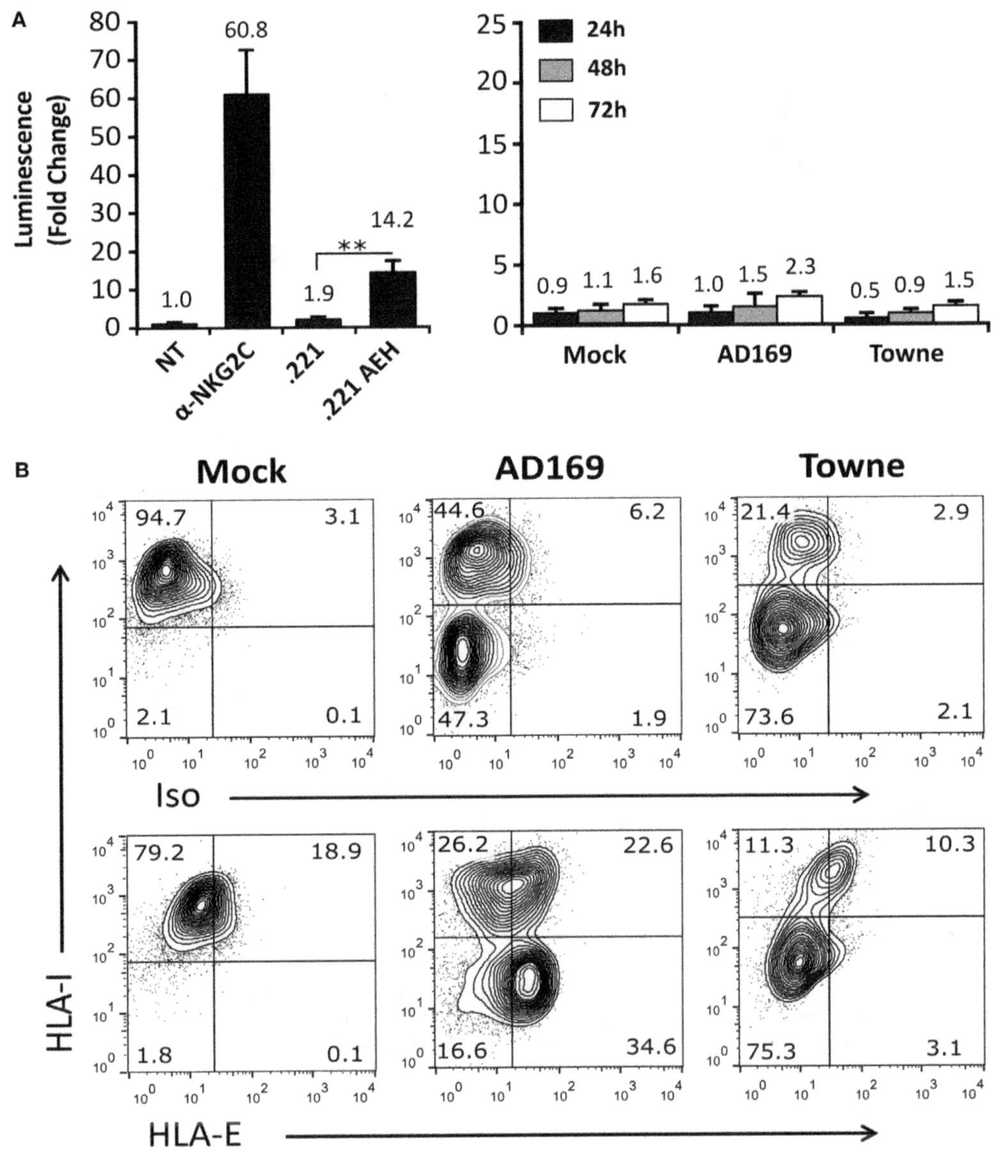

FIGURE 4 | Cells infected with AD169 or Towne human cytomegalovirus (HCMV) strains do not activate the Jurkat-NKG2C+ reporter independently of HLA-E expression. **(A)** HFFF cells infected with AD169 or Towne HCMV strains (multiplicity of infection = 10) were cocultured at 24, 48, and 72 h postinfection with Jurkat-NKG2C+ cells electroporated with the reporter plasmid. The average infection rate for each experiment was >50%. As a control, reporter activation following CD94/NKG2C receptor engagement by specific monoclonal antibodies or HLA-E was assessed in parallel, as described in **Figure 2**. Data correspond to three independent experiments (mean ± SD). **(B)** Total HLA-I and HLA-E expression at 72 h postinfection in mock-treated and HCMV-infected HFFF cells. Statistically significant differences are indicated (***p* < 0.01).

Jurkat-NKG2C+ cells with HFFF or MRC5 fibroblasts following infection with common HCMV laboratory strains (i.e., AD169, Towne), regardless of their ability to preserve surface HLA-E expression in infected cells.

Among variables that may potentially condition the expression of CD94/NKG2C ligand(s) by HCMV-infected cells, genomic differences between viral strains and changes associated to *in vitro* passage of the virus were considered potentially important. In fact, cells infected with two different clinical isolates and, to a lesser degree, with the endotheliotropic TB40/E strain, triggered Jurkat-NKG2C+ activation. Yet, this response was not inhibited by NKG2C- nor HLA-E-specific blocking mAbs and, furthermore, was independent of CD94/NKG2C expression. Experiments were carried out to explore the molecular basis of this dominant effect, which might potentially mask CD94/NKG2C-specific signaling. Pretreatment of Jurkat-NKG2C+ with a calcineurin inhibitor (FK506) hampered Luc expression triggered by anti NKG2C mAb or .221-AEH cells, but did not prevent the response of Jurkat-NKG2C+ cells to HCMV clinical isolates (data not shown), pointing out that a non-conventional activation of the reporter plasmid was induced under these experimental conditions. Further studies are required

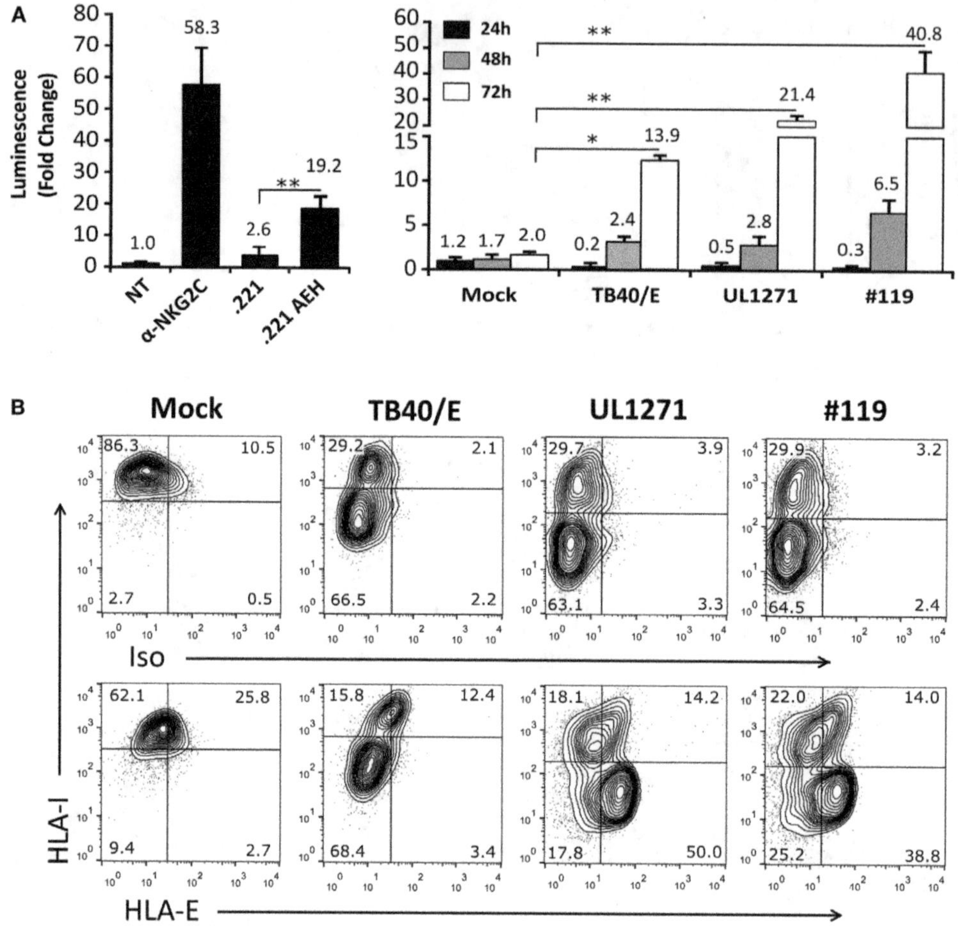

FIGURE 5 | Activation of Jurkat-NKG2C+ reporter cells by HFFF infected with human cytomegalovirus TB40/E or clinical isolates is independent of HLA-E expression. **(A)** The Jurkat-NKG2C+ reporter was cocultured with HFFF cells at different time-points (24–72 h) postinfection with TB40/E or two different clinical isolates (UL1271 and #119) (multiplicity of infection = 10). Data correspond to three independent experiments (mean ± SD). Average infection rate for each experiment was >50%. **(B)** Total HLA-I and HLA-E surface expression was monitored at 72 h postinfection. Statistically significant differences are indicated (*$p < 0.05$; **$p < 0.01$).

to assess the possibility of controlling this drawback, eventually re-designing the system.

NKG2C+ NK cells were reported to be activated upon interaction with primary aortic endothelial cells infected by the VHL/E HCMV strain (47); nevertheless, no data supporting an involvement of the receptor (e.g., blocking with anti-CD94 or -NKG2C mAbs) were provided. In our hands, Jurkat-NKG2C+ did not detectably respond to HCMV-infected MRC5 fibroblasts nor to human umbilical vein endothelial cells. Yet, the possibility that expression of putative CD94/NKG2C ligands might vary depending on the infected cell lineage is not ruled out.

Our observations should be discussed in the framework of previous indications supporting a CD94/NKG2C receptor interaction with infected cells. In this regard, NKG2C+ NK cells were shown to undergo *in vitro* expansion upon coculture with AD169 or Towne-infected fibroblasts (22, 23). The effect was dependent on cytokines (i.e., IL-12, IL-15) and was prevented by blocking anti-CD94, -NKG2C, or -HLA-E mAbs. The response was observed only in samples from some HCMV+ blood donors displaying baseline NKG2C^{bright} NK cell expansions, indicating that these experimental conditions promoted their proliferation, but did not strictly reproduce the differentiation process induced *in vivo* by primary HCMV infection. Of note, *in vitro* expansion was prevented upon infection by an HCMV BACmid mutant lacking the whole set of viral genes (i.e., US2-US11), which target HLA-I molecules, thus preserving their expression in infected cells (22). These experiments were interpreted as an indication of a relatively weak stimulation *via* CD94/NKG2C that was overridden by inhibitory KIR-HLA-I interactions which, by contrast, did not impair CD16-triggered NK cell effector functions against cells infected by the same virus (28).

On the other hand, a relation of NKG2C gene copy number with the magnitude of NKG2C+ NK cell expansion has been reported in HCMV+ healthy blood donors, in children with congenital infection and in renal transplant recipients (9, 14, 32, 48). Moreover, CD94/NKG2C surface expression levels and

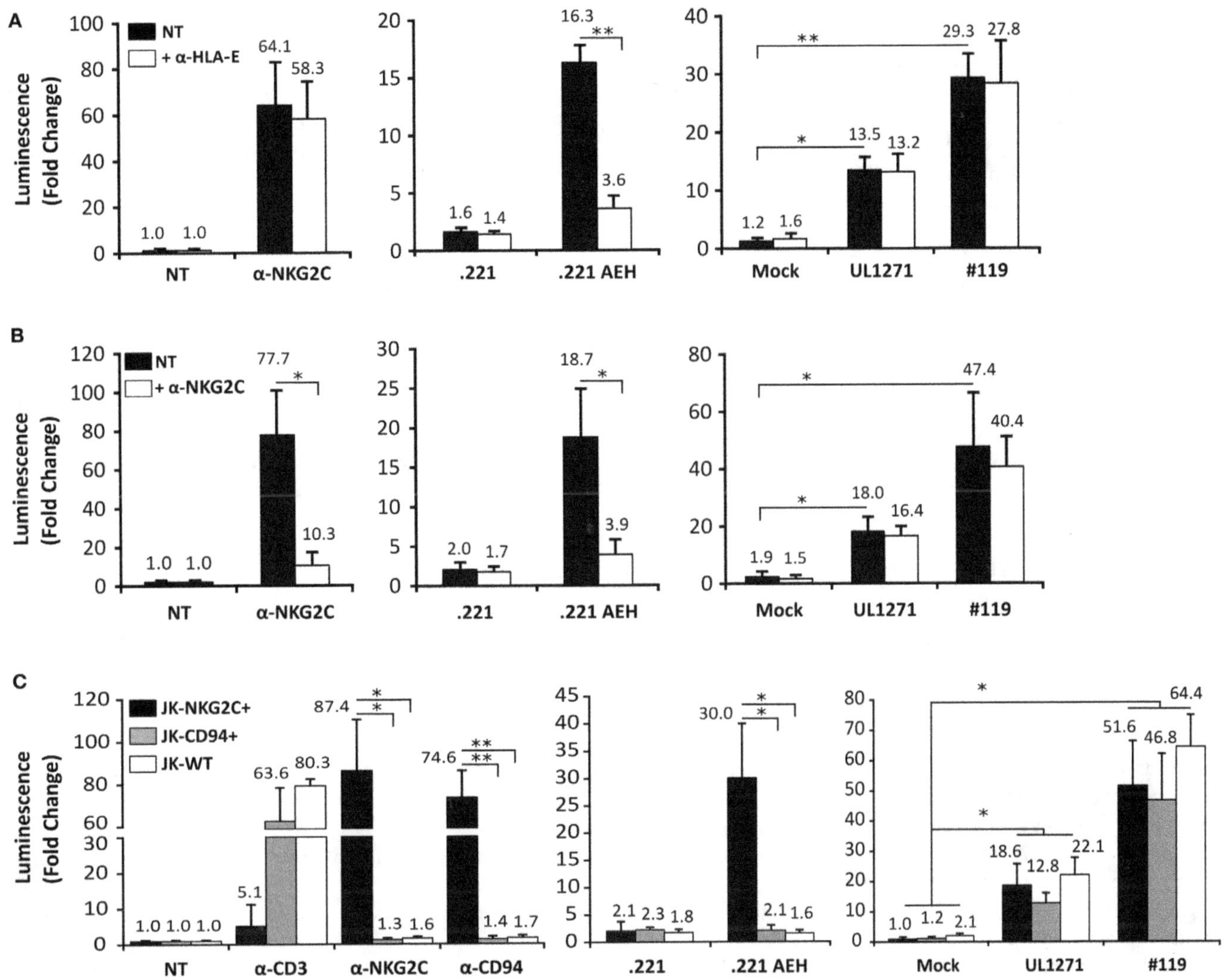

FIGURE 6 | Jurkat-NKG2C+ activation by cells infected with human cytomegalovirus (HCMV) clinical isolates is independent of the CD94/NKG2C receptor. **(A)** Response of Jurkat-NKG2C+ reporter to HFFF infected by HCMV clinical isolates or .221-AEH cells preincubated with the 3D12 anti-HLA-E monoclonal antibody (mAb) (5 μg/ml). **(B)** The blocking effect of an anti-NKG2C mAb (clone MAB1381) (5 μg/ml) on the response of Jurkat-NKG2C+ cells to stimulation with either plate-bound anti-NKG2C mAb, .221-AEH cells or HFFF infected by HCMV clinical isolates (white bars) was assessed in parallel. Data correspond to three independent experiments (mean ± SD); the average infection rate for each experiment was >50%. **(C)** Jurkat-NKG2C+ (black bars), Jurkat-CD94+ (gray bars), and wild-type Jurkat (WT) (white bars) were stimulated with 1 μg/ml of plate-bound anti-CD3, -NKG2C, -CD94 mAbs, or cocultured with .221 or .221-AEH cells (negative and positive controls). Confluent HFFF cells infected with UL1271 and #119 clinical isolates (multiplicity of infection = 10) for 72 h were cocultured with the different Jurkat cell lines (NKG2C+, CD94+, and WT) previously electroporated with the reporter plasmid. Data correspond to three independent experiments (mean ± SD); the average infection rate for each experiment was >50%. Statistically significant differences are indicated (*$p < 0.05$; **$p < 0.01$).

activation upon engagement by HLA-E were also greater in NKG2C^{bright} NK cells from *NKG2C*$^{+/+}$ than *NKG2C*$^{+/del}$ individuals (9). These data supported that subtle differences in CD94/NKG2C surface density may quantitatively influence adaptive NK cell differentiation/activation in response to HCMV, thus providing another indirect indication for a relatively low avidity of the hypothetical receptor–ligand interaction. The increased surface expression of CD94/NKG2C in differentiated adaptive NK cells (NKG2C^{bright}) from HCMV+ subjects as compared to the profile detected in non-infected individuals (NKG2C^{dim}) is also in line with this view. Of note, a limited affinity and/or reduced expression levels of the putative CD94/NKG2C ligand in infected cells may presumably render this pathway particularly sensitive to viral immune evasion mechanisms, eventually accounting for the undetectable activation of NKG2C+ NK cell effector functions and of Jurkat-NKG2C+ cells, in that case, independently of HLA-specific inhibitory receptors.

Despite the negative results obtained with the reporter system, the existence of specific ligand(s) for CD94/NKG2C in HCMV-infected cells still remains a plausible hypothesis, compatible with the involvement of a viral glycoprotein or/and an HLA-E-peptide complex. The latter option is suggested by the blocking effect of an anti-HLA-E mAb in the expansion of NKG2C+ NK cells cocultured with infected fibroblasts (23). A first candidate to be considered is the UL40 leader sequence-derived peptide reported to promote HLA-E expression in HCMV-infected cells by a mechanism refractory to the viral US6 TAP inhibitor (34) and favored by the intrinsic HLA-E resistance to other viral molecules targeting HLA-I (49). UL40-dependent HLA-E expression in infected cells was shown to inhibit NK cell activation by engagement of the CD94/NKG2A inhibitory receptor, presumably contributing to immune evasion (34). The canonical UL40-derived nonamer (VMAPRTLIL) binding to HLA-E is identical to that displayed by some HLA-I leader sequences, predictably promoting a low affinity interaction with CD94/NKG2C (50). Heatley et al. (51) compared the ability of a panel of UL40-nonamer variants to stabilize HLA-E in RMA-S cells, assessing the interaction of the corresponding HLA-E/peptide complexes with CD94/NKG2A or CD94/NKG2C by surface plasmon resonance assays, as well as their recognition by NK cell subsets, respectively, displaying these NKR. Of note, quantitative differences in the interaction of distinct HLA-E-peptide complexes were noticed, but none appeared preferentially recognized by CD94/NKG2C.

It is conceivable that HLA-E expression by HCMV-infected cells, in which HLA-I ligands for inhibitory KIRs are downregulated, might promote specific activation of NKG2C+ NKG2A- NK cells. Supporting this hypothesis, Jurkat-NKG2C+ cells were activated by .221 cells displaying HLA-E bound to the synthetic AD169 UL40-derived peptide (VMAPRTLIL), shared by some HLA-I alleles. Nevertheless, no reporter response was perceived following interaction with AD169-infected cells expressing HLA-E, pointing out to the influence of additional factors, and indicating that experimental approaches based on loading target cells with synthetic peptides provide valuable information but do not precisely reflect the complexity of HCMV infection. A putative role of UL40 in NKG2C+ NK cell expansion, detected following coculture with AD169-infected cells, was addressed employing a UL40 deletion mutant generated in an AD169 BACmid (HB5) (22, 23). In this setting, cells infected with ΔUL40 HB5 retained the ability to promote *in vitro* expansion of NKG2C+ NK cells; however, a caveat for interpreting these experiments is the gene deletion encompassing US2-US6 introduced for the generation of the HB5 BACmid. In fact, HCMV impairs surface expression of HLA-I molecules but not their biosynthesis and, therefore, endogenous HLA-I leader sequences in the absence of US6 may be presented by HLA-E in HB5-infected cells, eventually competing with the UL40-derived nonamer, or even replacing it in ΔUL40 HB5-infected HFFF. On the other hand, NKG2C+ NK cell expansions detected in cocultures with Towne-infected fibroblasts (22) may indirectly provide a case against the involvement of HLA-E expression, which was downregulated consistent with the existence of a genomic deletion spanning the UL40 leader sequence of this HCMV strain (44). Of note, no relation of the HLA-E dimorphism with the expansion of NKG2C+ cells was noticed in previous studies (7, 9).

In summary, molecular evidence supporting that a specific interaction of CD94/NKG2C with infected cells drives the adaptive NK cell response to HCMV remains elusive, leaving unanswered key interrelated questions, particularly, the cellular mechanisms underlying the stable expansion of NKG2C^{bright} NK cells and the basis for the wide variability of this effect in HCMV+ individuals. Observations in immunocompromised patients suggested that this pattern of response may compensate an inefficient T-cell-mediated control of the primary infection, eventually determined in healthy individuals by viral/host genetic factors as well as by other circumstantial variables (e.g., age at infection, viral load, etc.).

We hypothesize that primary HCMV infection promotes the differentiation, proliferation, and survival of a pool of progenitors, possibly stemming from NKG2C^{dim} NKG2A- NK cells present in seronegative individuals. Increased surface expression of CD94/NKG2C presumably constitutes a key early event, facilitating activation of this NK cell subset upon interaction with low-avidity ligand(s) displayed by infected cells. Such NKG2C-driven selection is dependent on cytokines required for adaptive NK cell differentiation/expansion and may be tuned by specific KIR-HLA-I interactions, consistent with the oligoclonal KIR expression profile of adaptive NK cells (8, 52). Following a contraction phase after viral replication is controlled, a pool of long-lived NKG2C^{bright} NK cells with clonal expansion potential survive, and homeostatic proliferation contributes to their persistent increased numbers in the circulation. The process is reminiscent of the generation of memory cytotoxic T lymphocytes as proposed for the response of Ly49H in mice (53); in fact, responsiveness of differentiated NKG2C^{bright} NK cells to cytokines (e.g., IL-15, IL-2) becomes dependent on signaling by activating NKR (e.g., NKG2C or CD16). HCMV reactivation/reinfection as well as other infectious pathogens may boost antibody- and cytokine-dependent activation of adaptive NK cells, leading to their progressive acquisition of late differentiation features (e.g., FcRγ downregulation). Further efforts are warranted to understand how HCMV infection resets the NK-cell compartment homeostasis in some individuals, and the implications that such persistent reconfiguration of the NK cell compartment may have in the development of the immune response under different pathological conditions.

AUTHOR CONTRIBUTIONS

AP developed the reporter system, carried out the experimental work, and revised the manuscript. MC-G collaborated in the experimental development. DF and AA collaborated in the genetic analysis of HCMV as well as in critical discussion of the results. HH contributed with essential support for the analysis of HCMV clinical isolates as well as in critical discussion of the results. AM contributed to the design, follow-up, and interpretation of the results. ML-B proposed the experimental approach, contributed to the design, follow-up, and interpretation of results, and wrote the draft that was revised by all authors.

ACKNOWLEDGMENTS

The authors thank Gemma Heredia and Andrea Vera for technical help, and Ramón Gimeno for advice in retroviral vector expression.

REFERENCES

1. Lanier LL. NK cell recognition. *Annu Rev Immunol* (2005) 23:225–74. doi:10.1146/annurev.immunol.23.021704.115526
2. Moretta A, Bottino C, Vitale M, Pende D, Cantoni C, Mingari MC, et al. Activating receptors and coreceptors involved in human natural killer cell-mediated cytolysis. *Annu Rev Immunol* (2001) 19:197–223. doi:10.1146/annurev.immunol.19.1.197
3. Arase H, Mocarski ES, Campbell AE, Hill AB, Lanier LL. Direct recognition of cytomegalovirus by activating and inhibitory NK cell receptors. *Science* (2002) 296(5571):1323–6. doi:10.1126/science.1070884
4. Smith HR, Heusel JW, Mehta IK, Kim S, Dorner BG, Naidenko OV, et al. Recognition of a virus-encoded ligand by a natural killer cell activation receptor. *Proc Natl Acad Sci U S A* (2002) 99(13):8826–31. doi:10.1073/pnas.092258599
5. Sun JC, Beilke JN, Lanier LL. Adaptive immune features of natural killer cells. *Nature* (2009) 457(7229):557–61. doi:10.1038/nature07665
6. van der Ploeg K, Chang C, Ivarsson MA, Moffett A, Wills MR, Trowsdale J. Modulation of human leukocyte antigen-C by human cytomegalovirus stimulates KIR2DS1 recognition by natural killer cells. *Front Immunol* (2017) 8:298. doi:10.3389/fimmu.2017.00298
7. Gumá M, Angulo A, Vilches C, Gómez-Lozano N, Malats N, López-Botet M. Imprint of human cytomegalovirus infection on the NK cell receptor repertoire. *Blood* (2004) 104(12):3664–71. doi:10.1182/blood-2004-05-2058
8. Béziat V, Liu L, Malmberg JA, Ivarsson MA, Sohlberg E, Bjorklund AT, et al. NK cell responses to cytomegalovirus infection lead to stable imprints in the human KIR repertoire and involve activating KIRs. *Blood* (2013) 121:2678–88. doi:10.1182/blood-2012-10-459545
9. Muntasell A, López-Montañés M, Vera A, Heredia G, Romo N, Peñafiel J, et al. NKG2C zygosity influences CD94/NKG2C receptor function and the NK-cell compartment redistribution in response to human cytomegalovirus. *Eur J Immunol* (2013) 43:3268–78. doi:10.1002/eji.201343773
10. López-Botet M, Muntasell A, Vilches C. The CD94/NKG2C+ NK-cell subset on the edge of innate and adaptive immunity to human cytomegalovirus infection. *Semin Immunol* (2014) 26(2):145–51. doi:10.1016/j.smim.2014.03.002
11. Lee J, Zhang T, Hwang I, Kim A, Nitschke L, Kim M, et al. Epigenetic modification and antibody-dependent expansion of memory-like NK cells in human cytomegalovirus-infected individuals. *Immunity* (2015) 42(3):431–42. doi:10.1016/j.immuni.2015.02.013
12. Schlums H, Cichocki F, Tesi B, Theorell J, Béziat V, Holmes TD, et al. Cytomegalovirus infection drives adaptive epigenetic diversification of NK cells with altered signaling and effector function. *Immunity* (2015) 42(3):443–56. doi:10.1016/j.immuni.2015.02.008
13. Monsivais-Urenda A, Noyola-Cherpitel D, Hernández-Salinas A, García-Sepúlveda C, Romo N, Baranda L, et al. Influence of human cytomegalovirus infection on the NK cell receptor repertoire in children. *Eur J Immunol* (2010) 40(5):1418–27. doi:10.1002/eji.200939898
14. Noyola DE, Fortuny C, Muntasell A, Noguera-Julian A, Muñoz-Almagro C, Alarcón A, et al. Influence of congenital human cytomegalovirus infection and the NKG2C genotype on NK-cell subset distribution in children. *Eur J Immunol* (2012) 42:3256–66. doi:10.1002/eji.201242752
15. Noyola DE, Alarcón A, Noguera-Julian A, Muntasell A, Muñoz-Almagro C, García J, et al. Dynamics of the NK-cell subset redistribution induced by cytomegalovirus infection in preterm infants. *Hum Immunol* (2015) 76(2–3):118–23. doi:10.1016/j.humimm.2015.01.017
16. Della Chiesa M, Falco M, Podesta M, Locatelli F, Moretta L, Frassoni F, et al. Phenotypic and functional heterogeneity of human NK cells developing after umbilical cord blood transplantation: a role for human cytomegalovirus? *Blood* (2012) 119(2):399–410. doi:10.1182/blood-2011-08-372003
17. Foley B, Cooley S, Verneris MR, Pitt M, Curtsinger J, Luo X, et al. Cytomegalovirus reactivation after allogeneic transplantation promotes a lasting increase in educated NKG2C+ natural killer cells with potent function. *Blood* (2012) 119(11):2665–74. doi:10.1182/blood-2011-10-386995
18. Gumá M, Cabrera C, Erkizia I, Bofill M, Clotet B, Ruiz L, et al. Human cytomegalovirus infection is associated with increased proportions of NK cells that express the CD94/NKG2C receptor in aviremic HIV-1-positive patients. *J Infect Dis* (2006) 194(1):38–41. doi:10.1086/504719
19. Petitdemange C, Becquart P, Wauquier N, Béziat V, Debré P, Leroy EM, et al. Unconventional repertoire profile is imprinted during acute chikungunya infection for natural killer cells polarization toward cytotoxicity. *PLoS Pathog* (2011) 7(9):e1002268. doi:10.1371/journal.ppat.1002268
20. Béziat V, Dalgard O, Asselah T, Halfon P, Bedossa P, Boudifa A, et al. CMV drives clonal expansion of NKG2C+ NK cells expressing self-specific KIRs in chronic hepatitis patients. *Eur J Immunol* (2012) 42(2):447–57. doi:10.1002/eji.201141826
21. Bjorkström NK, Lindgren T, Stoltz M, Fauriat C, Braun M, Evander M, et al. Rapid expansion and long-term persistence of elevated NK cell numbers in humans infected with hantavirus. *J Exp Med* (2011) 208(1):13–21. doi:10.1084/jem.20100762
22. Gumá M, Budt M, Sáez A, Brckalo T, Hengel H, Angulo A, et al. Expansion of CD94/NKG2C+ NK cells in response to human cytomegalovirus-infected fibroblasts. *Blood* (2006) 107(9):3624–31. doi:10.1182/blood-2005-09-3682
23. Rölle A, Pollmann J, Ewen EM, Le VT, Halenius A, Hengel H, et al. IL-12-producing monocytes and HLA-E control HCMV-driven NKG2C+ NK cell expansion. *J Clin Invest* (2014) 124(12):5305–16. doi:10.1172/JCI77440
24. Muntasell A, Vilches C, Angulo A, López-Botet M. Adaptive reconfiguration of the human NK-cell compartment in response to cytomegalovirus: a different perspective of the host-pathogen interaction. *Eur J Immunol* (2013) 43(5):1133–41. doi:10.1002/eji.201243117
25. Magri G, Muntasell A, Romo N, Sáez-Borderías A, Pende D, Geraghty DE, et al. NKp46 and DNAM-1 NK-cell receptors drive the response to human cytomegalovirus-infected myeloid dendritic cells overcoming viral immune evasion strategies. *Blood* (2011) 117(3):848–56. doi:10.1182/blood-2010-08-301374
26. Zhang T, Scott JM, Hwang I, Kim S. Cutting edge: antibody-dependent memory-like NK cells distinguished by FcRgamma deficiency. *J Immunol* (2013) 190(4):1402–6. doi:10.4049/jimmunol.1203034
27. Moraru M, Black LE, Muntasell A, Portero F, López-Botet M, Reyburn HT, et al. NK cell and Ig interplay in defense against herpes simplex virus type 1: epistatic interaction of CD16A and IgG1 allotypes of variable affinities modulates antibody-dependent cellular cytotoxicity and susceptibility to clinical reactivation. *J Immunol* (2015) 195(4):1676–84. doi:10.4049/jimmunol.1500872
28. Costa-García M, Vera A, Moraru M, Vilches C, López-Botet M, Muntasell A. Antibody-mediated response of NKG2Cbright NK cells against human cytomegalovirus. *J Immunol* (2015) 194(6):2715–24. doi:10.4049/jimmunol.1402281
29. Wu Z, Sinzger C, Frascaroli G, Reichel J, Bayer C, Wang L, et al. Human cytomegalovirus-induced NKG2C(hi) CD57(hi) natural killer cells are effectors dependent on humoral antiviral immunity. *J Virol* (2013) 87(13):7717–25. doi:10.1128/JVI.01096-13
30. Liu LL, Landskron J, Ask EH, Enqvist M, Sohlberg E, Traherne JA, et al. Critical role of CD2 co-stimulation in adaptive natural killer cell responses revealed in NKG2C-deficient humans. *Cell Rep* (2016) 15(5):1088–99. doi:10.1016/j.celrep.2016.04.005
31. Rölle A, Halenius A, Ewen EM, Cerwenka A, Hengel H, Momburg F. CD2-CD58 interactions are pivotal for the activation and function of adaptive natural killer cells in human cytomegalovirus infection. *Eur J Immunol* (2016) 46:2420–5. doi:10.1002/eji.201646492
32. Redondo-Pachon D, Crespo M, Yelamos J, Muntasell A, Pérez-Sáez MJ, Pérez-Fernández S, et al. Adaptive NKG2C+ NK cell response and the risk of cytomegalovirus infection in kidney transplant recipients. *J Immunol* (2017) 198(1):94–101. doi:10.4049/jimmunol.1601236
33. López-Botet M, Vilches C, Redondo-Pachón D, Muntasell A, Pupuleku A, Yélamos J, et al. Dual role of natural killer cells on graft rejection and control of cytomegalovirus infection in renal transplantation. *Front Immunol* (2017) 8:166. doi:10.3389/fimmu.2017.00166
34. Tomasec P, Braud VM, Rickards C, Powell MB, McSharry BP, Gadola S, et al. Surface expression of HLA-E, an inhibitor of natural killer cells, enhanced by human cytomegalovirus gpUL40. *Science* (2000) 287(5455):1031. doi:10.1126/science.287.5455.1031
35. Wang EC, McSharry B, Retière C, Tomasec P, Williams S, Borysiewicz LK, et al. UL40-mediated NK evasion during productive infection with human cytomegalovirus. *Proc Natl Acad Sci U S A* (2002) 99(11):7570–5. doi:10.1073/pnas.112680099

36. Falk CS, Mach M, Schendel DJ, Weiss EH, Hilgert I, Hahn G. NK cell activity during human cytomegalovirus infection is dominated by US2-11-mediated HLA class I down-regulation. *J Immunol* (2002) 169(6):3257–66. doi:10.4049/jimmunol.169.6.3257
37. Pérez-Villar JJ, Carretero M, Navarro F, Melero I, Rodríguez A, Bottino C, et al. Biochemical and serologic evidence for the existence of functionally distinct forms of the CD94 NK cell receptor. *J Immunol* (1996) 157(12):5367–74.
38. Lee N, Goodlett DR, Ishitani A, Marquardt H, Geraghty DE. HLA-E surface expression depends on binding of TAP-dependent peptides derived from certain HLA class I signal sequences. *J Immunol* (1998) 160:4951–60.
39. Llano M, Lee N, Navarro F, García P, Albar JP, Geraghty DE, et al. HLA-E-bound peptides influence recognition by inhibitory and triggering CD94/NKG2 receptors: preferential response to an HLA-G-derived nonamer. *Eur J Immunol* (1998) 28(9):2854–63. doi:10.1002/(SICI)1521-4141(199809)28:09<2854::AID-IMMU2854>3.0.CO;2-W
40. Longo PA, Kavran JM, Kim MS, Leahy DJ. Transient mammalian cell transfection with polyethylenimine (PEI). *Methods Enzymol* (2013) 529:227–40. doi:10.1016/B978-0-12-418687-3.00018-5.:227-40
41. Atalay R, Zimmermann A, Wagner M, Borst E, Benz C, Messerle M, et al. Identification and expression of human cytomegalovirus transcription units coding for two distinct Fcgamma receptor homologs. *J Virol* (2002) 76(17):8596–608. doi:10.1128/JVI.76.17.8596-8608.2002
42. Muntasell A, Magri G, Pende D, Angulo A, López-Botet M. Inhibition of NKG2D expression in NK cells by cytokines secreted in response to human cytomegalovirus infection. *Blood* (2010) 115(25):5170–9. doi:10.1182/blood-2009-11-256479
43. Hedin KE, Bell MP, Kalli KR, Huntoon CJ, Sharp BM, McKean DJ. Delta-opioid receptors expressed by Jurkat T cells enhance IL-2 secretion by increasing AP-1 complexes and activity of the NF-AT/AP-1-binding promoter element. *J Immunol* (1997) 159(11):5431–40.
44. Bradley AJ, Lurain NS, Ghazal P, Trivedi U, Cunningham C, Baluchova K, et al. High-throughput sequence analysis of variants of human cytomegalovirus strains Towne and AD169. *J Gen Virol* (2009) 90(Pt 10):2375–80. doi:10.1099/vir.0.013250-0
45. Sinzger C, Schmidt K, Knapp J, Kahl M, Beck R, Waldman J, et al. Modification of human cytomegalovirus tropism through propagation in vitro is associated with changes in the viral genome. *J Gen Virol* (1999) 80(Pt 11):2867–77. doi:10.1099/0022-1317-80-11-2867
46. Wilkinson GW, Davison AJ, Tomasec P, Fielding CA, Aicheler R, Murrell I, et al. Human cytomegalovirus: taking the strain. *Med Microbiol Immunol* (2015) 204(3):273–84. doi:10.1007/s00430-015-0411-4
47. Djaoud Z, Riou R, Gavlovsky PJ, Mehlal S, Bressollette C, Gérard N, et al. Cytomegalovirus-infected primary endothelial cells trigger NKG2C+ natural killer cells. *J Innate Immun* (2016) 8(4):374–85. doi:10.1159/000445320
48. Goodier MR, White MJ, Darboe A, Nielsen CM, Goncalves A, Bottomley C, et al. Rapid NK cell differentiation in a population with near-universal human cytomegalovirus infection is attenuated by NKG2C deletions. *Blood* (2014) 124(14):2213–22. doi:10.1182/blood-2014-05-576124
49. Llano M, Gumá M, Ortega M, Angulo A, López-Botet M. Differential effects of US2, US6 and US11 human cytomegalovirus proteins on HLA class Ia and HLA-E expression: impact on target susceptibility to NK cell subsets. *Eur J Immunol* (2003) 33(10):2744–54. doi:10.1002/eji.200324182
50. Valés-Gomez M, Reyburn HT, Erskine RA, López-Botet M, Strominger JL. Kinetics and peptide dependency of the binding of the inhibitory NK receptor CD94/NKG2-A and the activating receptor CD94/NKG2-C to HLA-E. *EMBO J* (1999) 18(15):4250–60. doi:10.1093/emboj/18.15.4250
51. Heatley SL, Pietra G, Lin J, Widjaja JM, Harpur CM, Lester S, et al. Polymorphism in human cytomegalovirus UL40 impacts on recognition of human leukocyte antigen-E (HLA-E) by natural killer cells. *J Biol Chem* (2013) 288(12):8679–90. doi:10.1074/jbc.M112.409672
52. Muntasell A, Pupuleku A, Cisneros E, Vera A, Moraru M, Vilches C, et al. Relationship of NKG2C copy number with the distribution of distinct cytomegalovirus-induced adaptive NK cell subsets. *J Immunol* (2016) 196:3818–27. doi:10.4049/jimmunol.1502438
53. Sun JC, Lanier LL. NK cell development, homeostasis and function: parallels with CD8(+) T cells. *Nat Rev Immunol* (2011) 11(10):645–57. doi:10.1038/nri3044

17

In Vitro Culture with Interleukin-15 Leads to Expression of Activating Receptors and Recovery of Natural Killer Cell Function in Acute Myeloid Leukemia Patients

Beatriz Sanchez-Correa[1]*, Juan M. Bergua[2], Alejandra Pera[3,4], Carmen Campos[3], Maria Jose Arcos[2], Helena Bañas[2], Esther Duran[5], Rafael Solana[1,3] and Raquel Tarazona[1]

[1] Immunology Unit, University of Extremadura, Cáceres, Spain, [2] Department of Haematology, Hospital San Pedro de Alcantara, Cáceres, Spain, [3] Instituto Maimónides de Investigación Biomédica de Córdoba (IMIBIC), Reina Sofia University Hospital, University of Cordoba, Red Espanola de Investigacion en Patologia Infecciosa (REIPI), Córdoba, Spain, [4] Brighton and Sussex Medical School, University of Sussex, Brighton, United Kingdom, [5] Histology and Pathology Unit, Faculty of Veterinary, University of Extremadura, Cáceres, Spain

Edited by:
Gianfranco Pittari, Hamad Medical Corporation, Qatar

Reviewed by:
Mar Vales-Gomez, Consejo Superior de Investigaciones Científicas (CSIC), Spain
Nicholas D. Huntington, Walter and Eliza Hall Institute of Medical Research, Australia

***Correspondence:**
Beatriz Sanchez-Correa
beatrizsanchezcorrea@gmail.com

Despite recent progress in the therapeutic approach of malignant hemopathies, their prognoses remain frequently poor. Immunotherapy could open a new window of great interest in this setting. Natural killer (NK) cells constitute an important area of research for hematologic malignancies, because this subpopulation is able to kill target cells spontaneously without previous sensitization, representing a novel tool in the treatment of them. Abnormal NK cytolytic function is observed in several hematological malignancies, including acute myeloid leukemia (AML) and myelodysplastic syndromes. Several mechanisms are involved in this abnormal function, such as decreased expression of activating receptors, increased expression of inhibitory receptors or defective expression of NK cell ligands on target cells. New immunotherapies are focused in identifying factors that could increase the expression of these activating receptors, to counteract inhibitory receptors expression, and therefore, to improve the NK cell cytotoxic capacities against tumor cells. In this work, we analyze the effect of interleukin (IL)-15 on the expression of NK cell-activating receptors that play a crucial role in the lysis of blasts from AML patients. Our results showed that IL-15 increased the surface expression of NKp30 on NK cells from healthy donors and AML patients with the consequent improvement of NK cell cytotoxicity. Besides, the upregulation of NKp30 induced by IL-15 is associated with an improvement of NK-mediated myeloid dendritic cells (DCs) maturation. NK cells cultured with IL-15 showed an upregulation of NKp30, which is associated with an increase anti-tumor activity and with an improved maturation of immature DCs. In our *in vitro* model, IL-15 exerted a great activating stimulus that could be used as novel immunotherapy in AML patients.

Keywords: natural killer cell, acute myeloid leukemia, interleukin-15, NKp30, dendritic cells, natural killer cell degranulation

INTRODUCTION

Natural killer (NK) cells constitutively possess lytic machinery and are able to spontaneously lyse virally infected or malignant cells without any prior sensitization to antigen. Their function is tightly controlled through a balance of signals from activating and inhibitory receptors, by cytokines and chemokines (1–3), as well as by cross talk with other immune cells, such as dendritic cells (DCs) (4), effector CD4$^+$ T cells (5), and regulatory T cells (Tregs) (6, 7).

There are several major histocompatibility complex (MHC) class I-specific inhibitory receptors with different molecular structures and specificities for different alleles of human leukocyte antigen (HLA) class I molecules. These receptors act as negative regulators of NK cytotoxicity ensuring that normal autologous cells are preserved (4, 8, 9). The two main groups of inhibitory receptors are the killer Ig-like receptors (KIR) (8), which recognize polymorphic determinants of HLA-A, -B, or -C molecules and the heterodimeric receptors CD94-NKG2A/B, which recognize peptides derived from the leader sequences of different HLA class I molecules presented by the non-classical MHC class I molecule HLA-E (10–13). Normal autologous cells that express HLA class I molecules are protected from NK-mediated attack, but cells that have reduced expression because of malignant transformation become susceptible to NK cell attack. Loss or downregulation of a single HLA-I allele, a relatively frequent event in cancer, can be sufficient to make tumor cells susceptible to NK cell cytotoxicity (8, 9).

In the absence of inhibitory signals, NK cell cytotoxicity requires signaling through activating receptors upon interaction with their ligands on target cells. An ample set of activating receptors have been described including, among others, members of the C lectin-like family as NKG2D and members of the immunoglobulin superfamily as natural cytotoxicity receptors (NCRs) and DNAM-1 (also known as CD226) (4, 8, 9, 14). NKG2D ligands are a group of MHC class I-like molecules, the expression of which is induced by cellular stress. NKG2D ligands included MHC-I polypeptide-related sequences A (MICA) and B (MICB), and UL16-binding proteins (ULBP1–6) (15, 16). DNAM-1 specifically recognizes CD155 and CD112, two members of the nectin family, that are also expressed on different types of tumors (17–20).

Natural cytotoxicity receptors are major activating receptors involved in tumor cell detection and lysis. NCRs include NKp46 (21–23), NKp30 (24, 25), and NKp44 (26), which mediate cell lysis of many cancer cells. NKp30 and NKp46 are expressed both in resting and activated NK cells, whereas NKp44 expression is restricted to activated NK cells (4, 9, 14, 21, 27). NCR ligands include pathogen-associated as well as stress-related molecules. The identity of NCR ligands on tumors remains in part elusive. The nuclear factor HLA-B-associated transcript 3 (BAT3) and B7H6 have been described as NKp30 ligands (24, 28, 29). NKp46 and NKp44 were shown to interact with the viral hemagglutinin protein (24, 30–32). A truncated isoform of mixed lineage leukemia 5 is an activating ligand for NKp44 (33). By contrast, proliferating cell nuclear antigen is an inhibitory tumor ligand for NKp44 (34). No tumor ligands for NKp46 have been identified so far.

We have recently demonstrated that acute myeloid leukemia (AML) patients have depressed NK cell function as well as altered cytokine production (35–37). NK cell-mediated rejection of leukemic blasts may be limited by the reduced expression of NK cell-activating receptors such as DNAM-1, NKp46, and NKp30 observed in AML patients (36, 37). Besides, NKp46 downregulation has been associated with decreased survival in AML patients (38).

In the last years, several cytokines have been extensively studied as potential therapeutic agents to manipulate the immune response against malignant cells due to their capacity of stimulate cell growth and survival as well as increase the cytotoxicity or cytokine production to boost immune reactivity (39–42). So far, only a small number of cytokines have reached clinical use probably due to the complexity of cytokine network. Among these cytokines tested in different *in vitro* and *in vivo* settings, interleukin (IL)-2 and IL-15 should be highlighted (40, 41, 43, 44). IL-2, initially described as a T cell growth factor, promotes CD8$^+$ T cell and NK cell cytolytic activity and modulates T cell differentiation in response to antigen. Moreover, IL-2 is essential for the development and maintenance of Tregs that may represent a limitation for its use in patients with cancer. The major disadvantage of IL-2 is its toxicity, including severe capillary leak syndrome that can accompany this treatment (43). Recently, IL-15 has emerged as a potential immunotherapeutic candidate for the treatment of cancer. IL-2 and IL-15 are structurally related and have overlapping functions including their role in T cell proliferation, promotion of cytotoxic T cell differentiation, production of immunoglobulin by B cells, and generation, proliferation, and activation of NK cells. In contrast to IL-2, IL-15 is not required for the maintenance of Tregs and, based on preclinical studies, IL-15 causes less vascular capillary leak (42). These factors support the role of IL-15 for cancer immunotherapy, boosting both innate and adaptive immunity against tumors (45). However, so far, few clinical trials have analyzed the security and efficacy of IL-15 in cancer patients. Thus, a Phase I study (NCT01572493) assessing the safety and efficacy of IL-15 in adults with advanced malignancies has been suspended for undisclosed reasons. IL-15 is being tested as an immunological adjuvant to haploidentical NK cell transfer in AML patients (NCT02395822).

Due to the ability of NK cells to spontaneously kill tumor cells, this population represents an attractive tool for cancer immunotherapy (46–48). However, NK cell defective function in AML patients may limit tumor control. The possibility of manipulating NK cells by cytokines for therapeutic purposes open new area of research in cancer.

The biological effects of IL-15 on NK cells will depend on the direct effect on NK cells as well as by indirect consequences mediated by other cells stimulated by IL-15. Thus, in order to analyze the whole figure we have used peripheral blood mononuclear cell (PBMC) cultures stimulated with IL-15.

In this study, we aimed to assess (i) whether IL-15 induces and expansion of NK cells from healthy donors (HDs) and AML patients and its effect on the expression of activating receptors after short-term culture *in vitro*; (ii) whether IL-15 increases the cytotoxic activity of NK cells; and (iii) whether the maturation of immature DCs (iDCs) is enhanced by culture with IL-15-stimulated PBMCs.

MATERIALS AND METHODS

Patients and HDs

Peripheral blood mononuclear cells were obtained from 14 newly diagnosed AML patients (ranged 18–89 years) at the Hospital San Pedro de Alcántara (Cáceres, Spain) prior to any treatment and from 20 HD volunteers (ranged 20–60 years). The study was approved by the local Ethics Committee and samples collected after written informed consent in accordance with the Declaration of Helsinki. Diagnosis was established by cytological criteria based on the French–American–British classification.

The collected blood was drawn into heparinized tubes and processed using Ficoll-Hypaque gradients. PBMCs were recovered and cell count and viability analysis were performed. PBMCs were immediately used for experiments. For co-culture experiments, we selected those patients with lower percentage of leukemic blasts and to eliminate leukemic blasts an adherence step was included in the protocol prior to co-culture with DCs.

Plasma was obtained after centrifugation and stored at −80°C for measurement of cytokine levels. Plasma samples from HDs and from AML patients used in the study had not been previously thawed.

Monoclonal Antibodies (mAbs)

Natural killer cell percentage and phenotype were evaluated in PBMC obtained from AML patients and HDs. The following anti-human mAbs were used for flow cytometry: CD56-FITC (NCAM16.2), CD56-PE (MY31), CD14-PE (MØP9), CD3-PerCP (SK7) all from BD Biosciences (San Jose, CA, USA); CD56-PECy7 (B159), CD16-APC Cy7 (3G8), NKG2D-PE (1D11), CD226-PE (DX11), CD107a-FITC (H4A3), CD107b-FITC (H4B4), CD86-FITC (2331(FUN-1)), CD1a-FITC (HI149) all from BD Pharmingen (San Diego, CA, USA); and NKp30-PE (AF29-4D12), NKp30-APC (AF29-4D12), NKp46-PE (9E2), NKp46-APC (9E2), CD3-VioBlue (BW264/56) all from Miltenyi Biotec (Bergisch Gladbach, Germany). Prior to use, mAbs were titrated to establish optimal staining dilutions. Isotype-matched immunoglobulins were included in all experiments as negative controls. Mean relative fluorescence intensity was calculated by dividing the mean fluorescent intensity (MFI) of the relevant mAb by the MFI of its isotype control.

Cell Culture and Flow Cytometry Analyses of NK Cells

Peripheral blood mononuclear cells were cultured in complete medium (RPMI-1640 supplemented with 10% FCS, L-glutamine, sodium pyruvate, non-essential amino acids and penicillin/streptomycin, all from BioWhittaker, Verviers, Belgium) and were stimulated with 100 ng/mL of recombinant human (rh)IL-15 from Peprotech (Rocky Hill, NJ, USA) or 750 U/mL of rhIL-2 (National Cancer Institute, Frederick, MD, USA) (49). Cells were harvested after 48 h, and the frequency and NK receptor repertoire were assessed by multi-parameter flow cytometry using a FACScan cytometer and the CellQuest software (BD Biosciences) or MACsQuant cytometer and the MACSQuantify software (Miltenyi Biotec). NK cells were defined as $CD3^-CD56^+$ cells within the lymphocyte gate and the expression of activating receptors analyzed was referred to this population. Analysis of NK cells at various time points after cytokine activation was performed in order to select the best timing for functional analysis (data not shown).

Analysis of IL-15 in Plasma

Interleukin-15 concentrations in the plasma of patients and healthy controls were determined by enzyme-linked immunosorbent assay (ELISA) using a Human IL-15 ELISA Ready-SET-Go! Kit (eBioscience) according to the manufacturer's instructions. The minimum detectable levels were 8 pg/mL, and the standard curve range was 8–1,000 pg/mL.

Two independent sets of experiments were performed. No significant variations were observed among the experiments. Plate was read in an Infinite® 200 (Tecan, Switzerland) plate reader.

NK Cell Degranulation Assay

Cytokine-stimulated NK cells were tested in a degranulation assay against the NK cell-susceptible target cell line K562. The analysis of NK cell degranulation was performed by measuring the expression of CD107a/b after activation with target cells at ratio 1:1 in the presence of BD GolgiStop (BD Biosciences) and a mixture of FITC-labeled anti-CD107a and anti-CD107b mAbs. After 4 h, cells were stained with PE-labeled anti-CD56 and PercP-labeled anti-CD3 from BD Biosciences and analyzed by flow cytometry by measuring the frequency of CD107a/b expression on $CD3^-CD56^+$ NK cells. Spontaneous basal NK cell degranulation was always below 10%. Background expression of CD107a/b ($CD107a^+$ NK cells in medium only) was subtracted from expression with target cells.

Generation of DCs

Monocyte-derived iDCs were obtained by adherence of monocytes to plastic. Thus, PBMCs from HDs were resuspended at 5×10^6 cells/mL in complete medium and allowed to adhere for 2 h at 37°C in culture flasks. Then, the non-adherent cells were removed and the adherent cells, predominantly monocytes, were cultivated in complete medium supplemented with 50 ng/mL of recombinant human granulocyte-macrophage colony-stimulating factor (Peprotech) and 20 ng/mL of rhIL-4 (Peprotech). After 6 days, the percentage of iDCs was analyzed by flow cytometry. iDCs were defined as $CD14^-CD1a^+CD83^-CD86^-$ cells.

To generate mDCs, iDCs were plated either in the absence or in the presence of allogeneic PBMCs cells stimulated or not with rhIL15 or rhIL-2 at ratio 1:5. After 2 days, DCs were assessed for the expression of CD86. As positive control, optimal DC maturation was induced by *Escherichia coli* lipopolysaccharide (LPS) 1 µg/mL (serotype 055:B5, Sigma-Aldrich, St Louis, MO, USA).

Statistical Analysis

Statistical analysis was performed using SPSS Statistics version 19. Because the cell population counts were generally not normally distributed, we report medians and ranges for parameter values. Non-parametric statistical methods were used to analyze the data. Paired differences between NK cells from the same patient with or without stimulation by cytokines were tested using the Wilcoxon signed-rank test. For comparison of NK cells between HDs and AML patients, the exact Wilcoxon rank sum test was used.

RESULTS

Effect of IL-2 and IL-15 on NK Cell Expansion *In Vitro*

Natural killer cells were evaluated in HDs and AML patients at the time of diagnosis and prior to any treatment. In order to study the modulatory effect of IL-2 and IL-15 on NK cell proliferation, we cultured PBMCs from HDs and AML patients in presence of these cytokines. Our results showed that IL-15 but not IL-2 induced a significant increase (p = 0.016) in the percentage of NK cells from HDs after 48 h of culture (**Figures 1A,B**). By contrast, none of the cytokines had an effect on the percentage of NK cells from AML patients (**Figures 1C,D**).

Upregulation of NK Cell-Activating Receptors by IL-2 and IL-15 *In Vitro*

In order to assess the effect of IL-2 and IL-15 on the NK cell receptor expression, we cultured PBMCs from HDs and AML patients in presence of these cytokines.

In HDs (**Figure 2**, left panels), we observed a statistically significant upregulation of the NK cell-activating receptors NKp30 and NKG2D after culture of PBMCs with IL-2 (p = 0.004 and p = 0.013, respectively) (**Figures 2A,D**). IL-15 also induced a significant increase in the expression of NKp30 on NK cells surface (p = 0.0044) (**Figure 2A**), but the upregulation of NKG2D induced by IL-15 was not statistically significant (p = 0.08) (**Figure 2D**). No significant changes in the expression of NKp46 and DNAM-1 were observed (**Figures 2B,C**).

The analysis of the effect of IL-2 on NK cells from AML patients (**Figure 2**, right panels) only showed a statistically significant upregulation of DNAM-1 expression (p = 0.046) (**Figure 2C**), without significantly affecting the expression of the other receptors considered. By contrast, the incubation with IL-15 induced a significant increase of NKp30 and NKG2D (p=0.001 and p=0.028) (**Figures 2A,D**). A representative example of the effect of IL-15 is shown in Figure S1 in Supplementary Material.

Cytotoxic Activity of Cytokine-Stimulated PBMCs

Once we observed the effect of the different cytokines on NK cell receptor expression, we analyzed if these phenotypes correlated with the cytotoxic capacity of NK cells against K562, a susceptible target cell line. A representative example is shown in **Figure 3A**. Our results demonstrated that concomitant with the upregulation of the activating receptors following 48 h of

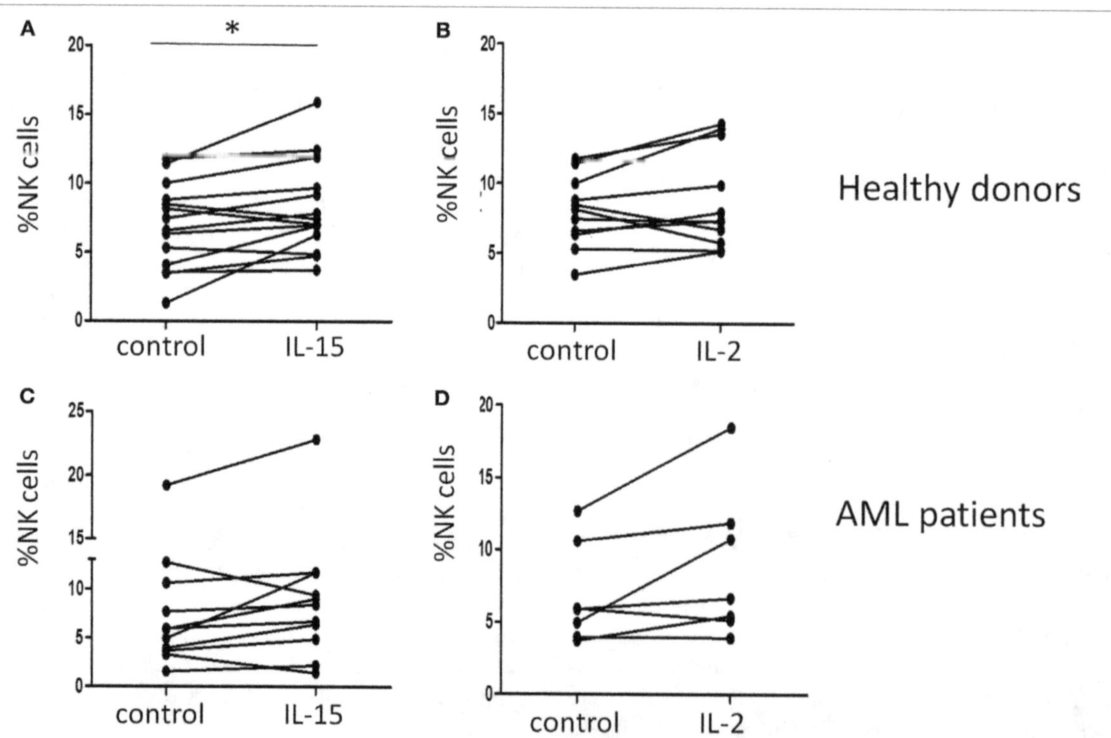

FIGURE 1 | Effect of interleukin (IL)-15 and IL-2 on natural killer (NK) cell expansion *in vitro*. The increase in the percentage of NK cells after incubation for 48 h with cytokines was analyzed by flow cytometry in healthy donors **(A,B)** and acute myeloid leukemia (AML) patients **(C,D)** in the presence of IL-15 **(A,C)** or IL-2 **(B,D)**.

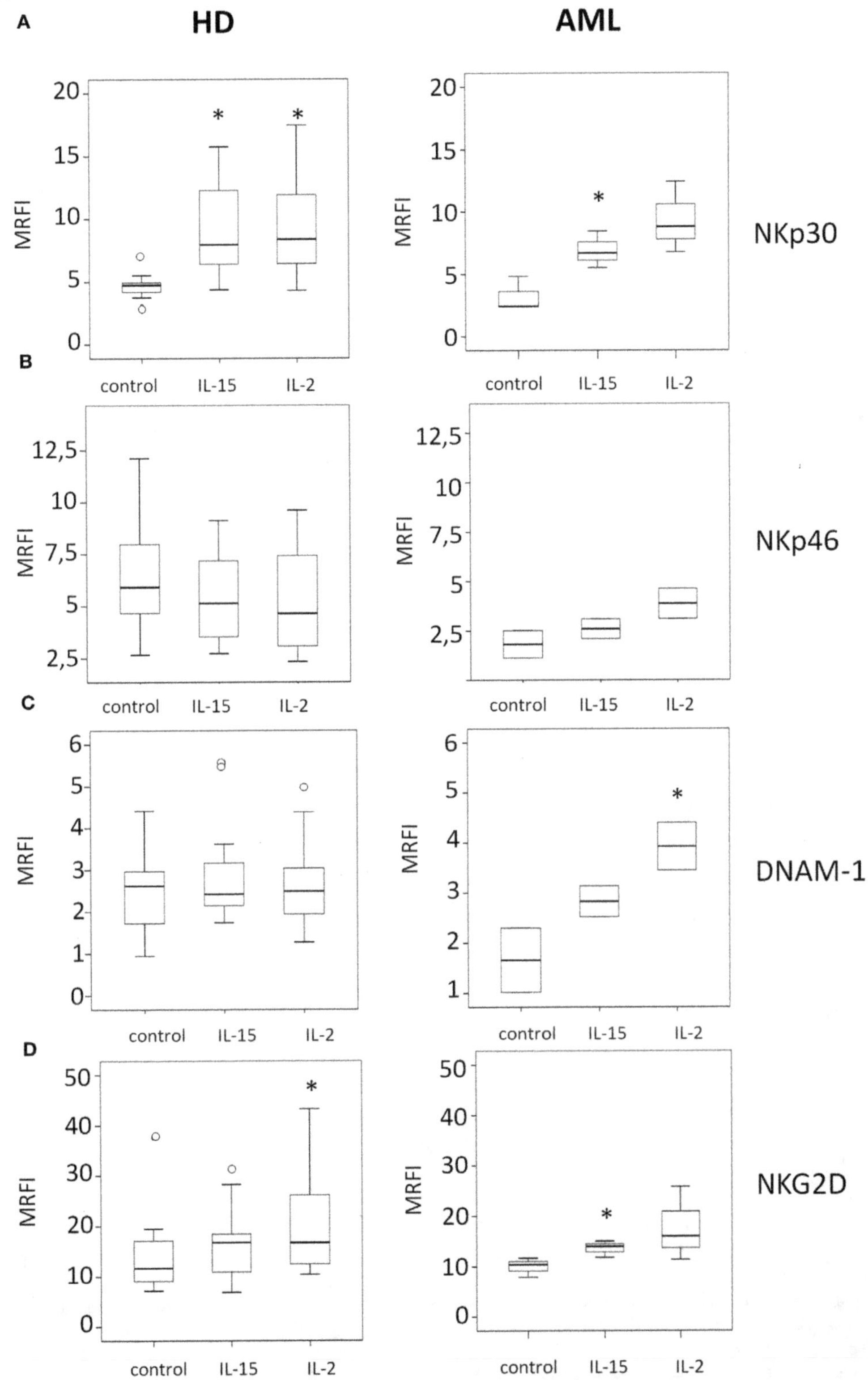

FIGURE 2 | Modulatory effect of interleukin (IL)-15 and IL-2 on natural killer cell activating receptor expression in healthy donors and acute myeloid leukemia patients. The expression of NKp30 **(A)**, NKp46 **(B)**, DNAM-1 **(C)**, and NKG2D **(D)** was analyzed by flow cytometry. NK cell receptor expression was analyzed after 48 h of culture with medium alone (control), IL-15, or IL-2. Left panels represent HDs and right panels represent AML patients. Mean relative fluorescence intensity (MRFI) was calculated by dividing the mean fluorescent intensity (MFI) of the relevant monoclonal antibody by the MFI of its isotype control. The lower boundary of the box indicates the 25th percentile and the upper boundary the 75th percentile. Bars above and below the box indicate the 90th and 10th percentiles. The line within the box marks the median. Circles (O) represent outliers values. (*$p < 0.05$, compared to control).

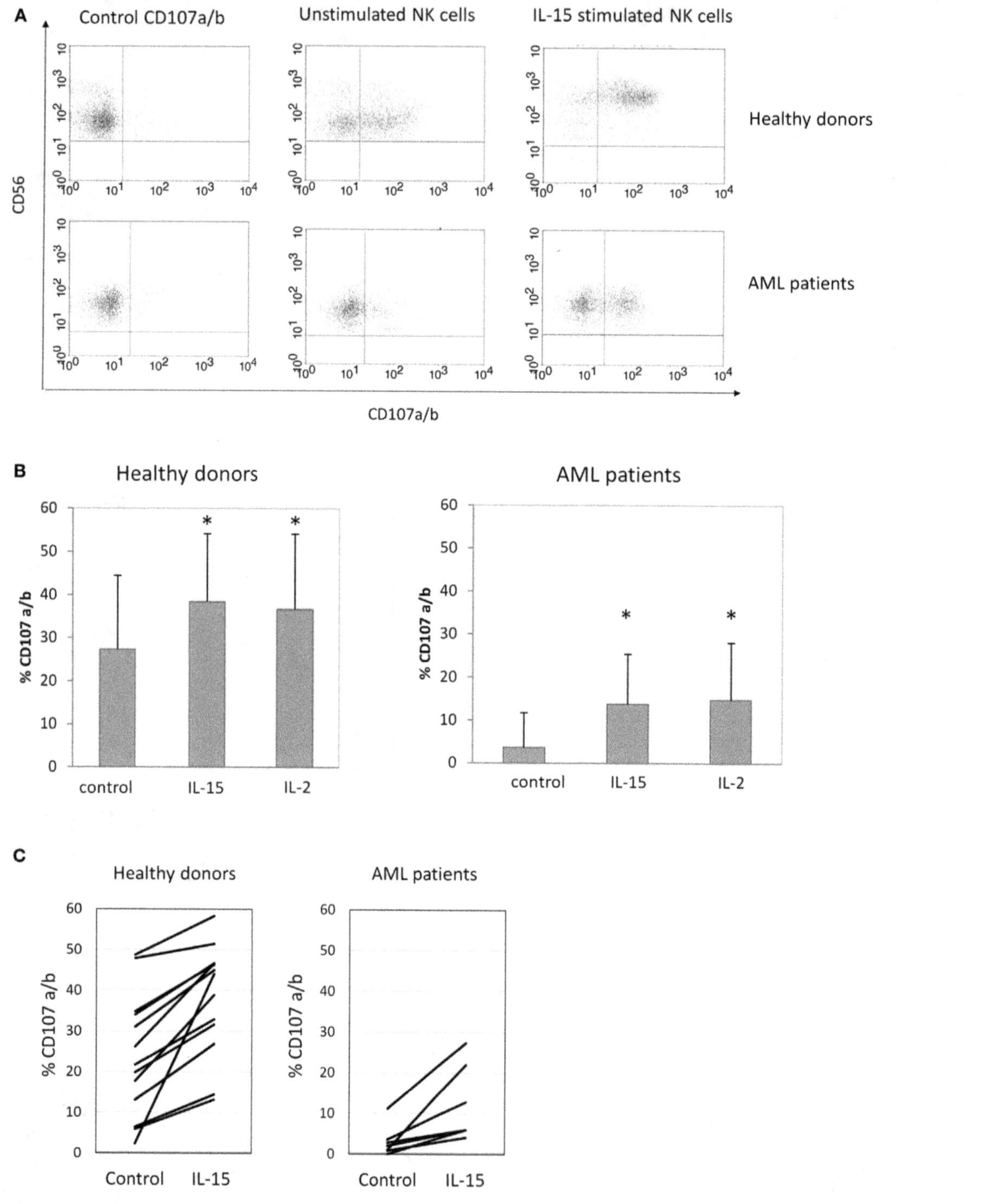

FIGURE 3 | Natural killer (NK) cell degranulation after cytokine stimulation. **(A)** Representative flow cytometry analysis of NK cell degranulation against K562 cells. **(B)** Column bars show the percentage of CD107a/b expression on NK cells from healthy donors (HDs) (left graph, $n = 16$) or acute myeloid leukemia (AML) patients (right graph, $n = 8$) in response to K562 cells. Error bars represent SD. **(C)** Individual representation of NK cell degranulation against K562 cells in HDs (right graph) and AML patients (right graph). NK cells were used after 48 h of culture with medium alone (control), interleukin (IL)-15, or IL-2 (*$p \leq 0.05$, compared to control).

IL-15 and IL-2 stimulation, there was a significant increase in NK cell cytotoxicity against K562 targets in both HDs and AML patients (**Figure 3B**). Although a high variability was observed, a positive effect of IL-15 on NK cell degranulation was found in all individuals analyzed but with different increases in degranulation (**Figure 3C**).

Effect of IL-15 and IL-2-Stimulated PBMCs on Maturation of iDCs

Recent studies demonstrated that human NK cells display the ability to kill or induce maturation of both autologous and allogeneic monocyte-derived iDCs This function depends on the engagement of NKp30 with its cellular ligands expressed by DCs. Considering that IL-15 and IL-2 have a potent effect in the expression of NKp30 on NK cell surface, we decided to study if these modifications affect iDC maturation mediated by NK cells. iDCs were generated as described in Section "Materials and Methods," harvested, counted and analyzed by FACS. After culture with cytokine-stimulated PBMCs, DC maturation was evaluated by the expression of CD86. Unstimulated PBMCs were used as control to compare the effect of IL-15 and IL-2 stimulation on PBMCs capacity to induce DC maturation. LPS-stimulated iDCs were used as positive control.

Our result showed that PBMCs from HDs and AML patients stimulated with IL-15 induced higher maturation of iDCs than unstimulated PBMCs although due to the high variability observed the differences were not statistically significant (**Figure 4A**; Figure S2). In response to cytokine-stimulated PBMCs, DCs displayed a onefold increase in the expression of CD86 compared to their response to unstimulated PBMCs (**Figure 4B**).

Presence of IL-15 in Plasma of HDs and AML Patients

Finally, we characterized the concentration of IL-15 in the plasma of HDs and AML patients. Our results showed that IL-15 was significantly higher in AML serum than in HDs ($p < 0.001$) (**Figure 5**). No correlation was observed between IL-15 levels in plasma and NKp30 expression on NK cells probably due to the high variability observed among donors (data not shown).

DISCUSSION

Acute myeloid leukemia is a heterogeneous disease that presents with different phenotypic and genotypic alterations in hematopoietic progenitors with the subsequent accumulation of immature hematopoietic stem cells that prevent the production of adequate amount of healthy hematopoietic cells. AML is more common in the elderly and co-morbidities and frailty often impact on patient tolerance to intensive treatment regimens. Thus, the prognosis of elderly AML patients remains poor, despite recent advances in the management and treatment options of AML patients including novel immunotherapies (50).

The role of NK cells against leukemia is supported by the discovery of NK cell spontaneous cytotoxicity against leukemia cell lines *in vitro* and the clinical benefits observed in KIR ligand mismatched allogeneic stem cell transplantation. Evidence for graft-versus-leukemia effect mediated by NK cells is observed in clinical studies with haploidentical donor transplants where the presence of alloreactivity due to KIR ligand mismatch was correlated with higher survival rates (51). NK cell-mediated cytotoxicity preserves healthy cells, and consequently, tumor control can be achieved in the absence of graft-versus-host disease (52). In addition, donor-derived NK cells have demonstrated to play a relevant role after hematopoietic stem cell transplantation (53).

Natural killer cells in AML patients habitually present defects in their cytotoxicity against autologous leukemic blasts probably as consequence of a reduced expression of activating receptors such as NCRs or DNAM-1 (36–38). Besides, the low expression of NKp46 and the NCR^{dull} (NKp46^{dull}NKp30^{dull}) phenotype have been associated with decreased survival in AML patients (38). NK cell phenotype and function at diagnosis of AML associate with clinical outcome. Thus, AML patients with low expression of

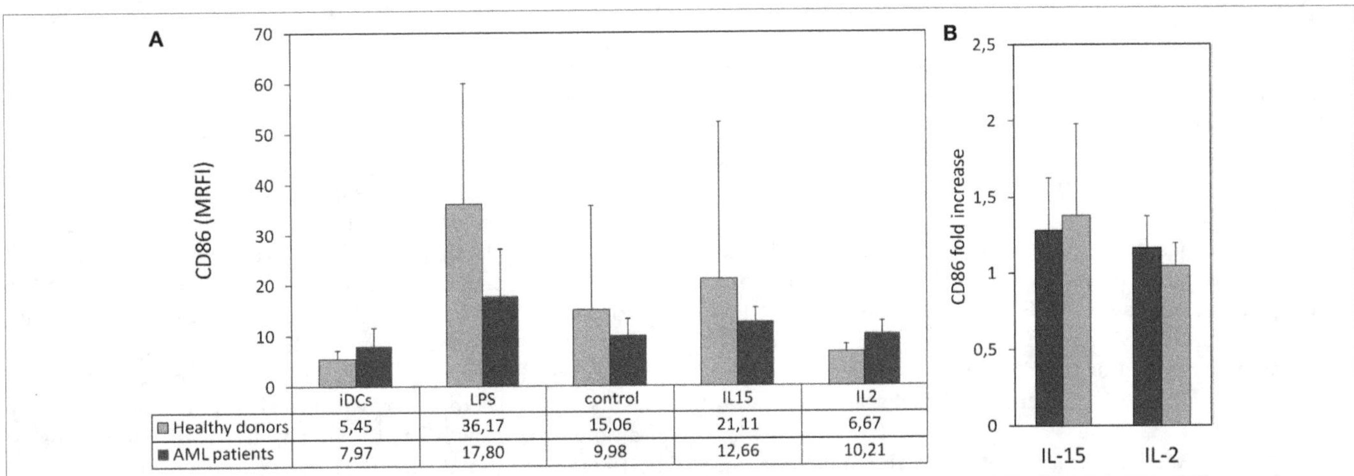

FIGURE 4 | Maturation of immature DCs (iDCs) co-cultured with cytokine-stimulated peripheral blood mononuclear cell (PBMC). Allogeneic iDCs were cultured in the presence cytokine-stimulated PBMC from healthy donors (HDs) or acute myeloid leukemia (AML) patients. Lipopolysaccharide (LPS) was used as positive control. Dendritic cell (DC) maturation was characterized by the expression of CD86. **(A)** Column bars show the MRFI of CD86 expression on DCs from HDs after culture in the presence of LPS- or cytokine-stimulated PBMC from HDs (light gray, $n = 6$) or AML patients (dark gray, $n = 4$). PBMCs were used after 48 h of culture with medium alone (control), interleukin (IL)-15 or IL-2. Error bars represent SD. **(B)** Comparative analysis of the effect of cytokines on PBMC-mediated DC maturation. Bars represent fold increase of CD86 expression on DCs cultured with PBMC treated with IL-15 or IL-2 relative to the expression in DCs co-cultured with unstimulated PBMCs.

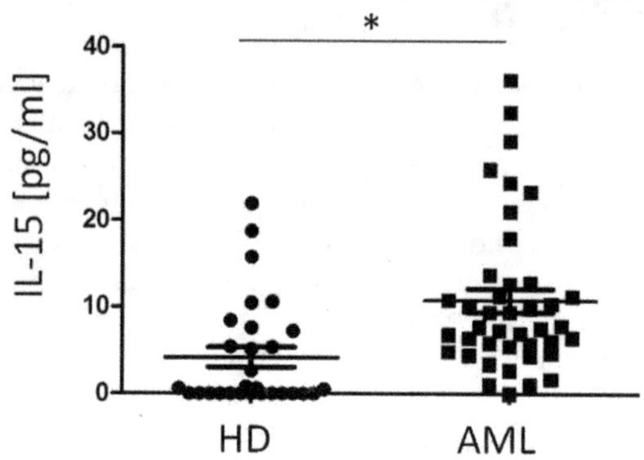

FIGURE 5 | Comparison of plasma IL-15 concentration in healthy donors (HDs) and acute myeloid leukemia (AML) patients. The horizontal bars represent the mean values. *$p \leq 0.05$.

activating receptors and decreased IFN-γ production had higher risk of relapse (54).

Previous reports demonstrated that IL-2 and IL-15 play an important role in the development, homeostasis, and function of T and NK cells. The administration of IL-2 as well as the adoptive transfer of IL-2-activated lymphocytes represented the first effective cancer immunotherapy for solid tumors such as melanoma. However, high doses of IL-2 were required to obtain therapeutic effect but a dose-related increase in toxicity was noted. Thus, IL-2 treatment was associated with adverse events including severe capillary leak syndrome that represent a major concern in IL-2 treatment. Besides IL-2 promotes the maintenance of Tregs that can reduce antitumor response (55). Since recombinant IL-2 was introduced, several clinical trials examining the role of IL-2 in preventing AML relapse have been developed. However, its use as monotherapy is not effective in terms of leukemia-free survival and overall survival (56). In a phase IV trial, AML patients in first complete remission received cycles of immunotherapy with histamine dihydrochloride and low dose of rIL-2 for 18 months to prevent leukemic relapse. The presence of high $CD56^{bright}$ NK cell counts and high expression of NKp30 or NKp46 on $CD16^+CD56^+$ NK cells independently predicted leukemia-free survival and overall survival in this clinical trial (57).

Other cytokines have been proposed for cancer treatment, and, in the last years, IL-15 has emerged as a potential immunotherapeutic candidate (43). Preclinical studies suggest that IL-15 may represent a more efficacious cytokine for cancer immunotherapy with less toxicity obtained with intermittent administration of IL-15 compared with daily administration (45, 58).

We have analyzed *in vitro* the effect of IL-15 and IL-2 on NK cell phenotype and function including NK cell cross talk with DCs. In our study, IL-2 was included as positive control since its role on NK cell activation both *in vitro* and *in vivo* has been extensively reported. Recently, IL-15 has been proposed as an ideal candidate for the expansion of NK cells *in vivo* since it does not promote expansion of Tregs (59). In our study, IL-15 but not IL-2 induced a significant expansion of NK cells in HDs after 48 h of culture. By contrast, NK cell expansion was very limited in AML patients showing higher variability.

Our results also show that short time culture with IL-15 induces upregulation of NKp30 and NKG2D on NK cells from AML patients in concordance with previous reports (60). Other studies have also shown an increment of DNAM-1 and NKp46 expression after culture with IL-15 (40, 60). We have also observed an increment of DNAM-1 and NKp46 but this increase did not reach statistical significance. The upregulation of activating receptor expression induced by IL-15 is related to the increased NK cell degranulation against K562 cell line in concordance with previous reports (60, 61). Altogether, these results suggest that AML-induced defective function of NK cells could be overcome by IL-15. Besides, we detect a significant increased production of IFNγ and granzymes A and B after culture of NK cells with IL-15 (data not shown). These observations are consistent with previous reports supporting the pivotal role of IL-15 in NK cell antitumor activity (40, 60, 61).

Natural killer cell-mediated induction of DCs maturation was mediated by NKp30 and cytokines released after NK cell activation such as TNF-α and IFN-γ (62, 63). We have also found that IL-15-activated PBMCs from AML patients and HDs have a higher capacity to support maturation of DCs than untreated or IL-2 treated PBMCs, suggesting that the increased expression of NKp30 after IL-15 culture improves the capacity of NK cells to collaborate in the maturation of DC.

These results suggest that IL-15 in addition to enhance NK cell cytotoxicity could also collaborate with the development of adaptive immunity by promoting NK cell-mediated maturation of DCs. Further analysis will be required to confirm this function.

Cytokines present in the tumor microenvironment modulate antitumor responses. In AML patients, cytokine profile at diagnosis is frequently aberrant and associates with pathogenesis, disease progression, and survival. We have previously shown that plasma levels of TNF-α, IL-6, and IL-10 are increased in AML patients. Low levels of IL-6 and high levels of IL-10 were associated with longer event-free survival and patient survival (35). In the present work, since IL-15 has demonstrated a role in NK cell activation, we analyzed IL-15 plasma levels in AML patients. We found higher plasma IL-15 levels in AML patients compared to HDs confirming previous results (64). In lymphoid leukemia, it has been described an increase of IL-15 and IL-5R (65). It has been also shown that IL-15 could act as a growth factor for a minor fraction of AML cell lines expressing IL-2Rβ/γ promoting their survival and proliferation (66). In addition, the finding that AML patients have higher levels of IL-15 in plasma than HDs together with the possibility that IL-15 may promote blast growth has to be considered in protocols using IL-15 as adjuvant. We were surprised by the contradictory results showing high levels of plasma IL-15 in AML patients, whereas NKp30 expression on NK cells is diminished. It has recently been described that maintained levels of IL-15 may induce NK cell exhaustion (67), and it has been suggested that for *ex vivo* expansion optimal dosing and timing of IL-15 is critical to get adequate NK cell activation (68). We can hypothesize that maintained levels of IL-15 induce changes on NK cell phenotype (e.g., activating receptor expression) that further contribute to the diminished NK cell activity observed in AML patients.

Thus, in spite of the encouraging data of IL-15 immunotherapy in murine models, together with its low toxicity in mice and primates that has led to the design of clinical trials in AML patients, the use of IL-15 *in vivo* as monotherapy or combined therapy in AML patients requires special caution.

A better understanding of IL-15/IL15R axis will allow the identification of novel therapeutic strategies directed to increase IL-15 immunomodulatory effect contributing to antitumor immune response but avoiding the promotion of leukemic cells survival. Our *in vitro* results support the relevance of IL-15 to induce functional active NK cells in AML patients with enhanced capacity to destroy leukemic cells and induce DCs maturation.

ETHICS STATEMENT

The study was approved by the local Ethics Committee and samples collected after written informed consent in accordance with the Declaration of Helsinki.

AUTHOR CONTRIBUTIONS

BS-C performed experiments. BS-C, CC, and AP analyzed data. JB, MJA, and HB selected the patients. BS-C, ED, RS, and RT designed the project and discussed data. BS-C and RT wrote the manuscript with support of all other co-authors.

ACKNOWLEDGMENTS

The following reagent was obtained through the NIH AIDS Reagent Program, Division of AIDS, NIAID, NIH: human rIL-2 from Dr. Maurice Gately, Hoffmann-La Roche Inc.

REFERENCES

1. Farag SS, Caligiuri MA. Human natural killer cell development and biology. *Blood Rev* (2006) 20:123–37. doi:10.1016/j.blre.2005.10.001
2. Long EO, Kim HS, Liu D, Peterson ME, Rajagopalan S. Controlling natural killer cell responses: integration of signals for activation and inhibition. *Annu Rev Immunol* (2013) 31:227–58. doi:10.1146/annurev-immunol-020711-075005
3. Vivier E, Tomasello E, Baratin M, Walzer T, Ugolini S. Functions of natural killer cells. *Nat Immunol* (2008) 9:503–10. doi:10.1038/ni1582
4. Moretta L, Bottino C, Pende D, Castriconi R, Mingari MC, Moretta A. Surface NK receptors and their ligands on tumor cells. *Semin Immunol* (2006) 18:151–8. doi:10.1016/j.smim.2006.03.002
5. Zingoni A, Sornasse T, Cocks BG, Tanaka Y, Santoni A, Lanier LL. Cross-talk between activated human NK cells and CD4+ T cells via OX40-OX40 ligand interactions. *J Immunol* (2004) 173:3716–24. doi:10.4049/jimmunol.173.6.3716
6. Ghiringhelli F, Menard C, Martin F, Zitvogel L. The role of regulatory T cells in the control of natural killer cells: relevance during tumor progression. *Immunol Rev* (2006) 214:229–38. doi:10.1111/j.1600-065X.2006.00445.x
7. Ghiringhelli F, Menard C, Terme M, Flament C, Taieb J, Chaput N, et al. CD4+CD25+ regulatory T cells inhibit natural killer cell functions in a transforming growth factor-beta-dependent manner. *J Exp Med* (2005) 202:1075–85. doi:10.1084/jem.20051511
8. Bottino C, Moretta L, Pende D, Vitale M, Moretta A. Learning how to discriminate between friends and enemies, a lesson from Natural Killer cells. *Mol Immunol* (2004) 41:569–75. doi:10.1016/j.molimm.2004.04.004
9. Moretta L, Bottino C, Pende D, Vitale M, Mingari MC, Moretta A. Different checkpoints in human NK-cell activation. *Trends Immunol* (2004) 25:670–6. doi:10.1016/j.it.2004.09.008
10. Borrego F, Ulbrecht M, Weiss EH, Coligan JE, Brooks AG. Recognition of human histocompatibility leukocyte antigen (HLA)-E complexed with HLA class I signal sequence-derived peptides by CD94/NKG2 confers protection from natural killer cell-mediated lysis. *J Exp Med* (1998) 187:813–8. doi:10.1084/jem.187.5.813
11. Braud VM, Allan DS, O'Callaghan CA, Soderstrom K, D'Andrea A, Ogg GS, et al. HLA-E binds to natural killer cell receptors CD94/NKG2A, B and C. *Nature* (1998) 391:795–9. doi:10.1038/35869
12. Carretero M, Palmieri G, Llano M, Tullio V, Santoni A, Geraghty DE, et al. Specific engagement of the CD94/NKG2-A killer inhibitory receptor by the HLA-E class Ib molecule induces SHP-1 phosphatase recruitment to tyrosine-phosphorylated NKG2-A: evidence for receptor function in heterologous transfectants. *Eur J Immunol* (1998) 28:1280–91. doi:10.1002/(SICI)1521-4141(199804)28:04<1280::AID-IMMU1280>3.0.CO;2-O
13. Iwaszko M, Bogunia-Kubik K. Clinical significance of the HLA-E and CD94/NKG2 interaction. *Arch Immunol Ther Exp (Warsz)* (2011) 59:353–67. doi:10.1007/s00005-011-0137-y
14. Bottino C, Moretta L, Moretta A. NK cell activating receptors and tumor recognition in humans. *Curr Top Microbiol Immunol* (2006) 298:175–82. doi:10.1007/3-540-27743-9_9
15. Champsaur M, Lanier LL. Effect of NKG2D ligand expression on host immune responses. *Immunol Rev* (2010) 235:267–85. doi:10.1111/j.0105-2896.2010.00893.x
16. Kasahara M, Yoshida S. Immunogenetics of the NKG2D ligand gene family. *Immunogenetics* (2012) 64:855–67. doi:10.1007/s00251-012-0638-9
17. Chan CJ, Martinet L, Gilfillan S, Souza-Fonseca-Guimaraes F, Chow MT, Town L, et al. The receptors CD96 and CD226 oppose each other in the regulation of natural killer cell functions. *Nat Immunol* (2014) 15:431–8. doi:10.1038/ni.2850
18. de Andrade LF, Smyth MJ, Martinet L. DNAM-1 control of natural killer cells functions through nectin and nectin-like proteins. *Immunol Cell Biol* (2014) 92:237–44. doi:10.1038/icb.2013.95
19. Fuchs A, Cella M, Giurisato E, Shaw AS, Colonna M. Cutting edge: CD96 (tactile) promotes NK cell-target cell adhesion by interacting with the poliovirus receptor (CD155). *J Immunol* (2004) 172:3994–8. doi:10.4049/jimmunol.172.7.3994
20. Pende D, Bottino C, Castriconi R, Cantoni C, Marcenaro S, Rivera P, et al. PVR (CD155) and Nectin-2 (CD112) as ligands of the human DNAM-1 (CD226) activating receptor: involvement in tumor cell lysis. *Mol Immunol* (2005) 42:463–9. doi:10.1016/j.molimm.2004.07.028
21. Bottino C, Biassoni R, Millo R, Moretta L, Moretta A. The human natural cytotoxicity receptors (NCR) that induce HLA class I-independent NK cell triggering. *Hum Immunol* (2000) 61:1–6. doi:10.1016/S0198-8859(99)00162-7
22. Mandelboim O, Porgador A. NKp46. *Int J Biochem Cell Biol* (2001) 33:1147–50. doi:10.1016/S1357-2725(01)00078-4
23. Sivori S, Vitale M, Morelli L, Sanseverino L, Augugliaro R, Bottino C, et al. p46, a novel natural killer cell-specific surface molecule that mediates cell activation. *J Exp Med* (1997) 186:1129–36. doi:10.1084/jem.186.7.1129
24. Kaifu T, Escaliere B, Gastinel LN, Vivier E, Baratin M. B7-H6/NKp30 interaction: a mechanism of alerting NK cells against tumors. *Cell Mol Life Sci* (2011) 68:3531–9. doi:10.1007/s00018-011-0802-7
25. Pende D, Parolini S, Pessino A, Sivori S, Augugliaro R, Morelli L, et al. Identification and molecular characterization of NKp30, a novel triggering receptor involved in natural cytotoxicity mediated by human natural killer cells. *J Exp Med* (1999) 190:1505–16. doi:10.1084/jem.190.10.1505
26. Cantoni C, Bottino C, Vitale M, Pessino A, Augugliaro R, Malaspina A, et al. NKp44, a triggering receptor involved in tumor cell lysis by activated human natural killer cells, is a novel member of the immunoglobulin superfamily. *J Exp Med* (1999) 189:787–96. doi:10.1084/jem.189.5.787
27. Moretta A, Bottino C, Vitale M, Pende D, Cantoni C, Mingari MC, et al. Activating receptors and coreceptors involved in human natural killer cell-mediated cytolysis. *Annu Rev Immunol* (2001) 19:197–223. doi:10.1146/annurev.immunol.19.1.197

28. Brandt CS, Baratin M, Yi EC, Kennedy J, Gao Z, Fox B, et al. The B7 family member B7-H6 is a tumor cell ligand for the activating natural killer cell receptor NKp30 in humans. *J Exp Med* (2009) 206:1495–503. doi:10.1084/jem.20090681
29. Pogge von Strandmann E, Simhadri VR, von TB, Sasse S, Reiners KS, Hansen HP, et al. Human leukocyte antigen-B-associated transcript 3 is released from tumor cells and engages the NKp30 receptor on natural killer cells. *Immunity* (2007) 27:965–74. doi:10.1016/j.immuni.2007.10.010
30. Arnon TI, Lev M, Katz G, Chernobrov Y, Porgador A, Mandelboim O. Recognition of viral hemagglutinins by NKp44 but not by NKp30. *Eur J Immunol* (2001) 31:2680–9. doi:10.1002/1521-4141(200109)31:9<2680::AID-IMMU2680>3.0.CO;2-A
31. Glasner A, Zurunic A, Meningher T, Lenac RT, Tsukerman P, Bar-On Y, et al. Elucidating the mechanisms of influenza virus recognition by Ncr1. *PLoS One* (2012) 7:e36837. doi:10.1371/journal.pone.0036837
32. Mandelboim O, Lieberman N, Lev M, Paul L, Arnon TI, Bushkin Y, et al. Recognition of haemagglutinins on virus-infected cells by NKp46 activates lysis by human NK cells. *Nature* (2001) 409:1055–60. doi:10.1038/35059110
33. Baychelier F, Sennepin A, Ermonval M, Dorgham K, Debre P, Vieillard V. Identification of a cellular ligand for the natural cytotoxicity receptor NKp44. *Blood* (2013) 122:2935–42. doi:10.1182/blood-2013-03-489054
34. Rosental B, Brusilovsky M, Hadad U, Oz D, Appel MY, Afergan F, et al. Proliferating cell nuclear antigen is a novel inhibitory ligand for the natural cytotoxicity receptor NKp44. *J Immunol* (2011) 187:5693–702. doi:10.4049/jimmunol.1102267
35. Sanchez-Correa B, Bergua JM, Campos C, Gayoso I, Arcos MJ, Banas H, et al. Cytokine profiles in acute myeloid leukemia patients at diagnosis: survival is inversely correlated with IL-6 and directly correlated with IL-10 levels. *Cytokine* (2013) 61:885–91. doi:10.1016/j.cyto.2012.12.023
36. Sanchez-Correa B, Gayoso I, Bergua JM, Casado JG, Morgado S, Solana R, et al. Decreased expression of DNAM-1 on NK cells from acute myeloid leukemia patients. *Immunol Cell Biol* (2012) 90:109–15. doi:10.1038/icb.2011.15
37. Sanchez-Correa B, Morgado S, Gayoso I, Bergua JM, Casado JG, Arcos MJ, et al. Human NK cells in acute myeloid leukaemia patients: analysis of NK cell-activating receptors and their ligands. *Cancer Immunol Immunother* (2011) 60:1195–205. doi:10.1007/s00262-011-1050-2
38. Fauriat C, Just-Landi S, Mallet F, Arnoulet C, Sainty D, Olive D, et al. Deficient expression of NCR in NK cells from acute myeloid leukemia: evolution during leukemia treatment and impact of leukemia cells in NCRdull phenotype induction. *Blood* (2007) 109:323–30. doi:10.1182/blood-2005-08-027979
39. Cany J, van der Waart AB, Spanholtz J, Tordoir M, Jansen JH, van der Voort R, et al. Combined IL-15 and IL-12 drives the generation of CD34-derived natural killer cells with superior maturation and alloreactivity potential following adoptive transfer. *Oncoimmunology* (2015) 4:e1017701. doi:10.1080/2162402X.2015.1017701
40. de Rham C, Ferrari-Lacraz S, Jendly S, Schneiter G, Dayer JM, Villard J. The proinflammatory cytokines IL-2, IL-15 and IL-21 modulate the repertoire of mature human natural killer cell receptors. *Arthritis Res Ther* (2007) 9:R125. doi:10.1186/ar2336
41. Fehniger TA, Cooper MA, Caligiuri MA. Interleukin-2 and interleukin-15: immunotherapy for cancer. *Cytokine Growth Factor Rev* (2002) 13:169–83. doi:10.1016/S1359-6101(01)00021-1
42. Steel JC, Waldmann TA, Morris JC. Interleukin-15 biology and its therapeutic implications in cancer. *Trends Pharmacol Sci* (2012) 33:35–41. doi:10.1016/j.tips.2011.09.004
43. Liao W, Lin JX, Leonard WJ. Interleukin-2 at the crossroads of effector responses, tolerance, and immunotherapy. *Immunity* (2013) 38:13–25. doi:10.1016/j.immuni.2013.01.004
44. Waldmann TA, Dubois S, Tagaya Y. Contrasting roles of IL-2 and IL-15 in the life and death of lymphocytes: implications for immunotherapy. *Immunity* (2001) 14:105–10. doi:10.1016/S1074-7613(09)00091-0
45. Berger C, Berger M, Hackman RC, Gough M, Elliott C, Jensen MC, et al. Safety and immunologic effects of IL-15 administration in nonhuman primates. *Blood* (2009) 114:2417–26. doi:10.1182/blood-2008-12-189266
46. Carotta S. Targeting NK cells for anticancer immunotherapy: clinical and preclinical approaches. *Front Immunol* (2016) 7:152. doi:10.3389/fimmu.2016.00152
47. Chester C, Fritsch K, Kohrt HE. Natural killer cell immunomodulation: targeting activating, inhibitory, and co-stimulatory receptor signaling for cancer immunotherapy. *Front Immunol* (2015) 6:601. doi:10.3389/fimmu.2015.00601
48. Iannello A, Thompson TW, Ardolino M, Marcus A, Raulet DH. Immunosurveillance and immunotherapy of tumors by innate immune cells. *Curr Opin Immunol* (2016) 38:52–8. doi:10.1016/j.coi.2015.11.001
49. Lahm HW, Stein S. Characterization of recombinant human interleukin-2 with micromethods. *J Chromatogr* (1985) 326:357–61. doi:10.1016/S0021-9673(01)87461-6
50. Almeida AM, Ramos F. Acute myeloid leukemia in the older adults. *Leuk Res Rep* (2016) 6:1–7. doi:10.1016/j.lrr.2016.06.001
51. Ruggeri L, Capanni M, Tosti A, Urbani E, Posati S, Aversa F, et al. Innate immunity against hematological malignancies. *Cytotherapy* (2002) 4:343–6. doi:10.1080/146532402760271127
52. Rezvani K, Rouce RH. The application of natural killer cell immunotherapy for the treatment of cancer. *Front Immunol* (2015) 6:578. doi:10.3389/fimmu.2015.00578
53. Pittari G, Fregni G, Roguet L, Garcia A, Vataire AL, Wittnebel S, et al. Early evaluation of natural killer activity in post-transplant acute myeloid leukemia patients. *Bone Marrow Transplant* (2010) 45:862–71. doi:10.1038/bmt.2009.265
54. Khaznadar Z, Boissel N, Agaugue S, Henry G, Cheok M, Vignon M, et al. Defective NK cells in acute myeloid leukemia patients at diagnosis are associated with blast transcriptional signatures of immune evasion. *J Immunol* (2015) 195:2580–90. doi:10.4049/jimmunol.1500262
55. Rosenberg SA. IL-2: the first effective immunotherapy for human cancer. *J Immunol* (2014) 192:5451–8. doi:10.4049/jimmunol.1490019
56. Buyse M, Squifflet P, Lange BJ, Alonzo TA, Larson RA, Kolitz JE, et al. Individual patient data meta-analysis of randomized trials evaluating IL-2 monotherapy as remission maintenance therapy in acute myeloid leukemia. *Blood* (2011) 117:7007–13. doi:10.1182/blood-2011-02-337725
57. Martner A, Rydstrom A, Riise RE, Aurelius J, Anderson H, Brune M, et al. Role of natural killer cell subsets and natural cytotoxicity receptors for the outcome of immunotherapy in acute myeloid leukemia. *Oncoimmunology* (2016) 5:e1041701. doi:10.1080/2162402X.2015.1041701
58. Romee R, Leong JW, Fehniger TA. Utilizing cytokines to function-enable human NK cells for the immunotherapy of cancer. *Scientifica (Cairo)* (2014) 2014:205796. doi:10.1155/2014/205796
59. Miller JS, Rooney CM, Curtsinger J, McElmurry R, McCullar V, Verneris MR, et al. Expansion and homing of adoptively transferred human natural killer cells in immunodeficient mice varies with product preparation and in vivo cytokine administration: implications for clinical therapy. *Biol Blood Marrow Transplant* (2014) 20:1252–7. doi:10.1016/j.bbmt.2014.05.004
60. Szczepanski MJ, Szajnik M, Welsh A, Foon KA, Whiteside TL, Boyiadzis M. Interleukin-15 enhances natural killer cell cytotoxicity in patients with acute myeloid leukemia by upregulating the activating NK cell receptors. *Cancer Immunol Immunother* (2010) 59:73–9. doi:10.1007/s00262-009-0724-5
61. Rettinger E, Kuci S, Naumann I, Becker P, Kreyenberg H, Anzaghe M, et al. The cytotoxic potential of interleukin-15-stimulated cytokine-induced killer cells against leukemia cells. *Cytotherapy* (2012) 14:91–103. doi:10.3109/14653249.2011.613931
62. Vitale M, Della CM, Carlomagno S, Pende D, Arico M, Moretta L, et al. NK-dependent DC maturation is mediated by TNFalpha and IFNgamma released upon engagement of the NKp30 triggering receptor. *Blood* (2005) 106:566–71. doi:10.1182/blood-2004-10-4035
63. Wehner R, Dietze K, Bachmann M, Schmitz M. The bidirectional crosstalk between human dendritic cells and natural killer cells. *J Innate Immun* (2011) 3:258–63. doi:10.1159/000323923
64. Kornblau SM, McCue D, Singh N, Chen W, Estrov Z, Coombes KR. Recurrent expression signatures of cytokines and chemokines are present and are independently prognostic in acute myelogenous leukemia and myelodysplasia. *Blood* (2010) 116:4251–61. doi:10.1182/blood-2010-01-262071
65. Xiong Y, Bensoussan D, Decot V. IL-15 as a potential target in leukemia. *Blood Lymphat Cancer* (2015) 2015(5):55–63. doi:10.2147/BLCTT.S78347
66. Meazza R, Basso S, Gaggero A, Detotero D, Trentin L, Pereno R, et al. Interleukin (IL)-15 induces survival and proliferation of the growth factor-dependent acute myeloid leukemia M-07e through the IL-2 receptor beta/gamma. *Int J Cancer* (1998) 78:189–95. doi:10.1002/(SICI)1097-0215(19981005)78:2<189::AID-IJC12>3.0.CO;2-6

67. Felices M, Lenvik A, Chu S, McElmurry R, Cooley S, Tolar J, et al. Continuous IL-15 signaling leads to functional exhaustion of human natural killer cells through metabolic changes that alters their in vivo anti-tumor activity. *Blood* (2016) 128:551.

68. Granzin M, Wagner J, Kohl U, Cerwenka A, Huppert V, Ullrich E. Shaping of natural killer cell antitumor activity by ex vivo cultivation. *Front Immunol* (2017) 8:458. doi:10.3389/fimmu.2017.00458

18

A Mature NK Profile at the Time of HIV Primary Infection is Associated with an Early Response to cART

Françoise Gondois-Rey[1], Antoine Chéret[2,3,4,5,6,7], Françoise Mallet[1], Ghislain Bidaut[8], Samuel Granjeaud[8], Camille Lécuroux[3,4,5], Mickaël Ploquin[6], Michaela Müller-Trutwin[6], Christine Rouzioux[7,9], Véronique Avettand-Fenoël[7,9], Andrea De Maria[10,11], Gilles Pialoux[12], Cécile Goujard[2], Laurence Meyer[13] and Daniel Olive[1]**

[1] CNRS, INSERM, Institut Paoli-Calmettes, CRCM, Immunity and Cancer Team, Aix Marseille Univ, Marseille, France, [2] APHP, Hôpital Bicêtre, Internal Medecin Unit, Le Kremlin-Bicêtre, France, [3] U1184, Paris-Sud Univ, Le Kremlin-Bicêtre, France, [4] CEA, DSV/iMETI, IDMIT, Fontenay-aux-Roses, France, [5] INSERM U1184, ImVA "Immunology of Chronic Viral Infections and Autoimmune Diseases", Le Kremlin-Bicêtre, France, [6] Institut Pasteur, HIV, Inflammation and Persistance Unit, Paris, France, [7] Virology Laboratory, APHP CHU Necker-Enfants Malades, Paris, France, [8] CNRS, INSERM, Institut Paoli-Calmettes, CRCM, CiBi Platform, Aix Marseille Univ, Marseille, France, [9] EA 7327 Paris Descartes University, Paris, France, [10] Dipartimento di Scienze della Salute and Centro di Eccellenza per la Ricerca Biomedica, Università di Genova, Genova, Italy, [11] Clinica Malattie Infettive, IRCCS Azienda Ospedaliera Universitaria San Martino – Istituto Nazionale per la Ricerca sul Cancro, Genova, Italy, [12] APHP, Hôpital Tenon, Infectious Diseases Unit, Paris, France, [13] INSERM, CESP U1018, Epidemiology and population health, APHP, Hôpital Bicêtre, Paris 11 Univ, Le Kremlin-Bicêtre, France

Edited by:
*Gianfranco Pittari,
Hamad Medical Corporation, Qatar*

Reviewed by:
*Anthony Jaworowski,
Burnet Institute, Australia
Miguel López-Botet,
Institut Hospital del Mar
d'Investigacions Mèdiques
(IMIM), Spain*

***Correspondence:**
*Françoise Gondois-Rey
francoise.gondois-rey@inserm.fr;
Daniel Olive
daniel.olive@inserm.fr*

Natural killer (NK) cells are major effectors of the innate immune response. Despite an overall defect in their function associated with chronic human immunodeficiency virus (HIV) infection, their role in primary HIV infection is poorly understood. We investigated the modifications of the NK cell compartment in patients from the ANRS-147-Optiprim trial, a study designed to examine the benefits of intensive combination antiretroviral therapy (cART) in patients with acute or early primary HIV infection. Multiparametric flow cytometry combined with bioinformatics analyses identified the NK phenotypes in blood samples from 30 primary HIV-infected patients collected at inclusion and after 3 months of cART. NK phenotypes were revealed by co-expression of CD56/CD16/NKG2A/NKG2C and CD57, five markers known to delineate stages of NK maturation. Three groups of patients were formed according to their distributions of the 12 NK cell phenotypes identified. Their virological and immunological characteristics were compared along with the early outcome of cART. At inclusion, HIV-infected individuals could be grouped into those with predominantly immature/early differentiated NK cells and those with predominantly mature NK cells. Several virological and immunological markers were improved in patients with mature NK profiles, including lower HIV viral loads, lower immune activation markers on NK and dendritic cell (DC), lower levels of plasma IL-6 and IP-10, and a trend to normal DC counts. Whereas all patients showed a decrease of viremia higher than 3 $\log_{10}$ copies/ml after 3 months of treatment, patients with a mature NK profile at inclusion

Abbreviations: PBMCs, peripheral blood mononuclear cells; NK cell, natural killer cell; PTC, posttreatment controller; KIR, killer-cell immunoglobulin-like receptor; HIV, human immunodeficiency virus; PMT, photo-multiplicators.

reached this threshold more rapidly than patients with an immature NK profile (70 vs. 38%). In conclusion, a better early response to cART is observed in patients whose NK profile is skewed to maturation at inclusion. Whether the mature NK cells contributed directly or indirectly to HIV control through a better immune environment under cART is unknown. The NK maturation status of primary infected patients should be considered as a relevant marker of an immune process contributing to the early outcome of cART that could help in the management of HIV-infected patients.

Keywords: HIV, primary infection, NK cells maturation, cART, memory-like NK

INTRODUCTION

Natural killer (NK) cells are one of the major innate immune components involved in the rapid response of the host to invading virus (1). Their function is probably crucial at the time of infection and can impact the quality of adaptive immune responses and the overall outcome of infections. NK cell activity is regulated by activating and inhibitory receptors, but their effector functions are intrinsically linked to their maturation (2). Cytolysis is the typical NK cell function, but NK cell also play an antiviral role through the release of soluble factors, such as IFN-γ and TNF-α (3), which activate T cells, macrophages, and dendritic cells (DCs). $CD56^{bright}$ NK cells are described as the progenitors of $CD56^{dim}$, the latter being endowed with the main NK cell effector functions (2). $CD56^{dim}$ cells sequentially progress from an immature population, characterized by a high degranulation and proliferation potential, to a terminally differentiated population, characterized by potent cytokine production at the expense of cell division and degranulation (4). Immature NK cells express NKG2A, a C-type lectin receptor forming an inhibitory heterodimer with CD94 to interact with HLA-E on target cells (5). HLA-E is a non-classical major histocompatibility complex (MHC) class I molecule whose expression is enhanced on infected cells through the presentation of viral peptides (6). During maturation, NKG2A loss is compensated by the acquisition of self-inhibitory killer-cell immunoglobulin-like receptor (KIR) expression, while CD57, a marker of senescence, is acquired (4). NKG2C/CD94 is the activating alternative receptor of HLA-E on NK cells (7). This receptor was initially described on a subset of NK cells expanded during CMV infection (8, 9), but recently, other viruses, including human immunodeficiency virus (HIV), were also shown to drive $NKG2C^{+}$ NK cell expansion, in the context of CMV co-infection (10, 11). The persistence of a $NKG2C^{+}CD57^{+}$ NK cell subset for more than 1 year after CMV or Hantavirus infection has led to the proposition that they are a memory-like form of NK cell (8, 12, 13). Therefore, the co-expression of CD56, CD16, NKG2A, NKG2C, and CD57 delineates sequential stages of the NK cell maturation process suggesting the acquisition of typical effector functions.

Natural killer cell functions are affected early after HIV infection (14). In addition, many modifications of the NK cell compartment, including decrease of $CD56^{bright}$, expansion of $CD56^{dim}$, and appearance of a functionally compromised subset of $CD56^{dim}$ expressing low levels of CD56 or CD16 were reported (15). Inversion of the ratio of NKG2A to NKG2C was described in primary HIV-infected patients (11, 16). In chronically infected patients, an overall increase of mature $CD57^{+}$ NK cells was observed (17). In cohort of individualist risk of HIV infection, NK cell activation at the time of primary infection has been both positively and negatively correlated with the risk of HIV acquisition (18, 19). Recently, a correlation was demonstrated between NK cell repertoire diversity, linked to progression to maturity, and increased susceptibility to HIV infection (20). Therefore, while NK maturation seems to be an interesting parameter in HIV infection, so far, the impact of the overall maturation of NK cells on the outcome of primary HIV-infection (PHI) remains elusive.

The Optiprim trial was designed to evaluate to what extent intensive antiviral therapy started during primary HIV infection contributes to a decrease in the size of HIV reservoirs and helps to achieve a so-called posttreatment controller (PTC) status (21). A sub-study was designed to investigate innate immune parameters. Considering the important role of NK cell maturation for their effector properties, we investigated the NK cell compartment with a combination of markers known to characterize sequential steps of the NK cell differentiation pathway. Thirty primary HIV-infected patients peripheral blood mononuclear cell (PBMC) samples were investigated at inclusion and 3 months after the onset of combination antiretroviral therapy (cART). Because cell populations expressing unexpected combinations of markers might be expanded in pathological conditions, new bioinformatics methods were applied to analyze multiparametric cytometry data in an unsupervised approach (22). This allowed the identification of a relationship between profiles of NK cells skewed to immaturity or maturity and virological and immune parameters reached naturally a few weeks after infection and after early cART.

SUBJECTS AND METHODS

Ethical Statement

All study participants provided written informed consent.

The study was approved by the Sud-Mediterranee-1 Ethics Committee and the French Health Products Safety Agency and complied with the Helsinki Declaration.

Study Population

Human immunodeficiency virus-1-infected subjects with PHI were included in a multicenter phase 3 randomized trial (ANRS-147 OPTIPRIM) (www.ClinicalTrials.gov, number NCT01033760). The endpoint was the impact of intensive vs. standard cART at month 24 on blood HIV-DNA levels. The results of this study have been published (21). Antoine Chéret was the Principal Investigator, Laurence Meyer the Methodological Investigator, and Christine Rouzioux the Virologist Investigator. We proposed a sub-study where the participants would give blood samples at day 0 (before cART initiation) and month 3. This sub-study, in which Daniel Olive and Françoise Gondois-Rey were the investigators, was designed to investigate parameters of innate immunity linked to cART efficacy. Among 90 patients in the main study, 30 patients were randomly included in this sub-study. This work shows original data on NK cells, DC, CMV, and plasma cytokines, and uses information from the main study. Patient characteristics are listed in **Table 1**.

Fifteen healthy donor samples were obtained from the French Blood Bank (EFS, Etablissement Français du Sang) as controls. PBMC samples were frozen and kept in liquid nitrogen until tested.

Flow Cytometry

Peripheral blood mononuclear cells were stained with multiparametric panels containing 9 or 12 fluorescent markers, respectively, designed to investigate NK and DC populations. The NK panel contained NKG2A-PacBlue (clone Z199, home-made), live–dead Aqua (Life Technology), CD57-FITC (Beckman Coulter; 1/30), NKG2C-PE (R&D; 1/40), CD14-PC5 (Beckman Coulter; 1/30), CD19-PC5 (Beckman Coulter; 1/30), CD56-PC7 (Beckman Coulter; 1/30), CD3-AF700 (BD Biosciences; 1/40), and CD16-APCH7 (BD Biosciences; 1/40). The DC panel contained live–dead Aqua (Life Technology), BDCA2-FITC (Miltényi; 1/30), CD123-PercpCy5.5 (BD Biosciences; 1/20), HLA-DR-ECD (Beckmann Coulter; 1/40), CD3-PC5 (BD Biosciences; 1/40), CD56-PC5 (Beckman Coulter; 1/30), CD19-PC5 (Beckman Coulter; 1/40), CD33-PC7 (BD Biosciences; 1/40), CD14-APCH7 (BD biosciences; 1/40), and CD16-AF700 (BD Biosciences; 1/40). Cells were incubated for 20 min at RT with reagents premixed in PBS, washed, and then fixed with 4% PFA. Data were acquired on a LSRII-SORP (BD Biosciences) equipped with four lasers (405 nm/100 mW, 488 nm/100 mW, 560 nm/50 mW, and 630 nm/40 mW). Photo-multiplicators were set using unstained and fully stained samples and linked to the cytometer standardization using the acquisition setting tool. Also, $7.74 \times 10^6 \pm 3 \times 10^5$ events were recorded. Compensations were performed with beads individually stained with corresponding reagents.

TABLE 1 | Patients' characteristics at inclusion.

Number of patients	30
Number of patients acutely infected	14
Time between estimated date of infection and enrollment (days)	34 (20–55)
Age (years)	39.4 (23–55)
Number of patients handled by intensive combination antiretroviral therapy	15
CD4 counts at T0 (count/μL)	550 (323–1,012)
CD8 counts at T0 (count/μL)	1,704 (417–8,157)
CD4 to CD8 ratio at T0	0.47 (0.08–1.32)
Human immunodeficiency virus (HIV)-RNA at T0 (log_{10} copies/mL)	5.51 (3.2–7)
HIV-DNA at T0 [log_{10} copies/10^6 peripheral blood mononuclear cell (PBMC)]	3.747 (2.78–4.68)
CD4 counts at M3 (count/μL)	656 (275–1,244)
CD8 counts at M3 (count/μL)	705 (371–1,282)
HIV-RNA at M3 (log_{10} copies/mL)	1.88 (1.3–3.42)
HIV-DNA at M3 (log_{10} copies/10^6 PBMC)	2.93 (2.25–3.62)

Mean values and (range) are indicated.

Flow Cytometry Data Analysis

Data were exported and analyzed with FlowJo (version 9-2, MacOS X). NK cells were gated as $CD3^-CD14^-CD19^-CD56^+CD16^+$, $CD3^-CD14^-CD19^-CD56^+CD16^-$, or $CD3^-CD14^-CD19^-CD56^-CD16^+$ (Figure S1 in Supplementary Material). Contaminating non-classical monocytes represented less than 0.11% of the gated events (not shown). The remaining data ($3.5 \times 10^5 \pm 2.5 \times 10^4$ events) were exported to create new files further subjected to automated gating. DC's were defined as $CD3^-CD14^-CD19^-CD56^-CD16^-HLA\text{-}DR^+$live cells. In the DC gate, the pDC were gated as $CD33^{low}BDCA2^+CD123^+$ cells, the mDC as $CD33^{high}BDCA2^-CD123^-$ cells.

Automatic Clustering

The flowClust implementation version 3.4.11on R version 3.1.2 under Linux Cent OS 6 was applied on the following parameters: CD56-PC7, CD16-APCH7, NKG2A-PacBlue, NKG2C-PE, CD57-FITC, to compute clusters as described (22). The number of clusters that fit the data optimally was estimated by computing flowClust on a predefined range and comparing them with the statistical criteria BIC and ICL (not shown). This number was estimated to be above 20; therefore, 27 clusters were computed in the events within the NK gate of the 60 samples (30 patients, T0, and M3) and one healthy donor. Two samples failed to be computed.

MFI values and event counts of the 1,593 clusters generated after computation were exported in an Excel table.

Multiparametric Data Management with MeV

MeV (version 4.9.0, https://sourceforge.net/projects/mev-tm4/) (23) was used to visualize and group multiparametric clusters using hierarchical clustering (HCL) of their centers of CD56, CD16, CD57, NKG2A, and NKG2C MFI. Euclidean distance and average linkage were chosen. Prior to MeV, centers were rescaled to adjust the 5 (respectively, 95) percentile of each dimension to −3 (respectively, +3) and normalized. The tree was cut interactively using objective MeV tools and color interpretation of the heatmap to define populations as homogeneous groups of clusters. Names were interactively given according to the comparisons of signatures between groups and the expression of known NK cell markers. The blocks were saved and imported in a spreadsheet program, leading to a matrix with a population identifier column associated to the initial count of events. The percentages of each population of each patient sample were summarized using pivot tables.

Virus Quantification

Human immunodeficiency virus-RNA was quantified in plasma by real-time RT-PCR with the Cobas TaqMan HIV1 v2.0 assay (Roche Diagnostics). Threshold values were arbitrarily given to samples below the threshold of the assay (20 RNA copies/mL). Total HIV-DNA was quantified by ultra-sensitive real-time PCR in PBMC using the Generic HIV-DNA assay from BioCentric (Bandol, France) as described (24).

Cytokine Quantification

Plasma Cytokines

IP-10 concentrations were determined in stored plasma or serum samples (−80°C) by specific enzyme-linked immunosorbent assay, human Quantikine CXCL10 (R&D Systems, Minneapolis, MN, USA) according to the manufacturers' instructions. Levels of IL-6 were measured in frozen plasma samples with specific ELISA assays (Human IL-6 Platinum ELISA, eBioscience). Samples with undetectable levels of IL-6 were arbitrarily attributed half the minimal detectable value (0.46 pg/mL).

Statistics

Statistical graphics were performed with Prism 6 software. The Kruskall–Wallis test followed by multiple comparison Dunn's posttest were used to compare variables between groups. Correlations were evaluated by using simple linear regression analysis and Spearman's rank correlation test.

RESULTS

Automatic Clustering Revealed NK Cell Differentiation Subtypes within HIV-Primary Infected Patients' PBMCs

We investigated the phenotypes of differentiation of NK cells with multiparametric cytometry using CD56, CD16, NKG2A, CD57, and NKG2C, known to delineate sequential stages of NK cell maturation (2, 4). In order to discover unexpected populations, multi-stained samples were analyzed through unsupervised computation of clusters, using a method previously validated (22). Computed clusters were visualized in a MeV heatmap according to their normalized MFI for the five markers and merged to identify populations as homogeneous groups of clusters (**Figure 1**). The tree was interactively cut to summarize the results in 19 groups, using objective tools of MeV and a subjective overview of the heatmap (**Figure 1**). Six of them including only a few clusters were excluded because of their rare representation. Thirteen groups were named according to comparisons of marker MFI and homogeneity with known NK population signatures, a process similar to manual gating of cytometry data visualized in dot plots (**Figure 1**). Accordingly, two groups of clusters expressing the highest levels of CD56, of NKG2A, no or low levels of CD16, a typical $CD56^{bright}$ signature, were thus named. Seven populations expressing medium levels of CD56 and high levels of CD16 were identified as $CD56^{dim}$. They included several groups differentially expressing NKG2A and CD57, two markers linked to immaturity or maturity, respectively, and NKG2C, a NK cell receptor expanded during CMV infections. Four populations expressing low levels of CD56 or CD16 previously described in HIV-infected patients as dysfunctional NK populations were identified (15). Finally, phenotypes of NK cells usually found in human PBMC, including populations specifically expanded during HIV infection, and new phenotypes characterized by unexpected combinations of markers were found in patients' samples using the automatic clustering and interactive merging approach. It should be emphasized that other analyses of the same data, based on the choice of higher or lower numbers of populations could have also been pertinent. As an example, the few clusters characterized by $CD56^{bright}CD16^{neg}NKG2C^{+}$ visible at the bottom of the tree were included in the $CD56^{bright}$ pool while other cuts of the tree could have separated this original combination of markers.

The Distribution of the Various Phenotypes of NK Cells Defined Groups of HIV-Primary Infected Patients with Different Maturation Profiles

We next determined how these NK populations were individually distributed among the 30 primary HIV-infected patients. The blocks of clusters were imported into a spreadsheet program, leading to a matrix with a population name, associated with initial counts of events and patient number. The frequencies of the 12 populations within the total NK cells of each patient were calculated and summarized using pivot tables (Table S1 in Supplementary Material). One population ($CD56^{neg}CD16^{+}CD57^{+}$), although present in different patients, represented less than 0.5% of total NK cells and was excluded from the analysis. Frequencies of the 12 remaining populations were visualized in a MeV heatmap and used to cluster patients (**Figure 2A**). Three groups of patients, named X, Y, and Z appeared on the map. Group X was characterized by the highest frequencies of $CD56^{dim}NKG2A^{+}$ and $CD56^{neg}CD16^{+}NKG2A^{+}$ NK cells. Group Y showed high frequencies of $CD56^{dim}NKGC2^{+}CD57^{+}$and $CD56^{dim}CD57^{+}$ NK cells. Group Z contained mainly a population of $CD56^{dim}$ $CD57^{-}NKG2A^{-}$ $NKG2C^{-}$ (**Figure 2A**).

We checked whether the frequencies of the most frequent NK cell populations directing the clustering (the four populations on top of the list) were significantly different between the three groups (**Figure 2B**). Group X showed significant higher frequencies of $CD56^{dim}NKG2A^{+}$ than Y and Z (34% vs., respectively, 13.8 and 14.3%). Group Y showed significant higher frequencies of $CD56^{dim}CD57^{+}$ NK cells than X and Z (21% vs., respectively, 2.9 and 7.2%), and higher frequencies of $CD56^{dim}NKG2C^{+}CD57^{+}$ than X and Z (17.8% vs., respectively, 2.9 and 0.9%). Group Z showed 42.6% of $CD56^{dim}$ $NKG2A^{-}NKG2C^{-}CD57^{-}$ while this phenotype represented only 12 or 16% of NK cells in, respectively, X and Y. Accordingly, the three groups of patients were significantly characterized by, respectively, high proportion of $CD56^{dim}NKG2A^{+}$ for group X, high proportion of $CD56^{dim}CD57^{+}$ NK cells for group

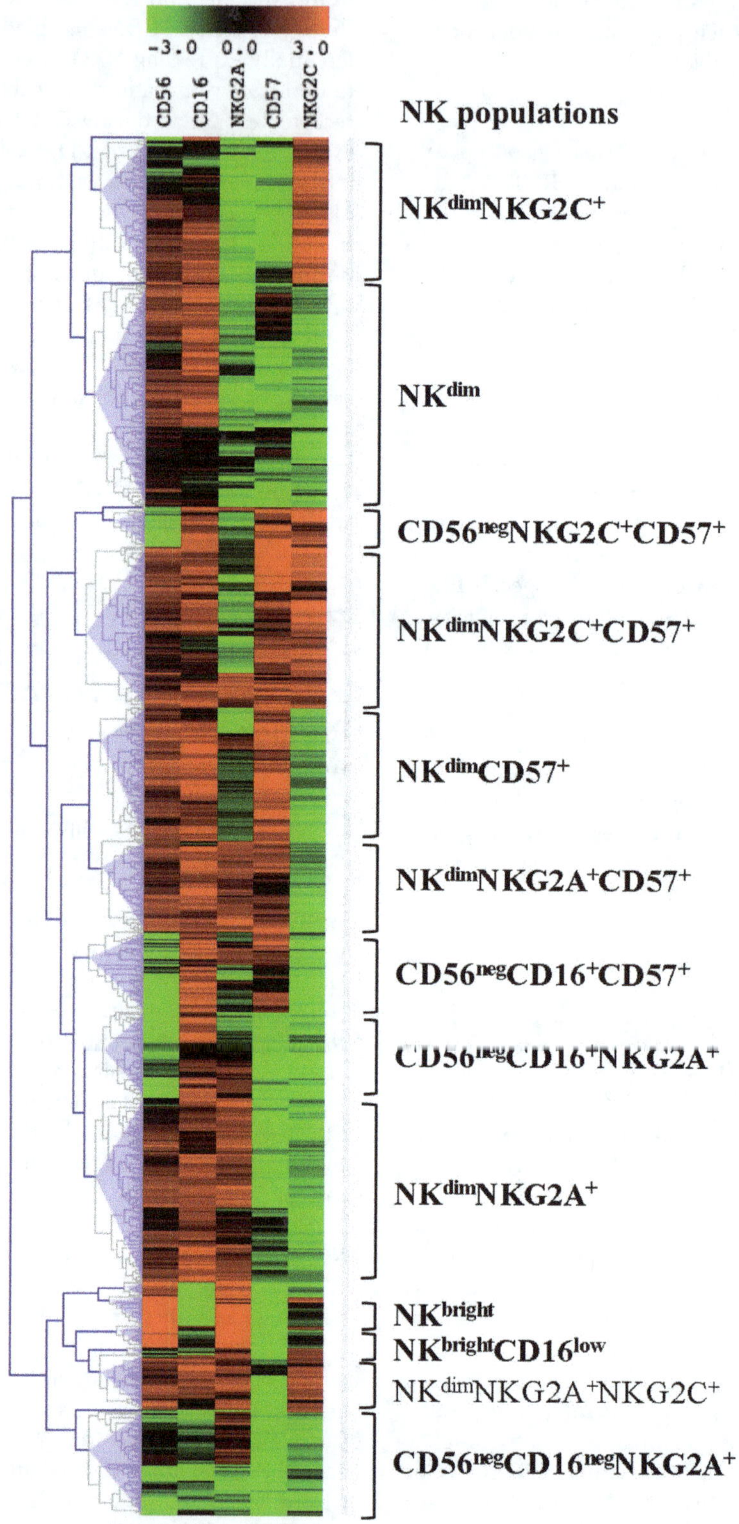

FIGURE 1 | Identification of natural killer (NK) cell populations. Visualization of the 756 clusters computed into the NK cells files from the 28 T0 patients' samples in a MeV heatmap. The clusters are visualized and merged according to their normalized MFI for the five markers shown on top using the MeV software hierarchical clustering tools. NK populations are defined as groups of merged clusters and annotated according to their signatures shown on the right.

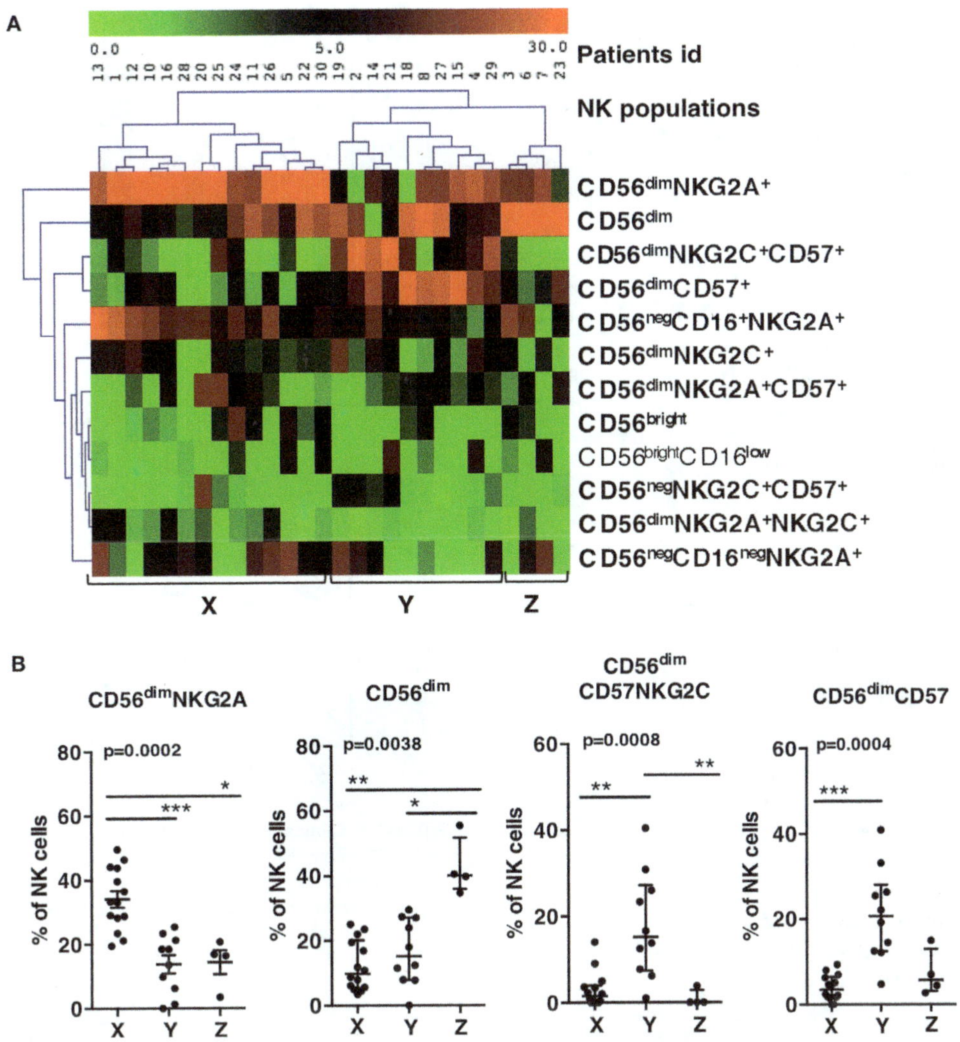

FIGURE 2 | Patients grouping according to the frequencies of groups of natural killer (NK) populations. (A) Double clustering of the 12 most frequent NK populations shown on right (lines) according to their percentages among NK cells of patients (columns). The squares display the respective frequencies according to the scale of color shown on top. **(B)** Patients X, Y, and Z display significant differences of percentages of CD56^{dim} NKG2A$^+$, CD56^{dim}, CD56^{dim}CD57$^+$NKG2C$^+$, and CD56^{dim}CD57$^+$ populations. Errors bars represent the median and the interquartile range. *p*-Values from Kruskall–Wallis test are indicated on top of groups, *p*-Values from Dunn's multiple comparison posttest on top of pairs: *<0.05; **<0.01; ***<0.001.

Y, and high proportion of CD56^{dim} NKG2A$^-$NKG2C$^-$CD57$^-$ for group Z.

NK Cell Profiles of Patients Correlated with HIV Viral Load at Inclusion

To evaluate the relationship between NK profiles of patients and HIV infection, the clinical and virological characteristics of patients within groups were compared at T0 (**Table 2**). Primary infection was defined by detectable plasma HIV-RNA and incomplete HIV-1 western blot, acute infection by one band or fewer (21) and early infection by more than one band. Four out of 14 patients from group X, 4 out of 10 patients of group Y, and all 4 patients of group Z were acutely infected (**Figure 3A**). The mean age of patients from each group was not significantly different (37, 40, and 42 years for, respectively, X, Y, and Z), nor was the mean of estimated time since infection and enrollment (36.5, 35.3, and 34.7 days for, respectively, X, Y, and Z) (**Table 2**). T-CD4 counts were similar for all groups, although group Z patients showed a trend to lower levels (**Table 2**). T-CD8 counts were significantly higher in X than Y and Z (mean of 2,148 counts/μL vs. 1,308 and 1,415 for, respectively, Y and Z) (**Table 2**).

A striking difference was observed when the viral load was compared between the three patient groups. Patients from groups X and Z displayed significantly higher mean HIV-RNA levels than patients from group Y (respectively, 5.77 $\log_{10}$ HIV-RNA copies/mL and 6.05 $\log_{10}$ HIV-RNA copies/mL vs. 4.88 $\log_{10}$ HIV-RNA copies/mL in group Y), while HIV-DNA were not significantly different despite a trend to lower levels for Y (**Figure 3B**). Thus, the major differences observed between the

TABLE 2 | Groups characteristics.

	Groups of patients		
Characteristics	**X**	**Y**	**Z**
Number of patients	14	10	4
Acute (% in the group)	28	40	100
Time between estimated date of infection and enrollment (days)	36 (23–55)	35 (22–46)	35 (32–41)
Age (years)	37 (23–64)	40 (23–62)	42 (24–55)
CD4 counts (count/μL)	549 (323–1,012)	584 (368–864)	430 (341–513)
CD8 counts (count/μL)	2,148 (502–8,157)	1,308 (417–2,716)	1,415 (1,140–1,966)
CD4 to CD8 ratio	0.41 (0.08–1.3)	0.57 (0.23–1.1)	0.31 (0.2–0.35)
Human immunodeficiency virus (HIV)-RNA (log_{10} cp/mL)	5.77 (4.6–7)	4.88 (3.2–5.7)	6.05 (5.6–7)
HIV-DNA (log_{10} cp/10^6 peripheral blood mononuclear cell)	3.8 (3.2–4.7)	3.5 (2.8–4.3)	3.9 (3.2–4.5)
Progenitors [% of natural killer (NK)]	7.7 (3–20)	4 (0–13.9)	6.3 (0.9–9.5)
Effectors (% of NK)	49.2 (29–66)	20.8 (8–28)	0
Intermediate (% of NK)	17.5 (3.5–29)	22.7 (10–39.5)	45.7 (40.5–56)
Mature (% of NK)	6.8 (0–16)	39 (21–54)	8.2 (2.7–15)
Dysfunctional (% of NK)	15 (5.7–41)	10.6 (3.2–21.6)	11.7 (2.3–16.3)

T0 mean values and (range) are indicated.

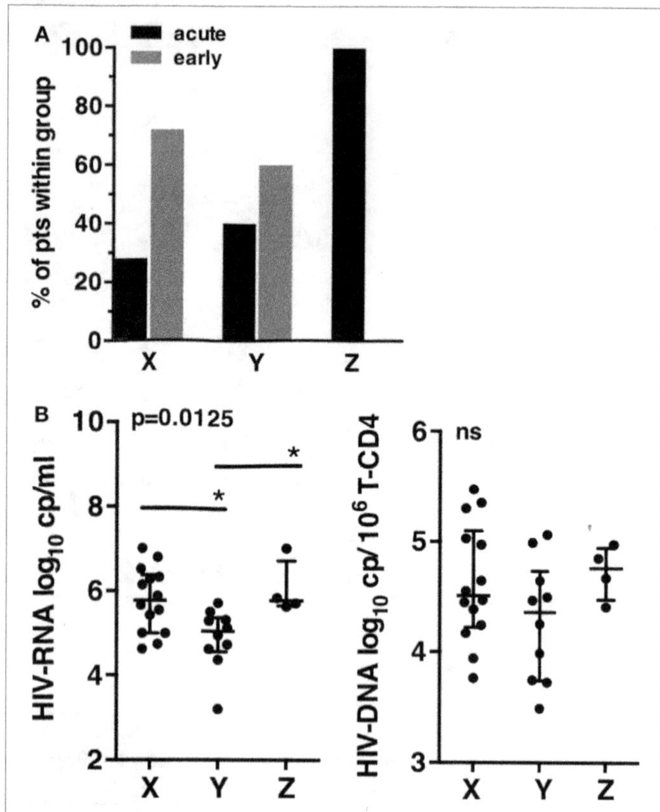

FIGURE 3 | Clinical and virological characteristics of patients groups at T0. (A) Primary infection status of patients according to the X, Y, and Z groups. **(B)** Human immunodeficiency virus (HIV) viral load. Plasma HIV-RNA (left), HIV-DNA copies associated to millions of T-CD4 cells (right). Error bars represent the median and interquartile range. *p*-Values from Kruskall–Wallis are indicated on top of groups, *p*-Values from Dunn's multiple comparison posttest on top of pairs: *<0.05; **<0.01; ***<0.001.

three groups of patients at inclusion highlighted the lower viral load of group Y, the group defined by an expansion of CD57$^+$ NK cells.

Patients with High Frequencies of Mature NK Cells Displayed Better Immunological Parameters at Inclusion

We next addressed the question of whether patients of group X, Y, and Z were different with respect to other major disease progression markers. We analyzed plasma inflammatory markers (IL-6, IP-10) (25, 26) and also focused on markers of innate immune activation and exhaustion, including CD38 expression on NK cells, CD86 expression on monocytes, PDL-1 expression on mDC (27), and pDC and mDC frequencies (28). These markers were all described to be linked to viral load and disease progression (29). They were compared between the groups of patients and some of them were compared to a group of 15 healthy donors (**Figure 4**).

Patients of group X displayed higher IL-6 and IP-10 plasma levels than patients of group Y (**Figure 4A**). CD38 expression on NK cells was significantly increased in all patients groups as compared to healthy donors. CD86 on monocytes and PDL-1 on mDC were significantly increased in patients X as compared to healthy donors, while patients Y showed a trend to lower increase and values of PDL-1 not different from that of healthy donors (**Figure 4B**). All patients groups showed significant decrease of pDC as compared to healthy donors, but patients Y showed a clear trend to higher values of pDC frequencies (**Figure 4C**). Only patients X and Z showed significant decrease of mDC frequencies while patients Y exhibited values similar to healthy donors (median of, respectively, 0.3 vs. 0.33%) (**Figure 4C**). Therefore, in addition to exhibit the lowest activation, Y patients displayed low exhaustion of DC, as shown by a trend to higher frequencies of pDC and frequencies of mDC not different from that of healthy donors (**Figure 4C**). As expected according to their high viremia levels, patients from groups X and Z displayed higher immune activation and exhaustion while patients Y exhibited a better immune status in PHI.

Patients with High Frequencies of Mature NK Cells Displayed a Lower Viral Load after 3 Months of cART

We then assessed the impact of the different virological and immunological status at inclusion on early cART outcome. While all patients showed a 3-log decrease of viral load after 3 months of cART (**Table 2**), we searched for those patients having reached a viral load below 50 copies/mL at M3. Notwithstanding the cART regimen, 6 out of 14 (57%) patients X, 1 out of 4 (25%) patients

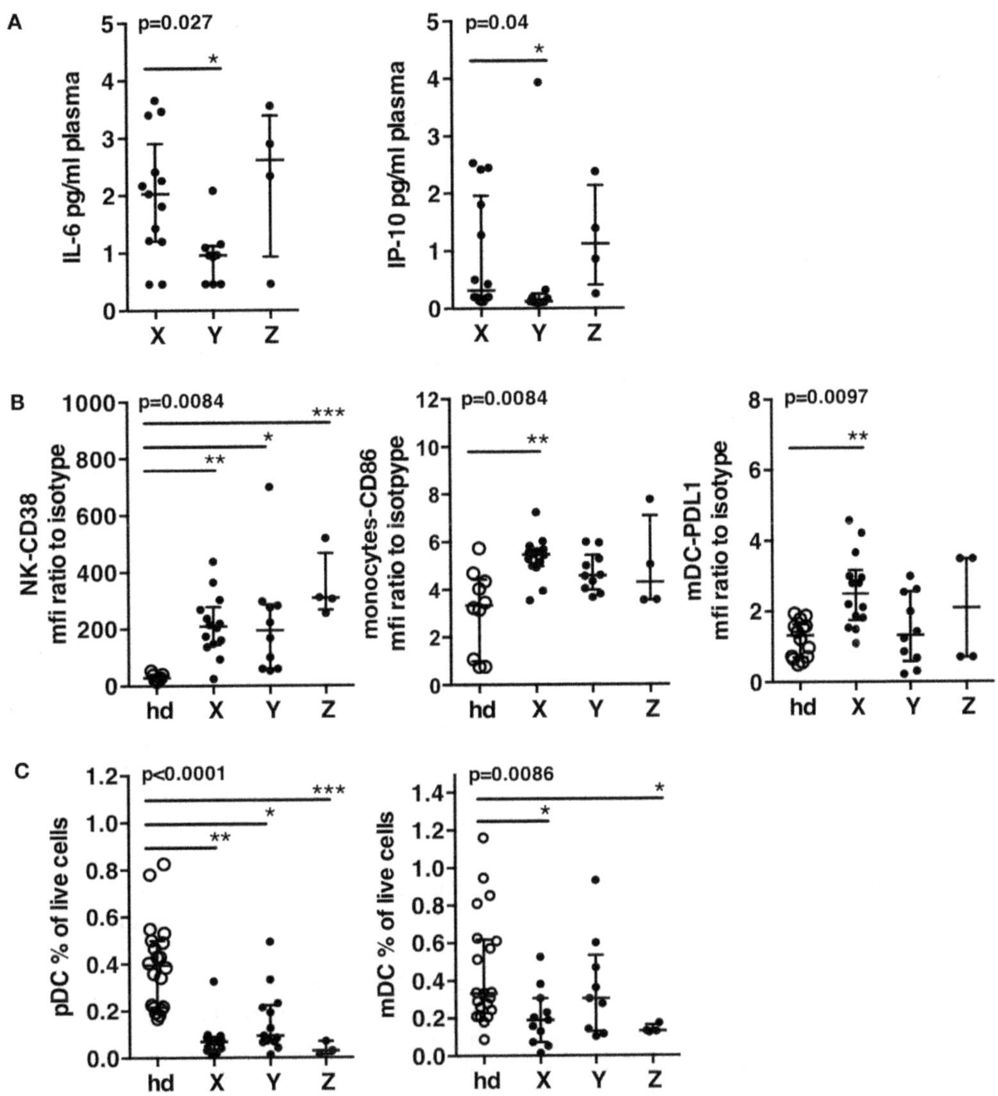

FIGURE 4 | Immune status of patient groups at T0. (A) Plasma levels of IL-6 and of IP-10. **(B)** Innate immune cells activation of patients groups and a group of 15 healthy donors. CD38 expression on natural killer cells, CD86 expression on monocytes, PDL-1 expression on mDC. **(C)** Dendritic cell exhaustion. Frequencies of pDC's and mDC's in the patients groups and in a group of 15 healthy donors. Errors bars represent the median and interquartile range. *p*-Values from Kruskall–Wallis test are indicated on top of groups, *p*-Values from Dunn's multiple comparison posttest on top of pairs: *<0.05; **<0.01; ***<0.001.

Z, and 7 out of 10 (70%) patients Y reached a viral load below 50 copies/mL at M3. Taken together, only 38% of patients lacking significant frequencies of $CD57^+$ NK cells (X + Z) reached a threshold of 50 $\log_{10}$ HIV-RNA copies/mL at M3 whereas 70% of patients with high frequencies of $CD57^+$ NK cells (Y) could reach it.

Early cART Modestly Modify the Frequencies of Mature $CD57^+$ NK Cells at M3

In order to evaluate the kinetics of NK cell maturation at a short time after infection, we compared the frequencies of $CD57^+$ NK cells at inclusion and after 3 months of cART in groups of patients X, Y, and Z (**Figure 5**). All $CD57^+$ NK cells were summed. HIV-RNA showed a mean decrease of 2 logs in both groups after 1 month of treatment, one additional log being lost during the third month (left graph). In the meantime, the frequency of CD57 NK cells remained unchanged in patients from group Y (47.5% at T0 vs. 48.9% at M3), while the frequencies of CD57 populations of patients from group X and Z significantly increased, or alternatively the proportion of immature NK cells decreased (13.9% at T0 vs. 20.5% at M3 for X; 11.7 vs. 19.4% for Z). Therefore, even if the frequencies of CD57 NK cells increased during the 3 months of cART, the values reached remained far from those of patients from group Y at T0 (47.5%).

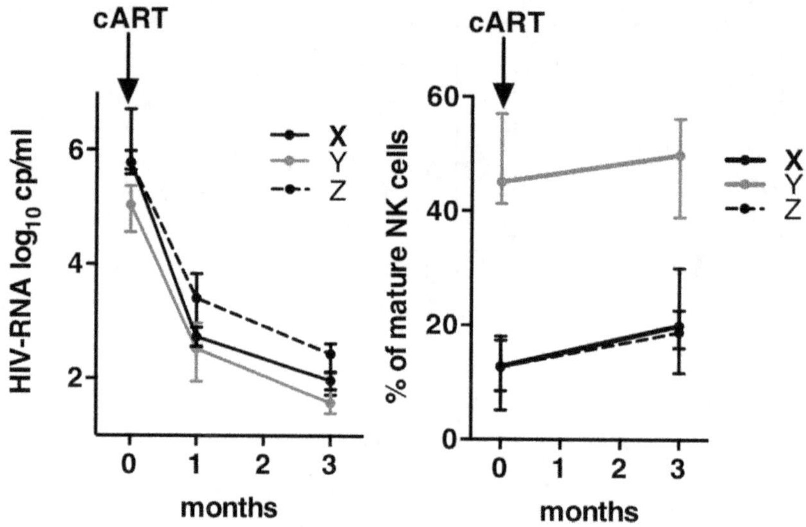

FIGURE 5 | Kinetics of viral load and of mature natural killer (NK) cells frequencies during 3 months of combination antiretroviral therapy (cART). Left graph shows the kinetics of human immunodeficiency virus-RNA between T0 and M3 after cART onset. Right graph shows the kinetics of the frequencies of mature NK cells between T0 and M3. Errors bars represent the median and interquartile range.

DISCUSSION

Natural killer cells are potent effectors of the innate immune system and key actors in the race engaged between the virus and the host. NK cell compartment is constituted by populations more or less advanced on the maturation pathway, which compose a unique landscape at an individual level. We studied the NK cell compartment of HIV-infected patients at the time of primary infection. In spite of limitations such as the number of patients and the lack of samplings before infection, our results globally support a link between NK cell compartment skewed toward maturation and decreased levels of viral load and immune activation at the time of the primary HIV infection.

While multiparametric cytometry allows deep investigation of human immune cells, discrete subsets resulting from unexpected combinations of markers are found by chance using classical manual analysis. Computation, through the consideration of all parameters at the same time allowed exploration of the full dataset and finding discrete subsets in an unsupervised approach (22). The description of the NK cells of patients into 12 subsets, whereas only six would have been searched according to previous reports (CD56^{bright}, four populations of CD56^{dim} and CD56^{neg}), provides an added value to the comprehension of the ongoing interplays between NK cell and HIV. One example is the finding of four different CD56^{neg} subsets and their clustering near four different NK cell populations, including CD56^{bright}. This result suggests that the defect affecting the NK cell can touch the lineage at different levels of maturation.

Among 12 NK populations, four co-expressed NKG2C in various combinations spanning the whole NK maturation process. The clustering of the CD56^{dim}NKG2A^{+}NKG2C^{+} subset near CD56^{bright}CD16^{low}, previously described at the transition between CD56^{bright} and CD56^{dim} (30), suggests they originate from the progenitor. NKG2C^{+} NK cell expansion was reported previously in HIV primary infection (10, 11, 16) and described during Hantavirus and Chikungunya acute infections (12, 31), always in the context of underlying CMV co-infection. The identification of discrete NKG2C^{+} NK cells subsets with a phenotype close to the NK cell progenitor suggests an ongoing generation during primary HIV infection.

The individual distribution of those 12 NK cell populations permits constitution of homogeneous and significantly different groups of patients. Group X included 14 patients whose NK cells were mainly composed of CD56^{dim}NKG2A^{+}, group Y included 10 patients with high frequencies of CD56^{dim}CD57^{+} and CD56^{dim}CD57^{+}NKG2C^{+}, and group Z included four patients whose NK cells displayed a CD56^{dim}NKG2A^{-}NKG2C^{-}CD57^{-} phenotype. To understand the role of those differences on HIV primary infection, clinical, virological, and immunological characteristics of the patients groups were compared. Kinetics of HIV infection could have been involved, but this hypothesis was not sustained by the comparison of the estimated time since infection or by the status of antibodies developed to HIV, defining an acute or early primary infection. The striking difference was the viral load at inclusion: patients with high frequencies of CD56^{dim}CD57^{+} NK cells (group Y) had significantly lower levels of HIV-RNA than patients with highest frequencies of CD56^{dim}NKG2A^{+} (group X) or CD56^{dim}NKG2A^{-}NKG2C^{-}CD57^{-} NK cells (group Z). Although the low number of patients limits the significance of some comparisons, patients Y, as expected according to their lower levels of HIV-RNA, showed lower immune activation and lower exhaustion of immune cells than patients X and Z. cART initiated in this better environment resulted in increased efficiency at M3.

Interestingly, one of the 10 patients showing high frequencies of mature NK cells at inclusion became later a PTC (21).

CD57 is a marker of senescence highly correlated to expression of self-KIR that identifies fully mature NK cells (4). Immature KIR$^-$ NK cells mostly express NKG2A to regulate their activity, only a small subset of NK cell co-express NKG2A and KIR. Accordingly, CD57$^+$ and NKG2A$^+$ NK cells were considered as covering, respectively, mature and immature NK populations. Accordingly, the balance of the NK cell compartment maturation is indeed the difference between the groups: NK cell of patients Y are mainly mature CD57$^+$ cells whereas most NK cells of patients X are immature. Thus, the NK cell maturation profile at the time of primary infection appears to be a pertinent marker of a better immune status and response to cART started at the primary infection.

The status of the NK cells before HIV infection is not known. Because NK cell maturation is a dynamic process, imbalance of the NK cell compartment toward maturity or immaturity could be a rapid early consequence of HIV infection. CD57$^+$ NK cells proliferate only slightly (4). The dynamics of their generation proposed by the follow-up of NKG2C subset during CMV reactivation in transplant patients (32) suggested that NKG2C$^+$CD57$^+$ resulted from the contraction of effector NKG2C$^+$ several months after control of CMV viremia. During the 3 months of cART, while HIV viremia was controlled, the frequencies of CD57$^+$ NK cells slightly increased in the Optiprim patients with low levels at inclusion but remained far below the values exhibited by patients with high frequencies at inclusion, suggesting that the dynamics of NK cell maturation is slower than that of viral load. An expansion of CD56^{dim} was reported during HIV primary infection (14). According to the inverse relationship between proliferation and maturation, this expansion should result in decreasing even more the frequencies of mature CD56^{dim}CD57$^+$. Taking into account the limited proliferation of mature CD56^{dim}CD57$^+$ and the slow dynamics of their generation, it is reasonable to speculate that the imbalance of the NK cell compartment toward maturation found at inclusion corresponded to its status at the time of HIV infection, whereas the imbalance toward immature NK cells might be a consequence of early NK cell expansion.

Accordingly, mature CD57$^+$ NK cell already present at the time of HIV infection must contribute better to immune control of the virus than CD56^{dim}NKG2A$^+$ NK cells does. As demonstrated in the NKG2C/CMV model, memory-like NKG2C$^+$CD57$^+$ NK cells expanded after CMV reactivation are potent producers of IFN-γ (8). Accumulation of NK cells expressing self-KIR was demonstrated in HIV PTCs of the Visconti study (33). Those NK cells were potent producers of IFN-γ upon stimulation with HIV-infected targets, suggesting that this function was crucial for virus control. IFN-γ can induce maturation and activation of T cells, DC, and macrophages (3), which cooperate in virus control. Indeed, DC frequencies in patients with high frequencies of mature CD57$^+$ NK cells were similar to healthy donors, suggesting indeed that innate immunity overall was involved.

Paradoxically, a prospective study on a cohort of prostitutes demonstrated an increased risk of HIV infection among those displaying highly diverse NK cell repertoire before infection, diversity being intrinsically linked to maturation (20). The high efficiency of NKG2A$^+$ NK cell to kill HIV-infected CD4-T cell targets *in vitro*, recently demonstrated (34) is an underlying mechanism that could be involved in the decreased susceptibility to HIV infection of prostitutes displaying an immature NK cell compartment before infection. Taken together, these observations suggest that if immature NK cells are efficient at the time of infection, once infection is established, other mechanisms involving mature CD57$^+$ NK cells contribute better to virus control.

Natural killer cell diversity increases with aging and the number of stimulations encountered, resulting in accumulation of mature NK cells heterogeneous with respects to functional activating and inhibitory receptors (20). Groups of patients showed similar mean age, their NK cell profiles should rather be attributed to individual histories resulting in a unique shaping of their NK cell repertoire. NKG2C$^+$CD57$^+$ NK cell could be considered as a particular subset of the mature CD57$^+$ NK cell compartment whose ligand is known. Increased expression of HLA-E on HIV-infected cells suggest a possible mechanism involving NKG2C+ NK cells (35). Other subsets of the mature CD57$^+$ pool might share unknown receptor specificities diversely able to target HIV-infected cells. Besides their functional receptors characteristics, NKG2C$^+$CD57$^+$ NK cells expanded after CMV infection are now considered as a memory-like or adaptive form of NK cells (8, 13). Recently, this subset has been characterized by a low expression of FcRγ (36) and enhanced potential for broad antiviral responses in the presence of virus-specific antibodies (37). While we cannot identify unambiguously adaptive NK cells, the CD57$^+$ NK cells found in patients partially controlling their viral load certainly overlap them. Irrespective of the various mechanisms possibly underlying the contribution of mature NK cell to HIV control, once infection is established, evaluation of this parameter during the primary HIV infection appears relevant in the search for prognostic markers to monitor HIV-infected patients.

AUTHOR CONTRIBUTIONS

AC, LM, and CR were the chief investigators of the OPTIPRIM study. AC, CG, and GP enrolled the patients. LM coordinated the data collection. DO and FG-R conceived and designed the innate immunity sub-study. FG-R and FM performed the multiparametric cytometry experiments. GB and SG developed bioinformatics tools of analysis. VA-F and CR were responsible for the virological investigations. CL, MM-T, and MP were responsible for immune investigations. FG-R, DO, AC, CR, and AM interpreted the data. FG-R generated the figures and tables. FG-R and DO wrote the paper. All authors reviewed, revised, and approved the final manuscript. AC, CR, LM, VA-F, and DO were part of the OPTIPRIM scientific committee.

ACKNOWLEDGMENTS

The authors thank Cyril Fauriat for helpful discussion. They are grateful to Marie-Laure Thibult for assistance with the use of the Cytometry Platform of the CRCM. The authors thank Bernard

ACKNOWLEDGMENTS

The authors thank Cyril Fauriat for helpful discussion. They are grateful to Marie-Laure Thibult for assistance with the use of the Cytometry Platform of the CRCM. The authors thank Bernard Chetrit of the Datacentre, Information technology and Scientific Computing (DISC) facility of the CRCM. The authors thank L. Leserman for editing the manuscript.

REFERENCES

1. Lanier LL. Evolutionary struggles between NK cells and viruses. *Nat Rev Immunol* (2008) 8(4):259–68. doi:10.1038/nri2276
2. Cooper MA, Fehniger TA, Caligiuri MA. The biology of human natural killer-cell subsets. *Trends Immunol* (2001) 22(11):633–40. doi:10.1016/S1471-4906(01)02060-9
3. Vivier E, Tomasello E, Baratin M, Walzer T, Ugolini S. Functions of natural killer cells. *Nat Immunol* (2008) 9(5):503–10. doi:10.1038/ni1582
4. Bjorkstrom NK, Riese P, Heuts F, Andersson S, Fauriat C, Ivarsson MA, et al. Expression patterns of NKG2A, KIR, and CD57 define a process of CD56dim NK-cell differentiation uncoupled from NK-cell education. *Blood* (2010) 116(19):3853–64. doi:10.1182/blood-2010-04-281675
5. Vales-Gomez M, Reyburn HT, Erskine RA, Lopez-Botet M, Strominger JL. Kinetics and peptide dependency of the binding of the inhibitory NK receptor CD94/NKG2-A and the activating receptor CD94/NKG2-C to HLA-E. *EMBO J* (1999) 18(15):4250–60. doi:10.1093/emboj/18.15.4250
6. Tomasec P, Braud VM, Rickards C, Powell MB, McSharry BP, Gadola S, et al. Surface expression of HLA-E, an inhibitor of natural killer cells, enhanced by human cytomegalovirus gpUL40. *Science* (2000) 287(5455):1031. doi:10.1126/science.287.5455.1031
7. Lanier LL, Corliss B, Wu J, Phillips JH. Association of DAP12 with activating CD94/NKG2C NK cell receptors. *Immunity* (1998) 8(6):693–701. doi:10.1016/S1074-7613(00)80574-9
8. Foley B, Cooley S, Verneris MR, Pitt M, Curtsinger J, Luo X, et al. Cytomegalovirus reactivation after allogeneic transplantation promotes a lasting increase in educated NKG2C+ natural killer cells with potent function. *Blood* (2012) 119(11):2665–74. doi:10.1182/blood-2011-10-386995
9. Guma M, Angulo A, Vilches C, Gomez-Lozano N, Malats N, Lopez-Botet M. Imprint of human cytomegalovirus infection on the NK cell receptor repertoire. *Blood* (2004) 104(12):3664–71. doi:10.1182/blood-2004-05-2058
10. Guma M, Cabrera C, Erkizia I, Bofill M, Clotet B, Ruiz L, et al. Human cytomegalovirus infection is associated with increased proportions of NK cells that express the CD94/NKG2C receptor in aviremic HIV-1-positive patients. *J Infect Dis* (2006) 194(1):38–41. doi:10.1086/504719
11. Brunetta E, Fogli M, Varchetta S, Bozzo L, Hudspeth KL, Marcenaro E, et al. Chronic HIV-1 viremia reverses NKG2A/NKG2C ratio on natural killer cells in patients with human cytomegalovirus co-infection. *AIDS* (2010) 24(1):27–34. doi:10.1097/QAD.0b013e3283328d1f
12. Bjorkstrom NK, Lindgren T, Stoltz M, Fauriat C, Braun M, Evander M, et al. Rapid expansion and long-term persistence of elevated NK cell numbers in humans infected with hantavirus. *J Exp Med* (2011) 208(1):13–21. doi:10.1084/jem.20100762
13. Cerwenka A, Lanier LL. Natural killer cell memory in infection, inflammation and cancer. *Nat Rev Immunol* (2016) 16(2):112–23. doi:10.1038/nri.2015.9
14. Alter G, Teigen N, Davis BT, Addo MM, Suscovich TJ, Waring MT, et al. Sequential deregulation of NK cell subset distribution and function starting in acute HIV-1 infection. *Blood* (2005) 106(10):3366–9. doi:10.1182/blood-2005-03-1100
15. Mavilio D, Lombardo G, Benjamin J, Kim D, Follman D, Marcenaro E, et al. Characterization of CD56-/CD16+ natural killer (NK) cells: a highly dysfunctional NK subset expanded in HIV-infected viremic individuals. *Proc Natl Acad Sci U S A* (2005) 102(8):2886–91. doi:10.1073/pnas.0409872102
16. Mela CM, Burton CT, Imami N, Nelson M, Steel A, Gazzard BG, et al. Switch from inhibitory to activating NKG2 receptor expression in HIV-1 infection: lack of reversion with highly active antiretroviral therapy. *AIDS* (2005) 19(16):1761–9. doi:10.1097/01.aids.0000183632.12418.33
17. Hong HS, Eberhard JM, Keudel P, Bollmann BA, Ballmaier M, Bhatnagar N, et al. HIV infection is associated with a preferential decline in less-differentiated CD56dim CD16+ NK cells. *J Virol* (2010) 84(2):1183–8. doi:10.1128/JVI.01675-09
18. Naranbhai V, Abdool Karim SS, Altfeld M, Samsunder N, Durgiah R, Sibeko S, et al. Innate immune activation enhances hiv acquisition in women, diminishing the effectiveness of tenofovir microbicide gel. *J Infect Dis* (2012) 206(7):993–1001. doi:10.1093/infdis/jis465
19. Scott-Algara D, Truong LX, Versmisse P, David A, Luong TT, Nguyen NV, et al. Cutting edge: increased NK cell activity in HIV-1-exposed but uninfected Vietnamese intravascular drug users. *J Immunol* (2003) 171(11):5663–7. doi:10.4049/jimmunol.171.11.5663
20. Strauss-Albee DM, Fukuyama J, Liang EC, Yao Y, Jarrell JA, Drake AL, et al. Human NK cell repertoire diversity reflects immune experience and correlates with viral susceptibility. *Sci Transl Med* (2015) 7(297):297ra115. doi:10.1126/scitranslmed.aac5722
21. Cheret A, Nembot G, Melard A, Lascoux C, Slama L, Miailhes P, et al. Intensive five-drug antiretroviral therapy regimen versus standard triple-drug therapy during primary HIV-1 infection (OPTIPRIM-ANRS 147): a randomised, open-label, phase 3 trial. *Lancet Infect Dis* (2015) 15(4):387–96. doi:10.1016/S1473-3099(15)70021-6
22. Gondois-Rey F, Granjeaud S, Rouillier P, Rioualen C, Bidaut G, Olive D. Multi-parametric cytometry from a complex cellular sample: improvements and limits of manual versus computational-based interactive analyses. *Cytometry A* (2016) 89(5):480–90. doi:10.1002/cyto.a.22850
23. Saeed AI, Bhagabati NK, Braisted JC, Liang W, Sharov V, Howe EA, et al. TM4 microarray software suite. *Methods Enzymol* (2006) 411:134–93. doi:10.1016/S0076-6879(06)11009-5
24. Bacchus C, Cheret A, Avettand-Fenoel V, Nembot G, Melard A, Blanc C, et al. A single HIV-1 cluster and a skewed immune homeostasis drive the early spread of HIV among resting CD4+ cell subsets within one month post-infection. *PLoS One* (2013) 8(5):e64219. doi:10.1371/journal.pone.0064219
25. Ploquin MJ, Madec Y, Casrouge A, Huot N, Passaes C, Lecuroux C, et al. Elevated basal pre-infection CXCL10 in plasma and in the small intestine after infection are associated with more rapid HIV/SIV disease onset. *PLoS Pathog* (2016) 12(8):e1005774. doi:10.1371/journal.ppat.1005774
26. Liovat AS, Rey-Cuille MA, Lecuroux C, Jacquelin B, Girault I, Petitjean G, et al. Acute plasma biomarkers of T cell activation set-point levels and of disease progression in HIV-1 infection. *PLoS One* (2012) 7(10):e46143. doi:10.1371/journal.pone.0046143
27. Trabattoni D, Saresella M, Biasin M, Boasso A, Piacentini L, Ferrante P, et al. B7-H1 is up-regulated in HIV infection and is a novel surrogate marker of disease progression. *Blood* (2003) 101(7):2514–20. doi:10.1182/blood-2002-10-3065
28. Sabado RL, O'Brien M, Subedi A, Qin L, Hu N, Taylor E, et al. Evidence of dysregulation of dendritic cells in primary HIV infection. *Blood* (2010) 116(19):3839–52. doi:10.1182/blood-2010-03-273763
29. Tomescu C, Liu Q, Ross BN, Yin X, Lynn K, Mounzer KC, et al. A correlate of HIV-1 control consisting of both innate and adaptive immune parameters best predicts viral load by multivariable analysis in HIV-1 infected viremic controllers and chronically-infected non-controllers. *PLoS One* (2014) 9(7):e103209. doi:10.1371/journal.pone.0103209
30. Beziat V, Duffy D, Quoc SN, Le Garff-Tavernier M, Decocq J, Combadiere B, et al. CD56brightCD16+ NK cells: a functional intermediate stage of NK cell differentiation. *J Immunol* (2011) 186(12):6753–61. doi:10.4049/jimmunol.1100330
31. Petitdemange C, Becquart P, Wauquier N, Beziat V, Debre P, Leroy EM, et al. Unconventional repertoire profile is imprinted during acute chikungunya infection for natural killer cells polarization toward cytotoxicity. *PLoS Pathog* (2011) 7(9):e1002268. doi:10.1371/journal.ppat.1002268
32. Lopez-Verges S, Milush JM, Schwartz BS, Pando MJ, Jarjoura J, York VA, et al. Expansion of a unique CD57(+)NKG2Chi natural killer cell subset during acute human cytomegalovirus infection. *Proc Natl Acad Sci U S A* (2011) 108(36):14725–32. doi:10.1073/pnas.1110900108
33. Scott-Algara D, Didier C, Arnold V, Cummings J-S, Boufassa F, Lambotte O, et al. Post-treatment controllers have particular NK cells with high anti-HIV capacity: VISCONTI study, abstract n° 52. *22nd Conference on Retroviruses and Opportunistic Infections (CROI 2015).* (2015).

34. Davis ZB, Cogswell A, Scott H, Mertsching A, Boucau J, Wambua D, et al. A conserved HIV-1-derived peptide presented by HLA-E renders infected T-cells highly susceptible to attack by NKG2A/CD94-bearing natural killer cells. *PLoS Pathog* (2016) 12(2):e1005421. doi:10.1371/journal.ppat.1005421
35. Nattermann J, Nischalke HD, Hofmeister V, Kupfer B, Ahlenstiel G, Feldmann G, et al. HIV-1 infection leads to increased HLA-E expression resulting in impaired function of natural killer cells. *Antivir Ther* (2005) 10(1):95–107.
36. Schlums H, Cichocki F, Tesi B, Theorell J, Beziat V, Holmes TD, et al. Cytomegalovirus infection drives adaptive epigenetic diversification of NK cells with altered signaling and effector function. *Immunity* (2015) 42(3):443–56. doi:10.1016/j.immuni.2015.02.008
37. Zhang T, Scott JM, Hwang I, Kim S. Cutting edge: antibody-dependent memory-like NK cells distinguished by FcRgamma deficiency. *J Immunol* (2013) 190(4):1402–6. doi:10.4049/jimmunol.1203034

19

A Two-Phase Expansion Protocol Combining Interleukin-15 and IL-21 Improves Natural Killer Cell Proliferation and Cytotoxicity against Rhabdomyosarcoma

Juliane Wagner[1,2,3], Viktoria Pfannenstiel[1,2,3], Anja Waldmann[4], Judith W. J. Bergs[5,6,7†], Boris Brill[4], Sabine Huenecke[1,2], Thomas Klingebiel[1], Franz Rödel[5,6,7], Christian J. Buchholz[7,8,9], Winfried S. Wels[4,6,7], Peter Bader[1,2] and Evelyn Ullrich[1,2,3]*

[1] Children's Hospital, Goethe University, Frankfurt am Main, Germany, [2] Division for Stem Cell Transplantation and Immunology, Department for Children and Adolescents Medicine, Hospital of the Goethe University Frankfurt, Frankfurt am Main, Germany, [3] LOEWE Center for Cell and Gene Therapy, Goethe University, Frankfurt am Main, Germany, [4] Georg-Speyer-Haus, Institute for Tumor Biology and Experimental Therapy, Frankfurt am Main, Germany, [5] Department of Radiotherapy and Oncology, Goethe University, Frankfurt am Main, Germany, [6] German Cancer Consortium (DKTK), Partner Site Frankfurt/Mainz, Frankfurt am Main, Germany, [7] German Cancer Research Center (DKFZ), Heidelberg, Germany, [8] German Cancer Consortium (DKTK), Partner Site Heidelberg, Heidelberg, Germany, [9] Molecular Biotechnology and Gene Therapy, Paul-Ehrlich-Institut, Langen, Germany

Edited by:
Gianfranco Pittari, Hamad Medical Corporation, Qatar

Reviewed by:
Nicolas Dulphy, Paris Diderot University, France
Raquel Tarazona, University of Extremadura, Spain

***Correspondence:**
Evelyn Ullrich
evelyn.ullrich@kgu.de

Rhabdomyosarcoma (RMS) is the most common soft tissue malignancy in children. Despite intensive research in recent decades the prognosis for patients with metastatic or relapsed diseases has hardly improved. New therapeutic concepts in anti-tumor therapy aim to modulate the patient's immune system to increase its aggressiveness or targeted effects toward tumor cells. Besides surgery, radiotherapy and chemotherapy, immune activation by direct application of cytokines, antibodies or adoptive cell therapy are promising approaches. In the last years, adoptive transfer of natural killer (NK) cells came into the focus of translational medicine, because of their high cytotoxic potential against transformed malignant cells. A main challenge of NK cell therapy is that it requires a high amount of functional NK cells. Therefore, *ex vivo* NK cell expansion protocols are currently being developed. Many culturing strategies are based on the addition of feeder or accessory cells, which need to be removed prior to the clinical application of the final NK cell product. In this study, we addressed feeder cell-free expansion methods using common γ-chain cytokines, especially IL-15 and IL-21. Our results demonstrated high potential of IL-15 for NK cell expansion, while IL-21 triggered NK cell maturation and functionality. Hence, we established a two-phase expansion protocol with IL-15 to induce an early NK cell expansion, followed by short exposure to IL-21 that boosted the cytotoxic activity of NK cells against RMS cells. Further functional analyses revealed enhanced degranulation and secretion of pro-inflammatory cytokines such as interferon-γ and tumor necrosis factor-α. In a proof of concept *in vivo* study, we also observed a therapeutic effect of adoptively transferred IL-15 expanded and IL-21 boosted NK cells in combination with image guided high precision radiation therapy using a luciferase-transduced RMS xenograft model. In summary, this two-phased feeder cell-free *ex vivo* culturing protocol combined efficient expansion and high cytolytic functionality of NK cells for treatment of radiation-resistant RMS.

Keywords: natural killer cells, radiotherapy, rhabdomyosarcoma, RH30 cells, RD cells, interleukin-15, interleukin-21

INTRODUCTION

With their ability to detect and directly destroy virally infected or malignant cells, natural killer (NK) cells form an important part of the first line defense of the immune system. They can be activated rapidly *via* germ-line encoded receptors that recognize the presence of stress ligands or absence of self-antigens on target cells (1–5).

In vivo development and survival of NK cells require cytokines (6–8). In this context, cytokines have been shown to activate NK cells potently during *ex vivo* expansion (9–12). The group of common γ-chain receptor cytokines encompassing interleukin (IL)-2, IL-4, IL-9, IL-15, and IL-21 has been studied intensively over the recent years. IL-2 and IL-15 have similar impacts on NK cells (13, 14). However, direct injection of IL-2 has been shown to be accompanied by severe side effects, such as vascular leak syndrome, activation-induced cell death, and strong induction of regulatory CD4^{pos} T cells, which did not occur after IL-15 administration (15, 16).

More recently, research has been focusing on IL-21 biology, but its effects on NK cell development are controversially discussed. IL-21 is known to be involved in the development and proliferation of NK cells from progenitor cells (17) and to induce receptor expression (18), interferon (IFN)-γ secretion and cytotoxicity (19). Conversely, IL-21 has also been reported to trigger apoptosis and to diminish IL-15-based benefits (20–22). These less favorable effects may be ascribed to the variability of experimental designs such as timing, cytokine concentration, additives, or accessory cells in culture as well as the developmental or maturation state and origin of NK cells. Of note, positive effects have been reported mostly upon cultivation of NK cells in the presence of auxiliary cells such as other peripheral blood mononuclear cells (PBMCs) (23), genetically modified feeder cells equipped with membrane-bound IL-21 (24, 25), or feeder-cell particles (26). The downside of these protocols is the necessity of elimination of hazardous cells, such as possibly graft-versus-host-disease (GvHD)-triggering cells or tumor-derived feeder cells, that might induce harmful side-effects *in vivo*. On the contrary, safer expansion strategies based on the exclusive application of cytokines, result in much lower absolute NK cell numbers (27). Thus, risk-free protocols for efficient expansion of functional NK cells are urgently needed.

Immune cell therapy is an effective anti-cancer strategy and hematopoietic stem cell transplantation (HSCT) has been shown to positively influence the outcome of patients with different hematologic diseases (28, 29). However, studies using HSCT did not achieve satisfactory improvement against high-risk rhabdomyosarcoma (RMS) (30–34). RMS is a rare malignant disease but the most common soft tissue cancer in children. The outcome of treatment for patients with stage IV RMS, relapsed or metastatic diseases arising from RMS, has scarcely improved during recent decades and, in general, is unfortunately considered to be poor even upon combination of surgery, chemotherapy, radiotherapy (RT), and HSCT (35, 36).

Natural killer cells are considered to potently initiate graft-versus-tumor (GvT) effects without provoking, but even preventing GvHD (37–41), a possible risk of HSCT (42–44).

Here, we present a two-phase protocol that combines IL-15-triggered NK cell expansion with an IL-21 boost to exert the stimulatory effects of both cytokines. To avoid contamination by other cell types, we employed enriched NK cells and circumvented the addition of any accessory or feeder cells. The NK cell product was characterized intensively in terms of proliferation, phenotype, and functionality. Finally, IL-15-expanded and IL-21-boosted NK cell products from different human donors were used for combined adoptive immune cell and radiation therapy in a xenograft model of RMS.

MATERIALS AND METHODS

Purification of Primary Human NK Cells

This study was approved (approval no. 329/10) by the Ethics Committee of the Goethe University Frankfurt (Frankfurt, Germany) and was performed in accordance with the Declaration of Helsinki with written informed consent given by every participant. NK cells were isolated from freshly generated donor buffy coats provided by the German Red Cross Blood Donation Service (DRK-Blutspendedienst Baden-Württemberg-Hessen, Frankfurt, Germany), using immunomagnetic negative selection (EasySep™ Human NK Cell Enrichment kit, StemCell Technologies, Vancouver, BC, Canada) according to the manufacturer instructions.

Briefly, PBMCs obtained from buffy coats by density gradient centrifugation were diluted to a final concentration of 100×10^6 cells/mL. Then, 50 µL of enrichment cocktail were applied per milliliter cell suspension and incubated for 10 min. Subsequently, 100 µL of microbead suspension were added and incubated for another 5 min. After incubation for 2.5 min in "The Big Easy" EasySep™ Magnet, the NK cell-enriched suspension was decanted into a new tube.

Cultivation of Primary Cells and Cell Lines

Purified primary NK cells were cultured at a concentration of 2×10^6 cells/mL in X-VIVO 10 medium (Lonza Group Ltd., Basel, CH) supplemented with 5% heat inactivated human fresh frozen plasma (FFP; provided by DRK-Blutspendedienst Baden-Württemberg-Hessen, Frankfurt, Germany) and 100 U/mL penicillin and 100 µg/mL streptomycin (Gibco, New York, NY, USA). Correlating to the batch, the cells were provided with IL-2 (ProleukinS, Novartis Pharmaceuticals, Horsham, UK), 100

U/mL (IL-2^{100}) or 1,000 U/mL (IL-2^{1000}), 10 ng/mL IL-15 (IL-15), 25 ng/mL IL-21 (both PeproTech, Rocky Hill, CT, USA) (IL-21) or combinations of those (IL-2^{100} + 15, IL-2^{100} + 15 + 21, IL-15 + 21). Every 3–4 days, half of the medium was replaced by fresh medium containing the corresponding cytokines or combinations. One batch was treated with IL-15 and boosted with IL-21 only 3–4 days before harvest and analysis (IL-15 + 21_{boost}).

Chronic myologenous erythroleukemia cell line K562 and RMS cell lines RH30 and RD (45) were purchased from the American Type Culture Collection (Manassas, VA, USA). Cells were maintained in Roswell Park Memorial Institute (RPMI) 1640 medium supplemented with 10% heat inactivated fetal calf serum (Invitrogen, Paisley, UK), 100 U/mL penicillin and 100 μg/mL streptomycin. The cells were splitted twice a week.

GFP/luciferase-expressing RD cells (RD$^{GFP/Luc}$) were generated *via* lentiviral transduction using vector particles pseudotyped with vesicular stomatitis virus G protein that were produced using the transfer plasmid pSEW-luc2, which encodes firefly luciferase and enhanced green fluorescent protein linked *via* a 2A peptide (46). GFP positive cells were enriched by fluorescence activated cell sorting (FACS) using a FACSAria II™ device (BD Biosciences, San Jose, CA, USA). Culture conditions for transduced cells were the same as for non-transduced cells.

Flow Cytometry

In order to check the quality of enriched NK cells and to monitor the phenotype of *ex vivo* expanded NK cells, samples were analyzed with a FACSCanto 10c™ system (BD Biosciences). Post-harvesting cells were resuspended in FACS buffer containing CellWASH (BD Biosciences), 0.5% bovine serum albumin (Sigma Aldrich, Taufkirchen, Germany) and 0.01% NaN_3 (0.1 M, Sigma Aldrich).

Intracellular staining was accomplished using formaldehyde (AppliChem GmbH, Darmstadt, Germany) for fixation and 90% methanol for membrane perforation.

The following antibodies were used: CD3-APC (#UCHT1), TRAIL-R-APC [#DJR2-4(7-8)], FAS-BV421 (#DX2), CD56-FITC (clone #HCD56), FAS-L-PE (#NOK-1), TRAIL-PE (#RIK-2), CD19-PerCP (#HIB19), CD16-PE/Cy7 (#3G8) all from Biolegend (San Diego, CA, USA); CD3-V450 (#UCHT1), CD19-V450 (#HIB19), CD14-V450 (#MφP9), CD45-BV510 (#HI30), NKp30-AF488 (#P30-15), DNAM-1-FITC (#DX11), NKp44-PE (#P44-8.1) CD45-APC (#2D1), CD137/4-1BB-APC (#4B4-1), CD107a-APC/H7 (#H4A3), IFN-γ-FITC (#B27), pAKT-AF647 (#F29-763), pERK1/2-AF647 (#20A), from BD Biosciences; CD56-APC/AF700, NKG2D-APC (#ON72), CD11a/LFA-1-FITC (#25.3) from Beckman Coulter Immunotech (Brea, CA, USA); CD45-PE (#HI30) from Invitrogen (Carslbad, CA, USA); and NKp46-APC (#9E2), KIR2D-FITC (#NKVFS1), CD158e/k-PE (#5.133), NKG2A-APC (#Z199) from Miltenyi Biotec (Bergisch-Gladbach, Germany). Depending on the panel Zombie Violet Fixable Viability Kit (BioLegend), 7AAD (BD Biosciences) or strongly diluted DAPI were used for live/death discrimination.

Data were acquired on a FACSCanto 10c™ instrument (BD Biosciences, San Jose, CA, USA) and analyzed using Flowjo (Tree Star Inc., Ashland, OR, USA).

Cytotoxicity Assay

To investigate the killing capacity of *ex vivo* expanded NK cells a FACS-based cytotoxicity assay was employed. NK effector cells were harvested after 6 days of cytokine stimulation. Target cells were harvested and stained for 5 min with Celltrace CFSE (Molecular Probes, Eugene, OR, USA) in a final concentration of 5 μM. After all cells had been washed with Dulbecco's Phosphate Buffered Saline (DPBS, Gibco), they were resuspended in X-VIVO 10 medium supplemented with 5% heat inactivated human FFP, 100 U/mL penicillin and 100 μg/mL streptomycin. NK cells and target cells were combined in a U-bottom 96-well plate at effector to target (E:T) ratios of 1:1, 5:1, and 10:1, adjusted to 25,000 target cells per well in a total volume of 200 μL. After 5 h of co-incubation the supernatant was removed, cells were harvested and resuspended in a 1:6,000 dilution of DAPI for live/death discrimination. From each well, the same amount of target cells was acquired using a FACSCanto 10c™ device. Samples exclusively containing target cells served as spontaneous lysis controls. Spontaneous lysis was subtracted from each sample to obtain specific lysis values. All experiments were conducted in triplicates for each NK cell donor.

To address additional antibody-dependent cellular cytotoxicity (ADCC) by NK cells, in a separate experiment, 10 μg/mL of anti-ErbB2 antibody Trastuzumab (Herceptin, ROCHE, Mannheim, Germany) were added and cytotoxicity compared to NK cell cytotoxicity in the absence of Trastuzumab.

Conjugation capacity of stimulated NK cells was addressed by staining NK cells with Celltrace CFSE, while target cells were stained with Celltrace Calcein Violet AM for 20 min at 4°C. After intensive washing, effector and target cells were co-incubated for 0–90 min, then shortly vortexed and fixed with 1–2% formaldehyde. Flow cytometry data were acquired on a FACSCanto 10c™ instrument.

Due to limited availability of NK cell numbers, cytotoxicity and conjugation assays were performed with cells from other donors than were used for proliferation assays.

Functional Activity and Degranulation Assay

Degranulation potential of cytokine stimulated NK cells was assessed as described (47), with cells harvested on day 6 of cultivation. Cells were washed and resuspended in fresh X-VIVO 10 medium supplemented with 5% heat-inactivated human FFP, 100 U/mL penicillin, and 100 μg/mL streptomycin. After 1 h, cells were incubated with anti-human CD107a, followed by an additional hour of incubation with GolgiStop™ (BD Biosciences). Cells were washed, blocked with human IgG and stained with Zombie Violet™ Fixable Viability Kit for live/death discrimination. Post washing, cells were stained for CD45, CD56, and CD16, fixed with formaldehyde solution (2% final concentration) and permeabilized with saponin buffer [0.2% saponin, 1% bovine serum albumin (both Sigma Aldrich) in DPBS]. In the end, cells were stained intracellularly with anti-human IFN-γ, washed, and measured by flow cytometry.

For Phosflow analysis freshly harvested cells were stained on the surface for CD45, CD3, and CD56, fixed and permeabilized with 90% methanol and intracellularly stained for pAKT, pERK1/2, and pMAPK. Flow cytometry data acquisition was performed on a FACSCanto 10c™ instrument.

Due to limited availability of NK cells numbers, degranulation assays were performed with cell preparations from different donors than proliferation or cytotoxicity assays.

Immunoblot Analyses

Western blotting was deployed for the assessment of production and release of apoptosis-mediating perforin and granzyme B. NK cells and culture supernatants were harvested on day 6 of cytokine stimulation. Cells were lysed using RIPA buffer supplemented with cOmplete™ Protease Inhibitor Cocktail (ROCHE) followed by sonification. Protein concentrations were determined *via* Bradford assay (Protein Assay Dye Reagent Concentrate, Bio-Rad, Munich, Germany). Separation of proteins was accomplished by SDS-PAGE followed by semi-dry blotting onto polyvinylidenfluoride membranes. Before antibody application, membranes were blocked with 5% skim milk powder in DPBS. Mouse monoclonal antibodies against human perforin (1:1,000, LifeSpan BioSciences, Seattle, WA, USA), granzyme B (1:200, #2C5, Santa Cruz, Heidelberg, Germany) and a rabbit antibody against human γ-tubulin (1:2,000, Sigma Aldrich) were used as primary antibodies. HRP conjugated rabbit anti-mouse IgG (1:15,000, Sigma Aldrich) and goat-anti rabbit (1:16,000, Sigma Aldrich) served as secondary antibodies. The antigen–antibody complexes were detected using an ECL-chemiluminescence system (Pierce™ ECL, Thermo Scientific, Waltham, MA, USA) according to the manufacturer's instructions, and visualized using X-ray films (Fujifilm, Tokyo, Japan).

Due to limited availability of NK cell numbers, immunoblot assays were performed with cells from other donors than used for proliferation or functional assays.

Cytometric Bead Array

Cytokine secretion was examined by cytometric bead array analyses (CBA) on supernatants of stimulated NK cells on the sixth day of cultivation using BD CBA Flex Sets for IFN-γ, tumor necrosis factor (TNF)-α, macrophage inflammatory protein (MIP)-1α, monocyte chemoattractant protein (MCP)-1, IL-8, IL-10, and granulocyte-macrophage colony-stimulating factor (GM-CSF) (BD Biosciences). The tests were performed according to the manufacturer's instructions using a mixture of PE-conjugated antibodies against the cytokines listed above. Data were acquired with the BD FACSVerse™ Bioanalyzer and were quantitated using the FCAP Array™ software (v3.0.1; BD Biosciences).

Murine RMS Xenograft Model and Treatment Protocol

The *in vivo* experiments were approved by the government committee (Regierungspräsidium Darmstadt, Darmstadt, Germany) and were conducted in accordance to the requirements of the German Animal Welfare Act.

Female, 12- to 16-week-old NOD/SCID/IL-2Rγc$^{-/-}$ mice (NSG) mice were injected subcutaneously with 10^5 luciferase expressing RD$^{GFP/Luc}$ cells. After visual detection of tumor nodes about 3 weeks post-cell injection, mice were imaged by performing a Cone-Beam CT (CBCT) operating at 65 kV, 0.5 mA and irradiated while immobilized with 2.5% isoflurane anesthesia (AbbVie, Wiesbaden, Germany) using a Small Animal Radiation Research Platform (SARRP, Xstrahl Ltd., Camberley, UK). CBCT images were transferred to MuriPlan™ Software and individual isocenters were selected for targeted radiation therapy applying a two-field geometry. Fractionated single doses of 2.5 Gy using a 1–10-mm collimated beam operating at 175 kV, 15 mA were applied four times a week to reach a total dose of 27.5 Gy. Post-termination of RT, adoptive transfer of IL-15 expanded and IL-21 boosted NK cells was accomplished in three injections, once a week. NK cells were purified from 11 buffy coats as described above and maintained separately during cultivation in 25 cm^2 suspension cell culture flasks (Cellstar, Frickenhausen, Germany) using the IL-15 + 21$_{boost}$ protocol as described before. After 10–11, 17–18, and 24–25 days of *ex vivo* expansion and stimulation, NK cells were harvested, washed, and pooled. The NK cells were injected intravenously *via* the tail vein with 10^7 NK cells in a total volume of 100 µL per mouse. Due to limited availability of NK cell numbers, *in vivo* application was performed with cells from other donors than used for proliferation or functional assays.

Tumor growth was monitored by caliper measurements and bioluminescence imaging (BLI) using an IVIS Lumina II system (Perkin Elmer, Waltham, MA, USA). For the latter method, mice were anesthetized by isoflurane inhalation and subcutaneously injected with 150 µg of *in vivo* grade VivoGlo™ luciferin (Promega, Madison, WI, USA) dissolved in 100 µL DPBS per mouse. Images were acquired after an incubation time of 15 min. BLI data analysis was performed using Living Image® software (Perkin Elmer).

Statistical Analyses

Results were analyzed using repeated measures one-way ANOVA with the Geisser–Greenhouse correction. For *in vivo* experiments, two-way ANOVA was used to compare the tumor-growth curves of different treatment groups. Statistical calculations were performed using GraphPad Prism v6 (GraphPad, La Jolla, CA, USA), and p values <0.05 were considered statistically significant.

RESULTS

IL-15-Driven NK Cell Expansion Is Not Impaired by a Short-term Boost with IL-21

The primary goal of *ex vivo* expansion of NK cells is to yield high cell numbers for adoptive transfer with an appropriate quality and optimal cytotoxic functionality of the NK cell product. To this end, NK cells were isolated by an immunomagnetic negative selection (as described in Section "Materials and Methods") and expanded *ex vivo* under feeder-cell free cultivation conditions addressing the impact of different cytokines. NK cells were cultured with IL-2^{100} (100 U/mL), IL-2^{1000} (1000 U/mL), IL-15 (10 ng/mL), IL-21 (25 ng/mL), combinations of these cytokines or IL-15 and a 3-day IL-21 boost (IL-15 + 21$_{boost}$). Purity of the enriched NK cells was determined and cell numbers as well as viability were monitored over a period of 4–6 weeks (**Figures 1A–D**).

The average content of starting material after NK cell purification consisted of 89% NK cells. Purity of the CD3^{neg}CD56^{pos} NK cells decreased slightly on day 3, but increased afterward during

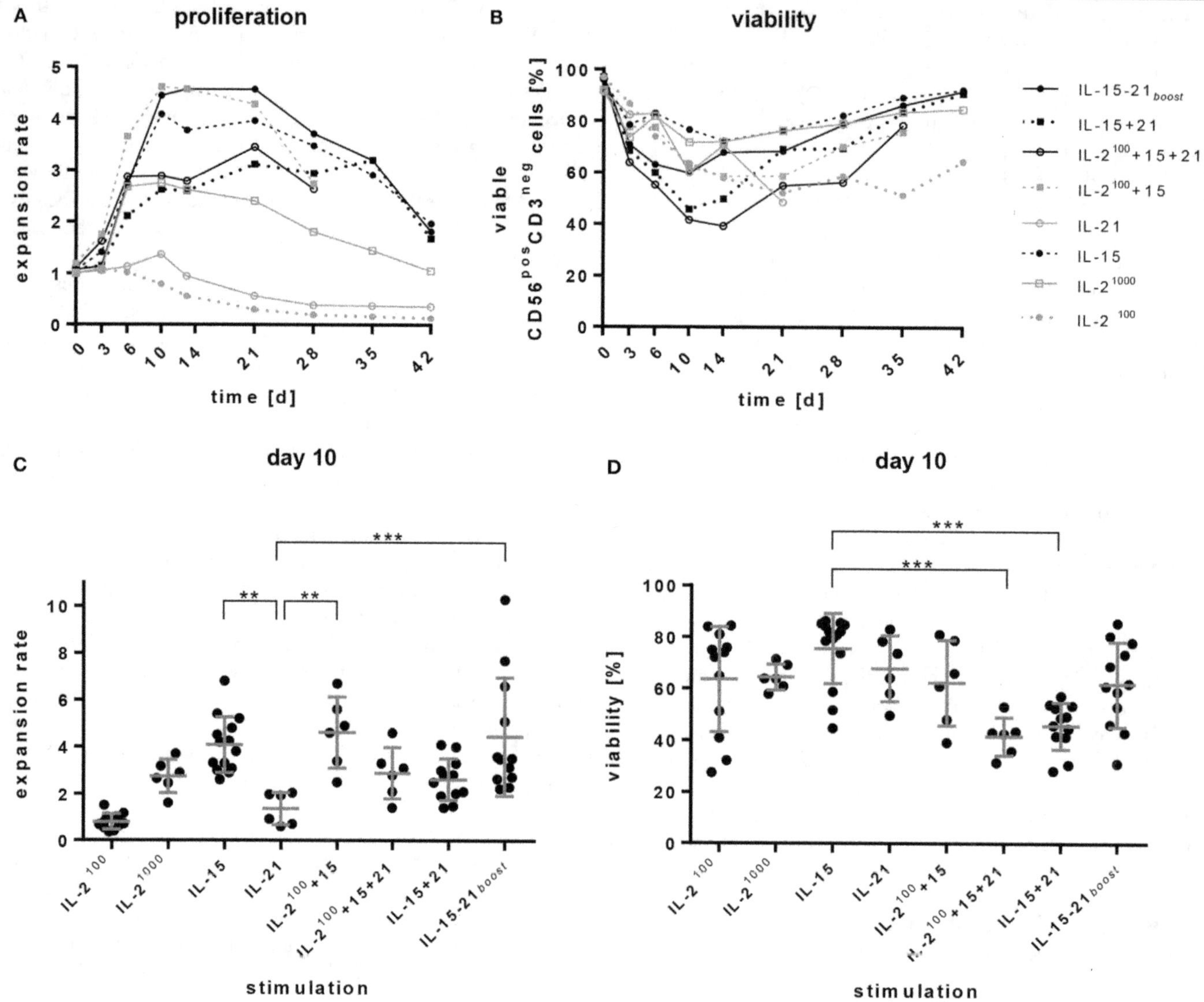

FIGURE 1 | Expansion and viability of natural killer (NK) cell products. Purified CD56^{pos}CD3^{neg} cells were cultivated for 4–6 weeks in the presence of different cytokines or cytokine combinations. **(A)** Proliferation is presented as expansion rate. Here, the mean values of 6–14 donors are shown over time. **(B)** Viability of CD56^{pos}CD3^{neg} NK cells was analyzed by DAPI staining at the indicated time points. **(C,D)** show the donor dependent distribution of values on day 10 for proliferation and viability, respectively. Lines represent mean values and SDs. Significant differences are indicated by asterisks (**$p < 0.01$, ***$p < 0.005$, one-way ANOVA). Although not indicated in the graph, IL-2^{100} and IL-21 alone induced significantly smaller expansion rates than all other protocols.

the culturing process. With protocols using IL-15, IL-15 + 21, and IL-15 + 21$_{boost}$, the purity reached over 95% from day 10 on. In cultures containing exclusively IL-2 also CD3^{pos}, cells expanded and diminished NK cell purity especially at late culture time points after 2–3 weeks (data not shown).

Remarkably, all protocols involving the addition of IL-15, such as IL-15 alone, IL-2^{100} + 15, and IL-15 + 21$_{boost}$, led to a high increase in the number of NK cells over the first 10 days reaching a plateau until day 21. Permanent stimulation with IL-21 (IL-2^{100} + 15 + 21 and IL-15 + 21) also provoked rapid but less distinct expansion. In general, all expansion rates decreased slowly after 3 weeks of *ex vivo* cultivation (**Figure 1A**).

High levels of IL-2 (IL-2^{1000}) evoked proliferation of NK cells during the first 6 days, but average expansion rates declined subsequently. Cultivation in the presence of low IL-2 levels (IL-2^{100}) or IL-21 alone did not induce proliferation, instead NK cells died (**Figures 1A–D**).

Irrespective of donor-dependent differences, stimulation with IL-15, IL-2^{100} + 15, and IL-15 + 21$_{boost}$ performed better than all other stimulation protocols. Of note, permanent exposure to IL-21 dampened IL-15-driven expansion, while a short boost with IL-21 did not disturb proliferation of NK cells, but, in some cases, even increased the expansion rate. With the IL-15 + 21$_{boost}$ protocol, expansion rates ranged between 2- and 10-fold depending on the donor, exhibiting an average 4.5-fold increase on day 10 of culture (**Figure 1C**).

Percentages of viable cells decreased during the first days, but then recovered upon further stimulation (**Figure 1B**).

Comparisons on day 10 of cultivation indicated significantly higher frequencies of viable cells in the product of IL-15-stimulated cells than in products obtained from expansion protocols with permanent addition of IL-21. Nevertheless, a short boost with IL-21 did not significantly affect cell viability compared to IL-15 mono treatment (**Figure 1D**).

IL-21 Triggers a Mature Phenotype of NK Cells

Next, we investigated changes in the phenotype and subset composition of NK cells triggered by cytokine-induced *ex vivo* expansion. To monitor their maturation state, NK cells were analyzed for the distribution of the CD56^{high}CD16^{neg} and CD56^{dim}CD16^{pos} subset over 4–6 weeks (**Figures 2A,B**). As all stimulation protocols induced an upregulation of the NK cell marker CD56, only a discrimination of CD16^{pos} and CD16^{neg} NK cells was implemented for further analysis of the maturation state.

Depending on the expansion protocol, the frequency of CD16^{pos} NK cells was reduced. Only protocols with permanent IL-21 exposure (IL-21, IL-15 + 21, and IL-2^{100} + 15 + 21) maintained a mature phenotype (**Figure 2A**). This effect became more pronounced the longer the *ex vivo* expansion endured (**Figures 2A,B**). In parallel with the decrease in mature phenotype, the proportion of less mature CD16^{neg} NK cells was increased (**Figure 2B**).

Furthermore, cytokine stimulation induced upregulation of activating and inhibitory receptors on the NK cell surface (**Figure 2C**). Interestingly, different cytokine combinations led to a significant increase of apoptosis inducing TRAIL or FAS ligand (FAS-L) in line with enhanced expression of TRAIL-R and FAS. We further observed that the presence of IL-21 had an additive effect with IL-15 regarding an enhanced surface expression of TRAIL, TRAIL-R, DNAM-1, FAS, FAS-L, NKp46, but also inhibitory NKG2A. For NKG2D, NKp30, NKp44 and inhibitory KIR2D and CD158 e/k (KIR3DL1/DL2), IL-21 counteracted the up-regulating effect of IL-15.

TRAIL, DNAM-1, NKp46, and NKG2A were rapidly upregulated upon short exposure to IL-21 if cells were pre-stimulated with IL-15 (IL-15 + 21$_{boost}$) and reached expression levels similar to those of NK cells stimulated permanently with IL-15 and IL-21 (IL-15 + 21). By contrast, changes in the levels of other surface markers were less rapid, showing significant differences for cells stimulated following the different protocols.

Almost all CD16^{neg} NK cells expressed two of the three activating receptors, NKG2D, DNAM-1, and NKp44, only a small fraction of these cells expressed only one (NKG2D) and an even smaller fraction none. After 6 days of NK cell expansion, almost all CD16^{neg} NK cells co-expressed all three receptors (Figure S1 in Supplementary Material, left panels). Among the CD16^{pos} fraction, half of it expressed only NKG2D and half co-expressed NKG2D and DNAM-1. After expansion with IL-15 alone or an additional short-term IL-21 boost, on day 6 of culturing, half of the CD16^{pos} fraction expressed all three receptors, and half co-expressed two. Permanent exposure to IL-21 further increased the fraction of CD16^{pos} NK cells that co-expressed all three receptors. Altogether, no prominent difference in the co-expression of NKG2D, NKp44, and DNAM-1 was observed upon expansion with the three different protocols (Figure S1 in Supplementary Material, right panel).

To analyze the kinetics of the receptor expression induced by the different expansion protocols based on the use of IL-15 and/or additional IL-21, flow cytometry data were acquired at three time points, before stimulation, early at day 6, and late at day 18 (Figure S2 in Supplementary Material). Permanent presence of IL-21 diminished the expression of NKG2D, NKp30, and NKp44 on the NK cell surface. In contrast, it had hardly any influence, when applied after an IL-15-induced expansion phase. Expression of NKp46 was only slightly reduced by long-term IL-21 treatment. The higher expression level of DNAM-1, aroused by permanent presence of IL-21, stayed stable over time (Figure S2A in Supplementary Material).

Inhibitory receptors of the KIR family (KIR2D and CD158e/k/KIR3D) did not show any relevant changes during NK cell expansion, while NKG2A was strongly upregulated during NK cell expansion, independent from the cytokines used (Figure S2B in Supplementary Material).

Besides DNAM-1, expression of the adhesion molecules LFA-1 (CD11a) and 4-1BB (CD137) was addressed. Expression of LFA-1 was early increased under all tested cytokine conditions. In contrast, 4-1BB (CD137) was upregulated only after a longer culture period (Figure S2C in Supplementary Material). Accordingly, no prominent differences in conjugate formation were observed comparing the NK cells of all three tested cytokine expansion protocols (Figure S2D in Supplementary Material).

Short Stimulation with IL-21 Increases NK Cell Cytotoxicity

Besides an increase in the yield of of NK cells, another aim of the *ex vivo* expansion is to obtain a therapeutic cell product that mediates optimal anti-tumor activity. In case of NK cell immunotherapeutics, cytotoxicity is a valuable indicator of functionality. Accordingly, cytotoxicity of cytokine-expanded NK cells was tested against targets such as the erythroleukemia cell line K562 and the RMS cell lines RD and RH30. The RD cell line represents the embryonal subtype of RMS while RH30 represents the alveolar subtype, which is more difficult to treat.

A short boost with IL-21 increased the cytotoxic effect of NK cells toward all three cell lines compared to stimulation with IL-15 alone or permanent exposure to IL-21 (**Figure 3**). An increase in specific lysis of approximately 10% was achieved by an IL-21 boost for all three cell lines as compared to sole IL-15 stimulation.

Cytotoxicity of expanded NK cells decreased with longer culture periods and reduced from 60% on day 6 to 40% on day 10 and 14 (E:T = 10:1). Still cytotoxicity of continuously expanded cells was superior to cytotoxicity of NK cells that were cryoconserved after 6 days of cytokine expansion (Figure S3A in Supplementary Material).

Despite the decrease in CD16^{pos} NK cell numbers, no decrease of ADCC was observed with NK cells expanded with the IL-15 or IL-15 + 21$_{boost}$ protocols compared to NK cells expanded with the IL-15 + 21 protocol in an assay against ErbB2^{pos} RD cells supplemented with anti-ErbB2 antibody Trastuzumab. All expansion protocols resulted in an additional lysis of about 5 to 10%

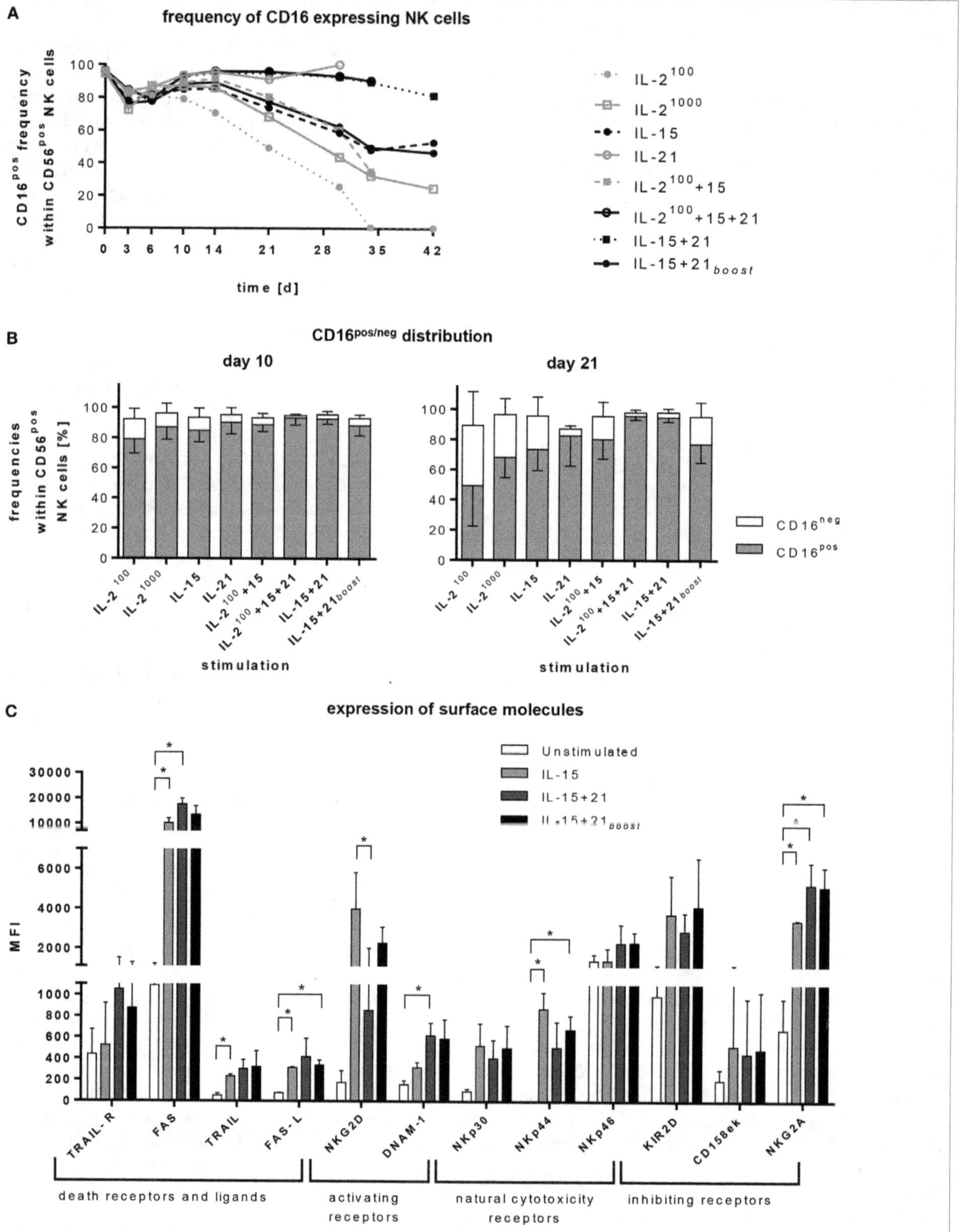

FIGURE 2 | Regulation of surface markers. Purified CD56^{pos}CD3^{neg} cells were cultivated for 4–6 weeks with different cytokines or cytokine combinations. **(A)** At indicated time points, frequencies of CD16^{pos} natural killer (NK) cells were assessed by flow cytometry. The graph represents means of 5–11 independent donors. **(B)** The graphs show the distribution of CD16^{pos} cells and their CD16^{neg} counterpart within CD56^{pos} NK cells on days 10 and 21. Bars represent mean values of 5–11 independent donors, lines indicate SDs. **(C)** Expression of various activating and inhibitory receptors after 6 days of cultivation is shown for selected protocols. Bars represent mean values of three independent donors, lines indicate SDs. Significant differences are indicated by asterisks (*$p < 0.05$, one-way ANOVA).

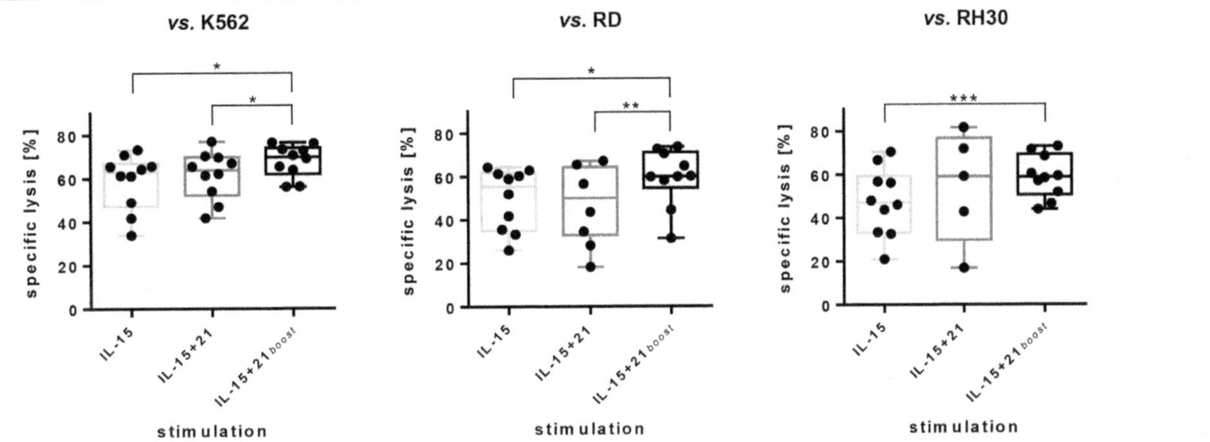

FIGURE 3 | Cytotoxic capacity of *ex vivo* expanded natural killer (NK) cells against erythroleukemia cell line K562 and RMS cell lines RD and RH30. NK cells were expanded for 6 days utilizing the stimulation protocols indicated. Data are given as mean values and SDs obtained with E:T ratios of 10:1 after 5 h of co-incubation. Specific lysis was calculated by substracting spontaneous lysis values from frequencies of dead target cells. Spontaneous lysis was determined from target cells cultured without NK cells. Here, combined results from independent donors are shown ($n = 10$; $n = 5$ for NK cells expanded with IL-15 + 21 vs. RH30). Assays were performed in tripliates for each donor. Asterisks indicate significant differences (*$p < 0.05$, **$p < 0.01$, ***$p < 0.005$; vs. K562 and vs. RD: one-way ANOVA; vs. RH30: Student's *t*-test).

by ADCC at an E:T ratio of 10:1 (Figure S3B in Supplementary Material).

The IL-21-Dependent Increase in NK Cell Cytotoxicity Is Based on Elevated Degranulation

To execute cytotoxicity upon activation by target cell recognition, one basic mechanism of NK cells is to degranulate and release cytolytic enzymes. In order to assess the impact of IL-21 to the cytolytic capacity and the underpinning mechanism, we stained for CD107a (lysosomal-associated membrane protein 1), which is located in the membranes of lytic granules and is expressed on the NK cell surface upon degranulation when lytic granules fuse with the outer membrane (47, 48).

Exposure to IL-21 during cultivation significantly increased the number of degranulating cells compared to IL-15 mono-treatment. This effect was rapidly induced after a short boost with IL-21. Permanent stimulation with IL-21 evoked similar results, suggesting that degranulation activity is maintained in the continuous presence of IL-21 (**Figure 4A**).

To further investigate if granule exocytosis is associated with cytolytic functionality, the expression of the pore-forming protein perforin and the serine protease granzyme B was addressed by immunoblotting. Indeed, expression and release of both proteins were induced strongly upon continuous stimulation with IL-21, but not subsequent to a short boost with IL-21 (**Figures 4B–D**). In particular, permanent presence of IL-21 in the culture provoked roughly a doubling of the relative expression of granzyme B and perforin in comparison to short-term or no IL-21 (**Figures 4C,D**).

In accordance with the slower induction of *de novo* production of granzyme B and perforin after short-term compared with continuous exposure to IL-21, apoptosis-inducing enzymes were released to a lesser extent despite similar degranulation levels were observed under both conditions (**Figures 4A,B**).

To further evaluate if this increase of degranulation was correlated with an increase of PI3K pathway signaling, we measured the phosphorylation of AKT and ERK1/2. Both were upregulated using IL-15 for expansion and even more with additional IL-21 (Figure S4 in Supplementary Material).

IL-21 Exposure Increases Cytokine Release by NK Cells

Natural killer cells can mediate direct cytotoxicity but also have an immunoregulatory function. The latter is achieved by secretion of cytokines. Hence, we analyzed the release of different cytokines after *ex vivo* expansion of NK cells using cytometric bead arrays (**Figure 5**).

Secretion of TNF-α was induced slightly by IL-15 and increased significantly upon stimulation with IL-21. Remarkably, after short-term exposure to IL-21, the TNF-α level was even higher in comparison with a continuous exposure to IL-21. A similar effect was observed for GM-CSF, although to a lesser extent. IFN-γ concentrations were significantly elevated with both protocols that utilized IL-21 causing a 10-fold increase in IFN-γ levels compared to IL-15 alone. Similarly, MIP-1α was secreted to a higher extent in the presence of IL-21 compared with IL-15 alone, independent of the duration of administration (**Figure 5**).

Levels of MCP-1 were similar for all three stimulation protocols, while IL-8 release was gradually more induced by IL-21 as compared to IL-15 alone and was further increased depending on the duration of IL-21 exposure. Also the levels of anti-inflammatory IL-10 were increased depending on the duration of IL-21 presence. In contrast, IL-15 application alone almost prevented secretion of IL-10 by NK cells (**Figure 5**).

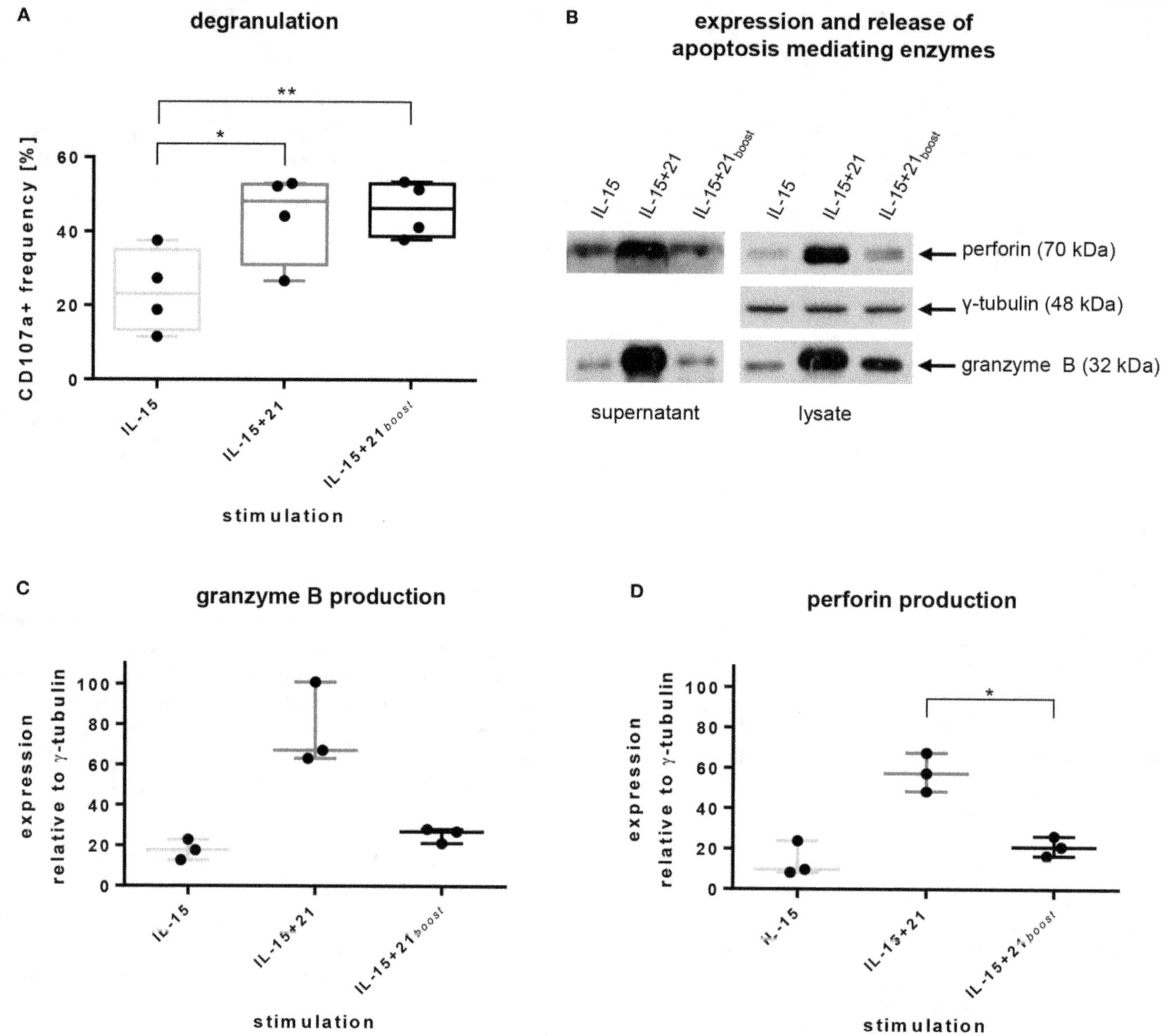

FIGURE 4 | Natural killer (NK) cell degranulation capacity, intracellular production, and release of apoptosis-mediating enzymes induced by selected *ex vivo* expansion protocols. Purified NK cells were kept in culture for 6 days using indicated protocols. **(A)** CD107a levels on the surface of cells were acquired by flow cytometry. Bars express mean values from four independent donors, lines indicate SDs. **(B)** Exemplary immunoblot analysis of supernatants and lysates of NK cells from one representative donor after stimulation with selected *ex vivo* expansion protocols. Proteins were detected using monoclonal antibodies against human perforin, γ-tubulin, and granzyme B. **(C,D)** Mean values and SDs of intracellular production of granzyme B and perforin obtained from three independent donors. Values are calculated in relation to γ-tubulin expression. Asterisks indicate significant differences (*$p \leq 0.05$, **$p \leq 0.01$, one-way ANOVA).

IL-21-Boosted Stimulation Enables *In Vivo* Cytotoxicity of NK Cells following Adoptive Transfer in a Xenograft Model

Given that NK cells cultured with the IL-15 + 21_{boost} protocol exhibited the highest expansion rate and cytotoxic activity toward different target cells, we finally aimed to validate the *in vivo* effectiveness of NK cells expanded with the two-phase protocol in a RMS xenograft model. To this end, NOD/SCID/IL-2Rγc^{-} (NSG) mice were subcutaneously injected with the RD$^{GFP/Luc}$ cell line and after establishing tumors, mice were locally irradiated by an image guided high precision local RT (**Figure 6A**). Subsequently, adoptive transfer of IL-15 + 21_{boost} continuously expanded NK cells was performed by three consecutive injections of 10^7 NK cells. Tumor burden was monitored by caliper measurement (not shown) or BLI analysis of eight mice per group over 79 days (**Figure 6B**). Animals treated with RT monotherapy displayed a significant decrease in luciferase intensity indicating a growth retardation (**Figures 6C,D**). Following combined RT and NK cell therapy, the cytostatic effect was even more pronounced as compared to untreated or irradiated tumor-bearing controls (**Figures 6B–E**).

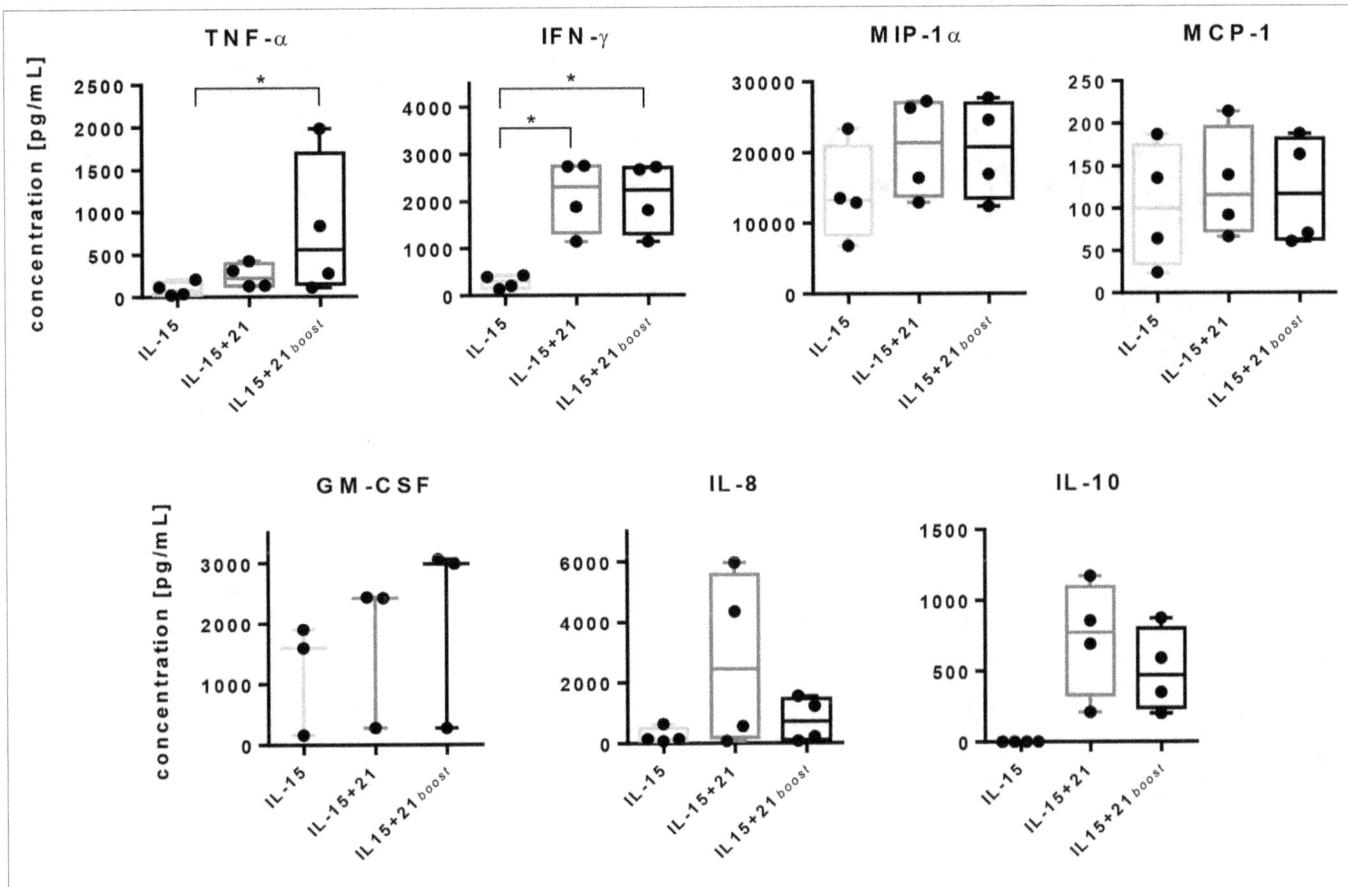

FIGURE 5 | Release of cytokines by purified natural killer (NK) cells upon stimulation with selected expansion protocols. Purified NK cells were cultured using different stimulation protocols. On day 6, release of tumor necrosis factor (TNF)-α, IFN-γ, macrophage inflammatory protein (MIP)-1α, monocyte chemoattractant protein (MCP)-1, granulocyte-macrophage colony-stimulating factor (GM-CSF), IL-8, and IL-10 were assayed by cytometric bead array analyses. Graphs show mean values and SDs obtained from four independent donors, except GM-CSF, which was measured for three independent donors. Asterisks indicate significant differences (*$p < 0.05$, one-way ANOVA).

DISCUSSION

Immunotherapy represents a promising approach against malignant diseases. In this context, NK cells have been explored for several years because they exhibit prevailing anti-tumor activity (9, 49–54). A major advantage of NK cell-based immunotherapy is the possibility to employ these cells in an allogeneic or haploidentical setting (6, 55, 56) without causing, or even preventing, GvHD (37–41). Numerous attempts have been made to expand NK cells efficiently for adoptive cell transfer focusing on different aspects such as high yields, efficient activation, cytotoxic potential, and/or good manufacturing practice (GMP) adequacy (27). Unfortunately, there are only limited publications available that state expansion rates after stimulating NK cells with cytokines only. Especially protocols using IL-21 on purified NK cells are quite rare. Wendt et al. did not mention absolute NK cell numbers but found an increased proliferation upon an IL-2 + 21 expansion over 72 h as shown by [^{3}H] thymidin incorporation. Researchers around Koehl et al. repeatedly reported expansion rates between four and five times after 12–14 days of culturing with high dose IL-2^{1000}. These results are similar to ours obtained with IL-15 or IL-15 + 21_{boost}, although our IL-2^{1000} expansion did not exceed a three times increase in our tests. Compared to that, protocols allowing additional accessory PBMCs resulted in an expansion rate of 23-fold by IL-15 (57), while under addition of IL-21 only a 3.7-fold expansion rate was reached (23). Addition of gene modified feeder cells resulted in several hundreds of multiplications. When IL-21 was added to such culturing conditions, an expansion of up to 2.7×10^{11} was reached after 46 days (24). The use of membrane bound IL-21 and 4-1BB ligand still led to an expansion of more than 10^5-fold (26).

In order to avoid contamination by possible GvHD causing cell subsets or genetically modified, tumor-derived feeder cells, we used purified NK cells for defining efficient expansion protocols. The major goal of the present study was to establish a feeder-cell free protocol for efficient expansion of primary NK cells and the improvement of their cytolytic capability to target highly aggressive and radiation-resistant RMS.

Natural killer cells were purified from buffy coats and stimulated with selected common γ-chain cytokines in different combinations. In the case of IL-2 stimulation, this study confirmed that the number of NK cells decreased under a low level of IL-2

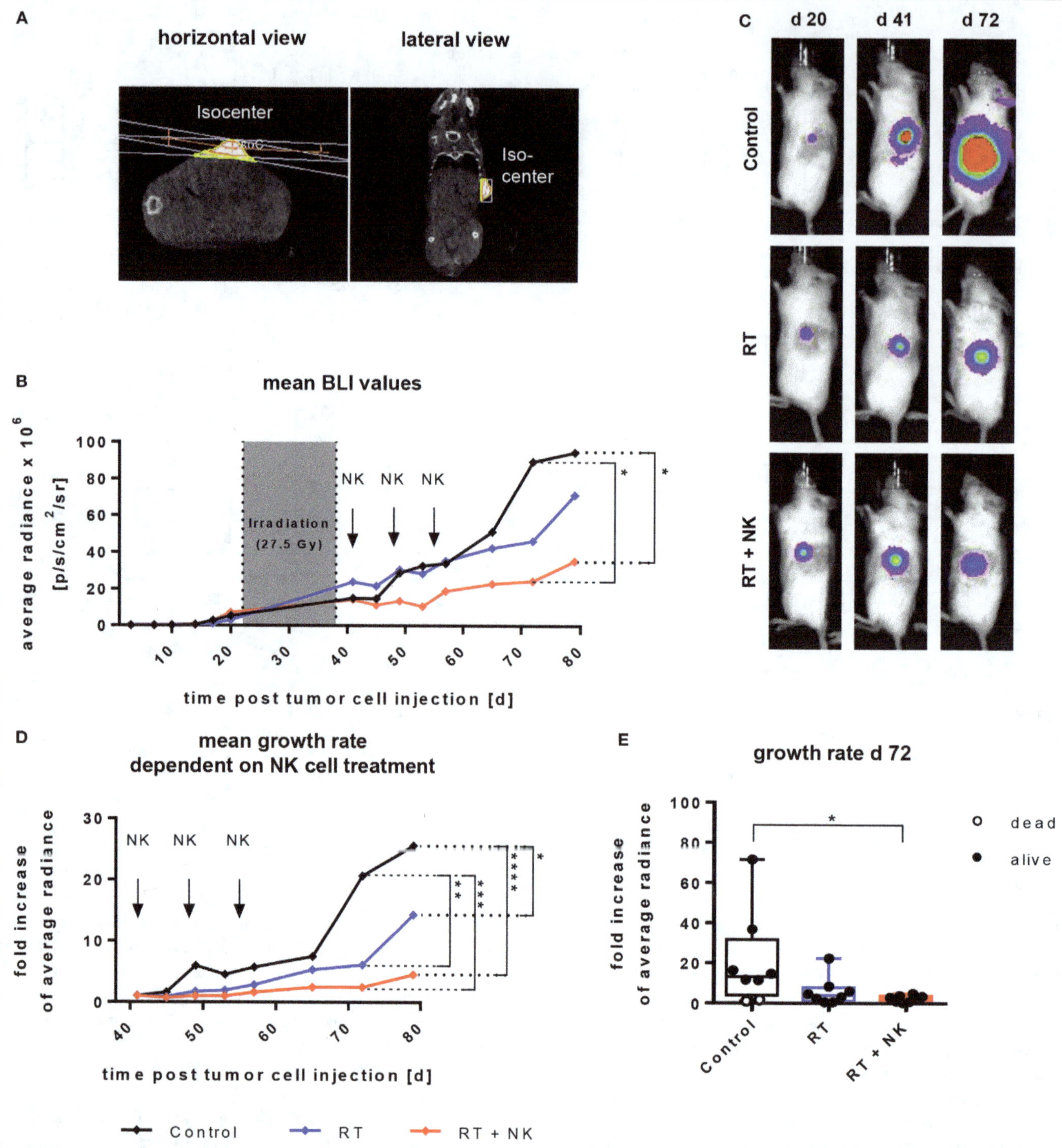

FIGURE 6 | Adoptive transfer of *ex vivo* IL-15-expanded and IL-21 boosted natural killer (NK) cells into RMS-bearing NSG mice subsequent to radiotherapy (RT). Mice were subcutaneously injected with 10^5 RD cells and 3 weeks later, tumor-bearing mice underwent local RT, followed by adoptive transfer of 10^7 NK cells by three weekly injections. NK cells were generated using the IL-15 + 21_{boost} protocol. **(A)** Exemplary CT-image guided planning of high precision tumor irradiation with a two-field geometry and RT isocenter presented in a horizontal and lateral view. **(B)** Time course of tumor growth, showing mean average radiance values from eight mice per group. The gray area indicates the period of RT, arrows indicate time points of NK cell administration. Significant differences are given as asteriks (*$p \leq 0.05$). **(C)** Pictures from bioluminescence imaging (BLI) of exemplary mice from each treatment group obtained before starting treatment (day 20), after termination of RT (day 41), and at the end of the experiment (day 72). **(D)** Kinetics of relative tumor growth rates normalized to tumor size at onset of NK cell treatment and **(E)** at day 72 after tumor inoculation. Displayed are mean values from eight mice per group and single values for each mouse **(E)**. Dead mice are shown as empty circles, asterisks indicate significant differences (*$p \leq 0.05$, **$p \leq 0.01$, ****$p \leq 0.001$, two-way ANOVA).

(100 U/mL) and increased only slightly upon administration of a high concentration of IL-2 (1000 U/mL) (**Figure 1A**) (14). However, as low IL-2 levels were shown to reduce CD3^{pos} T cell outgrowth from contaminating cells remaining after NK cell purification (58), combinations of a low IL-2-dosis with other cytokine stimulants more specifically improve NK cell expansion (59, 60). Of note, all IL-15-based expansion protocols, which were tested in this study, induced an increase in the amount of NK cells in the early expansion phase. In contrast, long-term presence of IL-21 negatively influenced NK cell proliferation capacity, while a short-time IL-21 boost up to 3 days enhanced proliferation and NK cell cytotoxicity (**Figures 1A,C** and **3**). In fact, the effects of IL-21 in NK cell proliferation are controversial. Immobilized or membrane-bound IL-21 has a positive effect on NK cell expansion (25, 61). Also, the longer survival of human NK cells following combined IL-21 and IL-2 stimulation has been reported (18). Conversely, an IL-21-induced apoptosis resulting in limited life span of NK cells has been described (20, 62). Overall, IL-21 appears to play an essential role in NK cell activation and cytotoxicity (21, 62).

Here, we observed a decline in the viability of NK cells after long-term exposure to IL-21 when added to other cytokines. Based on this observation, we developed a two-phase protocol first using IL-15 for the optimal expansion of NK cells, complemented by short-term addition of IL-21 to support proliferation and enhance the cytotoxic potential (IL-15 + 21_{boost}; **Figures 1A,B** and **3**). Regarding the cytokine-induced expansion of NK cells, it is important to note that most published and also our own results demonstrate a strong donor-dependent variability. It will be of high interest and potential clinical relevance to address and better understand underlying mechanisms of a donor-dependent variation in cytokine-induced NK cell response.

Cytokine stimulation and *ex vivo* expansion modulate the phenotype and function of NK cells (52). CD56 surface expression has been reported to be reduced during the purification process, down to complete absence of CD56 from the cell surface, but to be upregulated upon *in vitro* stimulation (63). Accordingly, we also found an increased expression of NK cell activating receptors upon cytokine stimulation (**Figure 2**). The CD16-expressing NK cell population was diminished during stimulation with IL-2 or IL-15 (**Figure 2A**), as previously reported to be mediated by metalloproteinase ADAM17 (64–66). In contrast, IL-21 maintained surface expression of CD16, hence, increasing the frequency of the more mature CD16^{pos} NK cell population (**Figure 2B**).

In parallel with the changes in CD16 expression, apoptosis-mediating receptors and ligands such as TRAIL, FAS, and FAS-L were upregulated upon cytokine-induced expansion. In this respect, we observed a marked increase in FAS expression triggered by all protocols (**Figure 2C**), confirming recent reports (67, 68). Usually, this mechanism is employed for the killing of tumor cells but also plays a part in the achievement of lymphocyte homeostasis following immune responses against infections (69). Herein, we observed TRAIL, FAS-L, DNAM-1, and NKp46 to be expressed strongly on the NK cell surface under IL-21 stimulation when compared with IL-15 stimulation alone (**Figure 2C**). In contrast, the expression of NKG2D, NKp30, and NKp44 was reduced under the influence of IL-21. It has been reported that NKp44 is downregulated by IL-21 treatment post-IL-15 stimulation (70, 71). Also, a decreased expression of NKp44 and NKG2D was attributed to IL-21-mediated inhibition of DAP10 and DAP12 (71, 72). In contrast, our results showed that the downstream signaling of DAP10 *via* the PI3K pathway was activated by IL-15, and even increased upon usage of IL-21 for NK cell expansion (Figure S4 in Supplementary Material). Upregulation of AKT phosphorylation is known to correlate with proliferation and cell survival. The even more activated ERK1/2 is associated with NK cell cytotoxicity, mediated by perforin and granzyme B mobilization (73, 74).

The maturation and activation state of NK cells in the final product is important, but the ability to carry out cytolytic functions is crucial. In line with that, we analyzed specific killing of the K562 erythroleukemia cell line as well as rhabdomyosarcoma cell lines RD and RH30. We observed an increase of cytotoxicity upon using the IL-15 + 21_{boost} protocol compared to IL-15 mono-treatment (**Figure 3**). For repeated NK cell applications, we tested continuously expanded cells and compared their cytotoxicity to that of cryopreserved cells. In line with reports on IL-15-stimulated NK cells (75), in our study, we observed a reduced cytotoxic capacity of cryopreserved cytokine-stimulated compared to continuously expanded NK cells (Figure S3A in Supplementary Material) and therefore performed analysis with the latter if several rounds of NK cell application were necessary. Another advantage of NK cell expansion using a feeder cell free IL-15 + 21_{boost} protocol will be that it easily fulfils the GMP criteria required for clinical application. Previous reports showed that IL-15-stimulated NK cells may attack various RMS cell lines more efficiently than unstimulated NK cells. Moreover, these investigations indicated that DNAM-1 and NKG2D may comprise initiators of cytotoxicity for resting NK cells, while killing by IL-15-stimulated NK cells involves additional factors including NKp30 and NKp46 (76). In our study, IL-21 enhanced NK cell cytotoxicity compared to IL-15 mono-treatment, increased DNAM-1 but reduced NKG2D expression. Although, LFA-1, among others, has been shown to be involved in the degranulation process (77) and we observed an increase in degranulating cells after prolonged IL-21 exposure or an IL-21 boost, we did not notice any difference in the expression levels of CD11a between the different stimulation protocols and also no prominent differences in the ability to form target cell conjugates (Figures S2C,D in Supplementary Material).

As previous reports showed that IL-21 induced killing can be independent from death receptor expression (62), but mediated by perforin (78), we investigated the release of lytic granules as one main mechanism mediating NK cell cytotoxicity. Indeed, IL-15 + 21_{boost}-stimulated NK cells showed an increased exposure of CD107a (**Figure 4A**), which is known to be an indicator of NK cell activity and degranulation (48). We further found that IL-21 strongly induced synthesis and release of granzyme B and perforin, but only after prolonged exposure. This is in accordance with reports on an increased production of perforin and IFN-γ in cytotoxic CD8^{pos} T cells from HIV patients upon IL-21 stimulation (79). Moreover, upregulation of intracellular perforin was reported for CD56^{pos} cells from HIV-infected individuals (80), from stimulated PBMCs (18) and from patients with malignant melanoma upon IL-21 stimulation (81). Moreover, the increased activation of AKT and ERK signaling that we observed with IL-21 has been reported to be accompanied by intracellular perforin and granzyme B redirection (73).

Although cytokine release follows pathways distinct from lytic granule exocytosis, both mechanisms often are regulated similarly (82). In parallel to the release of lytic granules, secretion of TNF-α and IFN-γ was elevated after IL-21 stimulation (**Figure 5**). In the context of an immunocompetent host, NK cells not only contribute to the immune response by direct cytotoxicity but also as mediators at an intersection between innate and adaptive immunity (83, 84). Surprisingly, secretion of IL-10 was strongly induced by IL-21, but not by IL-15. As IL-10 is reported to play a crucial role in immune suppression, its function following IL-21 stimulation remains to be evaluated. Nevertheless, opposing effects were reported for IL-12, which was shown to activate NK cells and trigger pro-inflammatory T cell responses (85, 86), but also to induce IL-10 secretion (87).

Finally, we addressed the *in vivo* anti-tumor efficacy of IL-15 + 21_{boost} expanded NK cells if combined with ionizing radiation in a RMS xenograft model. Adoptive NK cell immunotherapy in combination with RT has been addressed only in a few studies so far. Ames et al. recently reported that NK cell inoculation was effective at targeting cancer cells with a stem cell phenotype from a variety of solid malignancies, which was most effective when combined with RT before application (88). It has further been shown that a combined transfer of IL-12 + 15 + 18 activated NK cells and high dose (5 Gy) synchronous irradiation resulted in a growth retardation of RMA-S lymphoma and B16-RAE-1ε tumor cells (84). Consequently, in our proof of principle approach, we performed a multimodal treatment with a preceding local RT followed by adoptive transfer of IL-15 + 21_{boost} expanded NK cells. Here, we confirmed that combined NK cell immunotherapy and RT was superior to RT monotherapy in terms of growth retardation, which reaches a level of significances at 7.5 weeks. Moreover, for that purpose, a mouse CT-image guided and individually planed local irradiation procedure was applied using a SARRP with a fractionated daily 2.5 Gy RT protocol that to higher extent reflects the clinical situation where patients were treated with single 1.8 to 2 Gy fractions. Notably, combined modality treatment failed to cure the tumor burden (**Figures 6B,D**) indicating the necessity for advanced protocols with several repeated NK transfers or additional *in vivo* cytokine injections to further increase therapeutic efficacy. This will be addressed in future investigations.

In conclusion, we presented an optimized protocol for *ex vivo* NK cell expansion that combines the positive effects of both IL-15 and IL-21 on proliferation and activity of NK cells and that offers an ideal expansion protocol also under GMP conditions due to the absence of feeder cells. Additionally, our findings demonstrate the high *in vitro* and *in vivo* antitumor efficacy of IL-15 + 21_{boost} expanded NK cells, which may become useful for the development of innovative combined modality treatment strategies for radiation-resistant RMS.

ETHICS STATEMENT

This study was carried out in accordance with the recommendations of the Ethics Committee of the Goethe University Frankfurt (Frankfurt, Germany). All subjects gave written informed consent in accordance with the Declaration of Helsinki. This study was carried out in accordance with the recommendations of German Animal Welfare Act. The protocol was approved by the government committee (Regierungspräsidium Darmstadt, Darmstadt, Germany).

AUTHOR CONTRIBUTIONS

JW, EU, FR, and PB designed the project. JW, VP, JB, AW, SH, and BB performed experiments and analyzed data. JW, EU, WW, FR, PB, CB, TK, and SH discussed data. JW and EU wrote the manuscript with support from all other co-authors. All authors agreed to be accountable for the content of the work.

ACKNOWLEDGMENTS

The authors thank Halvard Bönig from the BSD Frankfurt, Germany, for providing healthy donor blood samples, Ralf Schubert and Petra Schoen for kind support in performing the CBA, Anke Muth for her support in the generation of the transfer plasmid for GFP-luciferase and the vector particles, Sara Tognarelli, Julius Oppermann, Annekathrin Heinze, Keven Hörster, Karina Winterling, Catharina Ettingshausen, and Sarah Kehr for their assistance within the experiments, Katja Stein and Bushra Rais for technical support as well as Sandra Kunz for excellent secretarial assistance.

REFERENCES

1. Moretta A, Bottino C, Vitale M, Pende D, Cantoni C, Mingari MC, et al. Activating receptors and coreceptors involved in human natural killer cell-mediated cytolysis. *Annu Rev Immunol* (2001) 19:197–223. doi:10.1146/annurev.immunol.19.1.197
2. Garrido F, Ruiz-Cabello F, Cabrera T, Pérez-Villar JJ, López-Botet M, Duggan-Keen M, et al. Implications for immunosurveillance of altered HLA class I phenotypes in human tumours. *Immunol Today* (1997) 18:89–95. doi:10.1016/S0167-5699(96)10075-X
3. Bauer S, Groh V, Wu J, Steinle A, Phillips JH, Lanier LL, et al. Activation of NK cells and T cells by NKG2D, a receptor for stress-inducible MICA. *Science* (1999) 285:727–9. doi:10.1126/science.285.5428.727
4. Pende D, Spaggiari GM, Marcenaro S, Martini S, Rivera P, Capobianco A, et al. Analysis of the receptor-ligand interactions in the NK-mediated lysis of freshly isolated myeloid or lymphoblastic leukemias. Evidence for the involvement of the poliovirus receptor (CD155) and Nectin-2 (CD112). *Blood* (2004) 105:2004–9. doi:10.1182/blood-2004-09-3548.Supported
5. Ullrich E, Koch J, Cerwenka A, Steinle A. New prospects on the NKG2D/NKG2DL system for oncology. *Oncoimmunology* (2013) 2:e26097. doi:10.4161/onci.26097
6. Miller JS, Soignier Y, Panoskaltsis-mortari A, Mcnearney SA, Yun GH, Fautsch SK, et al. Successful adoptive transfer and in vivo expansion of human haploidentical NK cells in patients with cancer. *Cancer* (2005) 105:3051–7. doi:10.1182/blood-2004-07-2974.Supported

7. Fujisaki H, Kakuda H, Shimasaki N, Imai C, Ma J, Lockey T, et al. Expansion of highly cytotoxic human natural killer cells for cancer cell therapy. *Cancer Res* (2009) 69:4010–7. doi:10.1158/0008-5472.CAN-08-3712
8. Noguchi M, Yi H, Rosenblatt HM, Filipovich AH, Adelstein S, Modi WS, et al. Interleukin-2 receptor gamma chain mutation results in X-linked severe combined immunodeficiency in humans. *Cell* (1993) 73:147–57. doi:10.1016/0092-8674(93)90167-O
9. Miller JS, Klingsporn S, Lund J, Perry EH, Verfaillie C, McGlave P. Large scale ex vivo expansion and activation of human natural killer cells for autologous therapy. *Bone Marrow Transplant* (1994) 14:555–62.
10. Brehm C, Huenecke S, Quaiser A, Esser R, Bremm M, Kloess S, et al. IL-2 stimulated but not unstimulated NK cells induce selective disappearance of peripheral blood cells: concomitant results to a phase I/II study. *PLoS One* (2011) 6:e27351. doi:10.1371/journal.pone.0027351
11. Suck G, Koh MBC. Emerging natural killer cell immunotherapies: large-scale ex vivo production of highly potent anticancer effectors. *Hematol Oncol Stem Cell Ther* (2010) 3:135–42. doi:10.1016/S1658-3876(10)50024-4
12. Huenecke S, Zimmermann SY, Kloess S, Esser R, Brinkmann A, Tramsen L, et al. IL-2-driven regulation of NK cell receptors with regard to the distribution of CD16+ and CD16- subpopulations and in vivo influence after haploidentical NK cell infusion. *J Immunother* (2010) 33:200–10. doi:10.1097/CJI.0b013e3181bb46f7
13. Giri JG, Ahdieh M, Eisenman J, Shanebeck K, Grabstein K, Kumaki S, et al. Utilization of the beta and gamma chains of the IL-2 receptor by the novel cytokine IL-15. *EMBO J* (1994) 13:2822–30.
14. Carson WE, Giri JG, Lindemann MJ, Linett ML, Ahdieh M, Paxton R, et al. Interleukin (IL) 15 is a novel cytokine that activates human natural killer cells via components of the IL-2 receptor. *J Exp Med* (1994) 180:1395–403. doi:10.1084/jem.180.4.1395
15. Assier E, Jullien V, Lefort J, Moreau JL, Di Santo JP, Vargaftig BB, et al. NK cells and polymorphonuclear neutrophils are both critical for IL-2-induced pulmonary vascular leak syndrome. *J Immunol* (2004) 172:7661–8. doi:10.4049/jimmunol.172.12.7661
16. Waldmann TA, Lugli E, Roederer M, Perera LP, Smedley JV, Macallister RP, et al. Safety (toxicity), pharmacokinetics, immunogenicity, and impact on elements of the normal immune system of recombinant human IL-15 in rhesus macaques. *Blood* (2011) 117:4787–95. doi:10.1182/blood-2010-10-311456
17. Parrish-Novak J, Dillon SR, Nelson A, Hammond A, Sprecher C, Gross JA, et al. Interleukin 21 and its receptor are involved in NK cell expansion and regulation of lymphocyte function. *Nature* (2000) 408:57–63. doi:10.1038/35040504
18. Skak K, Frederiksen KS, Lundsgaard D. Interleukin-21 activates human natural killer cells and modulates their surface receptor expression. *Immunology* (2008) 123:575–83. doi:10.1111/j.1365-2567.2007.02730.x
19. Park Y-K, Shin D-J, Cho D, Kim S-K, Lee J-J, Shin M-G, et al. Interleukin-21 increases direct cytotoxicity and IFN-γ production of ex vivo expanded NK cells towards breast cancer cells. *Anticancer Res* (2012) 32:839–46.
20. Li Q, Ye LJ, Ren HL, Ting HY, Li J, Shi JL, et al. Multiple effects of IL-21 on human NK cells in ex vivo expansion. *Immunobiology* (2015) 220:876–88. doi:10.1016/j.imbio.2015.01.009
21. Brady J, Carotta S, Thong RPL, Chan CJ, Hayakawa Y, Smyth MJ, et al. The interactions of multiple cytokines control NK cell maturation. *J Immunol* (2010) 185:6679–88. doi:10.4049/jimmunol.0903354
22. Kasaian MT, Whitters MJ, Carter LL, Lowe LD, Jussif JM, Deng B, et al. IL-21 limits NK cell responses and promotes antigen-specific T cell activation: a mediator of the transition from innate to adaptive immunity. *Immunity* (2002) 16:559–69. doi:10.1016/S1074-7613(02)00295-9
23. Choi I, Yoon SR, Park SY, Kim H, Jung SJ, Jang YJ, et al. Donor-derived natural killer cells infused after human leukocyte antigen-haploidentical hematopoietic cell transplantation: a dose-escalation study. *Biol Blood Marrow Transplant* (2014) 20:696–704. doi:10.1016/j.bbmt.2014.01.031
24. Granzin M, Stojanovic A, Miller M, Childs R, Huppert V, Cerwenka A. Highly efficient IL-21 and feeder cell-driven ex vivo expansion of human NK cells with therapeutic activity in a xenograft mouse model of melanoma. *Oncoimmunology* (2016) 5:e1219007. doi:10.1080/2162402X.2016.1219007
25. Denman CJ, Senyukov VV, Somanchi SS, Phatarpekar PV, Kopp LM, Johnson JL, et al. Membrane-bound IL-21 promotes sustained ex vivo proliferation of human natural killer cells. *PLoS One* (2012) 7:e30264. doi:10.1371/journal.pone.0030264
26. Oyer JL, Pandey V, Igarashi RY, Somanchi SS, Zakari A, Solh M, et al. Natural killer cells stimulated with PM21 particles expand and biodistribute in vivo: clinical implications for cancer treatment. *Cytotherapy* (2016) 18:653–63. doi:10.1016/j.jcyt.2016.02.006
27. Granzin M, Ullrich E, Wagner J, Köhl U, Cerwenka A, Huppert V. Shaping of natural killer cell antitumor activity by ex vivo cultivation. *Front Immunol* (2017) 8:458. doi:10.3389/fimmu.2017.00458
28. Röllig C, Bornhäuser M, Kramer M, Thiede C, Ho AD, Krämer A, et al. Allogeneic stem-cell transplantation in patients with NPM1-mutated acute myeloid leukemia: results from a prospective donor versus no-donor analysis of patients after upfront HLA typing within the SAL-AML 2003 trial. *J Clin Oncol* (2015) 33:403–10. doi:10.1200/JCO.2013.54.4973
29. Czerw T, Labopin M, Gorin N-C, Giebel S, Blaise D, Meloni G, et al. Long-term follow-up of patients with acute myeloid leukemia surviving and free of disease recurrence for at least 2 years after autologous stem cell transplantation: a report from the Acute Leukemia Working Party of the European Society for Blood and Marr. *Cancer* (2016) 122:1880–7. doi:10.1002/cncr.29990
30. Koscielniak E, Klingebiel TH, Peters C, Hermann J, Burdach ST, Bender-Götze C, et al. Do patients with metastatic and recurrent rhabdomyosarcoma benefit from high-dose therapy with hematopoietic rescue? Report of the German/Austrian Pediatric Bone Marrow Transplantation Group. *Bone Marrow Transplant* (1997) 19:227–31. doi:10.1038/sj.bmt.1700628
31. Carli M, Colombatti R, Oberlin O, Stevens M, Masiero L, Frascella E, et al. High-dose melphalan with autologous stem-cell rescue in metastatic rhabdomyosarcoma. *J Clin Oncol* (1999) 17:2796–803. doi:10.1200/jco.1999.17.9.2796
32. Dantonello TM, Int-Veen C, Harms D, Leuschner I, Schmidt BF, Herbst M, et al. Cooperative trial CWS-91 for localized soft tissue sarcoma in children, adolescents, and young adults. *J Clin Oncol* (2009) 27:1446–55. doi:10.1200/JCO.2007.15.0466
33. Peinemann F, Kröger N, Bartel C, Grouven U, Pittler M, Erttmann R, et al. High-dose chemotherapy followed by autologous stem cell transplantation for metastatic rhabdomyosarcoma – a systematic review. *PLoS One* (2011) 6:e17127. doi:10.1371/journal.pone.0017127
34. Thiel U, Koscielniak E, Blaeschke F, Grunewald TGP, Badoglio M, Diaz MA, et al. Allogeneic stem cell transplantation for patients with advanced rhabdomyosarcoma: a retrospective assessment. *Br J Cancer* (2013) 109:2523–32. doi:10.1038/bjc.2013.630
35. Yamazaki F, Osumi T, Shigematsu N, Morioka H, Shimada H. Successful treatment of metastatic rhabdomyosarcoma with radiochemotherapy and allogeneic hematopoietic stem cell transplantation. *Jpn J Clin Oncol* (2015) 45:225–8. doi:10.1093/jjco/hyu189
36. Hiniker SM, Donaldson SS. Recent advances in understanding and managing rhabdomyosarcoma. *F1000Prime Rep* (2015) 7:59. doi:10.12703/P7-59
37. Asai O, Longo DL, Tian ZG, Hornung RL, Taub DD, Ruscetti FW, et al. Suppression of graft-versus-host disease and amplification of graft-versus-tumor effects by activated natural killer cells after allogeneic bone marrow transplantation. *J Clin Invest* (1998) 101:1835–42. doi:10.1172/JCI1268
38. Ruggeri L, Capanni M, Urbani E, Perruccio K, Shlomchik WD, Tosti A, et al. Effectiveness of donor natural killer cell alloreactivity in mismatched hematopoietic transplants. *Science* (2002) 295:2097–100. doi:10.1126/science.1068440
39. Olson JA, Leveson-Gower DB, Gill S, Baker J, Beilhack A, Negrin RS. NK cells mediate reduction of GVHD by inhibiting activated, alloreactive T cells while retaining GVT effects. *Blood* (2010) 115:4293–301. doi:10.1182/blood-2009-05-222190
40. Meinhardt K, Kroeger I, Bauer R, Ganss F, Ovsiy I, Rothamer J, et al. Identification and characterization of the specific murine NK cell subset supporting graft-versus-leukemia- and reducing graft-versus-host-effects. *Oncoimmunology* (2015) 4:e981483. doi:10.4161/2162402X.2014.981483
41. Ullrich E, Salzmann-Manrique E, Bakhtiar S, Bremm M, Gerstner S, Herrmann E, et al. Relation between acute GVHD and NK cell subset reconstitution following allogeneic stem cell transplantation. *Front Immunol* (2016) 7:595. doi:10.3389/fimmu.2016.00595
42. Negrin RS. Graft-versus-host disease versus graft-versus-leukemia. *Hematology Am Soc Hematol Educ Program* (2015) 2015:225–30. doi:10.1182/asheducation-2015.1.225
43. Ferrara JLM, Levine JE, Reddy P, Holler E. Graft-versus-host disease. *Lancet* (2009) 373:1550–61. doi:10.1016/S0140-6736(09)60237-3

44. Blazar BR, Murphy WJ, Abedi M. Advances in graft-versus-host disease biology and therapy. *Nat Rev Immunol* (2012) 12:443–58. doi:10.1038/nri3212
45. Hinson AR, Jones R, Crose LE, Belyea BC, Barr FG, Linardic CM. Human rhabdomyosarcoma cell lines for rhabdomyosarcoma research: utility and pitfalls. *Front Oncol* (2013) 3:183. doi:10.3389/fonc.2013.00183
46. Abel T, El Filali E, Waern J, Schneider IC, Yuan Q, Münch RC, et al. Specific gene delivery to liver sinusoidal and artery endothelial cells. *Blood* (2013) 122(12):2030–8. doi:10.1182/blood-2012-11-468579
47. Tognarelli S, Jacobs B, Staiger N, Ullrich E. Flow cytometry-based assay for the monitoring of NK cell functions. *J Vis Exp* (2016) e54615. doi:10.3791/54615
48. Alter G, Malenfant JM, Altfeld M. CD107a as a functional marker for the identification of natural killer cell activity. *J Immunol Methods* (2004) 294:15–22. doi:10.1016/j.jim.2004.08.008
49. Ljunggren HG, Malmberg KJ. Prospects for the use of NK cells in immunotherapy of human cancer. *Nat Rev Immunol* (2007) 7:329–39. doi:10.1038/nri2073
50. Koehl U, Kalberer C, Spanholtz J, Lee DA, Miller JS, Cooley S, et al. Advances in clinical NK cell studies: donor selection, manufacturing and quality control. *Oncoimmunology* (2016) 5:e1115178-1-11. doi:10.1080/2162402X.2015.1115178
51. Terme M, Ullrich E, Delahaye NF, Chaput N, Zitvogel L. Natural killer cell-directed therapies: moving from unexpected results to successful strategies. *Nat Immunol* (2008) 9:486–94. doi:10.1038/ni1580
52. Vivier E, Tomasello E, Baratin M, Walzer T, Ugolini S. Functions of natural killer cells. *Nat Immunol* (2008) 9:503–10. doi:10.1038/ni1582
53. Pietra G, Manzini C, Rivara S, Vitale M, Cantoni C, Petretto A, et al. Melanoma cells inhibit natural killer cell function by modulating the expression of activating receptors and cytolytic activity. *Cancer Res* (2012) 72:1407–15. doi:10.1158/0008-5472.CAN-11-2544
54. Morvan MG, Lanier LL. NK cells and cancer: you can teach innate cells new tricks. *Nat Rev Cancer* (2015) 16:7–19. doi:10.1038/nrc.2015.5
55. Bachanova V, Burns LJ, McKenna DH, Curtsinger J, Panoskaltsis-Mortari A, Lindgren BR, et al. Allogeneic natural killer cells for refractory lymphoma. *Cancer Immunol Immunother* (2010) 59:1739–44. doi:10.1007/s00262-010-0896-z
56. Stern M, Passweg JR, Meyer-Monard S, Esser R, Tonn T, Soerensen J, et al. Pre-emptive immunotherapy with purified natural killer cells after haploidentical SCT: a prospective phase II study in two centers. *Bone Marrow Transplant* (2013) 48:433–8. doi:10.1038/bmt.2012.162
57. Iliopoulou EG, Kountourakis P, Karamouzis MV, Doufexis D, Ardavanis A, Baxevanis CN, et al. A phase I trial of adoptive transfer of allogeneic natural killer cells in patients with advanced non-small cell lung cancer. *Cancer Immunol Immunother* (2010) 59:1781–9. doi:10.1007/s00262-010-0904-3
58. Besser MJ, Schallmach E, Oved K, Treves AJ, Markel G, Reiter Y, et al. Modifying interleukin-2 concentrations during culture improves function of T cells for adoptive immunotherapy. *Cytotherapy* (2009) 11:206–17. doi:10.1080/14653240802590391
59. Claus M, Greil J, Watzl C. Comprehensive analysis of NK cell function in whole blood samples. *J Immunol Methods* (2009) 341:154–64. doi:10.1016/j.jim.2008.11.006
60. De Maria A, Bozzano F, Cantoni C, Moretta L. Revisiting human natural killer cell subset function revealed cytolytic CD56(dim)CD16+ NK cells as rapid producers of abundant IFN-gamma on activation. *Proc Natl Acad Sci U S A* (2011) 108:728–32. doi:10.1073/pnas.1012356108
61. Li XM, He CH, Liu CZ, Ma J, Ma P, Cui HL, et al. Expansion of NK cells from PBMCs using immobilized 4-1BBL and interleukin-21. *Int J Oncol* (2015) 47:335–42. doi:10.3892/ijo.2015.3005
62. Brady J, Hayakawa Y, Smyth MJ, Nutt SL. IL-21 induces the functional maturation of murine NK cells. *J Immunol* (2004) 172:2048–58. doi:10.4049/jimmunol.172.4.2048
63. Kloess S, Huenecke S, Piechulek D, Esser R, Koch J, Brehm C, et al. IL-2-activated haploidentical NK cells restore NKG2D-mediated NK-cell cytotoxicity in neuroblastoma patients by scavenging of plasma MICA. *Eur J Immunol* (2010) 40:3255–67. doi:10.1002/eji.201040568
64. Harrison D, Phillips JH, Lanier LL. Involvement of a metalloprotease in spontaneous and phorbol ester-induced release of natural killer cell-associated Fc gamma RIII (CD16-II). *J Immunol* (1991) 147:3459–65.
65. Grzywacz B, Kataria N, Verneris MR. CD56dimCD16+ NK cells downregulate CD16 following target cell induced activation of matrix metalloproteinases. *Leukemia* (2007) 21:356–9. doi:10.1038/sj.leu.2404499
66. Romee R, Foley B, Lenvik T, Wang Y, Zhang B, Ankarlo D, et al. NK cell CD16 surface expression and function is regulated by a disintegrin and metalloprotease-17 (ADAM17). *Blood* (2013) 121:3599–608. doi:10.1182/blood-2012-04-425397
67. Medvedev AE, Johnsen A-C, Haux J, Steinkjer B, Egeberg K, Lynch DH, et al. Regulation of Fas and Fas-ligand expression in NK cells by cytokines and the involvement of Fas-ligand in NK/LAK cell-mediated cytotoxicity. *Cytokine* (1997) 9:394–404. doi:10.1006/cyto.1996.0181
68. Lin S-J, Wang L-Y, Huang Y-J, Kuo M-L. Effect of interleukin (IL)-12 and IL-15 on apoptosis and proliferation of umbilical cord blood mononuclear cells. *Bone Marrow Transplant* (2001) 28:439–45. doi:10.1038/sj.bmt.1703182
69. Felices M, Lenvik TR, Ankarlo DEM, Foley B, Curtsinger J, Luo X, et al. Functional NK cell repertoires are maintained through IL-2Rα and Fas ligand. *J Immunol* (2014) 192:3889–97. doi:10.4049/jimmunol.1302601
70. Lim D-P, Jang Y-Y, Kim S, Koh SS, Lee J-J, Kim J-S, et al. Effect of exposure to interleukin-21 at various time points on human natural killer cell culture. *Cytotherapy* (2014) 16:1419–30. doi:10.1016/j.jcyt.2014.04.008
71. de Rham C, Ferrari-Lacraz S, Jendly S, Schneiter G, Dayer J-M, Villard J. The proinflammatory cytokines IL-2, IL-15 and IL-21 modulate the repertoire of mature human natural killer cell receptors. *Arthritis Res Ther* (2007) 9:R125. doi:10.1186/ar2336
72. Burgess SJ, Marusina AI, Pathmanathan I, Borrego F, Coligan JE. IL-21 down-regulates NKG2D/DAP10 expression on human NK and CD8+ T cells. *J Immunol* (2006) 176:1490–7. doi:10.4049/jimmunol.176.3.1490
73. Wei S, Gamero AM, Liu JH, Daulton AA, Valkov NI, Trapani JA, et al. Control of lytic function by mitogen-activated protein kinase/extracellular regulatory kinase 2 (ERK2) in a human natural killer cell line: identification of perforin and granzyme B mobilization by functional ERK2. *J Exp Med* (1998) 187:1753–65. doi:10.1084/jem.187.11.1753
74. Jiang K, Zhong B, Gilvary DL, Corliss BC, Hong-Geller E, Wei S, et al. Pivotal role of phosphoinositide-3 kinase in regulation of cytotoxicity in natural killer cells. *Nat Immunol* (2000) 1:419–25. doi:10.1038/80859
75. van Ostaijen-ten Dam MM, Prins HJ, Boerman GH, Vervat C, Pende D, Putter H, et al. Preparation of cytokine-activated NK cells for use in adoptive cell therapy in cancer patients: protocol optimization and therapeutic potential. *J Immunother* (2016) 39:90–100. doi:10.1097/Cji.0000000000000110
76. Boerman GH, van Ostaijen-ten Dam MM, Kraal KCJM, Santos SJ, Ball LM, Lankester AC, et al. Role of NKG2D, DNAM-1 and natural cytotoxicity receptors in cytotoxicity toward rhabdomyosarcoma cell lines mediated by resting and IL-15-activated human natural killer cells. *Cancer Immunol Immunother* (2015) 64:573–83. doi:10.1007/s00262-015-1657-9
77. Nielsen N, Ødum N, Ursø B, Lanier LL, Spee P. Cytotoxicity of CD56 bright NK cells towards autologous activated CD4 + T cells is mediated through NKG2D, LFA-1 and TRAIL and dampened via CD94/NKG2A. *PLoS One* (2012) 7:e31959. doi:10.1371/journal.pone.0031959
78. Ma H-L, Whitters MJ, Konz RF, Senices M, Young DA, Grusby MJ, et al. IL-21 activates both innate and adaptive immunity to generate potent antitumor responses that require perforin but are independent of IFN-γ. *J Immunol* (2003) 171:608–15. doi:10.4049/jimmunol.171.2.608
79. White L, Krishnan S, Strbo N, Liu H, Kolber MA, Lichtenheld MG, et al. Differential effects of IL-21 and IL-15 on perforin expression, lysosomal degranulation, and proliferation in CD8 T cells of patients with human immunodeficiency virus-1 (HIV). *Blood* (2007) 109:3873–80. doi:10.1182/blood-2006-09-045278
80. Strbo N, de Armas L, Liu H, Kolber MA, Lichtenheld M, Pahwa S. IL-21 augments natural killer effector functions in chronically HIV-infected individuals. *AIDS* (2008) 22:1551–60. doi:10.1097/QAD.0b013e3283089367
81. Davis ID, Brady B, Kefford RF, Millward M, Cebon J, Skrumsager BK, et al. Clinical and biological efficacy of recombinant human interleukin-21 in patients with stage IV malignant melanoma without prior treatment: a phase IIa trial. *Clin Cancer Res* (2009) 15(6):2123–9. doi:10.1158/1078-0432.CCR-08-2663
82. Reefman E, Kay JG, Wood SM, Offenhäuser C, Brown DL, Roy S, et al. Cytokine secretion is distinct from secretion of cytotoxic granules in NK cells. *J Immunol* (2010) 184(9):4852–62. doi:10.4049/jimmunol.0803954

83. Lodolce JP, Burkett PR, Boone DL, Chien M, Ma A. T cell-independent interleukin 15Ralpha signals are required for bystander proliferation. *J Exp Med* (2001) 194:1187–94. doi:10.1084/jem.194.8.1187
84. Moretta L, Ferlazzo G, Bottino C, Vitale M, Pende D, Mingari MC, et al. Effector and regulatory events during natural killer–dendritic cell interactions. *Immunol Rev* (2006) 214:219–28. doi:10.1111/j.1600-065X.2006.00450.x
85. Trinchieri G. Interleukin-12 and its role in the generation of TH1 cells. *Immunol Today* (1993) 14:335–8. doi:10.1016/0167-5699(93)90230-I
86. Manetti R, Parronchi P, Giudizi MG, Piccinni MP, Maggi E, Trinchieri G, et al. Natural killer cell stimulatory factor (interleukin 12 [IL-12]) induces T helper type 1 (Th1)-specific immune responses and inhibits the development of IL-4-producing Th cells. *J Exp Med* (1993) 177:1199–204. doi:10.1084/jem.177.4.1199
87. Mehrotra PT, Donnelly RP, Wong S, Kanegane H, Geremew A, Mostowski HS, et al. Production of IL-10 by human natural killer cells stimulated with IL-2 and/or IL-12. *J Immunol* (1998) 160:2637–44.
88. Ames E, Canter RJ, Grossenbacher SK, Mac S, Smith RC, Monjazeb AM, et al. Enhanced targeting of stem-like solid tumor cells with radiation and natural killer cells. *Oncoimmunology* (2015) 4:e1036212. doi:10.1080/2162402X.2015.1036212

Permissions

All chapters in this book were first published in TNKCRLIAAE, by Frontiers; hereby published with permission under the Creative Commons Attribution License or equivalent. Every chapter published in this book has been scrutinized by our experts. Their significance has been extensively debated. The topics covered herein carry significant findings which will fuel the growth of the discipline. They may even be implemented as practical applications or may be referred to as a beginning point for another development.

The contributors of this book come from diverse backgrounds, making this book a truly international effort. This book will bring forth new frontiers with its revolutionizing research information and detailed analysis of the nascent developments around the world.

We would like to thank all the contributing authors for lending their expertise to make the book truly unique. They have played a crucial role in the development of this book. Without their invaluable contributions this book wouldn't have been possible. They have made vital efforts to compile up to date information on the varied aspects of this subject to make this book a valuable addition to the collection of many professionals and students.

This book was conceptualized with the vision of imparting up-to-date information and advanced data in this field. To ensure the same, a matchless editorial board was set up. Every individual on the board went through rigorous rounds of assessment to prove their worth. After which they invested a large part of their time researching and compiling the most relevant data for our readers.

The editorial board has been involved in producing this book since its inception. They have spent rigorous hours researching and exploring the diverse topics which have resulted in the successful publishing of this book. They have passed on their knowledge of decades through this book. To expedite this challenging task, the publisher supported the team at every step. A small team of assistant editors was also appointed to further simplify the editing procedure and attain best results for the readers.

Apart from the editorial board, the designing team has also invested a significant amount of their time in understanding the subject and creating the most relevant covers. They scrutinized every image to scout for the most suitable representation of the subject and create an appropriate cover for the book.

The publishing team has been an ardent support to the editorial, designing and production team. Their endless efforts to recruit the best for this project, has resulted in the accomplishment of this book. They are a veteran in the field of academics and their pool of knowledge is as vast as their experience in printing. Their expertise and guidance has proved useful at every step. Their uncompromising quality standards have made this book an exceptional effort. Their encouragement from time to time has been an inspiration for everyone.

The publisher and the editorial board hope that this book will prove to be a valuable piece of knowledge for researchers, students, practitioners and scholars across the globe.

List of Contributors

Kattria van der Ploeg, Chiwen Chang, Martin A. Ivarsson, Ashley Moffett and John Trowsdale
Department of Pathology, University of Cambridge, Cambridge, UK

Mark R. Wills
Department of Medicine, University of Cambridge, Cambridge, UK

Wilford Goh and Nicholas D. Huntington
The Walter and Eliza Hall Institute of Medical Research, Parkville, VIC, Australia
Department of Medical Biology, University of Melbourne, Melbourne, VIC, Australia

Julia Pollmann
Research Group Innate Immunity, German Cancer Research Center (DKFZ), Heidelberg, Germany

Alexander Rölle
Clinical Cooperation Unit Applied Tumor-Immunity, German Cancer Research Center (DKFZ), National Center for Tumor Diseases (NCT), Heidelberg, Germany

Maike Hofmann
Faculty of Medicine, Department of Medicine II, University Hospital Freiburg, University of Freiburg, Freiburg im Breisgau, Germany

Adelheid Cerwenka
Research Group Innate Immunity, German Cancer Research Center (DKFZ), Heidelberg, Germany Medical Faculty Mannheim, Division of Immunbiochemistry, University Heidelberg, Heidelberg, Germany

Silke Heidenreich and Nicolaus Kröger
Department of Stem Cell Transplantation, University Medical Center Hamburg-Eppendorf, Hamburg, Germany

Amy K. Erbe, Wei Wang, Patrick K. Reville and Jacquelyn A. Hank
Department of Human Oncology, University of Wisconsin, Madison, WI, United States

Lakeesha Carmichael, KyungMann Kim, Eneida A. Mendonca and Yiqiang Song
Department of Biostatistics and Medical Informatics, University of Wisconsin, Madison, WI, United States

Wendy B. London
Dana-Farber Cancer Institute/Boston Children's Cancer and Blood Disorder Center, Harvard Medical School, Boston, MA, United States

Arlene Naranjo
COG Statistics and Data Center, Department of Biostatistics, University of Florida, Gainesville, FL, United States

Fangxin Hong
Department of Biostatistics, Harvard University, Dana Farber Cancer Institute, Boston, MA, United States

Michael D. Hogarty
Children's Hospital of Philadelphia, University of Pennsylvania School of Medicine, Philadelphia, PA, United States

John M. Maris
Provenance Biopharmaceuticals, Carlisle, MA, United States

Julie R. Park
Seattle Children's Hospital/University, Seattle, WA, United States
University of Washington, Seattle, WA, United States

M. F. Ozkaynak
New York Medical College, Valhalla, NY, United States

Jeffrey S. Miller
Department of Medicine, University of Minnesota, Minneapolis, MN, United States

Andrew L. Gilman
Levine Children's Hospital, Charlotte, NC, United States

Brad Kahl
Department of Medicine, Washington University, St. Louis, MO, United States

Alice L. Yu
Department of Pediatrics, Hematology/Oncology, Moores Cancer Center, University of California San Diego, San Diego, CA, United States
Institute of Stem Cell and Translational Cancer Research, Chang Gung Memorial Hospital, Taoyuan, Taiwan

Paul M. Sonde
Department of Human Oncology, University of Wisconsin, Madison, WI, United States
Department of Pediatrics, University of Wisconsin-Madison, Madison, WI, United States

Mirte Post, Angelica Cuapio, Markus Osl, Dorit Lehmann, Ulrike Resch and Erhard Hofer
Institute of Vascular Biology and Thrombosis Research, Center of Physiology and Pharmacology, Medical University of Vienna, Vienna, Austria

David M. Davies and Amit Nathwani
Department of Oncology, University College London Cancer Institute, London, UK

Martin Bilban
Department of Laboratory Medicine, Medical University of Vienna, Vienna, Austria

Bernhard Schlechta and Wolfgang Eppel
Department of Obstetrics and Gynecology, Medical University of Vienna, Vienna, Austria

Dagmar Stoiber
Ludwig Boltzmann Institute of Cancer Research, Vienna, Austria
Institute of Pharmacology, Center of Physiology and Pharmacology, Medical University of Vienna, Vienna, Austria

Jan Spanholtz
Glycostem Therapeutics, Oss, Netherlands

Emilio Casanova
Ludwig Boltzmann Institute of Cancer Research, Vienna, Austria
Institute of Physiology, Center of Physiology and Pharmacology, Comprehensive Cancer Center, Medical University of Vienna, Vienna, Austria

Simona Carlomagno
Dipartimento di Medicina Sperimentale, Università degli Studi di Genova, Genova, Italy

Michela Falco, Maria Bono
Istituto Giannina Gaslini, Genova, Italy

Claudia Alicata, Alessandro Moretta and Simona Sivori
Dipartimento di Medicina Sperimentale, Università degli Studi di Genova, Genova, Italy
Centro di Eccellenza per le Ricerche Biomediche, Università degli Studi di Genova, Genova, Italy

Lucia Garbarino and Michela Mazzocco
S.C. Laboratorio di Istocompatibilità e IBMDR, E.O. Ospedali Galliera, Genova, Italy

Lorenzo Moretta
Dipartimento di Immunologia, IRCCS Ospedale Bambin Gesù, Roma, Italy

Bercelin Maniangou, Nolwenn Legrand, Catherine Willem, Gaëlle David and Christelle Retière
Etablissement Français du Sang Pays de la Loire, Nantes, France
CRCINA, INSERM U1232 CNRS, Université d'Angers, Université de Nantes, Nantes, France

Mehdi Alizadeh
Laboratoire de Recherche et Développement, EFS Rennes, Rennes, France

Ulysse Guyet, Eric Charpentier and Alexandre Walencik
L'institut du thorax, INSERM, CNRS, UNIV Nantes, Nantes, France

Katia Gagne
Etablissement Français du Sang Pays de la Loire, Nantes, France
CRCINA, INSERM U1232 CNRS, Université d'Angers, Université de Nantes, Nantes, France
Laboratoire de Recherche et Développement, EFS Rennes, Rennes, France
Laboratoire d'Histocompatibilité, EFS Nantes, Nantes, France
LabeX Transplantex, Université de Strasbourg, Strasbourg, France

Sabine Huenecke, Claudia Cappel, Verena Pfirrmann, Emilia Salzmann-Manrique, Sibille Betz, Eileen Keitl, Julia Banisharif-Dehkordi, Shahrzad Bakhtiar, Christoph Königs, Andrea Jarisch, Jan Soerensen, Thomas Klingebiel, Peter Bader and Melanie Bremm
Clinic for Pediatric and Adolescent Medicine, University Hospital, Frankfurt, Germany

Ruth Esser
GMP Development Unit, Hannover Medical School, Institute of Cellular Therapeutics, Hannover, Germany

Evelyn Ullrich
Clinic for Pediatric and Adolescent Medicine, University Hospital, Frankfurt, Germany
LOEWE Center for Cell and Gene Therapy, Goethe University, Frankfurt, Germany

Antje Isernhagen
Institute of Cellular and Molecular Immunology, University Medical Center Göttingen, Göttingen, Germany

Dörthe Malzahn and Heike Bickeböller
Institute of Genetic Epidemiology, University Medical Center Göttingen, Göttingen, Germany

Ralf Dressel
Institute of Cellular and Molecular Immunology, University Medical Center Göttingen, Göttingen, Germany
Institute of Genetic Epidemiology, University Medical Center Göttingen, Göttingen, Germany
DZHK (German Center for Cardiovascular Research), Partner Site Göttingen, Göttingen, Germany

Dominik Schmiedel and Ofer Mandelboim
Faculty of Medicine, The Lautenberg Center for General and Tumor Immunology, The BioMedical Research Institute Israel-Canada, The Hebrew University Hadassah Medical School, Jerusalem, Israel

María Delso-Vallejo, Jutta Kollet and Volker Huppert
Miltenyi Biotec GmbH, Bergisch-Gladbach, Germany

Ulrike Koehl
Hannover Medical School, Institute for Cellular Therapeutics, IFB-Tx, Hannover, Germany

Matilde Risti and Maria da Graça Bicalho
LIGH - Immunogenetics and Histocompatibility Laboratory, Federal University of Paraná, Curitiba, Brazil

Markus Granzin
Clinical Research, Miltenyi Biotec Inc., Gaithersburg, MD, USA

Juliane Wagner and Evelyn Ullrich
Division for Stem Cell Transplantation and Immunology, Department for Children and Adolescents Medicine, Hospital of the Goethe University, Frankfurt, Germany
LOEWE Center for Cell and Gene Therapy, Cellular Immunology, Goethe University, Frankfurt, Germany

Ulrike Köhl
Institute of Cellular Therapeutics, Integrated Research and Treatment Center Transplantation, Hannover Medical School, Hannover, Germany

Adelheid Cerwenka
Innate Immunity Group, German Cancer Research Center, Heidelberg, Germany
Division of Immunbiochemistry, Medical Faculty Mannheim, Heidelberg University, Heidelberg, Germany

Volker Huppert
R&D Reagents, Miltenyi Biotec GmbH, Bergisch Gladbach, Germany

Aldi Pupuleku and Marcel Costa-García
Department of Experimental and Health Sciences, University Pompeu Fabra, Barcelona, Spain

Domènec Farré
Immunology Unit, Department of Biomedical Sciences, Medical School, University of Barcelona, Barcelona, Spain

Hartmut Hengel
Institute of Virology, Albert Ludwigs University of Freiburg, Freiburg, Germany
Faculty of Medicine, Albert Ludwigs University of Freiburg, Freiburg, Germany

Ana Angulo
Immunology Unit, Department of Biomedical Sciences, Medical School, University of Barcelona, Barcelona, Spain
Institut d'Investigacions Biomèdiques August Pi i Sunyer, Barcelona, Spain

Aura Muntasell
Hospital del Mar Medical Research Institute (IMIM), Barcelona, Spain

Miguel López-Botet
Department of Experimental and Health Sciences, University Pompeu Fabra, Barcelona, Spain
Hospital del Mar Medical Research Institute (IMIM), Barcelona, Spain

Beatriz Sanchez-Correa and Raquel Tarazona
Immunology Unit, University of Extremadura, Cáceres, Spain

Juan M. Bergua, Maria Jose Arcos and Helena Bañas
Department of Haematology, Hospital San Pedro de Alcantara, Cáceres, Spain

Alejandra Pera
Instituto Maimónides de Investigación Biomédica de Córdoba (IMIBIC), Reina Sofia University Hospital, University of Cordoba, Red Espanola de Investigacion en Patologia Infecciosa (REIPI), Córdoba, Spain
Brighton and Sussex Medical School, University of Sussex, Brighton, United Kingdom

Carmen Campos
Instituto Maimónides de Investigación Biomédica de Córdoba (IMIBIC), Reina Sofia University Hospital, University of Cordoba, Red Espanola de Investigacion en Patologia Infecciosa (REIPI), Córdoba, Spain

Esther Duran
Histology and Pathology Unit, Faculty of Veterinary, University of Extremadura, Cáceres, Spain

Rafael Solana
Immunology Unit, University of Extremadura, Cáceres, Spain
Instituto Maimónides de Investigación Biomédica de Córdoba (IMIBIC), Reina Sofia University Hospital, University of Cordoba, Red Espanola de Investigacion en Patologia Infecciosa (REIPI), Córdoba, Spain

Françoise Gondois-Rey, Françoise Mallet and Daniel Olive
CNRS, INSERM, Institut Paoli-Calmettes, CRCM, Immunity and Cancer Team, Aix Marseille Univ, Marseille, France

Antoine Chéret
APHP, Hôpital Bicêtre, Internal Medecin Unit, Le Kremlin-Bicêtre, France
U1184, Paris-Sud Univ, Le Kremlin-Bicêtre, France
CEA, DSV/iMETI, IDMIT, Fontenay-aux-Roses, France
INSERM U1184, ImVA "Immunology of Chronic Viral Infections and Autoimmune Diseases", Le Kremlin-Bicêtre, France
Institut Pasteur, HIV, Inflammation and Persistance Unit, Paris, France
Virology Laboratory, APHP CHU Necker-Enfants Malades, Paris, France

Ghislain Bidaut and Samuel Granjeaud
CNRS, INSERM, Institut Paoli-Calmettes, CRCM, CiBi Platform, Aix Marseille Univ, Marseille, France

Camille Lécuroux
U1184, Paris-Sud Univ, Le Kremlin-Bicêtre, France
CEA, DSV/iMETI, IDMIT, Fontenay-aux-Roses, France
INSERM U1184, ImVA "Immunology of Chronic Viral Infections and Autoimmune Diseases", Le Kremlin-Bicêtre, France

Mickaël Ploquin and Michaela Müller-Trutwin
Institut Pasteur, HIV, Inflammation and Persistance Unit, Paris, France

Christine Rouzioux, Véronique Avettand-Fenoël
Virology Laboratory, APHP CHU Necker-Enfants Malades, Paris, France
EA 7327 Paris Descartes University, Paris and France

Andrea De Maria
Dipartimento di Scienze della Salute and Centro di Eccellenza per la Ricerca Biomedica, Università di Genova, Genova, Italy
Clinica Malattie Infettive, IRCCS Azienda Ospedaliera Universitaria San Martino - Istituto Nazionale per la Ricerca sul Cancro, Genova, Italy

Gilles Pialoux
APHP, Hôpital Tenon, Infectious Diseases Unit, Paris, France

Cécile Goujard
APHP, Hôpital Bicêtre, Internal Medecin Unit, Le Kremlin-Bicêtre, France

Laurence Meyer
INSERM, CESP U1018, Epidemiology and population health, APHP, Hôpital Bicêtre, Paris 11 Univ, Le Kremlin-Bicêtre, France

Sabine Huenecke and Peter Bader
Children's Hospital, Goethe University, Frankfurt am Main, Germany
Division for Stem Cell Transplantation and Immunology, Department for Children and Adolescents Medicine, Hospital of the Goethe University Frankfurt, Frankfurt am Main, Germany

Anja Waldmann and Boris Brill
Georg-Speyer-Haus, Institute for Tumor Biology and Experimental Therapy, Frankfurt am Main, Germany

Judith W. J. Bergs and Franz Rödel
Department of Radiotherapy and Oncology, Goethe University, Frankfurt am Main, Germany
German Cancer Consortium (DKTK), Partner Site Frankfurt/Mainz, Frankfurt am Main, Germany
German Cancer Research Center (DKFZ), Heidelberg, Germany

Thomas Klingebiel
Children's Hospital, Goethe University, Frankfurt am Main, Germany

Christian J. Buchholz
German Cancer Research Center (DKFZ), Heidelberg, Germany
German Cancer Consortium (DKTK), Partner Site Heidelberg, Heidelberg, Germany
Molecular Biotechnology and Gene Therapy, Paul-Ehrlich-Institut, Langen, Germany

Winfried S. Wels
Georg-Speyer-Haus, Institute for Tumor Biology and Experimental Therapy, Frankfurt am Main, Germany
German Cancer Consortium (DKTK), Partner Site Frankfurt/Mainz, Frankfurt am Main, Germany
German Cancer Research Center (DKFZ), Heidelberg, Germany

Ulrike Koehl
Hannover Medical School, Institute of Cellular Therapeutics, Hannover, Germany
Institute of Clinical Immunology, University Hospital and Medical Faculty, University of Leipzig, Leipzig, Germany
Fraunhofer Institute for Cell Therapy and Immunology, Leipzig, Germany

Antoine Toubert
Institut National de la Santé et de la Recherche Médicale (INSERM) U1160; Université Paris Diderot, Sorbonne Paris Cité, Paris, France
Laboratoire d'Immunologie et Histocompatibilité, Hôpital Saint-Louis, Paris, France

Gianfranco Pittari
Department of Medical Oncology, National Center for Cancer Care and Research, Hamad Medical Corporation

Juliane Wagner, Viktoria Pfannenstiel and Evelyn Ullrich
Children's Hospital, Goethe University, Frankfurt am Main, Germany
Division for Stem Cell Transplantation and Immunology, Department for Children and Adolescents Medicine, Hospital of the Goethe University Frankfurt, Frankfurt am Main, Germany
LOEWE Center for Cell and Gene Therapy, Goethe University, Frankfurt am Main, Germany

Index

www.ingramcontent.com/pod-product-compliance
Lightning Source LLC
LaVergne TN
LVHW080947210926
848485LV00004B/789